Frequency Standards and Metrology

A. De Marchi (Ed.)

Frequency Standards and Metrology

Proceedings of the Fourth Symposium,
Ancona, Italy, September 5 - 9, 1988

With 271 Figures

Springer-Verlag Berlin Heidelberg GmbH

Professor Dr. Andrea De Marchi
Dept. of Electronics and Automatics, Università degli Studi di Ancona,
Via delle Brecce Bianche, I-60131 Ancona, Italy

SPONSORED AND SUPPORTED BY THE FOLLOWING ORGANIZATIONS
National Research Council (*Canada*)
National Bureau of Standards (*USA*)
Centre National de la Recherche Scientifique (*France*)
Consiglio Nazionale delle Ricerche (*Italy*)
Istituto di Radioastronomia del CNR (*Italy*)
Telespazio SpA (*Italy*)

ISBN 978-3-642-74503-4 ISBN 978-3-642-74501-0 (eBook)
DOI 10.1007/978-3-642-74501-0

Originally published by Springer-Verlag Berlin Heidelberg New York in 1989
Softcover reprint of the hardcover 1st edition 1989

2156/3150-543210 – Printed on acid-free paper

Preface

Since the previous Symposium, several exciting new developments and advances have occurred in the field of frequency standards and metrology. These include the first results on the long-term stability of a millisecond Pulsar, for which data integrated over several years now show a stability of around 10^{-14}.

Improvements in the understanding of various biases in Cesium beam standards promise accuracies in the low 14s for primary standards and in the low 13s for short commercial tubes, for which long term stabilities in low 14s have already been shown to be obtainable by accuracy improvement. Beams using optical pumping for state selection and for detection have been operated with excellent results, and more are being realized.

Other new frequency standards which have appeared include a macroscopic rf trap with Mercury ions, which performs in the low 15s in one day, the sub-millimeter metastable Magnesium beam, which has shown a short term stability in the low 12s in one second and promises an accuracy of 10^{-19}, and the cold Hydrogen masers, which have such high stabilities that they cannot be measured with existing local oscillators.

Prospects for future developments include laser manipulation of neutrals and spectroscopy of single ions at rest in a trap. Both these groups of techniques have great potential for unprecedented accuracy and short- and long-term stability, and new superior frequency standards are expected to be realizable in this way in the not too distant future.

Stability and reproducibility have been advanced at optical frequencies with lasers stabilized on atomic or molecular resonances, but the phase-coherent connection to the established microwave standards that are to be used in the field is still missing. More work is needed in this direction.

These and other promised advances underline once more the need for better oscillators with outstanding short-term stability, needed to measure performances and to be master oscillators that do not degrade the stability of the system. Striking results have been obtained in this area with superconducting cavity oscillators and with line-narrowed lasers, but more work is necessary.

Ancona, January 1989 *A. De Marchi*

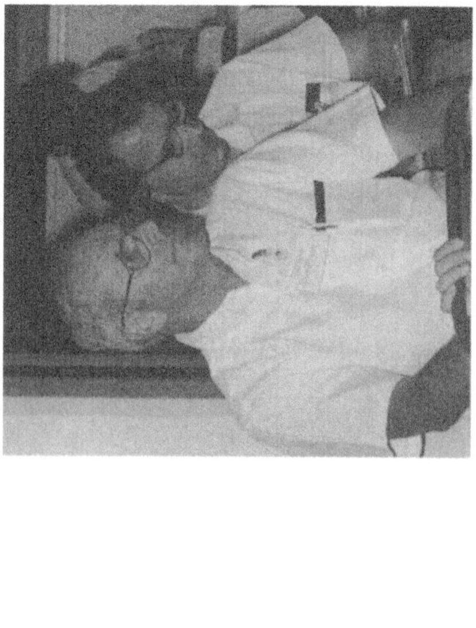

Opening Remarks by the Chairman

In opening this Symposium I want to first thank all the Institutions and the people who have made it possible.

The University of Ancona and Prof. Bruni, our "Rettore" or Dean in particular, have obviously played an important role, and it must be recognized as a merit to have been able to see the relevance of this meeting even if the scientific content is still beyond the reach of this young University. I am sure that his welcoming you here means a lot for the direction this Institution should take.

A particular recognition goes to the National Bureau of Standards, without whose hefty financial help this meeting would not have been possible, but I also want to mention the contributions of the National Research Council of Canada and the Centre National de la Recherche Scientifique of France, as foreign help.

In Italy we received the support of the Consiglio Nazionale delle Ricerche and its Istituto di Radioastronomia, of Telespazio SpA and of course of the University of Ancona. To this the help received in all forms from the Dipartimento di Elettronica e Automatica must be added.

The list continues with all the scientists who have helped me put together the scientific program, with my fair ladies of the Cooperativa Interlingua, without whom I'd be in high waters, and with all of you who have come to participate, some at their own expenses, and are going to make an exciting Symposium. Unfortunately some have not been able to make it. We miss them.

This is the fourth time that we get together to discuss our work at the frontier of frequency standard work, metrology, and physics, where advances in one field carry immediate relevance for the others. We have done so in 1971 in Quebec, Canada, at Forêt Montmorency, where Jacques Vanier was our host, in 1976 in Colorado, USA, where the organizer was Helmut Hellwig, and in 1981 in Savoie, France, where Claude Audoin took the trouble and the honor. Each time the meeting was a success, and I hope we can stand up to the record this time. The attendance has been growing: from 65 to 95 to 125. Here we are 165 from 17 countries. The number of talks and posters has grown too, probably as much as it can grow for a week long program. This means that our field is becoming increasingly important all over the world, and it is certainly due to the awesome precision which can be obtained in Frequency Metrology. However it poses some difficult problems that we will have to discuss and solve if we want to preserve the spirit of a select, yet friendly group, which has been the key to success of these Symposia. It is partly because of this that we publish here a Charter, as a guideline for future meetings.

It was a great pleasure for me to seek and accept the challenge of organizing this Fourth Symposium on Frequency Standards and Metrology. I wish all of us a very fruitful week, and hope everything goes well.

Before starting the Opening Session let me ask Claude Audoin to address you too as the Chairman of the last Symposium.

Claude!

A. De Marchi

Opening Remarks by Claude Audoin

Ladies and Gentlemen:
On behalf of the Committees for the Fourth Symposium on Frequency Standards and Metrology, I am very pleased to express my warmest thanks to Professor Bruni, Rettore dell' Universita di Ancona, for his hospitality and to welcome the attendees of this meeting.

Our Italian friends have expressed the wish to organize a meeting of this series since a long time, tracing back to the time when the Aussois meeting was in preparation. We are very grateful to Andrea De Marchi who has coordinated the initiatives and who has taken the challenge of organizing this Symposium.

I have to say that the success of the preceding meetings of this series has in fact encouraged scientists from different countries to consider the organization of a Symposium on Frequency Standards and Metrology. Jacques Vanier, Helmut Hellwig and myself are very happy of this healthy situation. However, we are extremely attached to preserve the spirit of the preceding meetings, which we believe has greatly contributed to the presence, here, of a large number of very distinguished physicists. This led us to elaborate a Charter describing the requirements which must be fulfilled to organize a conference under the name "Symposium on Frequency Standards and Metrology". This Charter creates a Governing Board gathering the organizers of all prior Symposia of the series. This Board has the power of decision to approve or to disapprove the organization of a "Symposium on Frequency Standards and Metrology". The text of this Charter will be published in the Proceedings of this conference.

Since our last meeting, substantial changes and progress occurred in our field of interest. The definition of the meter has been changed, being now related to that of the S.I. second. Knowledge of the properties of various types of frequency standards has made significant progress. Cryogenic active masers have been realized, mercury ion frequency standards contribute currently to improve the realization of the USNO time scale, the optically pumped cesium beam frequency standard has become a reality, and so on. The most fascinating advance is certainly related to the spectacular progress made in the observation and the manipulation of stored ions and atoms. I would personally appreciate that the impact that this will have on a next generation of microwave and optical frequency standards be discussed here.

I have no doubt that the outstanding quality of the speakers and of the attendees invited from all over the world by the Program Committee, and that the excellent housing conditions will result in an exciting, fruitful and friendly meeting. Let us start the scientific part of the meeting.

Contents

Part I	**Opening Session**

New Looks at Old Ideas
By N.F. Ramsey 2

Frequency Metrology and the Fundamental Physical Constants
By B.W. Petley and J.L. Flowers (With 2 Figures) 10

Coherent Spectroscopy on a Single Atomic System at Rest
in Free Space III
By H. Dehmelt 15

Part II	**Use and Availability of the Frequency Reference**

Use and Availability of the Frequency Standards
By M. Granveaud 22

Advanced Applications of Frequency Standards
By S. Leschiutta (With 2 Figures) 23

In Search of the Best Clock – an Update
By D.W. Allan (With 6 Figures) 29

Comparison of Primary Frequency Standards
By B. Guinot and J. Azoubib (With 5 Figures) 37

Part III	**Established Microwave Standards**

Established Microwave Frequency Standards
By H. Hellwig 44

A Point of View on the Accuracy Performances and on the Prospects of
Improvement of Primary Cesium Beam Frequency Standards
By C. Audoin 46

The Accuracy of Commercial Cesium Beam Frequency Standards
By A. De Marchi (With 2 Figures) 52

Reproducibility of H Maser Standards
By S.B. Crampton, J.J. Krupczak, and Phat Vu (With 2 Figures) 57

Advances in the Theory of Gas-Cell Atomic Frequency Standards
By J.C. Camparo (With 2 Figures) 62

Part IV **New Standards**

Presentation of the Session Entitled "New Standards"
By C. Audoin 70

High Accuracy Spectroscopy of Stored Ions
By D.J. Wineland, W.M. Itano, J.C. Bergquist, J.J. Bollinger, F. Diedrich, and S.L. Gilbert (With 1 Figure) 71

Mg Beam Frequency Standard
By A. Godone, E. Bava, and C. Novero (With 4 Figures) 78

Part V **Cold H Masers**

Introduction to Cold Hydrogen Masers
By S.B. Crampton (With 1 Figure) 86

The Cold Hydrogen Maser
By R.F.C. Vessot, E.M Mattison, R.L. Walsworth, and I.F. Silvera (With 3 Figures) 88

Performance of the UBC Cryogenic Hydrogen Maser
By M.D. Hürlimann, W.N. Hardy, M.E. Hayden, and R.W. Cline (With 6 Figures) 95

Quantum Scattering of Atomic Hydrogen from the Surface of Liquid Helium
By J.T.M. Walraven and J.J. Berkhout (With 2 Figures) 102

Part VI **Optically Pumped Cesium**

Influence of Low Magnetic Field on Pumping Efficiency in an Optically Pumped Cesium Beam Resonator
By G. Théobald, P. Cérez, N. Dimarcq, and V. Giordano (With 6 Figures) 110

An Optically Pumped Primary Frequency Standard
By R.E. Drullinger, J.H. Shirley, D.J. Glaze, and L. Hollberg 116

Design of an Optically Pumped Cs Laboratory Frequency Standard
By E. de Clercq, A. Clairon, B. Dahmani, A. Gérard, and P. Aynié (With 7 Figures) 120

Preliminary Results and Some Considerations on Optically Pumped Cesium Beam Frequency Standard
By Wang Yiqiu, Xie Linzhen, Wang Depei, and Yao Shutong (With 5 Figures) 126

Experimental Results on an Optically Pumped Cs Frequency Standard
By S. Ohshima, Y. Nakadan, T. Ikegami, and Y. Koga (With 4 Figures) 132

Temperature Measurements of Laser Cooled Cesium Atoms
By C. Salomon and J. Dalibard (With 4 Figures) 137

Part VII **Local Oscillators**

Local Oscillators – Introductory Remarks
By V.F. Kroupa 144

Stability of Frequency Locked Loops
By F.L. Walls (With 3 Figures) 145

1/f Noise in Oscillators: Theoretical and Experimental Progresses
By J.-J. Gagnepain (With 5 Figures) 150

RF Local Oscillators for Future Frequency Standards
By R.J. Besson 157

Ultra-Stable Cavity-Stabilized Lasers with Subhertz Linewidth
By D. Hils and J.L. Hall (With 7 Figures) 162

Low-Noise and Stable LO's for Deep Space Applications
By L. Maleki (With 2 Figures) 174

Part VIII **Frequency Synthesis**

Progress in Infrared Frequency Synthesis
By F. Bertinetto 180

Coherent Frequency Measurements of the Hfs-Resolved Methane Line
By G. Kramer, C.O. Weiss, and B. Lipphardt (With 5 Figures) 181

Phase-Locked Frequency Chains to 130 THz at NRC
By B.G. Whitford (With 3 Figures) 187

Direct Comparison of the Characteristics of Optical and Hydrogen Frequency Standards
By S.N. Bagayev, V.P. Chebotayev, V.M. Klementyev, B.A. Timchenko, and V.F. Zakharyash (With 4 Figures) 191

Part IX **Near IR Frequency Standards**

The Physics of Optical Frequency Standards Using Saturation Methods
By C.J. Bordé (With 10 Figures) 196

Double-Mode He-Ne/CH_4 Laser Stabilized by Methane Resonances with Relative Width of 10^{-11}
By M.A. Gubin, A.V. Nikulchin, and D.A. Tyurikov (With 3 Figures) .. 206

State-of-the-Art for High Accuracy Frequency Standards in the 28 THz Range Using Saturated Absorption Resonances of OsO_4 and CO_2
By A. Clairon, O. Acef, C. Chardonnet, and C.J. Bordé (With 13 Figures) 212

Part X Solid State Lasers

Solid State Lasers – Introduction
By H.R. Telle 224

Narrow-Linewidth Operation of Diode-Laser-Pumped Nonplanar Ring Oscillators
By A.C. Nilsson, T. Day, A.D. Farinas, E.K. Gustafson, and R.L. Byer (With 4 Figures) 225

Optical Stabilization of Semiconductor Lasers
By L. Hollberg (With 3 Figures) 231

Frequency Stabilization of Highly Coherent AlGaAs Diode Lasers
By A.M. Akulshin, V.V. Nikitin, V.A. Sautenkov, V.V. Vasiliev, V.L. Velichansky, E.K. Yurkin, and A.S. Zibrov (With 4 Figures) 236

Frequency Control of Semiconductor Lasers
By M. Ohtsu, K. Kuboki, C.H. Shin, and M. Murata (With 3 Figures) .. 242

Part XI Containment of Neutrals

Recent Progress in the Control of the Motion of Atomic Beams by Laser Fields
By V.I. Balykin, V.S. Letokhov, Y.B. Ovchinnikov, A.I. Sidorov, and S.V. Shul'ga (With 6 Figures) 248

Atoms Laser-Cooled Below the Doppler-Cooling Limit
By P.D. Lett, C.I. Westbrook, R.N. Watts, S.L. Rolston, P.L. Gould, H.J. Metcalf, and W.D. Phillips (With 4 Figures) 264

High-Contrast, High-Resolution Single Recoil Component Ramsey Fringes in Ca
By J. Helmcke, J. Ishikawa, and F. Riehle (With 4 Figures) 270

Obtaining of Supernarrow Saturated Absorption Resonances on Cold Particles and Their Use for Optical Frequency Standards
By S.N. Bagayev, V.P. Chebotayev, A.K. Dmitriyev, A.E. Om, Y.V. Nekrasov, and B.N. Skvortsov (With 4 Figures) 277

Laser Cooling in Calcium and Magnesium Atomic Beams
By N. Beverini, E. Maccioni, D. Pereira, F. Strumia, and G. Vissani (With 2 Figures) 282

Part XII Trapped Ions

Introduction to the Session on Trapped Ions
By H. Dehmelt (With 1 Figure) 286

Hg^+ Single Ion Spectroscopy
By J.C. Bergquist, F. Diedrich, W.M. Itano, and D.J. Wineland
(With 3 Figures) 287

Lifetime Measurements of Metastable States in Ions
By G. Werth (With 2 Figures) 293

Nonlinear Effects in Trapped Three-Level Ions
By H. Gilhaus, T. Sauter, R. Blatt, W. Neuhauser, and P.E. Toschek
(With 10 Figures) 300

New Frequency Standards Based on Yb^+
By R. Blatt, R. Casdorff, V. Enders, W. Neuhauser, and P.E. Toschek
(With 5 Figures) 306

Laser Spectroscopy of a Single Barium Ion Using "Shelving"
By W. Nagourney, Nan Yu, and H. Dehmelt (With 4 Figures) 312

Frequency Standards Utilizing Penning Traps
By J.J. Bollinger, S.L. Gilbert, W.M. Itano, and D.J. Wineland
(With 2 Figures) 319

XII.1 General Physics

The Interferometric Detection of Gravitational Waves – a Status Report
By A. Brillet 326

Relativity, Pulsar Time, and Atomic Time
By J.H. Taylor (With 2 Figures) 332

Testing Spatial Isotropy Using Optically-Pumped ^{199}Hg and ^{201}Hg
By S.K. Lamoreaux (With 2 Figures) 338

XII.2 Quantum Physics

Squeezed States, Interferometric Limits and Back-Action Evasion
By R.E. Slusher, A. La Porta, B. Yurke, and P. Grangier (With 4 Figures) 343

Mass Ratio Spectroscopy and the Proton's Atomic Mass
By R.S. Van Dyck, Jr., F.L. Moore, D.L. Farnham, and P.B. Schwinberg
(With 4 Figures) 349

The Physics of Circular Atoms and the Measurement of the Rydberg
Constant
By M. Gross, J. Hare, P. Goy, and S. Haroche (With 3 Figures) 356

Determination of the Rydberg Constant by Doppler-Free Two-Photon Spectroscopy of Atomic Hydrogen Rydberg States
By M. Allegrini, F. Biraben, B. Cagnac, J.C. Garreau, and L. Julien (With 4 Figures) 362

Part XIII Poster Sessions

The Primary Cesium Atomic Clocks of the PTB
By A. Bauch and T. Heindorff (With 3 Figures) 370

Metrological Parameters of the Primary Cesium Beam Frequency Standards in the National Time and Frequency Standard of the USSR
By Y.G. Abashev, V.N. Baryshev, G.A. Yolkin, and Y.N. Yakovlev (With 4 Figures) 374

Work on Cesium Standards at NRC
By J.-S. Boulanger and R.J. Douglas 379

Optical Pumping with Finite-Linewidth Lasers: Predictions for the Optically Pumped Cesium Beam Frequency Standard
By P. Tremblay and C. Jacques (With 5 Figures) 382

Work on Primary Cesium Beam Frequency Standard at CRL
By K. Nakagiri, M. Shibuki, H. Okazawa, M. Aida, and N. Kotake (With 1 Figure) 386

Improvement and Reevaluation of Chinese Primary Frequency Standard
By Yang Xiaoren, Wu Changhua, and Li Mingshou (With 4 Figures) ... 389

Preliminary Results on Small Optically Pumped Cesium Resonator
By P. Thomann, H. Schweda, and G. Busca (With 3 Figures) 392

Determining the Velocity Distribution in Cesium Beams from the Response to a Microwave Pulse
By G.D. Rovera, A. Premoli, and A. De Marchi (With 5 Figures) 395

Physical Problems Determining the Reliability of the Classical Microwave Frequency Standards
By G. Busca, H. Schweda, P. Thomann, and L.G. Bernier 398

Difference Frequency in Two Isotopes Atomic Beam Magnetic Resonance Apparatus as Candidate for Frequency Standard
By K.W. Radecki (With 1 Figure) 400

Studies and Developments of H-Masers at LHA
By R. Barillet, P. Lesage, P. Petit, J. Viennet, and C. Audoin 402

Hydrogen Maser Auto-Tuning Systems
By D. Morris (With 2 Figures) 404

Work on Hydrogen Maser
By Cheng Yumin (With 1 Figure) 406

Advanced Atomic Hydrogen Masers
By H. Peters, B. Owings, and P. Koppang (With 1 Figure) 408

Recent Progress in Oscillating Compact Hydrogen Masers
By H.T.M. Wang and R.R. Hayes (With 1 Figure) 410

Shapes of Storage Bulbs for Hydrogen Masers for Maximum Filling Factor
By J.E.B. Ponsonby (With 1 Figure) 412

H Maser and Rb Work in Romania
By O. Gheorghiu, L. Giurgiu, M. Dinca, C. Szekely, and C. Mandache (With 2 Figures) 414

Collisions of Atoms in Microwave and Laser Fields
By B.J. Verhaar and H.T.C. Stoof 416

Temperature of Mercury Ions Stored in an R.F. Cylindrical Trap
By C. Meis, M. Jardino, B. Gely, and M. Desaintfuscien (With 1 Figure) 418

A High Stability Oscillator Based on a Sapphire Loaded Superconducting Cavity Resonator
By S.K. Jones, D.G. Blair, and A. Giles (With 3 Figures) 420

Recent Scientific Results at the Istituto di Radioastronomia
By R. Ambrosini (With 2 Figures) 425

The Time and Frequency System at the Matera Space Geodesy Center
By A. Cenci, L. Garramone, G. Bianco, B. Pernice, and G. Sylos Labini (With 2 Figures) 427

Optical Pumping of Rubidium Vapor with Laser-Diodes: Application to Coherent Optical Communications
By M. Têtu, N. Cyr, B. Villeneuve, V. Giordano, S. Thériault, and M. Breton (With 3 Figures) 431

Improvements in Short- and Long-Term Frequency Stabilities of Diode Laser Pumped Rubidium Atomic Clock
By M. Hashimoto and M. Ohtsu (With 2 Figures) 434

Efficient Frequency Noise Reduction of GaAlAs Laser Diodes by Negative Electronic Feedback
By H.R. Telle and B. Lipphardt (With 2 Figures) 436

Some Applications of Extended-Cavity Semiconductor Lasers
By M. de Labachelerie, K. Diomande, and P. Cérez 439

Ultra-Narrow Linewidth of Optically Self-Locked Diode Lasers
By C. Bréant, P. Laurent, and A. Clairon (With 5 Figures) 441

Manipulation of Metastable Helium with a LNA Laser
By R. Kaiser, N. Vansteenkiste, E. Arimondo, A. Aspect, C. Cohen-Tannoudji, M. Karrais, and H. Haberland (With 2 Figures) 445

Quantitative Study of Laser Cooling in a Penning Trap
By W.M. Itano, L.R. Brewer, D.J. Larson, J.J. Bollinger, S.L. Gilbert, and D.J. Wineland 447

Laser Cooling and Resonant Transport of MG^+ Ions in a Penning Trap
By Jin Yu, F. Plumelle, and M. Desaintfuscien (With 1 Figure) 449

Trapped Yb^+ as a Potential Optical Frequency Standard
By G.P. Barwood, A.S. Bell, P. Gill, and H.A. Klein (With 3 Figures) .. 451

Kinetic Energy and Spatial Distribution of Ion Clouds in Paul Traps
By R. Blatt, I. Siemers, M. Schubert, T. Sauter, and W. Neuhauser (With 2 Figures) 453

Tuning Characteristic of 543 nm Internal Mirror He-Ne Laser for Frequency Stabilization
By T. Tako, M. Yamashita, and M. Suzuki (With 2 Figures) 455

Double-Mode CO_2 Laser with Complex Resonator for Frequency Standards and Superhigh Resolution Spectroscopy
By V.Y. Kurochkin, V.N. Petrovskiy, E.D. Protsenko, and A.N. Rurukin 457

Laser Frequency Measurement at NPL
By D.J.E. Knight, K.I. Pharaoh, and M.J. Padgett 459

Saturation Spectroscopy with the NH_3 Laser
By K.J. Siemsen, A.A. Madej, and J. Reid (With 1 Figure) 461

OsO_4 Molecular Beam for High Accuracy Frequency Standard
By A. Godone, M.P. Sassi, E. Bava, and C. Caldera 463

He-Ne Laser at 612 nm Stabilized to $^{127}I_2$ Using FM Spectroscopy
By F. Bertinetto, G.B. Picotto, P. Cordiale, and S. Fontana (With 1 Figure) 465

Frequency Stabilization of a 543.5 nm Wavelength He-Ne Laser to an Iodine Absorption Line
By U. Brand and J. Helmcke (With 2 Figures) 467

Measurement of Isotope Shift in Optical Transitions of Atomic Oxygen
By G.M. Tino, K. Ernst, A. Sasso, and M. Inguscio (With 3 Figures) ... 469

Selective Electron Attachment of SF_6 Molecules in Collision with Xe(nf) Rydberg Atoms in a R.F. Quadrupole Trap and Correlative Effects on SF_6^- Ions Lifetime
By S. Rajab Pacha, G. Brincourt, R. Catella, Y. Zerega, and J. Andre (With 1 Figure) 472

Closing Remarks 475

List of Participants 477

Index of Contributors 481

Part I

Opening Session

New Looks at Old Ideas

N.F. Ramsey

Lyman Physics Laboratory, Harvard University, Cambridge, MA 02138, USA

1. Introduction

The title for my talk, New looks at Old Ideas, was originally chosen by Andrea De Marchi and the conference Board of Directors. Initially I refused to give a talk under this title because such a talk might sound too much like a typical older person who says most new ideas are just old ideas warmed over. Later I realized that there was much to be said on the subject of New Looks at Old Ideas so I agreed to use that title, but I wish to emphasize at the outset that in so doing I am by no means belittling the importance and the novelty of the great new ideas in frequency standards and metrology. In fact there are so many in both areas that to avoid diffuseness I shall concentrate my comments on the field of frequency standards.

2. Use of Coherent Radiation

An old and new idea in frequency standards is the use of coherent radiation. Even the simplest tuned oscillator automatically provides coherent radiation since all the photons are driven by the same device. The study of spectroscopy with coherent radiation, however, was a much later development since the known spectral lines were for a long time beyond the reach of any coherent laboratory source. The first use of coherent radiation was in the crude 1933 microwave spectroscopy experiment of Cleeton and Willams[1] which studied the inversion spectrum of ammonia. However, there were no subsequent experiments by similar methods for the next 22 years.

The first extensive use of coherent radiation in spectroscopy came with the invention of the molecular beam magnetic resonance method by Rabi[2] in 1937. The initial experiments[2] with this method were primarily thought of as resonance measurements of nuclear spin precession frequencies, but Kellogg, Rabi, Ramsey and Zacharias[3] soon observed structure to their resonances and gave their paper the subtitle The Radiofrequency Spectrum of H_2. The primary emphasis was on the low frequency of the spectra rather than on the coherence of the radiation. In view of subsequent developments in the field, a better and more general subtitle to the first paper would have

Frequency Standards and Metrology Editor: A. De Marchi

been Spectroscopy With Coherent Radiation. However, the coherency of the radiation was recognized and used in both theoretical calculations and experiments. This was particularly true of Ramsey's separated oscillatory field method[4] which is now extensively used as a frequency standard[5,6] and on which the definition of the second is based.

The invention of the maser by Townes[7] and of the hydrogen maser by Kleppner and Ramsey[8] increased the significance of coherency. However, coherent radiation could not be used at optical frequencies until the invention of the laser by Townes and Schawlow[9] and even then further developments, such as two photon spectroscopy, were required before precision spectroscopy with coherent radiation became possible at optical frequencies. By now the progress has been such that almost all new devices proposed for accurate frequency standards depend on spectroscopy with coherent electromagnetic radiation.

3. Atomic and Molecular Beams

The first atomic beam experiment was that of Dunoyer[5,10] in 1911 to show that atoms travelled in straight lines. The technique owes much to improvements by Stern[5,11] and his associates but it was the invention of the magnetic resonance method by Rabi[2] and the development of that method into radiofrequency spectroscopy by Rabi and his associates[2,3,5] that made atomic beams important in frequency standards and metrology. Developments of atomic cesium frequency and time standards were initiated by Lyons, Kusch, Essen, Parry and others[6] and the initial atomic beam frequency standards have subsequently been markedly improved by contributions from many scientists and engineers. The Cs frequency standard is now widely used and is the basis for the international definition of the second. Further improvements to the atomic beam standard are currently being developed including laser detection, laser state selection, use of reversed beams and use of different atoms such as Mg. Many of the laser devices that are promising as frequency standards are also in part dependent on molecular beams.

4. Elimination of First Order Doppler Effects

In the earliest precision spectroscopy , the effects of the first order Doppler shift in broadening spectral lines were the principal limitations to be overcome. Schuler[12] and others in 1926 used cooled discharge tubes as sources of their spectra to reduce the Doppler broadening. Beginning in 1932 Jackson and Kuhn[13] and others reduced the first order Doppler effects by designing their sources so the emitting or absorbing atoms were in a directed atomic beam which could be viewed transversely to reduce the atoms' velocity components in the viewing direction.

In the early low frequency molecular beam magnetic resonance[2,3] experiments the first order Doppler shift was reduced by using standing waves, but at higher frequencies there was serious Doppler broadening. The separated oscillatory field method[4] removed the first order Doppler shifts in atomic beam spectroscopy provided the two coherent oscillatory fields had the same phase. Atomic beams with separated oscillatory fields achieved sufficient accuracy that it became apparent that for any other atomic frequency standard to be competitive its first order Doppler shift would have to br removed.

In most microwave absorption devices, such as the optically pumped ^{87}Rb frequency standard[14], the average velocities and thereby the first order Doppler effects are reduced by the frequent collisions with a buffer gas as discussed by Dicke[15]. Unfortunately the frequent collisions of the radiating atoms with the buffer gas atoms also produce frequency shifts which limit the accuracy.

In the hydrogen maser[8], where the atoms are stored in a bottle for a relatively long time, the first order Doppler effect is eliminated by the fact that the average velocity of the atoms is approximately zero due to the collisions with the surfaces of the storage bottle. Since the collisions with the walls are much less frequent than collisions with a buffer gas the frequency shifts are correspondingly reduced, but there still remain small shifts due to the collisions with the walls of the containing bottle. In some more recent optically pumped frequency standards the hydrogen maser practice of omitting a buffer gas has been followed.

Even after the invention of the laser[9], highly accuracy measurements at optical frequencies were not possible until methods were devised to overcome the large effects of the first order Doppler shift. This was effectively accomplished with either saturated absorption spectroscopy or two photon spectroscopy which employ two laser beams at the same frequency going in opposite directions so that the first order Doppler shifts cancel[16,17].

5. Cooling

The idea of cooling in spectroscopy has been around a long time. Schuler[12], for example, in 1926 used a cooled hollow cathode gaseous discharge source to reduce the effects of Doppler broadening in spectroscopy Molecular beam sources were frequently cooled[3, 5, 18] to liquid nitrogen temperatures or below[19] to reduce the numbers of rotational states and to obtain narrower resonance lines by keeping the molecules in the apparatus for longer times.

The big revolution in cooling, however, was the invention of laser cooling by Hansch and Schawlow[20] and Wineland and Dehmelt[17, 20]; the "new look" was a great new invention far more ingenious than the original "old idea" for cooling. All methods of precision spectroscopy, after overcoming other limitations, such as first order Doppler effects, are ultimately limited by the frequency shifts and broadenings due to the second order Doppler shift. Since that shift is due to to the relativistic slowing of any moving clock, it can be overcome only by slowing the atoms. Atoms and ions have now been laser cooled to a few micro-K, where the Doppler shifts are completely negligible. Furthermore, when cooling is combined with trapping, very weak traps can be used. It is highly probable that future dramatic improvements in frequency standards will involve laser cooling

6. Trapping

In the earliest spectroscopic experiments the atoms were "trapped" in the tube where the discharge took place, but this trapping was merely to retain the atomic concentration and it provided no benefit in narrowing the spectral lines. In the microwave absorption and optically pumped spectrometers the atoms were confined both by the cell and by the buffer gas with the latter diminishing the first order Doppler broadening as discussed above.

The original atomic beam experiments provided no trapping, which was both an advantage and a disadvantage. The spectral lines were not broadened or shifted by collisions of the radiating atom with either buffer gas atoms or the walls of the containing vessel. On the other hand, by the Heisenberg uncertainty principle, the observed resonance lines were broadened by the shortness of the time required for the atoms to pass through the oscillatory field region. The first efforts to overcome this disadvantage were atomic beam resonance experiments with stored atoms by Kleppner and Ramsey[22]. Although narrower resonances were found in this way the most significant consequence of the experiment was that it lead to the invention of the hydrogen maser[8] . In that maser the H atoms are trapped by a teflon coated bottle for 10 seconds, which is 10,000 times longer than the passage time in a typical atomic beam apparatus. The new look looks for such physical traps are the development of entirely new surfaces, such as surfaces of superfluid helium[23, 24].

Electromagnetic ion traps have also been around for a long time. The highly successful Penning[25] trap was first used in 1936 and the focussing principles for the Paul[26] alternating gradient field traps for ions were invented in 1953. But successful spectroscopy with such trapped ions only began with the experiments of Dehmelt[27] and his associates. Even then the accuracy of the experiments was relatively low due to the high kinetic energy of the trapped ions, which made the second order Doppler shifts large. The application of the previously discussed laser cooling techniques[20,21] completely changed the situation and extremely narrow resonances have been observed by Wineland, Dehmelt[29] and others[17, 28, 29,] . Typical kinetic energies of ions in traps now correspond to temperatures of a few micro-K and precise measurements can be made with only a single ion[17]. The principal remaining difficulties of this promising method are the necessity of using a low ion density to avoid space charge effects which in turn reduces the signal to noise ratio.

Likewise principles for focussing neutral atoms with radiofrequency trapping magnetic fields have been known since Paul's[30] work in 1950. However, such traps were originally thought to be of little value in atomic spectroscopy due to the distortion of the atomic energy levels by the strong and highly inhomogeneous fields required for the trapping of neutral atoms. Now with laser cooling the trapping fields can be much weaker. There have been a variety of successful experiments in slowing or stopping atoms in a beam. Due to the Doppler shift, the resonance frequencies of the atoms change as they slow down so either the energies of the atomic states or the frequency of the oscillator must be changed. The former has been accomplished by Phillips[31, 32] and others[17, 31] and the latter, usually called chirping, by Hall[33] and others[31]. The slowed atoms can be retained a relatively long time by optical molasses[34] or they can be trapped by inhomogeneous laser beams that depend on dipole forces[34], on scattering forces[36] or on stimulated molasses cooling[37]. Based on these principles there are a number successful laser traps by Chu[34], Phillips[35], Pritchard[36], Wiemann[36], Cohen-Tanoudji[37] and others[17, 31]. The principal problems are the distortions of the atomic energy levels by the laser trapping fields, which often make it necessary to turn off the trapping laser beams while the atomic spectrum is observed.

7. Separated or Successive Oscillatory Fields

The separated oscillatory field resonance method was invented by Ramsey[4] to obtain narrower resonances and to permit the molecular beam resonance method to work well at higher frequencies. It was later generalized[4] to more than two separated oscillatory fields and to fields that were successive in time rather than space. The first atomic beam Cs clocks developed by Zacharias[5,6], Essen and Parry[5, 6] utilized the separated oscillatory field method and it has been the basis of almost all subsequent atomic beam standards. Although the separated field method was originally planned for molecular beam experiments at radiofrequencies it has subsequently been adapted[4] to other devices including lasers by[4] Baklanov, Dubetsky, Chebotsev, Bergquist, Lee, Hall, Salour, Cohen-Tannoudji, Borde, Hansch and others.

8. Fountain Experiments

My next new look at an old idea is one that failed as an old idea but now with a "new look" appears very promising. In 1953 Zacharias proposed obtaining very narrow spectral lines with what he called a fountain experiment. He proposed using a vertical beam and atoms which were so far on the slow side of the Maxwellian velocity distribution that they would fall under gravity after rising a few meters and would be detected after they fell. Coherent separated oscillatory fields would excite the atoms on the way up and on the way down so that the observed resonance would be of the order of 1 Hz in width corresponding to the approximately 1 second rise and fall time. Although this method had not yet worked at the time my molecular beam book[5] was published, it appeared so promising for the future that I described it in detail in the book. Unfortunately, the method failed due to the number of atoms at very slow speeds being far below the Maxwell predictions, as a result of the non-equilibrium conditions in the source slits. With laser cooling, new looks at this old idea are possible and several groups are undertaking promising fountain experiments with laser cooling.

9. Use of Higher Frequencies

The idea of using higher frequencies in coherent resonance spectroscopy has been around for a long time and new possibilities were opened in 1939 when the molecular beam magnetic resonance method was extended from radiofrequencies to microwaves. But sources of coherent radiation at optical frequencies became available only after the invention of the laser. Likewise, from the earliest days of radiofrequency oscillators, there have been techniques for multiplying and dividing frequencies; but ingenious new inventions have been required to do so at optical frequencies.

10. Feedback and High Q Cavities

From the earliest days of radio and electronic circuits, negative feedback and high Q circuits have been used, but currently there is a new look at these old ideas as they are applied to lasers, including diode lasers. Optical or electrical negative feedback from a high Q cavity markedly diminishes the laser noise and increases its short term stability. Another valuable new look is the use of superconducting cavities for both high Q and high stability.

11. Astronomical Time and Frequency Standards

The earliest time and frequency standards were based on periodic astronomical phenomena. The development of the concept of time probably originated from the obvious periodic changes of days to nights, due to the rotation of the earth. For many years the second was defined in terms of astronomical periodicities: first in terms of the rotation frequency of the earth and later on the earth's orbit around the sun. Only in 1967, long after the development of accurate atomic clocks, was the second finally defined in terms of an atomic frequency.

Now there is a great new look at the old idea of astronomical time and frequency standards. Some of the millisecond pulsars appear to be so stable as to be promising long term frequency standards. If future observations confirm their stability and if enough stable pulsars can be discovered to permit effective cross checks of their relative stabilities, future definitions of the second might once again be based on an astronomical periodicity, but atomic clocks will still be required for the measurements.

12. Conclusion

Many of what are now be considered "old ideas" were at an earlier time "new looks" at still older ideas. Furthermore, the "new looks" often require more ingenuity and creativity than the original "old ideas". With the combinations of new ideas and new looks at old ideas that now exist in frequency standards and metrology, the fields are dynamic and challenging. The discussions at this meeting will help shape the future.

REFERENCES

1. C. E. Cleeton and N. H. Williams: Phys. Rev.,45, 234, (1934).
2. I. I. Rabi, J. R. Zacharias, S. Millman and P. Kusch: Phys. Rev., 53, 318 (1938).
3. J. M. B. Kellogg, I. I. Rabi, N. F. Ramsey and J. R. Zacharias: Phys. Rev., 56, 728 (1939) and 57, 677 (1949).
4. N. F. Ramsey: Phys. Rev., 76, 996 (1949), 78, 695 (1950) and Physics Today, 33(7), 25 (1980).
5. N. F. Ramsey: Molecular Beams (Oxford University Press, England, !956 and 1985).
6. N. F. Ramsey: Journ. of Res. of N. B. S., 88, 301 (1983).
7. J. P. Gordon, H. Z. Zeiger and C. H. Townes: Phys. Rev., 95, 282 (1954) and 99, 1264 (1955).

8. H. M. Goldenberg, D. Kleppner and N. F. Ramsey: Phys. Rev. Lett., 8, 361 (1960) and Phys. Rev., 126, 603 (1962) and 138A, 972 (1965).
9. A. L. Schawlow and C. H. Townes: Phys. Rev., 112, 1940 (1958).
10. L. Dunoyer, Le Radium: 8, 142 (1911).
11. O. Stern and F. Knauer: Zeits. f. Physik, 39, 751 (1926), 39, 764 (1926) and 53, 766 (1929).
12. H. Schuler: Zeits. f. Pysik, 35, 323 (1926).
13. L. Jackson and H. Kuhn: Nature, 134, 25 (1934).
14. A. Kastler: Journ. de Physique, 11 , 225 (1950).
15. R. H. Dicke:Phys. Rev. 89, 472 (1953).
16. T. W. Hansch and Y. R. Shen: Laser Spectroscopy, 7 (Plenum Press, 1985).
17. D. J. Wineland and W. Itano: Physics Today, 40(6), 34 (1987) and Science, 237, 612 (1987).
18. O. Stern, R. O Frisch and I. Estermann: Zeits. f. Physik, 85, 4 and17 (1933).
19. R. F. Code and N. F. Ramsey: Rev. SC. Instr., 42, 896 (1971) and A4, 1945 (1972).
20. T. W. Hansch and A. L. Schawlow: Opt. Commun. Nederlands 13, 68 (1975).
21. D. Wineland and H. Dehmelt: Bull. Am. Phys. Soc., 18, 1521 (1973) and 20, 60, 61, 637 (1975).
22. D. Kleppner, N. F. Ramsey and P. Fjelstadt: Phys. Rev. Lett., 1, 232 (1958).
23. W. N. Hardy and M. Morrow: J. Phys. (Paris) Colloq, 42, C8-171 (1981).
24. R. L. Wadsworth, I. F. Silvera, R. F. C. Vessot, et al: Phs. Rev. A34, 2550 (1986).
25. F. M. Penning: Physica 3, 873 (1936).
26. H. Friedburg and W. Paul: Naturwissenschaft 37, 20 (1950), 38, 159 (1951) and Z. f. Phys. 130, 493 (1951).
27. H. Dehmelt: Adv.At. Mol. Phys. 3, 53 (1967).
28. D. J. Wineland, W. M. Itano, et al Phys. Rev. A20, 1521 (1979), A25, 35 (1982) and A36, 428 (1987).
29. W. Neuhauser, M. Hohenstatt, P. Toschek and H. Dehmelt: Phys. Rev. Lett. 41, 233 (1978) and Phys. Rev. A22, 1137 (1980).
30. W. Paul, H. Steinwedel and M. Raether: Z. f. Naturforsch. 8a, 448 (1953), 38, 159 (1951) and Z. f. Phys. 140, 262 (1955).
31 Abstracts for Contributed Posters at Eleventh Intl. Conf. on Atomic Phys. (1988)..
32 W. D. Phillips, et al: J. Opt. Soc. Am. B2, 1751 (1985) and reference 31.
33. J. Hall, R. N. Watts and C. E Wieman: Opt. Letts. 11, 291 (1986).
34. S. Chu, et al: Phys. Rev. Lett. 55, 48 (1985) and 57, 314 (1986).
35. W. D. Phillips and H. J. Metcalf: Phys. Rev. Lett. 54, 992 (1985) and Sc. American 50 (March 1987).
36. D. Pritchard, C. Wieman, et al: Phys. Rev Lett. 57, 310 (1986).

Frequency Metrology and the Fundamental Physical Constants

B.W. Petley and J.L. Flowers

National Physical Laboratory, Teddington, Middlesex, TW11 0LW, UK

1. Introduction

The fundamental physical constants of the twentieth century have conventionally comprised the group: c, e, m_e, m_p, h, N_A, k and G. This group involves all of the base SI units save the candela. Frequency measurements play an increasingly important part in determinations of their values[1-3] in SI units.

2. c and the Rydberg Constant

Over the centuries the frequency associated with the measurements of c has gradually increased (Fig. 1). If we assume that there is a causal relationship between them (possibly as a reflection of the growth of frequency technology), we can see that, by the year 2000 AD, one might expect to measure c at a frequency of around 10^{16} Hz, which is beyond that of the hydrogen Lyman-alpha line, with a fractional accuracy of better than 10^{-11}!

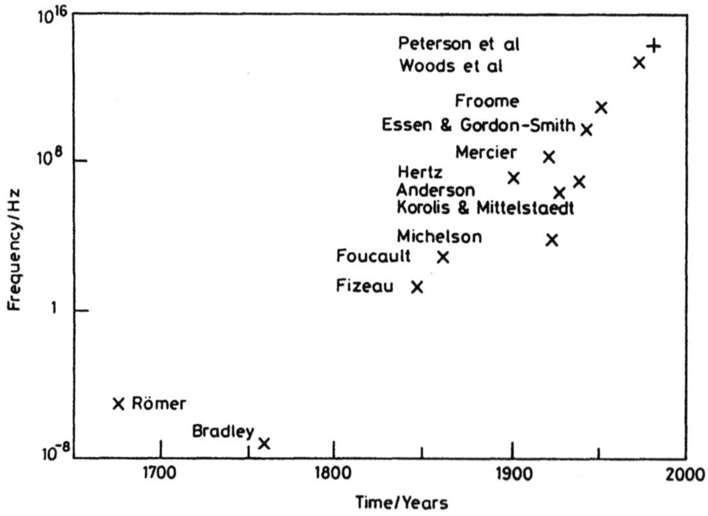

Figure 1. The frequency associated with measurements of c over the centuries; (+) - highest frequency measured directly to date.

Frequency Standards and Metrology Editor: A. De Marchi

The measurements of the Rydberg constant $R_\infty = \mu_0^2 c^3 m_e e^4/8h^3$, have often been essentially limited by the accuracy of the realisation of the metre. Although undesirable, such difficulties need have little effect on the contribution that measurements of R_∞ make to physics since R_∞ is divided out when one intercompares measurements on H, D, He^+, Ps, etc (provided, of course, that the same length standard is used for each of the measurements).

3. Accuracy and Frequency

Along with the increased frequency, the measurements of the velocity of light also improved greatly in accuracy over the centuries and so one might expect that measurement accuracy would also be correlated with frequency. The values of the quantities included by Cohen and Taylor[2] in the 1986 CODATA review of the best values of the fundamental physical constants (and some more recent work) are shown as a function of frequency in Fig. 2. Broadly their accuracy correlates with frequency, improving roughly as $(\text{frequency})^{1/3}$, with an intercept depending on the epoch.

The correlation of measurement accuracy with frequency is partly due to the fact that the quality factor, or 'Q' (i.e. frequency/linewidth - which tends to increase with frequency), is very important in frequency metrology. That is we can only estimate the centre of a resonance to a certain fraction of the

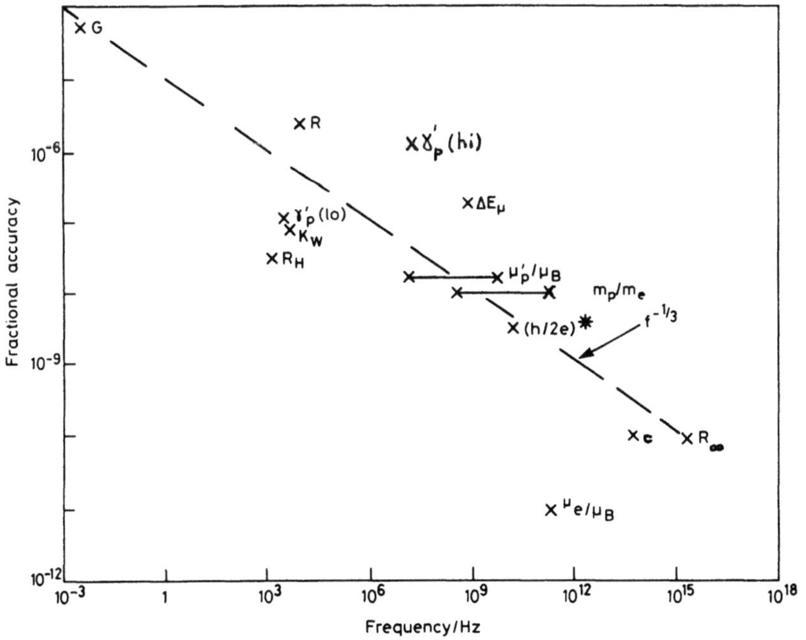

Figure 2. The precision of present-day measurements of the fundamental physical constants as a function of frequency; (*) value in terms of maintained volt.

line-width. The accuracy achieved in a particular case depends on how well the theory of the lineshape is understood.

The advent of laser fringe-counting interferometers has greatly facilitated the length measurements required for measurements of the fundamental physical constants. The fringe counting speed varies from about one x-ray fringe per second in the measurements of the lattice parameter of silicon, to around 10 MHz in the measurements of the acceleration due to gravity. The NPL gas constant measurements[4] employed an acoustic interferometer whose length was measured by fringe counting while the ultrasonic transducer frequency was also measured. Similarly in the NBS[5] measurement of R the acoustic resonances of a spherical cavity were excited at around 15 kHz. Incidentally, its linear dimensions were checked by exciting the cavity microwave resonances.

4. Fine Structure and Hyperfine Splittings

The quantity cR_∞ characterises the frequency of transitions between the main electronic levels in atoms and ions, and there are other important quantities which involve the fine structure constant α (that is $\mu_0 ce^2/2h$ or $1/137{\cdot}01..$), as well. These are the fine-structure transitions and the hyperfine transitions, which are characterised by $\alpha^2 cR_\infty$, and Lamb shifts characterised by $\alpha^3 cR_\infty$. In these and other cases, higher order correction terms are involved, most of which involve term series in powers of α. Since α is a dimensionless quantity, comparisons between different types of measurements form an important meeting ground between theory and experiment and these provide increasingly stringent tests of Quantum Electro-Dynamics (QED).

The hydrogen maser measurement of the ground-state splitting of atomic hydrogen has long provided us with the most accurately measured physical quantity and, but for wall-shift effects, might well have served to define the second. Unfortunately, although it may be expressed in terms of the R_∞ and α and correction terms involving powers of α, the theory remains incomplete because the effect of the proton electronic polarisability, nuclear structure and recoil cannot be computed with certainty.

By using the correction terms known at that time Cohen and Taylor (1973)[3] obtained the value of α^{-1} of $137.03597(22)[1 + \delta_N]$, from the hydrogen maser frequency where δ_N is due to the polarisability etc of the proton. Their value may be compared with the later value of 137.035 9914(11) which is obtained from value of g_e-2, measured by Van Dycke, Schwinberg and Dehmelt[6], by using the associated theory of Kinoshita[7]. It is clear from this that the polarisability of the proton is small and may well be zero.

Atomic hydrogen is one of the simplest systems and it is surely one of the paradoxes of metrology that one of the most accurately measured quantities is not calculable. This provides a major challenge for future theoreticians rather than the experimenter. It reminds us that it may take some time for theoretical physicists to meet the challenge provided by the spectacular accuracies that will be achievable with ion traps and atom traps.

5. The SI Units and the Quantised Hall Resistance

The quantised Hall resistance[8] R_H, is measured by a DC method, and already this measurement of h/e^2 is limited by our ability to realise the ohm with adequate accuracy - indeed an increasing number of national standards laboratories are using R_H to monitor the stability of their ohm standards, and the Josephson effect $2e/h$ to monitor the stability of their volt. The calculable capacitor realisation of the ohm is often made using an angular frequency of 10^4 radians per second for the AC bridges so that 10^{-8} F corresponds to 10^4 Ω etc.

Our knowledge of many of the fundamental physical constants is limited by the accuracies of 'absolute' realisations of the ampere, ohm, volt and watt. Direct and indirect measurements via the fundamental physical constants have shown that the disseminated electrical units are variously in error at the 1.5-16 x 10^{-6} level, and consequently they, along with the IPTS-68 practical temperature scale, will be changed from 1st January 1990.

6. Dimensionless Quantities

There is an important group of dimensionless quantities which are beautifully determined via various combinations of the spin precession and cyclotron frequencies of electrons, muons, positrons and protons. These include μ_p/μ_N, m_p/m_e[9], μ_p/μ_B, g_e-2 for the proton electron and positron, and $g_\mu-2$, for the muon.

The lepton magnetic moment anomalies may be calculated with high accuracy, which thereby provides one of the triumphs QED. The proton magnetic moment by contrast is not yet amenable to calculation beyond the 0.1% level of accuracy. It has a diminishing role in evaluations of the fundamental physical constants.

7. The Pace of Metrology

As in other parts of metrology, the pace of technological change has been reflected by a tenfold improvement of the accuracy with which the fundamental physical constants are known every decade or so[10] - however despite this we cannot catch up until frequency metrology reaches a natural accuracy limit!

Today we may use particular combinations of the fundamental physical constants to maintain many of the SI units, indeed they are involved with all of the base units save the kilogram. Constants such as the velocity of light, the Josephson effect 2e/h, and the gyromagnetic ratio of the proton enable length, voltage and flux density respectively to be measured in terms of a frequency - which provides the basis of Quantum Metrology. Once the Rydberg constant is measureable directly in terms of frequency such measurements are likely to make formidable demands on theory. The identity of the fundamental constants associated with matter and anti-matter will be investigated shortly as well.

For most quantities the ultimate accuracy limit lies many decades away. Already Josephson effects (involving $2e/h$) have been used at the 10^{-15} level of accuracy[9] to investigate by a null method the gravitational acceleration of microwave photons in falling through a height of a few millimetres[11]. Although it is rarely a limiting factor, frequency measurement plays a key part in measurements of combinations of all of the fundamental physical constants (with the possible exception of direct measurements of the Avogadro constant). There can be no doubt that such measurements at the highest levels of accuracy will continue to be a challenging and rewarding field of endeavour.

8. References

1. B W Petley: *The Fundamental Physical Constants and the Frontier of Measurement*, (Adam Hilger, London, 1988)
2. E R Cohen and B N Taylor: Revs. Mod. Phys. 59, 1121-48 (1987)
3. E. R. Cohen and B. N. Taylor: J. Phys. Chem. Ref. Data, 2, 662-734 (1973); and E. R. Cohen, in *Quantum Metrology and Fundamental Physical Constants*, eds. P. H. Cutler and A. A. Lucas (Plenum, London, 1983)
4. A. R. Colclough, T. J. Quinn and T. R. J. Chandler: Proc. R. Soc. (Lond.): A368 165,(1979)
5. M. R. Moldover, J. P. Trusler, T. J. Edwards, J. B. Mehl and R. S. Davis: Phys. Rev. Lett. 60, 249 (1988)
6. R. S. Van Dycke,jr., P. B. Schwinberg and H. A. Dehmelt: Phys. Rev. Lett. 59, 26 (1987)
7. T. Kinoshita and W. B. Lindquist: Phys. Rev. D. 27, 867, 887 (1983)
8. K. von Klitzing, G. Dorda and M Pepper: Phys. Rev. Lett. 45, 494, (1980)
9. R. S. Van Dycke, jr., F. L. Moore, D. L. Farnum, and P. B. Schwinberg: Bull. Am. Phys. Soc. 31, 244 (1986)
10. B. W. Petley: in *Physics in a Technological World*, ed. A. P. French (American Institute of Physics, New York: 1988)
11. A. K. Jain, J. E. Lukins and J -S Tsai: Phys. Rev. Lett 58, 1165 (1987)

1 September 1988

Coherent Spectroscopy on a Single Atomic System at Rest in Free Space III

H. Dehmelt

Department of Physics, University of Washington, Seattle, WA98195, USA

1. Introduction

When I gave the first talk under this title in 1976 at the second of these symposia (below referred to as II), serious mono-particle spectroscopy had just arrived on the scene with a precision measurement of the g-factor of an individual electron by the Seattle group. This experiment, published under the title

Axial, magnetron, cyclotron and spin-cyclotron-beat frequencies measured on single electron almost at rest in free space (geonium),

already made use of an early rf variant of sideband cooling. At II I also discussed the 1973-75 abstracts

Proposed $10^{14}\Delta\nu < \nu$ Laser Fluorescence Spectroscopy on Thallium$^+$ Mono-ion Oscillator I-III,

which went on to 'electron shelving' and laser side band cooling. Prior to 1973, impressive spectral resolution had been attained in Seattle, NASA, Bell, and Bonn-Mainz studies of hfs resonances in clouds of trapped He^+, H_2^+, Hg^+ ions in Paul rf traps, and of cyclotron and spin resonances in clouds of electrons in Penning traps. Historically, spectra of electrons trapped in vacuum were observed first. A 1961 progress report to the NSF detailed the observation of cyclotron, magnetron and axial resonances on small clouds in a Penning trap in Seattle. Thus many of the techniques and procedures now widely used for trapped charged and neutral atoms were first proposed or exercised for electrons. Even so, no individual **neutral** atom has been isolated to date. In 1981, at the third symposium (or III), the Heidelberg/Seattle group reported the first mono-ion **optical** spectroscopy experiment, carried out in 1980 with an individual Ba^+ ion in a newly pioneered 1mm miniature Paul rf quadrupole trap with MHz oscillation frequencies for good resolution of side bands for easy zero-point confinement. At III Boulder also reported success with 1981 work on a single Mg^+ ion in a 1cm Penning trap with a magnetron frequency <30kHz and little resolution of sidebands.

2. Mono-Electron RF Spectra

After continuous detection of an individual electron had been demonstrated at Seattle in 1973, it was still necessary to develop the

Frequency Standards and Metrology Editor: A. De Marchi

compensated Penning trap and to invent the continuous Stern-Gerlach effect before cyclotron and spin-resonances could be observed on an individual electron. The width of the cyclotron motion side band of the spin resonance, at $\nu_s - \nu_c$, was reduced by us from the ≈500Hz value reported at II to 3Hz in 1987. The g-factor follows from $\nu_s - \nu_c$ as

$$\tfrac{1}{2}g \equiv \nu_s/\nu_c = [(\nu_s - \nu_c)/\nu_c] + 1.$$

The latest value, measured both for electron and positron by the Seattle group, has an estimated error of only 4 parts in 10^{12} and is the most precisely determined parameter of an elementary particle. The value has withstood such tests as changing the sign of the magnetic bottle and drastically altering the trap electrode structure undertaken as good experimental practice. Our result puts stringend bounds on new theories put forth by Salam and others, which, beyond the standard model, view leptons as composed of three smaller spin-½ fermions. In fact, based on our g-value certain composite models discussed by Brodsky and Drell yield a new 10^4 times smaller value for the Radius of the electron

$$R < 10^{-20}\ \text{cm}.$$

This agrees with a $|g-2|$ vs $R/\lambda\!\!\!^{-}_C$ plot for p, ^3H and ^3He, all with $I=\tfrac{1}{2}$.

It is certainly clear from the close agreement of our data with qed theory that the physical electron cannot be very different from the Dirac point electron and must be very small. This implies that its constituents must be very massive, in fact $\approx 10^{10}$ more massive than an electron, so that their Schrödinger Zitter-Bewegung orbits fall within the spatial extension of the electron. I have extrapolated this model by postulating that the electron constituents have even more massive sub-constituents and so on, but not ad infinitum. Rather, after a finite number of rungs on the ladder the progression stops with the 'cosmon', the most massive particle or largest eigenvalue of the mass-energy operator ever realized in the universe. A tightly bound cosmon/anticosmon pair of zero total relativistic energy was created from the metastable vacuum in a random, spontaneous quantum jump, not unlike that recently observed in Ba^+. The 'cosmonium atom' introduced here is merely a modernized version of Lemaître's 'l'atome primitif'. The subsequent ultra-rapid breakup of its components, cosmon and anticosmon, by cascading down the ladder of quantized mass-energy values, then initiated the big bang. No point particles appear in this description of a universe likewise finite in the large.

Further progress in our g-factor experiments hinges on substantial reduction of the magnetic bottle necessitated by the continuous Stern-Gerlach effect and on the control of cavity shifts of the cyclotron resonance. The relativistic cyclotron resonance has recently been analyzed /1/ at Seattle and on this basis a Seattle/Harvard group has obtained a resonance with a bottle 10 times weaker than in the experiments described in II. The very steep low frequency edge realized here may allow us to resolve the spin fine structure of the cyclotron resonance and therefrom ascertain the spin state. At the same time, the spin-resonance would be more than an order of magnitude narrower.

3. Two-Electron Geonium

The 'ground state' of 2e-geonium, two electrons at rest at a distance d apart on the axis of the Penning trap where trap forces balance Coulomb repulsion was briefly described in Seattle in 1979. For our ν_z=60MHz trap $d = (e^2/2\pi^2 m\nu_z{}^2)^{1/3}$ = 15.3µm. Realization that metastable states of this system might be nearly immune to trap cavity shift has greatly increased interest in them. In particlular, the state with one electron in (mnkq)=(-½0kq) and the other electron in (mnkq)=(-½1kq) should exhibit a **metastable** cyclotron excitation. Metastability of the out-of-phase or antisymmetric classical z-oscillation near 60MHz of the two electrons, which must be excited parametrically, is under study in Seattle experiments.

4. Mono-Ion Optical Spectra

Today, at this symposium, I am still unable to announce that the goals of high resolution mono-ion optical spectroscopy, as formulated by me at II and III, have been reached. No experimental work on quadrupole shift-free $^1S_0$ - $^3P_0$ forbidden transitions in Thallium$^+$-like ions with 10^{-18} resolution has been reported sofar. However, these experiments of the past few years stand out. In 1984 the Seattle optical spectroscopy group observed the $6^2S_{1/2}$ - $5^2D_{3/2}$ 2-photon transition at 2.07µm in Ba$^+$ side band-free in the Lamb-Dicke regime. In 1986 the **shelved electron** scheme and quantum jumps between **electronic** levels were demonstrated in Seattle and Hamburg for this ion, and for Hg$^+$ in Boulder. Simultaneously or soon after, as proposed at III, forbidden transitions from the $S_{1/2}$ ground state to a metastable $D_{5/2}$ state were induced by laser radiation for both ions. For Ba$^+$ and Hg$^+$ the wavelength are 1.762 µm and 0.2815 µm and the realized upper **electronic** state life times are 30 sec and 0.1 sec respectively. This almost certainly guarantees experimental line widths equaling the natural ones of 5milliHz and 1.6Hz, when optical flywheel sources of sufficient short term stability are available for scanning. While such narrow widths have not yet been reached, excellent progress has been made at Boulder, where admirable flywheel oscillators have been perfected. For Hg$^+$ in Boulder, where in place of the original 1cm Penning trap described in II and III a miniature rf trap very similar to that developed in Heidelberg has now been adopted, this has produced a spectrum of 5 sidebands only 30kHz wide and spaced at ≈1.5 MHz. Spectra of forbidden Ba^+ lines have also been recorded in Seattle.

One way to solve the flywheel oscillator problem may be development of a stored ion laser, in which in distinction from current lasers, and as in the H-maser, the oscillation frequency is determined by atoms and not by the cavity, as proposed at Seattle in 1966 and later. In fact, as proposed in 1986, it may be possible to operate such an ultra-narrow stored ion laser on the forbidden $5D_{5/2}$ - $6S_{1/2}$ transition just mentioned. To this end, a linear string, compare Straubel 1955 /2/, of Ba^+ ions would be laser cooled and confined in a 1-meter rf race track trap developed for this purpose at Seattle in 1969. The ≈200,000 laser cooled ions would collect at the bottom of the ≈2m long potential valley and would arrange themselves, 10μm apart, like a closed string of pearls in the only Stark and 2nd Doppler shift-free multi-ion configuration described sofar. One straight section of the race track would serve as laser. As in the shelved electron experiment, but by more intense 455nm $S_{1/2}$ - $P_{3/2}$ excitation, the ions in the other straight section would rapidly be pumped into the $D_{5/2}$ level in a manner quite analogous to Maiman's original three-level Ruby laser. Light pressure would cause the ring of ions to slowly rotate, thereby circulating them from the pumping into the lasing section. Stark quenching, by applying an rf voltage across opposite trap electrodes, could greatly increase the emission rate, and facilitate laser action.

Along somewhat similar lines, but hopefully much easier to realize, a close-by identical but excited ion has been proposed /3/ as a source of ultrasharp λ=1.762μm radiation. In Heidelberg in 1980 it had been demonstrated, that two Ba^+ ions could be squeezed together until their Coulomb repulsion balances the trapping forces, which was 4.3μm in this case. So it seems possible, to push the two oscillating quadrupoles together to <<λ into the near-zone where excitation exchange proceeds much faster than emission into space. Conveniently, Boulder and San Jose groups have recently demonstrated two ions aligned on the axis of an rf trap. In more detail, the Seattle experiment /3/ will make use of the shelved electron scheme in the following way: Laser-cool two Ba^+ ions to equilibrium distance of ≈8μm on the axis of the trap, and identify them as ion A and ion B visually or photoelectrically by their location on the z-axis. Briefly turn on a 455nm spectral lamp until one of the ions, say the A-ion, becomes invisible by being pumped into the metastable $D_{5/2}$ level. Focus on the visible, spatially well resolved B-ion, and turn off both the blue and the red illumniation lasers for ≈15sec. Then turn them back on again and check if by chance the excitation has been transferred to the B-ion and it is now in the $D_{5/2}$ level and dark, while the A-ion is bright. When the two ions are pushed together to ≈λ/8 = 0.22μm, by turning on a much stronger trapping field during the excitation exchange period, one might be able to detect excitation transfer in >10% of the attempts. Tuning of the atoms relative to each other makes use of the ac Stark effect produced by the rf trapping field. A 0 - 10mV/cm variable dc field in the z-direction displaces the ions axially, and causes them to see different rf fields, which detunes them relative to each other. Stark-quenching of the metastable level can greatly accelerate the experiment. In this way it may be possible to demonstrate a resonant transfer response of extraordinary sharpness, namely approaching twice the natural width of the $D_{5/2}$ level, 11 milliHz.

Zero-point confinement of the electron in its cyclotron motion, $\langle n \rangle \approx 0.16$, in geonium has been routine since 1979. Also, the earliest analysis of sideband cooling in 1978 already showed, that when the secular oscillation frequency $\omega_v > \gamma$, the width of the optical transition used for cooling, the average oscillation quantum number in the trapping well will quickly drop to

$$v \approx (3/16)(\gamma/\omega_v)^2 \ll 1.$$

This clearly implied that it is possible to realize frequent random zero-point confinement, and confirmed my estimate of a cooling limit $v \approx 4/5$ for Thallium$^+$ announced at the 2nd of these symposia in 1976. Now the Boulder group reports zero-point confinement for one more degree of freedom but not yet for the third one and the micromotion of an Hg^+ ion in their rf trap. In a later talk here we hope to hear more about this advance. It seems that the title

Coherent Spectroscopy on a Single Atomic System at Rest in Free Space

of my talks at the last 3 of these symposia already sounds quite realistic now!

References to most of the work touched upon here can be found in earlier articles /4-7/ and the proceedings /8/ of II - IV.

Diana Dundore read the manuscript. The National Science Foundation supports the SINGLE ATOMIC/ELEMENTARY PARTICLE AT REST IN SPACE PROJECTS.

5. References

1. Hans Dehmelt, Richard Mittleman and Yuan Liu: Proc. Natl. Acad. Sci. USA 85 (1988)
2. H. Straubel: Naturwissenschaften 42, 506 (1955)
3. Hans Dehmelt and Warren Nagourney: Proc. Natl. Acad. Sci. USA 85 (1988)
4. Hans Dehmelt: Physica Scripta T22, 102 (1988)
5. W. Itano, J. Bergquist, and D. Wineland: Science 237, 612 (1987)
6. Annales de Physique, No. 6, 10 (1985)
7. Hans Dehmelt: in Advances in Laser spectroscopy, ed. by F. T. Arecchi, F. Strumia & H. Walther (Plenum, New York, 1983) p.153
8. Journal de Physique, Colloque C8, Supplement to No. 12, 42 (1981)

Part II

Use and Availability of the Frequency Reference

Use and Availability of the Frequency Standards

M. Granveaud

Laboratoire Primaire du Temps et des Fréquences, Observatoire de Paris, 61 av. de l'Observatoire, F-75014 Paris, France

The atomic time history began officially 21 years ago with the cesium based definition of the second. Since 1967, progress have been made on various aspects of the time/frequency metrology and on the use of the atomic frequency standards.

The primary laboratory cesium standards have improved their capacity to realize the definition of the second by one order of magnitude : from 10^{-12} in 1967, the accuracy has decreased to 10^{-13} and less for a few standards in 1988. Much is hoped, in the accuracy domain, from the optical pumping technics.

Another fundamental aspect is concerned with the T/F standards intercomparisons. Use of artificial satellites has brought a real breakthrough : one order of magnitude has been gained in precision and accuracy on the T/F operational intercomparisons. Through the Global Positioning System, two frequency standards running on two continents are currently compared at the level of 3×10^{-14} over time intervals of one week.

Finally, the most impressive feature of atomic frequency standards especially rubidium and cesium standards, lies in their standardization as routine reference devices. Twenty years ago, they were essentially devoted to metrology, research work and army tasks ; nowadays, beside these traditional applications, cesium standards are used in various fields : for example, in geodesy to locate oil deposits ; in navigation to measure ranges ; in telecommunications to ensure syntonization of digital data. In 20 years, the needs of the users have largely increased approaching the possibilities of the best T/F standards. It is time for a new generation !

Frequency Standards and Metrology Editor: A. De Marchi

Advanced Applications of Frequency Standards

S. Leschiutta

Dipartimento di Elettronica, Politecnico di Torino,
corso Duca Abbruzzi, 24, I-10129 Torino, Italy

1 Introduction

Countless are nowadays the applications of Time/Frequency methods and systems to science and technology: the daily production of more than 300 000 quartz units, with precisions ranging from 10E-5 to 10E-10, barely satisfies the needs of factories making wristwatches, cars, PCs, dishwashers, etc..

Applications to navigation and timekeeping being well known, this review paper will focus on some less known cases in which atomic frequency standards can or must be used. These cases are satellite positioning and modern digital and optical communications; some unusual applications will be quoted only, while in the last section , an attempt will be performed to forecast future needs and uses.

Nevertheless it can be of interest to know that any airliner, for communication, navigation, housekeeping or just entertainment, is fitted with not less than one hundred frequency sources. On ground, to support the safe flight of that plane, an equivalent number of frequency standards are on service, some of which must be, in some cases, atomic ones.

2 Satellite Localisation and Positioning

All the used satellite positioning systems, both for near-earth and deep space spacecrafts, are based on Time and Frequency.

The position of a satellite, i.e. the knowledge of its orbit , is needed for a number of reasons, from the simple stationkeeping, when a spacecraft must be kept in some specific position and orientation to serve a given Earth area or to fulfill some electromagnetic compatibility requirements, to the cases in wich a specific orbit must be followed in order to enable a mission to be performed (the satellite, e.g., carries an altimeter with a resolution and accuracy of a meter).

The most important instance for the implied metrological requirements, is when the orbit itself is the quantity of interest, as happens for the geodetical and some geophysical satellites; in these latter cases an accuracy of the order of a centimeter is needed.

Table I presents the most used systems, gathered following the phenomenon on wich the specific system is based ; needless to say that the accuracy in ranging is related to the accuracy of the frequency standards used. For laser ranging, atomic frequency standards must be adopted: an accuracy of E-10 - E-11 in needed on the field (the environment of a laser station is usually far more severe compared to that of a metrologic laboratory).

In recent years a new method of ranging took widespread use.This method, called PSRN (Pseudo Random Noise Code) was pioneered in the 60' by JPL for lunar ranging. The method can be easily understood with reference to Fig.1.

Frequency Standards and Metrology Editor: A. De Marchi

TABLE I SATELLITE POSITIONING METHODS AND SYSTEMS

TWO WAY PROPAGATION TIME		
	RADIO OR OPTICAL PULSES	RADAR - LASER RANGING
	MODULATION OF A CARRIER	SIDE TONE RANGING
	PSEUDO RANDOM NOISE CODES	
	ONE-WAY	GPS-GLONAS-STARFIX
	TWO-WAY	PRARE-MITREX-GEOSTAR-LOCSTAR
FREQUENCY VARIATIONS - DOPPLER		
	ONE-WAY	TRANSIT-ARGOS-TSIKADA-DORIS
	TWO-WAY	some proposals
PHASE INTERPHEROMETRY		QUASI-VLBI

in the last column, some existing systems are listed, for references /1/

From an atomic frequency standard a binary code is derived , whose properties ,seen in a time window wider of some specific value, are those of a white noise limited in band. But well inside that value, the code is stricty deterministic, in the sense that repetes itself. The code is sent , via a microwave carrier to the spacecraft, where it is trasponded back to the earth station. In a correlator are sent the received code and a suitably delayed replica of the trasmitted code. When cross-correlation is found, the locally introduced delay is in principle equal to the total round-trip time.

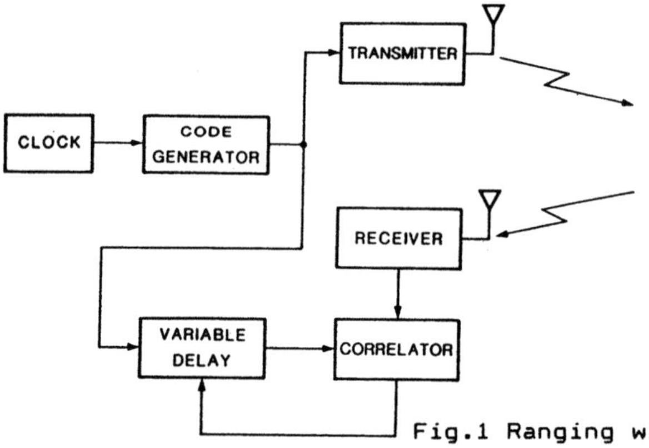

Fig.1 Ranging with a Pseudo Random Noise Code

For deep space operations, i.e. with the spacecraft at some A.U. from the Earth, Hydrogen masers are needed. Moreover if more than one receiving station is involved, it is easy to calculate that the masers must be "Syntonised" to about 10E-13 and "synchronised" within 10 - 30 ns, for accurate and smooth operations.

Let us assume a "chip" length (the duration ot an element inside the code) of 100 ns, a range of 10 A.U. leading to a round trip time of about 10 000 s , a reasonable margin between the drift or the absolute time errors of the local masers and the chip duration; the above mentioned orders of magnitude are obtined.

Frequency variations methods based on the Doppler effect, are very effective and vere historically important because they were unique, for the first fifteen years of the space age, in providing accurate orbital data.

Derivation of the needed accuracies and stabilities are not strigthforward as in the previous cases. It can be shown indeed that the two most stringent requirements came fron the time tagging of the data and from the frequency stability of the two involved oscillator, one on ground , the other spaceborne.

As regards time tagging, let us assume that the speed of the spacecraft is 5 km/s and that an istantaneous position must be time tagged with a maximum error of 10 cm ; a maximum error with respect to UTC of 20 us can be tolerated. But if the time-tagging is made on board and if the clock is corrected once a week, the average fractional frequency error of the spaceborne clock must be less than 3E-11.

As regards the stability problem during the transit of the satellite occurring between t1 and t2, the error in range difference versus the <u>combined</u> frequency drift of the two oscillators during t2-t1 is given by the relation :

$$N = \text{number of beating cycles} = (f_r - f_s)(t_2 - t_1) + f_s(r_2 - r_1)/c$$

where r1 and r2 are the ranges at t1 and t2, respectively, and fs and fr are the frequencies of the satellite and of the receiving station. The transit duration spans from 10 to 1000 s , the frequencies are in the range 150 - 2000 MHz and consequently in those time integration intervals, relative (two oscillators are involved) instabilities in the 1E-11 - 1E-13 are called for range difference errors of 1-10 cm.

The most accurate positioning method is the satellite laser ranging; accuracies of few centimeters in orbit determination are achieved for satellites up to 7000 km such LAGEOS. In June 1988 the European Space Agency, mainly with support of France and Italy , has launched a new experiment called LASSO (LAser Synchronisation from Stationary Orbit) with two aims : nanosecond accuracy in timing and centimeter accuracy in orbit determination at geostationary orbit. This positioning accuracy is needed for the determination of low-order armonics of the geopotential.

The trend of satellite-position accuracy versus time is given in Fig.2 for the different methods ; the figures near the Doppler method are the number of harmonics of the geopotential that were used in the calculations.

To reach an accuracy around one centimeter, two major advances are needed:

- reliable two-color lasers, possibly spaceborne, to overcome the tropospheric delay problems, and
- a time interval counter, with sub-picosecond resolution and accuracy as a streak camera, but with a range up to one second.

To summarize, it can be pointed out that all the recent advances in DEEP SPACE NAVIGATION, PLATE TECTONICS, OCEAN ALTIMETRY AND CIRCULATION, GEODYNAMICS, GEOKINEMATICS, GEODESY, GRAVITY FIELD, via space techniques, were made possible by the use of frequency standards. The same applies for Navigation.

Fig.2 Satellite ranging accuracy versus time

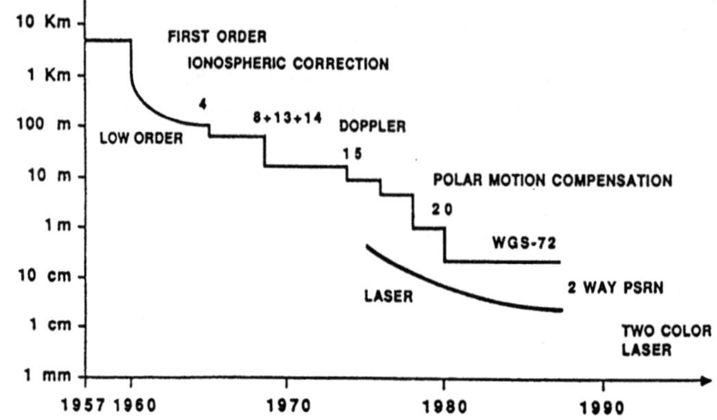

3 Telecommunications

A major evolution is under way in the whole realm of telecommunication; the well esthablished analog techiques, used since the inception of telecommunication with wires or wireless for voice, still pictures and TV, are with an increasing pace replaced by numeric techniques.

The new systems, globally called ISDN (Integrated Sevices Digital Network) are fully digital and " transparent" to the signals , i.e. the system is completely indifferent to the carried signal that can be phone, fax, telex, data, TV, etc.

Any signal is sampled, each sample is digitized and transmitted as a series of " 0 " and " 1 ". In other terms, the signal is converted in a stream of pulses grouped in bounches, called frames. Consequently to recover the digital meaning of any pulse inside a frame, two informations are needed : the beginning of a frame and the position inside the frame .

Beginning means that some synchronisation is needed, position requires the availability of a frequency standards plus a counter . Consequently the same frequency must be available at the two terminals of any link and synchronized clocks must be available in order to recover the information.

These syntonisation and synchronisation needs are provided with a number of methods ans systems /2/, not to be described here .

Here it is sufficient to state that inside a nation a number of clock networks is formed. Inside each network , usually covering one defined geographic area, three to four levels of mutually interconnected frequency standards\clocks are set up.

The main pourpose for this interconnections is to maintain some specific limits to the phase departures between the clocks inside a given level and between the various levels. For instance , for the international networks, the CCITT (International Consultative Committee for Telegraphy and Telephony) dictates /2/ that a specific parameter called TIE time interval error, or total integrated error) must follow the limitation, given in nanoseconds :

$$TIE \leq (0.01\, t + 3000)\ ns, \text{ with } t, \text{ integration time, in seconds.}$$

In order to admit a decoding error every fifteen days, a caesium frequency standards must used ; moreover since in telecommunications the overall relaibilitity is a must, some redundancy is required and consequently at least three frequency standards are used.

Three different criteria are followed in establishing such clock networks.

In the first case a pure hierarchy is adopted with master-slave relations between the various levels.

In a more subtle approach, all the clocks belonging to a specific level - of the same nominal quality by definition - partecipate in sort of internal "voting" and each clock corrects himself with a weigth calculated by a suitable algoritm. Being the premium the stability, usually the weigth reflects the past behaviuor of the interested clock, one possible criterion being the number of frequency corrections that were required in the previous months in order to maintain the phase of that clock inside some given limits.

In the latter instance, sintonisation nor sinchronisation are feasible and consequently each network is time independent ; the system is called "plesiochronous" , but to allow interconnection between the networks the abovementioned requirements must be met and all the clocks at the higher level are by definition caesium standards.

4 Some Unusual Applications

Time and Frequency are the physical quantities that can be measured with higher accuracies at a given costs and efforts and are unique in solving strange problems. No wonder if one finds frequency standards and clocks in very unusual applications. A list of some cases , not all requiring atomic devices, is here presented:

- the docking of a Very Large Crude Carrier (a millimetric doppler radar with extremely high spectral purity: the ship speed is less than 1mm/s ....),
- power dispatching on electrical power lines,
- real time power losses measurement along electrical power lines (looking the phase difference at the ends of the line),
- personal navigation with a credit-card size device (based on the differential absoption of two millimeter wave carriers, one of wich is tuned on the Oxigen absorption line) /3/,
- real time management of high pressure gas-lines,
- car navigation and localisation using Pseudo-random noise codes in hyperbolic mode.

5 Future Applications and Requirements

The most demanding present application of Time/Frequency methods, at least for measurement times between 10 and 10.000 s is Very Large Base Interferometry ; H masers only can provide 10E-15 stability in that time range.

Better results could be reached with wider baselines, not available on Earth ; consequently Space VLBI experiments are under study, with one station on Earth, the second spaceborne.

Two projects now are considered /4/, RADIOASTRON in the USSR, QUASAT inside ESA (European Space Agency) with some USA participation.

Expected improvements are in :

astronomy : motion of extragalactic masers with orbital parallax methods, independent evaluation of the Hubble constant,
astrometry : better links between frame references,
geosciences: better results , mostly for geodynamics.

Very challenging goals are to be met for Space-VLBI :

- stability of the frequency reference on ground : integration times up to 10.000 s are planned at 22 GHz,
- two-way phase transfer from ground to an orbiting satellite at 77.000 km,
- real time measurement of the baseline, with an accuracy better of 10E-9.

More orbiting radiotelescopes, with common frequency references, could pave the road to fundamental science researches /5/, such as the detection of gravitational waves, the establishement in space of man-made reference frames, whose eventual rotation cold be measured, using the Sagnac effect, with a sensitivity of about 2E-12 rad/s.

Another challenging applications will be the millimeter, submillimeter or infrared satellite-satellite links, for communication, tracking, positioning, and in general data relay .

To achieve the needed sensitivity and controlled bandwhidts, the use of heterodyne techniques is considered, with the consequent need of stable and precise frequency sources in those ranges , plus the availability of frequency synthesizers.

Coming to Earth-bound applications , there is the need for an inexpensive frequency standard, with the accuracy for life of a rubidium device (about E-10 - E-11) but with the MTBF (medium time between failures) of a crystal standard (some 150 000 hours); proposed applications are rho-rho navigation and the BITS (Building Integrated Timing System) , a concept evaluated in the USA, in wich all the time disciplined devices inside a building or a limited area, are synchronized by a local clock.

To summarize, it can be stated that atomic frequency standards and clocks will continue to play a fundamental role in Science and Technology ; these devices are indeed instrumental for any progress in Fundamental Constants,
Metrology, Astronomy and Astrophysics, Geosciences, Communication , electronic instrumentation , and Navigation on land,seas, air and space.

References:

1 - S.Leschiutta, U.Pisani:Uses of Precise Time and Frequency, 1er Forum Europeen Temps-Frequence,p.279-283,1987.
2 - R.D.Calvert : Timing Requirements of future ISDNs , 2nd European Frequency and Time Forum, p.71-98, 1988.
3 - M.Lanchais : The INDEX new compass and new media system, Proc. ITTT Seminar (ISBN 0-86050-187-6) ,p.241-252,1987.
4 - I.Feyes et Al.: On astrometric and geodynamics aspects of Space VLBI,Intercosmos Scientific Conference, Szentendre, Ungary, 1987.
5 - dr.Robert Vessot, personal communication.

In Search of the Best Clock – an Update

D.W. Allan

Time and Frequency Div., National Institute of Standards and Technology,
Boulder, CO 80303, USA

Because of the increased need for better clock performance than is currently available, this paper addresses some fundamental questions regarding clock metrology. Heretofore, most work has focussed on improving the clocks to meet the increased need. Though this is fundamental, we will show that significant gains are also available through the algorithms (computational methods for optimally combining the information) which process the readings of the clocks and through international comparisons now available via satellite. Proper algorithms for processing seem to be more important than the proportionate attention generally given them. In fact, to date, the only way we have been able to investigate some of the outstanding time predictability in long-term of the millisecond pulsar, PSR 1937+21, is by using such optimization algorithms.

Introduction

In the search for the best clock, we explicitly mean the clock with the best long-term frequency stability, which in turn can be shown to lead to the minimum long-term time dispersion [1]. This paper shows that such a clock is not a single physical device, but rather an optimal computed combination of information from a set of clocks and frequency standards. There are three important steps in time and frequency metrology for producing a best clock. <u>Step one</u> is to have an accurate primary frequency standard; the better the frequency accuracy, the better will be the achievable long-term frequency stability. <u>Step two</u> is to have a precise and accurate method of accumulating the time intervals defined by the available frequency standards. <u>Step three</u> is the optimum use by some computer algorithm of the time differences among all the clocks involved and of the frequency calibrations with the primary frequency standards. The deviations of the optimally computed time and frequency are less than those of the best contributing member clock [2-4].

There are now sufficient data on the long-term stability of clocks from high accuracy satellite time and frequency comparison techniques that we can view most of the best clocks in the world as if they were in the same laboratory. Therefore, we are now at a point in international time and frequency metrology where we have

Frequency Standards and Metrology Editor: A. De Marchi
© Springer-Verlag Berlin, Heidelberg 1989

never been before. We can combine the time differences and frequency calibrations from the best clocks and frequency standards in the world without long-term deterioration and generate a "world's best clock" using an optimization algorithm. The clocks studied are primary cesium beam frequency standards, hydrogen masers, mercury ion frequency standards and various commercial atomic clocks. All of the primary timing centers' outputs, which are available to NBS via the Global Positioning System's common-view measurements, were used in the study. The millisecond pulsar signal from PSR 1937+21 was used as well. The efforts to search for the "best clock" in this study have the advantage that post analysis better helps us characterize and deal with systematic deviations.

2. Current Clock Performance and Some Future Considerations

The stochastic behaviors of some of the main frequency references are plotted in Figure 1 [5]. The frequency instabilities resulting from systematic and environmental perturbations are also of basic importance. The causes of these instabilities have to be considered in order to approach "best clock" performance [6,7].

Figure 2 illustrates an idea that uses technology which is nearly available and which needs a digital servo or computer processor to appreciate the theoretical estimation of frequency stability of 1 part in 10^{16} for an averaging time of about

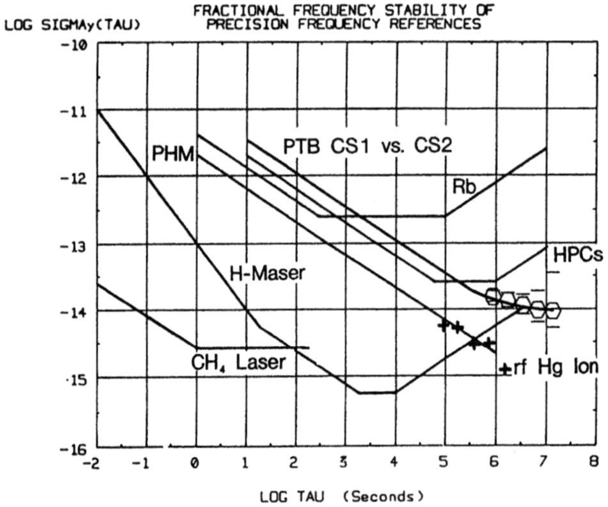

Figure 1. A $\sigma_y(\tau)$ versus τ plot of the measured stability of some of the best frequency references. The hexagons are for PTB CS1 versus PTB CS2. Rb is for a commercial rubidium gas-cell frequency standard. Better long-term stability performance has been observed [10,13]. The "HPCs" stand for a high performance commercial cesium-beam frequency standard and "PHM" for a passive hydrogen maser. The "+" symbols are compliments of the USNO for their rf trapped mercury-ion standard versus their active hydrogen maser, M19, for 36 days of data. "H-Maser" stands for an active hydrogen maser with the measurement system bandwidth, f_h, set at 500 Hz; the stability shown is $1/\sqrt{2}$ times the stability between a pair -- under the assumption of equal contribution. "CH_4" is for methane-stabilized laser.

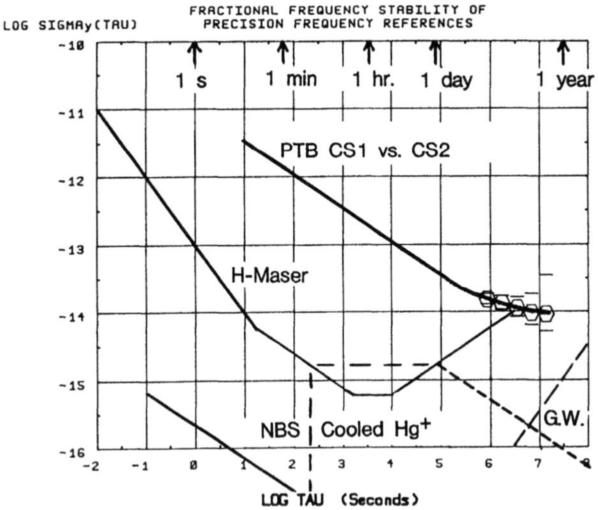

Figure 2. An illustration of a digital and/or computer servo technique to achieve a stability of one part in 10^{16} at $\tau = 1$ year.

1 year. The idea is to use very high accuracy (10^{-16} or better) frequency standards that will soon be available in the optical or near optical region of the spectrum. A frequency divider and/or multiplier chain would be needed for operation for a few hundred seconds once a day in order to calibrate the rate of an active hydrogen maser, which could then carry the memory of that frequency day by day. If the uncertainties in each calibration were limited primarily by the instability of the maser, and these uncertainties were random and uncorrelated from day to day, then the maser could be made to track the frequency of the optical standard (at least in a software sense). The knowledge of the optical standard's frequency would then improve as $\tau^{-1/2}$ down to the accuracy limit of the optical standard itself.

3. Ensemble Time

We give below some examples of applications of ensemble time concepts.

A. The NBS ensemble algorithm from which UTC(NBS) is derived features two weighting factors per clock [2]. In theory, the algorithm's free-running output, NBS(AT1), should have better short-term and long-term stability than any member clock(s) in short-term and/or in long-term stability. Another feature of the NBS algorithm is that a clock's weighting factors are adaptive, conforming to the performance of each clock as it ages -- sometimes getting better and sometimes getting worse toward the end of clock life. This algorithm's output has been tested using other time scale data and using simulated data. The tests appear to be consistent with the theory.

B. Recently, a global Kalman ensemble time was generated using the data from all the principal timing centers available via GPS [8]. Figure 3 is a plot of the

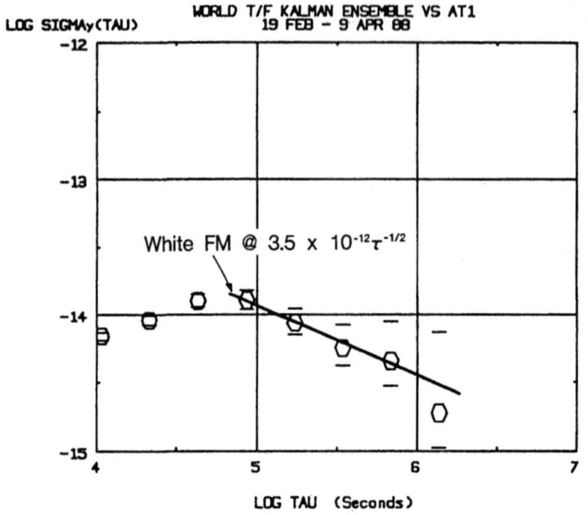

Figure 3. Frequency stability of NBS(AT1) or UTC(NBS) with respect to a Global ensemble taken via GPS in common-view [8]. The Kalman smoother characteristics cause the stability values to be too optimistic at τ = 1/8, 1/4, 1/2 day.

frequency stability between this ensemble time and NBS(AT1). For sample times shorter than one day the time/frequency Kalman smoother biases the stability values which are too low. For τ greater than or equal to one day the plotted values obviously fit the theoretical curve very well.

C. The Global Position System is anticipating the use of ensemble time both at the Master Control Center and for the satellite vehicle (SV) clocks. To show the quality of such an approach we employed the NBS time scale algorithm used in generating AT1 and applied it to the SV clocks [9]. This set of space clocks also gave us an independent reference significantly decoupled from the usual environmental disturbances. The stability plot of $\sigma_y(\tau)$-versus-τ of this space ensemble against NBS(AT1) gave rise to a performance which was better than that of most of the major timing centers in the world [9,10]. The long-term stability was well modeled by a flicker noise FM ($\sigma_y(\tau)$ was independent of the sample time, τ) for 10 days < τ < 1/2 year at a level of about 2×10^{-14}.

D. TTBIPM88 ensemble time, as generated by Guinot at BIPM, is based on a an algorithm with special weight functions for the set of about 160 clocks in order to meet the most stringent requirements for long-term stability, such as for the studies of millisecond pulsars [5]. Most of the clocks in TTBIPM88 are commercial cesium clocks, a few are primary cesium clocks, and a few are hydrogen masers. TTBIPM88 ensemble time along with the USNO master clock (with the coordinate time steering removed to make it an independent ensemble time), the compensated arrival times of the millisecond pulsar (PSR1937+21) and the NBS(AT1) ensemble time were used in an N-cornered-hat relationship to estimate the stability of each of the four. The estimated stabilities are plotted in Figure 4.

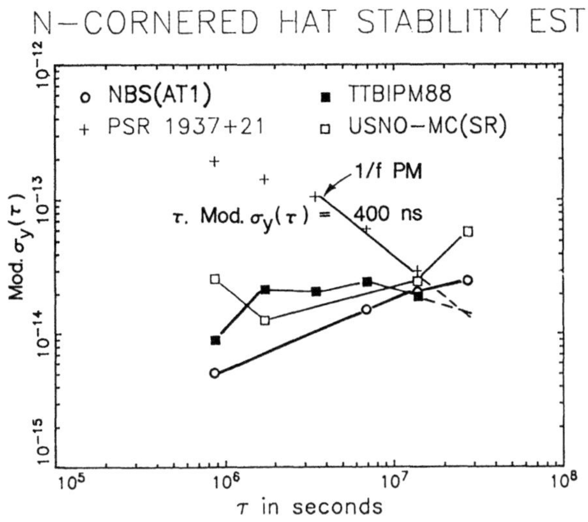

Figure 4. An N-cornered hat frequency stability estimate with four nominally independent contributing clocks: millisecond pulsar PSR 1937+21, NBS(AT1), TTBIPM88 and UTC(USNO SR) (The USNO master clock with steering removed). Because of correlations and apparent correlations, due to finite data lengths (this data set covered 1060 days), negative variances (not plotted) may result.

Because TTBIPM88 shows annual variations relative to some other time scales; for example, TAI, and because it so highly dependent in long term on the stabilities of the PTB primary cesium standards, a study was performed to determine whether long-term correlations in long term exist between the PTB Cs 1 and the PTB Cs 2 standards. Changes in humidity have been identified as one probable driving mechanism [10-12]. The correlation analysis was done by observing these two standards against other presumably independent time scales, and by using the cross-correlation technique used in reference [10]. Figure 5 is a time residual plot of

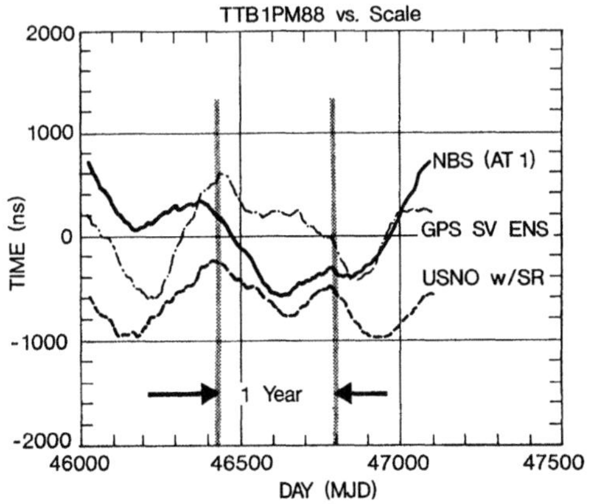

Figure 5. A plot of the time residuals of TTBIPM88 versus three other independent time scales - showing an apparent correlation in the annual variations (changes in slopes).

TTBIPM88 with respect to three other time scales and shows evidence of annual correlations. That one of the time scales was the GPS SV ensemble would rule out a humidity-coupling mechanism between it and the other two. Figure 6 shows the results of the cross-correlation analysis using NBS(AT1) and the GPS SV ensemble as the other independent scale to test for correlation between PTB Cs 1 and PTB Cs 2. The two independent reference scales were interchanged in their roles, and the correlation curve remained essentially the same as shown in Figure 6.

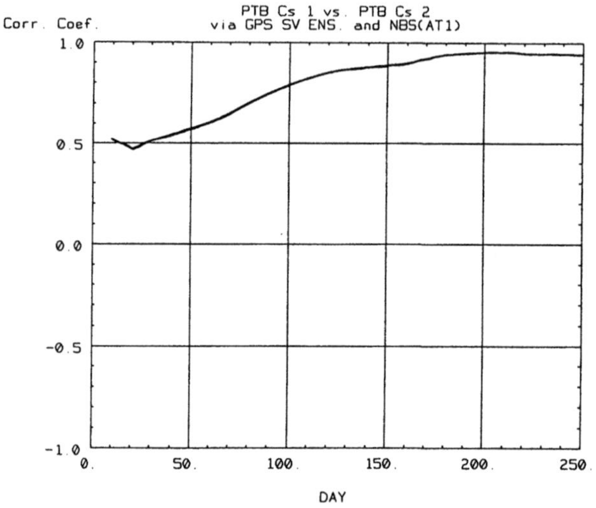

Figure 6. A cross-correlation analysis between PTB Cs 1, PTB Cs 2, NBS(AT1) and GPS SV ensemble time - showing evidence of annual correlations between the two primary cesium beam standards.

4. Conclusions

A few of the more significant time and frequency metrology advantages which arise from using appropriate processing algorithms have been demonstrated. Without the use of clock ensemble algorithms it would have been impossible to estimate the stability of the millisecond pulsar PSR 1937+21 at one year sample time at the current level (Mod.$\sigma_y(\tau = 360$ days) $\lesssim 1.5 \times 10^{-14}$). Within about 10 years we project that, using advanced clocks which are now being developed and the ensemble methods described here, the level of stability 1×10^{-15} for $\tau = 1$ year will be achieved. At this level of atomic-time stability there is a possibility for observing gravitational waves in comparisons with millisecond pulsars, that is, assuming that uncertainties in millisecond pulsar time due to earth orbit uncertainties are sorted out.

The time scale TTBIPM88 is a scale constructed in retrospect and appears to be the best long-term stable time scale yet constructed. In the long term PTB's primary cesium clocks are given high weights and are credited with making a major contribution toward this excellent long-term stability (sampling times of about a year and longer). We have shown some evidence that the random fluctuations between PTB Cs1 and Cs2 may not be random and uncorrelated; i.e., there may be some annual correlated movement of the two primary cesium beam standards, which in turn could induce an annual term in TTBIPM88. The size of the deviation seems to be of the order of 300 ns to 400 ns in half a year. This tentative conclusion needs further study for confirmation.

Better algorithms are being studied and insights into environmental effects on clocks are being gained. Future algorithms will need to make better use of information theory, maximum likelihood estimation techniques, Kalman filters, efficient statistical processing, models for systematic effects and of ways to compensate for environmental perturbations. With continued data sharing and international cooperative effort, all time scales stand to benefit. The output of such cooperation can be the "best clock" possible.

5. References

1. D. W. Allan, "Time and Frequency (Time-Domain) Characterization, Estimation, and Prediction of Precision Clocks and Oscillators," IEEE Transactions on Ultrasonics, Ferroelectrics, and Frequency Control, UFFC-34, 647-654, 1987.

2. D. W. Allan, J.E. Gray, and H.E. Machlan, "The National Bureau of Standards Atomic Time Scale: Generation, Stability, Accuracy and Accessibility," Chapter 9 in Time and Frequency: Theory and Fundamentals, ed. B.E. Blair, NBS Monograph 140 (Out of Print); D.W. Allan and M.A. Weiss, "The NBS Atomic Time Scale, AT1," NBS Technical Note 1316, 1988.

3. D. M. Granveaud and B. Guinot, "Atomic Time Scales," IEEE Trans. Instr. Meas., IM-21, 396, 1972.

4. B. Guinot, "Some Properties of Algorithms for Time Scales," Metrologie, 24, 195-198, 1987.

5. Private communication: Dr. DeBoer of the PTB, Dr. Winkler of the USNO Dr. Wineland of the NBS, Dr. Drullinger of the NBS, and Dr. B. Guinot of BIPM.

6. CCIR Report 364-5, "Performance of standard-frequency generators", Standards Frequencies and Time Signals, Vol. VII, 120-129, 1986.

7. F. L. Walls, "Frequency Stability of a Locked Oscillator," to be published in Proceedings of Fourth Symposium in Frequency Standards and Metrology, Ancona, Italy, September 6-9. 1988.

8. M.A. Weiss, "The Design of Kalman Smoothers for Global Positioning Data," submitted to CPEM Digest, 1988.

9. D.W. Allan and T.K. Peppler, "Ensemble Time and FRequency STability of GPS Satellite Clocks," to be published in Proc. of 42nd Annual Symposium on Frequency Control, Baltimore, MD, June 1-4, 1988.

10. D.W. Allan, "A Study of Long-Term Stability of Atomic Clocks," Proceedings of the Precise Time & Time Interval Planning Meeting, December 1987.

11. E. Bava, F. Cordara, V. Pettiti, and P. Tavella, "Analysis of the Seasonal Effects on a Cesium Clock to Improve the Long-Term Stability of a Time Scale, to be published in the Proceedings of the 19th Annual Precise Time and Time Interval Planning Meeting, Redondo Beach, CA, December 1-3, 1987.
12. J. E. Gray, H. E. Machlan and D. W. Allan, "The Effect of Humidity on Commercial Cesium Beam Atomic Clocks," to be published, Proc. of the 42nd Annual Symposium on Frequency Control, 1988.
13. F. Danzy and W. J. Riley, "Stability Test Results for GPS Rubidium Clocks," to be published in the Proceedings of the 19th Annual Precise Time and Time Interval Planning Meeting, Redondo Beach, CA, December 1-3, 1987.

Comparison of Primary Frequency Standards

B. Guinot and J. Azoubib

Bureau International des Poids et Mesures, F-92312 Sèvres Cedex, France

1. Introduction

This report covers the interval 1976.0 to present. It deals with all the data of primary frequency standards which have been made available to the Bureau International de l'Heure (BIH), then to the BIPM. A list of these standards, together with information, is given by Table I. In the following, the correction for the gravitational frequency offset is made: all the frequencies are expressed at sea level.

Table I. Primary frequency standards contributing data to BIH/BIPM since 1976

Laboratory	Designation of the standard	Interrog. length [cm]	Type of operation	Stated uncert. $[10^{-13}]$	Ref.
Communications Research Laboratory (Japan)	CRL-CS1	55	discontinuous, once or twice per year	1	(1)
National Bureau of Standards (USA)	NBS-6	374	discontinuous, once per year to once per 3 years	1	(2)
National Research Council (Canada)	NRC-CsV	210	primary clock since 1975 May 1	0.5	(3)
	NRC-CsVI A, B, C	100	primary clocks since 1979	0.7	(4)
Physikalisch-Technische Bundesanstalt (Fed. Rep. of Germany)	PTB-CS1	80	discontinuous until 1978 July, then primary clock	0.3	(5) (6)
	PTB-CS2	80	primary clock since 1985	0.2	(7)
Gosstandart (USSR)	MCsR 101 MCsR 102	?	discontinuous at 1 to 2 months interval	1	

2. The Common Reference for Comparisons, the Annual Frequency Variations

Some of the standards are operated in a discontinuous mode; their comparison requires the use of a common reference with the best possible long term stability. This

Frequency Standards and Metrology Editor: A. De Marchi

reference is here the time scale TT(BIPM88), which has been established by BIPM for the most demanding applications, in particular for millisecond pulsars. TT(BIPM88) has been computed according to the method described in (8), which aims to combine
- the stability of the International Atomic Time TAI up to about 3 years,
- the accuracy and therefore the stability over longer averaging times of NRC-CsV, PTB-CS1 and PTB-CS2.

In preparing TT(BIPM88), we were faced to the well known seasonal variation which appears between the time scales based on industrial cesium clocks and the primary standards (9,10). Let us consider the facts.

(a) In 1984, NRC-CsV had troubles. For the years 1979-1983, expressing the normalized frequency difference between NRC-CsV and PTB-CS1 by a constant and an annual sine wave gives, the time t being in years,

$$[y(\text{NRC-CsV}) - y(\text{PTB-CS1})] \times 10^{14} = \begin{array}{l} 0.7 - 0.6 \sin 2\pi t + 1.3 \cos 2\pi t \\ \pm 0.9 \pm 1.3 \qquad\qquad \pm 1.3 \end{array}$$

The agreement of the mean frequencies is good; the seasonal difference is small and not statistically significative.

(b) With the mean of NRC-CsV and PTB-CS1, one obtains the following annual term:

$$1976\text{-}78,\ [y(\text{TAI}) - y(\text{st.})] \times 10^{14} = 1.4\ (\pm 1.3) \sin 2\pi t + 5.0\ (\pm 1.3) \cos 2\pi t.$$

Later, by averaging the frequencies of PTB-CS1 and PTB-CS2:

$$1987,\ [y(\text{TAI}) - y(\text{st.})] \times 10^{14} = 1.3\ (\pm 0.8) \sin 2\pi t + 0.7\ (\pm 0.8) \cos 2\pi t.$$

(c) The sensitivity of industrial frequency standards to environmental conditions, especially to humidity, has been often reported (see, for instance, 11, 12, 13).

A way to reconcile these facts is to assume that the primary standards NRC-CsV, PTB-CS1 and PTB-CS2 are free from annual variations, that the annual variation lies in TAI, as a consequence of environmental effects on the numerous industrial cesium clocks which enter in its establishment, that the decrease of the amplitude of the annual variation between TAI and the primary standards reflects improvements of the condition of operation of the industrial cesium clocks and/or the new participation of clocks under different climates. But it is not a proof that the primary standards are not at all influenced by seasonal effects. By an analysis of the data of satellite-born clocks of the Global Positioning System (GPS), D.W. Allan from the National Bureau of Standards (NBS) came recently to the conclusion that the PTB primary clocks might have annual variations. The other standards listed in Table I are not accurate enough to bring any light on these matters. The question remains open. We can nevertheless settle an upper limit of the order of 3×10^{-14} (peak to peak) to the seasonal frequency variation (if any) of the best primary standards.

As NRC-CsV has not been used for evaluating TT(BIPM88) during intervals of malfunction in 1984/85 and 1987, this time scale has a frequency which remains close to that of PTB-CS1, as shown by Fig. 1. It is therefore interesting to check the stability of PTB-CS1 with respect to the quasi-independent time scale EAL established

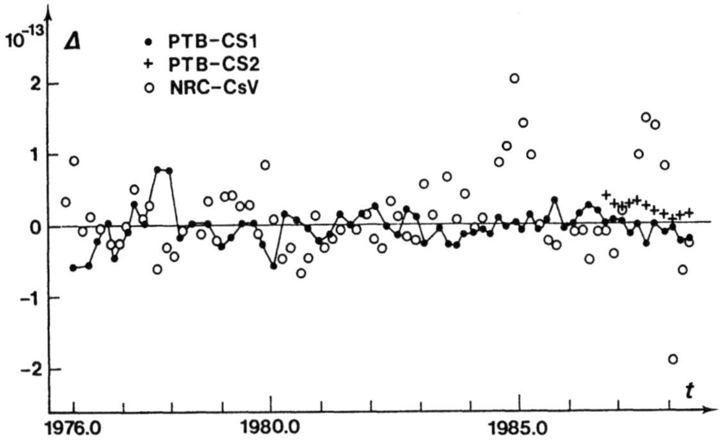

Fig. 1. Δ is y(standard) - y(TT(BIPM88))

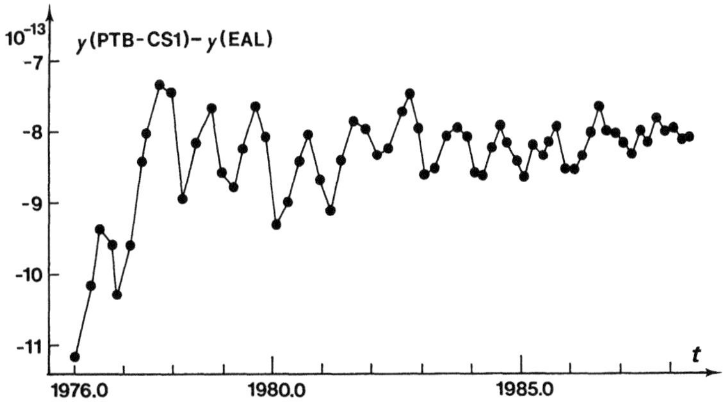

Fig. 2. Comparison of the frequency of PTB-CS1 and EAL

by BIH/BIPM with the data of 75(1976) to 160(1987) atomic clocks. As seen on Fig. 2, a change of frequency occurred in 1976/77, which is not explained. Subsequently, the long-term stability of the frequency difference is:

$$\sigma_y(1 \text{ year}) = 1.4 \times 10^{-14}, \qquad \sigma_y(2 \text{ years}) = 1.5 \times 10^{-14}.$$

No secular frequency change above a few units of 10^{-14} is observed. Such a result is compatible with the estimated stability of EAL (8) and accuracy of PTB-CS1. It is thus very probable that the secular frequency variation of TT(BIPM88) is also below a few units of 10^{-14}, since 1977.

On Fig. 1, the annual component with decreasing amplitude is clearly apparent. We have assumed in evaluating TT(BIPM88), that the NRC and PTB standards have no annual frequency variation.

3. Transfer of the Frequency Measurements to the Common Reference TT(BIPM88)

(a) Standards operated in a discontinuous mode. Usually the laboratories k provide F(UTC(k)) - F(standard) over a specified interval. The transfer to TAI, then to TT(BIPM88) involves long distance time comparisons and also the stability of UTC(k) and TAI. This problem has been considered in (14). At present, with the exception of the USSR Gosstandart, all the primary laboratories are linked by GPS and the noise introduced by the transfer to TT(BIPM88) is negligible. The Gosstandart provides the frequency of UTC(SU) on intervals ranging from a few days to one month. Unfortunately the noise of the time comparisons with other laboratories requires that the frequency difference F(TAI) - F(UTC(SU)) be estimated over 6 months.

(b) For the primary clocks, we have used frequency samples over 60 days, which are now practically free from time comparison errors. The PTB/NRC link was as follows:

1976 Jan. - 1978 May	LORAN-C
1978 May - 1980 Feb.	{ NRC-OP by SYMPHONIE (15) OP-PTB by LORAN-C
1980 Feb. - 1982 Aug.	SYMPHONIE
1982 Aug. - 1983 June	LORAN-C
1983 June - 1984 June	{ NRC-USNO by LORAN-C USNO-PTB by GPS
1984 June - present	GPS

Most of the time, satellites have provided a stable link between these laboratories.

4. Results

Figure 1 shows that NRC-CsV and PTB-CS1 agree normally quite well. The discrepancies in 1985/86 and 1987 are due to recognized defects in the functioning of NRC-CsV. PTB-CS1 and PTB-CS2 agree within the limits of their stated uncertainties; but there is a permanent offset of about 2×10^{-14}.

The frequencies of NRC-CsVI, A, B and C (Fig. 3) differ from the frequency of TT(BIPM87), in the average for 1980-1988, by respectively 4.1, - 1.1, 3.7 in 10^{-14}.

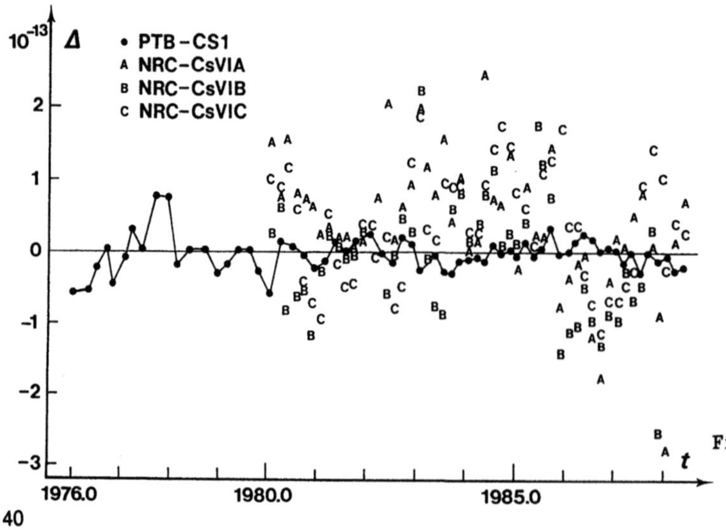

Fig. 3. Δ is y(standard) - y(TT(BIPM88))

The standard deviations of the 60-day frequency samples are 9.2, 8.9, 5.9 in 10^{-14}. Unfortunately, there are correlations both in time and among the standards.

Figure 4 shows the data for CRL-CS1 and NBS-6. The standard deviation of the frequency difference with TT(BIPM88), and practically with PTB-CS1 also, are: 12×10^{-14} for CRL-CS1; 11×10^{-14} for NBS-6. These figures agree with the stated uncertainties.

The results of the USSR standards (Fig. 5) raise more embarrassing questions. They show a good consistency, but an offset of 25×10^{-14} appeared in 1984 and persists, for which we have no explanation.

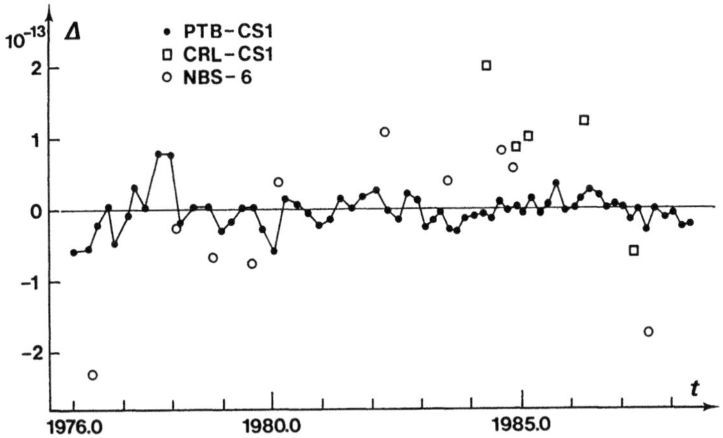

Fig. 4. Δ is y(standard) - y(TT(BIPM88))

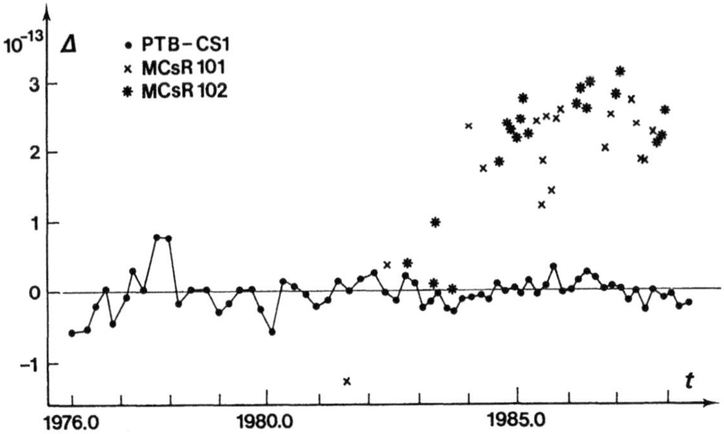

Fig. 5. Δ is y(standard) - y(TT(BIPM88))

5. Conclusion

For several years the agreement of NRC-CsV and PTB-CS1 has been most satisfactory. Later, PTB-CS2 confirmed the results of PTB-CS1. But the other standards which appeared did not reach a sufficient level of accuracy to provide useful comparisons. In addition, we are, for some intervals, left with a single laboratory which carries almost alone the weight of the accurate realization of the SI second, and of the frequency steering which ensures the accuracy of the unitary scale interval of TAI.

References

1. K. Nakagiri, M. Shibuki, S. Urabe, M. Ishizu, Y. Ohta, T. Morikawa, Y. Saburi: Proc. 38th annual Freq. Control Symp., 447 (1984)
2. D. Glaze, H. Hellwig, D. Allan, S. Jarvis: Metrologia 13, 17 (1977)
3. A. Mungall, C. Costain: Metrologia 13, 105 (1977)
4. A. Mungall, J.-S. Boulanger: IEEE Trans. IM-29, 291 (1980)
5. G. Becker: Metrologia 13, 99 (1977)
6. A. Bauch, H. de Boer, B. Fischer, T. Heindorff, P. Hetzel, R. Schröder: Actes du Congrès de Chronométrie, 59 (1986)
7. A. Bauch, K. Dorenwendt, B. Fischer, T. Heindorff, E.K. Müller, R. Schröder: IEEE Trans. IM-36, 613 (1987)
8. B. Guinot: Astron. Astrophys. 192, 370 (1988)
9. B. Guinot, J. Azoubib: IEEE Trans. IM-29, 226 (1980)
10. B. Guinot, M. Granveaud, J. Azoubib: J. Inst. of Electronics and Telecomm. Engineers, India, 27, 524 (1981)
11. S. Iijima, K. Fujiwara, H. Kobayashi, T. Kato: Annals Tokyo astr. Obs. 17, 50 (1978)
12. F. Cordara, V. Pettiti, P. Tavella: Proc. 2nd European Frequency and Time Forum, 469 (1988)
13. J.E. Gray, H.E. Machlan, D.W. Allan: Proc. 42th annual Freq. Control Symp. (1988)
14. J. Azoubib, M. Granveaud, B. Guinot: Metrologia 13, 87 (1977)
15. C.C. Costain et al.: Proc. 11th PTTI, 499 (1979)

Part III

Established Microwave Standards

Established Microwave Frequency Standards

H. Hellwig

National Institute of Standards and Technology,
(formerly the National Bureau of Standards), Gaithersburg, MD 20899, USA

Introduction

Atomic and molecular beam physics, which began with the experiment of O. Stern and W. Gerlach in 1921, is the foundation of today's atomic time and frequency standards. The following papers discuss progress in the area of microwave frequency standards and clocks. Microwave devices are defined, for our purposes here, as those atomic resonators which feature resonances in the microwave region of the electromagnetic spectrum. The combination of atomic and molecular beam physics with the idea to use an oscillator-driven magnetic field to induce transitions was born in the 1930's and led to molecular beam spectroscopy. The invention of efficient sources for microwave radiation in the 1940's as a result of the Second World War made practical devices possible. The manipulation of atomic excitations through optical pumping together with the invention of the maser-laser principle in the 1950's have combined to allow novel realizations of atomic resonators. The advent of solid state electronics permitted the construction of clocks and frequency standards of considerable reliability, together with highly accurate performance.

Although, during the last 40 years, numerous atoms and molecules with microwave resonances have been studied for frequency standard purposes in laboratory as well as in commercial prototype versions, only three types of atomic clocks have reached maturity. Respectively, they are based on the resonances of cesium at 9.2 GHz, rubidium at 6.8 GHz and hydrogen at 1.4 GHz. Although each of these three atoms has been used with a variety of experimental techniques, three particular realizations have dominated their use and have found their way into commercial practice: For cesium, it is the atomic beam using a Ramsey-type cavity and a surface-ionization hot-wire detector. For rubidium, it is a gas cell resonator filled with a suitable buffer gas; optical pumping is used for population inversion, and detection is accomplished with a photodetector. For hydrogen, it is the maser principle with atomic hydrogen confined to a coated storage vessel within a microwave cavity; detection of the microwave oscillation is performed using a superheterodyne receiver.

All three resonances found early commercial realizations. The cesium beam standard was the first commercially-produced device in the late 1950's. During the 1960's, rubidium gas cell and hydrogen maser frequency standards also became available from commercial sources. Today, commercial availability of all three device types is assured through more than one source for each type. The total number of devices in use worldwide has reached surprisingly large numbers. There are of the order of 10,000 devices in use for cesium as well as rubidium and of the order of 100 devices are deployed in the case of hydrogen masers. Their applications span a wide variety of uses ranging from fundamental spectroscopic experiments and tests of theories of relativity to commercial television and nuclear submarines. The majority of applications fall into two major areas: navigation and communication. In navigation, the precision timing available from atomic clocks is the main reason behind their applications. Time translates into distance via the speed of light, i.e., 3 nanoseconds are the equivalent of approximately one meter. Atomic clocks are at the heart of signal emissions used for navigation in both terrestrial (e.g., LORAN C) and space systems (NAVSTAR/GPS) as well as providing interpolation capability and autonomy to the navigating vehicles, whether they be stationary,

Frequency Standards and Metrology Editor: A. De Marchi

ground vehicles, ships, aircraft or spacecraft. In communication, the principal feature of the atomic standards is their high frequency accuracy and stability. For example, interface specifications at the major nodes of communication networks are now at the 10^{-11} accuracy level. Again, in some similarity to their deployment in navigation, the use of atomic frequency standards in communication is characterized by use as source of signal emissions in major networks (television, data communication, and voice), but also to provide interpolation capabilities and autonomy to local or regional networks and to individual users allowing them to bridge outages and to more easily phase in and out of major networks.

The cesium beam standard holds the record for accuracy with a current performance of parts in 10^{14}; also, the cesium beam principle has been used since 1967 to define the unit of time (length of the second). The active hydrogen maser is the record-holder for frequency stability featuring instabilities as low as parts in 10^{16} for sampling times of hours. Rubidium gas cell devices excel in utility, particularly because of their very compact size which is of the order of 100 cubic centimeters. It is interesting to note that the corresponding numbers, two decades ago, for all three areas of performance were about two orders of magnitude larger in each case. This historical development should give us confidence that further improvements are still possible. The following papers provide a discussion of some of these opportunities. However, we have to realize that these established devices are challenged. Technical competition has come for some time from the application of optical principles and optical transitions in atoms and molecules. More recently, a fourth type of microwave frequency standard has entered as a serious contender. It is based on the use of trapped ions. The U.S. Naval Observatory has operated mercury ion storage devices for over 2 years and currently has three such devices in use. They are commercial prototypes featuring mercury ions stored in an electromagnetic rf trap with optical pumping and optical detection of the resonance at 40.5 GHz. Optical pumping and microwave interrogation alternate to avoid light shifts. Preliminary results have shown a long term stability equal to that of the whole ensemble of cesium and hydrogen clocks at the USNO. Nevertheless, we have to realize that any new device, including the mercury ion frequency standard, in order to become a serious contender for commercial use has to meet two challenges: (a) it must equal or exceed the exquisite combination of size, cost and performance (frequency stability and accuracy) which hydrogen, rubidium and cesium deliver in unique ways, and (b) it must match the considerable level of technical refinement and proven reliability in the field of today's commercial devices.

A Point of View on the Accuracy Performances and on the Prospects of Improvement of Primary Cesium Beam Frequency Standards

C. Audoin

Laboratoire de l'Horloge Atomique,
Equipe de Recherche du CNRS associée à l'Université Paris-Sud,
Bât. 221, Université Paris-Sud, F-91405 Orsay Cedex, France

A status of the frequency accuracy claimed by the different laboratories operating primary cesium beam frequency and time standards is given. The main design choices made are summarized. Recent progress in the theoretical and experimental analysis of some frequency shifts is reported. Frequency shifts whose study require further attention are pointed out. Possible ways of improving the accuracy are summarized. It is concluded that research work on the behaviour of traditional primary cesium beam frequency standards must be actively pursued.

1. Presently Operated Primary Cesium Standards

All the standards considered in this review use magnets for the state selection of atoms as well as for the detection of the microwave resonance.

Most of the primary standards already in operation at the time of our 1981 Aussois meeting are still in service without major design changes. Two new standards have been made working, at VNIIFTRI, in USSR, in 1982 and at PTB, in FRG, in 1985. Table 1 gives the claimed accuracy figures of the twelve standards which, according to my knowledge, are presently available /1-8/. Five of them are operated as clocks /9/.

2. Main Features of the Standards

At the present time, primary cesium beam standards have a horizontal beam. In most of them, the direction of the beam can be reversed, providing the capability of measuring the cavity phase shift.

Two different types of magnet state selectors are used /10/. This leads to a beam cross section which is either rectangular or cirular. In the second case, the holes in the microwave cavity walls are smaller. Consequently, the probability that neighbouring $\Delta F = \pm 1$, $\Delta m = \pm 1$ transitions occur is reduced. Furthermore, the effect of the distributed cavity phase shift is also reduced and the requirements for the beam retrace upon beam reversal can be more easily fulfilled (except for the effect of the atom free fall). In all cases, the characteristics of the state selectors and the beam geometry determine the velocity range of atoms striking the detector. The mean value, $\bar{v}$, of the velocity is comprised between approximately 90 and 300 ms^{-1}.

Frequency Standards and Metrology Editor: A. De Marchi

Table 1. Accuracy figures of the primary cesium beam frequency standards considered. The date is that at which the accuracy has been evaluated.

Institution	Designation	10^{14} x Accuracy	Date
PTB	Cs 1	3	1986
	Cs 2	1.5	1988
NRC	Cs V	10	1986
	Cs VI-A	10	1988
	Cs VI-C	10	1988
Gostandart	MCsR 101	10	1987
	MCsR 102	10	1987
NBS	NBS-6	20	1987
CRL	Cs 1	7	1986
NRLM	NRLM-II	2.2	1985
NIM	Cs III	25	1987 (?)
SO		60	1987

The linewidth, $\Delta\nu$, of the Ramsey central line depends also on the separation L between the interaction regions. We have $\Delta\nu \simeq \bar{v}/2L$.

The atomic beam may be parallel to the small side of the waveguide cross section. Then the amplitude of the µw field varies abruptly at the beam entrance and exit. It follows that the Rabi pedestal shows pronounced sidelobes. In the second possible configuration, the beam is parallel to the large side of the waveguide cross section. The amplitude of the µw field varies smoothly, with a subsequent attenuation of the pedestal sidelobes /11/. In the second case, the static magnetic field is parallel to the beam direction instead of being perpendicular. This field is then created by a long solenoïd and is less sensitive to imperfections of the magnetic shield shape. Thus, the configuration where the atomic beam is parallel to the large side of the guide cross section shows advantages.

The separation between the two interaction regions takes noticeably different values,from 0.55 to 3.7 m, but without a correlation with the accuracy.

In many of the standards of Table 1, the atomic resonance is interrogated using a square-wave frequency modulation. In some of them, the beam tube response is processed digitally.The related advantages are described in the literature /12/,/13/.

3. Progress in the Knowledge of Frequency Shifts

3.1. Millman Effect and Measurement of the Static Magnetic Field

A theoretical analysis has clearly shown that the Millman effect, although observed for $\Delta m = \pm 1$ transitions cannot occur for $\Delta m = 0$ transitions /14/. Consequently, it cannot displace the F = 4, m = 0 $\longleftrightarrow$ F = 3, m = 0 reference transition.

Measurement of the frequency separation between purposely chosen microwave field dependent $\Delta m = 0$ transitions has been recommended for an accurate calculation of the second order Zeeman frequency shift /15/.

3.2. Distributed Cavity Phase Shift

The frequency shift related to the phase gradient in the interaction regions has been neatly measured /16/. The phase gradient was found about twice the theoretical value, which was attributed to wall roughness of the waveguide.

It has been shown that the phase gradient can be greatly reduced at a given place in a ring cavity /17/. Preliminary experimental results confirm the interest of replacing each of the Ramsey cavity end sections by such a ring /1/.

3.3. Frequency Pulling by Neighbouring Transitions

It has been shown that the frequency pulling due to asymmetries of the neighbouring $\Delta m = 0$ lines (and to a lesser extent to that of $\Delta m = \pm 1$ lines) may be, in fact, noticeably larger than routinely estimated in the past /11/. Furthermore, the value of the frequency shift oscillates when the value of the C-field is changed and it usually exists values of this field for which the frequency shift is equal to zero.

Measurements have shown that the frequency shift behaves as predicted theoretically /1/,/18/.

3.4. Influence of Modulation and Demodulation on Frequency Shifts

It has been shown that besides the amplitude of the µw field, the characteristics of its modulation modify the exact value of most of the frequency shifts /10/,/13/. This must be taken into consideration.

Processing of the third harmonic of the beam response gives the possibility of eliminating the pullings due to neighbouring transitions and to a cavity mistuning /13/.

3.5. Measurement of the Velocity Distribution of the Atoms Detected

Knowledge of the velocity distribution of the atoms detected in necessary to calculate most of the frequency shifts. It also enables i) to verify the beam retrace upon beam reversal and ii) to search for velocity distribution distorsions caused by Majorana transitions.

A new measurement method proved to be efficient. It consists in computing the Fourier transform of the variation of the beam tube response versus the microwave power /19/.

4. Frequency Shifts Which Require Further Attention

4.1. AC Stark Shift

It has been shown that the random electric field of the blackbody radiation displaces the frequency of the reference transition /20/. It is highly desirable that experiments be conducted in order to measure this effect.

4.2. Spin-exchange Frequency Shift

Collisions between alkali atoms cause a frequency shift due to spin exchange interactions /10/. The spin-exchange frequency shift cross section of the cesium atom is not known. Theoretical and experimental informations on this frequency shift occuring in a cesium beam are necessary.

4.3. RF Leakage

Frequency shifts due to RF leakage, both in the sections of cut-off waveguides placed on each side of the interaction regions and in the C-field region have been mentioned in several occasions. It would be useful to relate quantitatively these frequency shifts to the level of the stray RF field.

4.4. Majorana Transitions

Majorana transitions may be induced in regions where the adiabatic condition is not fulfilled. They cause a change of the velocity distribution of atoms which are able to strike the detector and, therefore, a change of the velocity dependent frequency shifts. Furthermore, it has been shown that a purposely created particular static field configuration gives unusual features and a frequency shift of the order of 10^{-11} /1/. To my knowledge, a satisfactory explanation of such a frequency shift is not yet available.

5. Possible Ways of Reducing the Frequency Shifts

Besides technical improvements of the present state of the art, several possibilities would change greatly the design of cesium beam frequency standards /10/,/21/.

Further increase of the value of the T_c-temperature of the new superconductive materials (preferably above room temperature !) would be highly beneficial to cesium beam frequency standards, to reduce the cavity phase shift and to create homogeneous and stable static magnetic fields.

Use of very slow atoms would reduce drastically a number of frequency shifts. This can be attempted by slowing down actively the cesium atoms, as described in other papers of this issue. However, the atomic beam intensity must remain sufficient to preserve a good short term frequency stability. On the other hand, the product of the atom density in the beam times the atom relative velocity must remain below a limit, to avoid the occurence of a spin exchange frequency shift.

Optical state preparation and detection offer very interesting perspectives that will be described in detail in other papers. It will be possible to decrease the frequency shifts related to C-field inhomogeneities, to take care more easily of the distributed cavity phase shift, to eliminate the pulling by neighbouring transitions and to suppress Majorana transitions. Furthermore, the increase of the frequency stability will render easier the measurement of some of the frequency shifts.

Continued progress in the design of electronic systems which generate the interrogation signal and which process the beam tube response are desirable to take full advantage of the expected improvement of accuracy.

6. Conclusions

It appears that, on the average, no spectacular progress has been made since 1981, as long as the accuracy figure of the ensemble of primary cesium beam frequency standards is concerned. Several reasons may be invoked to understand this state of affairs : some of the machines are getting old ; the burden of time keeping prevents scientists to improve their frequency standards ; emphasis has been placed on frequency shifts, such as that due to the cavity phase shift, which are real but others of at least the same importance, such as that due to neighbouring transitions, have received a sufficient attention only very recently. It is my opinion that, for reasons which are understandable and respectable, not enough research activity has been devoted to the improvement of accuracy. In this respect, it is striking to note that only one cesium beam apparatus, of the same class as the best listed in Table 1, has been operated for the purpose of doing experiments on causes of frequency shifts. This is the machine named Cs X of the PTB.

However, an accuracy of 1×10^{-14} is almost at hand in a very few of the primary standards. This is an accessible goal. It is highly desirable that a number of primary cesium beam frequency standards reach this level of accuracy in order to enable valuable comparisons. Eventual discrepancies at this level would, ounce understood, enable further progress of the accuracy and also of the long term frequency stability. We know that an improvement of the latter is absolutely necessary to measure the long term behaviour of the rotation of some pulsars /22/.

Application of optical methods to the cesium beam frequency standard is an attractive subject of work. However efforts devoted to the improvement of the traditional primary frequency and time standards should still be actively pursued at least until a detailed comparison of the two versions of this standard has been done at the level of 1×10^{-14}, or better.

Acknowledgements

The author is indebted to Drs. H. de Boer, J.S. Boulanger, I.K. Bakhaev, R. Drullinger, K. Nagakiri, Y. Koga, Yang Xiaoren and Zhai Zao-Cheng who supplied valuable informations which helped greatly the preparation of this paper.

References

1. A. Bauch, H. de Boer, K. Dorenwendt, B. Fischer, T. Heindorff, P. Hetzel and R. Schröder : PTB-Mitteilungen 98, 19 (1988)
2. A.G. Mungall, H. Daams, J.S. Boulanger and C.C. Costain : Journal de Physique 42, C8-233 (1981) ; J.S. Boulanger : private communication (1988)
3. Yu G. Abashev, L.P. Elkina and S.B. Pushkin : IEEE Trans. on Instr. and Meas. IM-36, 627 (1987)
4. L.L. Lewis, F.L. Walls and D.J. Glaze : Journal de Physique 42, C8-241 (1981) ; R. Drullinger : private communication (1988)
5. K. Nakagiri, M. Shibuki, H. Okazawa, J. Umezu, Y. Ohta and H. Saitoh : IEEE Trans. on Instr. and Meas., IM-36, 617 (1987)
6. Y. Nakadan and Y. Koga : IEEE Trans. on Instr. and Meas. IM-34, 133 (1986)
7. Yang Xiaoren : private communication (1988)
8. Zhai Zao-Cheng : private communication (1988)
9. B. Guinot : this issue
10. J. Vanier and C. Audoin : The Quantum Physics of Atomic Frequency Standards, Adam Hilger, in press
11. A. de Marchi: G.D. Rovera and A. Premoli, Metrologia 20, 37 (1984)
12. G. Becker: Metrologia 13, 99 (1977) ; C. Audoin, V. Candelier and J. Vanier : IEEE Trans. on Ultrasonics, Ferroelectronics and Frequency Control UFFC-34, 573 (1987)
13. A. de Marchi, G.D. Rovera and A. Premoli : IEEE Trans. on Ultrasonics, Ferroelectronics and Frequency Control UFFC-34, 582 (1987)
14. J. Vanier, A.G. Mungall and J.S. Boulanger: Metrologia 20, 101 (1984)
15. Y. Koga: Japanese Journal of Applied Physics 23, 97 (1984)
16. A. Bauch, T. Heindorff and R. Schröder: IEEE Trans. on Instr. and Meas. IM-34, 136 (1985)
17. A. de Marchi: Proc. of the 40th Annual Frequency Control Symposium, p. 441 (1986)
18. A. de Marchi, G.D. Rovera, R. Drullinger and D.A. Howe : Proc. of the 39th Annual Frequency Control Symposium, p. 3 (1985) ; A. de Marchi: Proc. of the 41th Annual Frequency Control Symposium, p. 53 (1987) ; A. de Marchi and R.E. Drullinger: Metrologia 24, 23 (1987)
19. J.S. Boulanger : Metrologia 23, 37 (1986)
20. W.M. Itano, L.L. Lewis and D.J. Wineland : Journal de Physique 42, C8-283 (1981)
21. D.J. Wineland, Metrologia 13, 121 (1977)
22. D.W. Allan, this issue

The Accuracy of Commercial Cesium Beam Frequency Standards *

A. De Marchi

Dipartimento di Elettronica e Automatica,
Universitá di Ancona, Ancona, Italy
*Work supported by the National Bureau of Standards,
Boulder, Colorado, and by the United States Naval Observatory,
Washington DC, USA

The accuracy of Cesium beam frequency standards has been studied extensively over the years. Most of the published work, however, was carried out on few devices, which have been built in a small number of national laboratories. This puts the Cs beam standard in a strange and unique position because it is the primary reference, and it is available to the users as a commercial unit, but the studies that should guarantee its accuracy are relative to standards which are substantially different from those that the users can directly access.

For this reason, and because commercial Cesium standards exist in a great number, are relatively inexpensive, and are very reliable, it is of interest to study their accuracy in the light of all new understanding.

The other important reason for this analysis is of course the fact that long term stability is intimately connected to accuracy through the fractional instability of the various frequency biases. In fact if a bias of a 10^{-12} magnitude can vary in long term by 10%, due to whatever reason, it will cause a long term frequency instability of 10^{-13}. Although some users may find little thrill in the quest for ever greater accuracy, thinking of it as something for the time scale Vestals, the same people are often likely to get excited by improvements in long term stability, and ought to give serious consideration to the statement reported above.

A well known analysis of an error budget for a commercial Cesium beam standard was published 24 years ago [1] and was actually more a design guideline for a small tube than an accuracy analysis for any particular standard. The design goal of an accuracy better than $1x10^{-13}$ was presented as being within reach. Magnetic field inhomogeneity and pulling by neighboring transitions (Rabi pulling) were correctly indicated as the biggest biases, while phase shifts were recognized as insignificant in well designed and realized cavities.

Yet standards designed after those guidelines appeared to have accuracies worse than predicted by one order of magnitude. More than a decade was spent in wondering why and how this should happen, and it is now clear that for scientists who were used to work on the accuracy of long laboratory standards it was natural to take assumptions on short tubes based on their experience. This has been misleading.

In fact both distributed and end to end cavity phase shift, which are indeed the most serious problems for long tubes, constitute less of a limitation in short tubes. For the distributed shift this is due to the fact that beam apertures don't need to be as wide in short tubes, therefore the beam is passed at the very end of the cavities, skimming the surface of the short circuit, and the spatial phase variation is then a factor of 5 smaller [2] than at half wavelength from the short, where the beam passes in long tubes. For the end-to-end one the reason is that it is much easier to symmetrize a short microwave structure than a long one. As a result, phase shift problems can be expected to be smaller than 10^{-13} in short tubes [4].

Nevertheless observed shifts of the order of 10^{-12} and drifts of 10^{-12}/year have been attributed to cavity phase shift [3], and the fact that beam reversal cannot be performed in a sealed tube has been often mentioned as the reason why such a device would never be fully evaluated. In fact it has been believed that the actual phase difference in a particular tube is impossible to measure, much less to compensate.

* Work supported by the National Bureau of Standards, Boulder, Colorado, and by the United States Naval Observatory, Washington DC. USA.

Frequency Standards and Metrology Editor: A. De Marchi

In this paper we speculate on the possibility of attempting such a compensation, to reduce the phase shift effects below 10^{-13}. We do this to show that this effect should not stand in the way of accuracy assessment for a commercial standard down to the 10^{-13} level. Moreover, in a standard using a short tube and good electronics, there are few other bias effects which intervene at that level. We will discuss them here with the aim of showing that they can be either corrected or evaluated in such a way that the overall accuracy could approach it.

A list of all effects known to produce frequency biases larger than 10^{-14} in a commercial standard is given in Table I. Their typical magnitude is indicated if it is at least 10^{-13}. In this case the evaluation uncertainty is reported as it can be obtained with good analysis or alignment. Additional information is also given as to whether the bias is velocity (power) dependent.

Some assumptions are made in this Table, which deserve a little explanation:
1) C-field homogeneity and cavity realization are assumed to be "state of the art".
2) The same is true for the servo loop (infinite gain double integration is assumed).
3) The RF Spectrum in particular is supposed to be clean.

Table I: Systematic effects in a Cesium beam standard.

Classification		Effect	Bias$\times10^{13}$	Uncert.$\times10^{-14}$
Physical Perturbations on Cesium atoms	*	Quantization H field	10^3	3
	μ	Doppler 1st order		
	μ	Doppler 2nd order	1	1
	*	Gravitational Red-shift		
	*	Black Body Radiation		
	?	Majorana Transitions		
	?	Spin Exchange		
Effects due to the interrogation method	μ	End-to-end Phase Shift	1	3
	μ	Distributed Phase Shift		
	μ	Cavity Pulling	10	3
	μ	Rabi Pulling	10	3
	μ	Bloch Siegert effect		
Problems from the Servo Electronics	μ	Integrator offset		
	μ	Slave Oscillator offset		
	μ	Mod/demod distortions	1	3
	μ	RF Spectrum asymmetry		

Orders of magnitude are for a small tube and are indicated only if $>10^{-13}$.
* Independent of atomic velocity and trajectory
μ Dependent on Microwave Power and Modulation Parameters
? Needs more study: not clear how important

Most of the biases depend on power. However most do so at a level which is irrelevant for this study. These include Doppler shifts and phase shifts [6].

It appears that only a few biases can cause problems at the stated target level of accuracy. In the following we will first discuss cavity phase shift, to show how it can be checked and corrected to the size indicated in the Table, then we will show how cavity pulling and Rabi pulling can be reduced to that size. For the case of field homogeneity instead we will refer to [1] for an illustration of how uncertainties can be assessed to be small enough.

A full analytical description of the Ramsey cavity was given in [5] with the help of the Scattering Matrix formalism. In that analysis it is shown that what is relevant to the end-to-end phase shift is the symmetry of the whole microwave structure including the feeding junction, not just of the two arms of the U-shaped cavity. Resonant modes can be sustained by the cavity that are either symmetric or antisymmetric with respect to the centre, and the feeding junction is designed to excite selectively either kind, but not the other. In this way the modes which do couple to the outside through the junction, which we call desired modes or D-modes, are very much overcoupled, and their loaded Q will be dominated by the external structure. The other kind of modes, which we call undesired modes or U-modes, couple only through the slight asymmetries of the structure and will therefore show a loaded Q close to the unloaded value. It is possible in principle to monitor how well the structure is symmetrized by looking at the sharp peaks of the U-modes in reflection from the outside of the cavity.

It is also shown in [5] that an asymmetry can be introduced in the microwave structure from the outside of the sealed tube through an asymmetry of the feed. This is possible if the distance between vacuum seal and junction is not too many times the penetration depth of the first superior asymmetric modes. In commercial tubes the microwave structure is usually a U-shaped standard X-Band waveguide shorted at the ends and fed in the centre with an E-plane Tee. The first superior asymmetric modes are the TE_{11} and the TM_{11}, which are degenerate, with a cutoff frequency of 16.2 GHz and a penetration depth of 4 mm at Cs frequency. Therefore if these modes were excited 25 mm from the junction, and all the energy were in them at that point, a fraction e^{-6} (or $2x10^{-3}$) of the field would be still in these modes at the junction. This would give a phase difference of about 10 μrad between the two arms, which means a shift of the clock frequency of 10^{-13}.

That this is true was shown recently by the author in the course of an experiment which will be reported in full when completed. The idea is to measure the frequency difference upon reversal of the section of the microwave structure external to the sealed tube. In the device studied this section has a cavity defining vane, which is asymmetric and therefore feeds the cavity via the mentioned asymmetric modes (Fig.1). This vane is at a distance of about 30 mm from the junction. The measured frequency difference upon reversal of the asymmetry is $3x10^{-13}$, i.e. half that much is the effect of the end-to-end phase difference introduced by the asymmetry of the feed.

It is interesting to point out that the position of the tuning screw does not seem to have a significant effect on the asymmetry, confirming that the relevant element is the vane.

The importance of the feed seems to be an additional problem, because now one has to worry not only about the asymmetry of the cavity, but about the feed too, and its "symmetric" arm at that!

On the other hand it should make it possible in principle to correct cavity asymmetries from the outside of the tube, and reduce the uncertainty on phase shift well below concern. The compensation can be performed by monitoring in reflection the size of the sharp U-modes, while modifying the asymmetry of the feed. Of course this would require a feeding microwave section designed for this purpose.

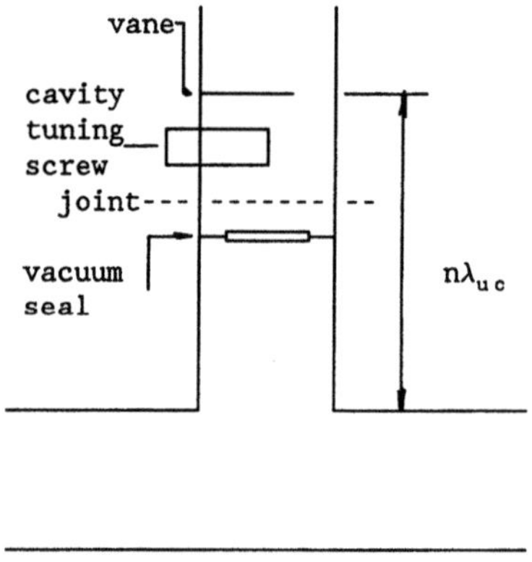

Fig.1 - Configuration of the feed in the tube used for the feed reversal experiment. λ_{uc} is the penetration depth of the first asymmetric superior modes.

Once the phase shift is taken care of, cavity pulling and Rabi pulling are left as the only effects to be evaluated for an accuracy assessment at the level of 10^{-13}.

The first one is essentially a power shift because it stems from the fact that different points of the atomic resonance are excited with different power. This gives a distortion of the response, which depends on Q_c and Q, quality factors of cavity and Ramsey line, on the cavity mistuning $\delta\nu_c=\nu_c-\nu$ from Cesium resonance frequency, on modulation parameters [6] and on microwave power. Below Optimum Power (PO) the tube response is enhanced on the side of the resonance closer in frequency to the cavity and decreased on the other side. Then the atomic resonance appears closer to the cavity resonance than it really is (cavity pulling). The opposite is true above

PO (cavity pushing). Therefore there is a point where the effect vanishes. This power setting is at PO if slow square wave frequency modulation is used [1], but can be different for other modulation schemes [6].

The expression for cavity pulling is:

$$\left.\frac{\delta\nu}{\nu}\right|_c = \Lambda_c \left(\frac{Q_c}{Q}\right)^2 \frac{\delta\nu_c}{\nu} \qquad (1)$$

where Λ_c is a factor which depends on power and modulation parameters. A typical curve of Λ_c versus power is shown in fig.2, as was calculated for an existing standard using actual modulation parameters. Since it is very unlikely that a cavity mistuning should occur greater than half the cavity width, the order of magnitude of cavity pulling should be expected to be $\delta\nu/\nu|_c < \frac{1}{2}\Lambda_c(Q_c/Q^2)$. In short tubes it is usually $Q \leq 3\times10^7$. With $Q_c < 10^4$ and $\Lambda_c(PO) < 0.2$, the upper limit is $\delta\nu/\nu|_c < 10^{-12}$. Therefore it would not be necessary to tune the cavity better than 10% of its half width for a bias smaller than 10^{-13}. However it must be pointed out that even at the power where it vanishes this effect causes power sensitivity proportional to $\delta\nu_c$. For long term stability purposes the latter can be often compensated by Rabi pulling, as illustrated below, but a bias from Rabi pulling must then be accepted. For accuracy purposes the cavity should be tuned better than that.

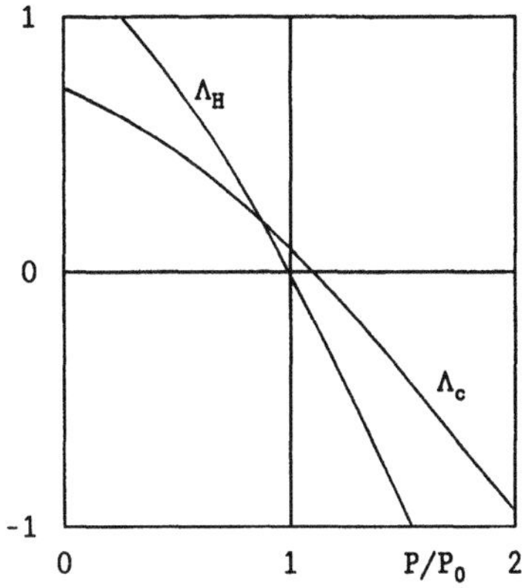

Fig.2 - Curves of Λ_H (slow square wave modulation) and Λ_c (sinewave with actual modulation parameters) versus microwave power.

As mentioned above the position of the tuning screw does not affect end-to-end phase shift noticeably and it is therefore a quite pure cavity tuning adjustment. Although this is very sensitive, in the configuration of the tubes which have been aligned in the course of this work, it can be done very well with some care and experience by maximizing the CW beam current signal at centre frequency. In order to judge how well the cavity is tuned it is then advisable to check the power sensitivity of the standard at high C-field, where Rabi pulling is insignificant, because then any power sensitivity is caused by cavity pulling.

As for Rabi pulling it will be argued here that, after these other effects are reduced as indicated, it too is easily reduced. In fact the power sensitivity of the standard is now caused solely by Rabi pulling, and if a C-field value is found for which it vanishes that will mean that the pulling is null.

That field values exist at which Rabi pulling vanishes was shown in [7] from a theoretical point of view, and experimentally in [8,9,10]. The reason for this is that the tails of Rabi lines are a $(\sin x/x)^2$ function integrated over the velocity distribution that atoms have in that hyperfine sublevel. Their slope is therefore oscillating with the distance from the centre frequency of the line. Since the two neighboring lines on the two sides of the clock transition have different shapes, due to different atomic velocity distributions, their two slopes at Cs frequency do not compensate and produce a baseline slope which distorts the line. Rabi pulling is the term used for the resulting shift in first harmonic loops. This baseline slope also has an oscillating behavior, therefore field values exist for which it vanishes. Both π and σ neighboring transitions contribute to this shift, and make it sometimes difficult to give a theoretical prediction of the shape of the Rabi pulling curve, but these zero crossing points can easily be found experimentally provided a good frequency reference is available. A long term stability of a few parts in 10^{14} is required of the latter to guarantee that Rabi pulling is smaller than 10^{-13}.

Operating the standard at one of these zero baseline slope points would now give accuracy in excess of 10^{-13} and high power insensitivity. Long term stability could be expected to reach the low 10^{-15} range. However, if a residual cavity mistuning were present, power insensitivity would not be guaranteed by operation at a zero baseline slope point because there would still be the power shift caused by cavity pulling. Power insensitivity would then be found at a C-field point close to the zero crossing of Rabi pulling, where the latter is such that its power dependence exactly compensates that of cavity pulling. Extremely good long term stability can still be obtained, but Accuracy will be limited to a few parts in 10^{13}. In fact a shift of $2x10^{-13}$ due to Rabi pulling causes a power shift of less than 10^{-13}/dB, which would compensate power sensitivity from cavity pulling as encountered if the cavity is tuned not better than 10% of one half width of the cavity. An accuracy of a few parts in 10^{13} can safely be stated in this way, but no more. To do better, the cavity tuning must be perfected separately, because power insensitivity is the only available criterion to judge precisely how close the C-field is to the zero crossing of Rabi pulling.

As a conclusion it appears that the following statements can be safely made:

1) Phase shift is not an obstacle toward accuracy assessment down to the 10^{-13} level in commercial Cesium beam frequency standards using short sealed tubes.
2) Accuracies of a few parts in 10^{13} can be obtained by just operating the standard at a C-field value at which power sensitivity is null. Extremely good long term stability and environmental insensitivity are expected in this case [8,9,10].
3) Accuracies in excess of 10^{-13} could be obtained by going through the illustrated procedure for phase shift compensation and tuning the cavity very carefully at the operating temperature before looking for the zero power sensitivity point.

It is also probably worth mentioning here that it is very advisable to do all the frequency measurements, necessary to perform the alignment procedure sketched above, with the standard already operating close to a zero crossing of power sensitivity. In this way the disturbance introduced by environmental sensitivity is minimized.

References

[1] J.H.Holloway, R.F.Lacey: "Factors which limit the Accuracy of Cesium atomic beam frequency standards"- Proc. 7th CIC (1964).

[2] R.F.Lacey: "Phase shift in microwave Ramsey structures"- Proc.22nd FCS (1968).

[3] D.W.Allan, H.W.Hellwig, S.Jarvis Jr., D.A.Howe, R.M.Garvey: "Some causes and cures of frequency instabilities (drift & noise) in Cesium beam frequency standards"- Proc. 31st FCS (1977).

[4] A.De Marchi, G.P.Bava: "On cavity phase shift in commercial Cesium beam frequency standards" - Metrologia, 20 (1984).

[5] A.De Marchi: "The effect of feeding junction asymmetries on end to end phase difference in Ramsey Cavities"- Proc. 2nd European Time & Frequency Forum, Neuchâtel (1988).

[6] A.DeMarchi, G.D.Rovera, A.Premoli: "Effects of servo loop modulation in atomic beam frequency standards employing a Ramsey cavity"- IEEE Trans. UFFC, 34 (1987).

[7] A.De Marchi, G.D.Rovera, A.Premoli: "Pulling by neighboring transitions and its effects on the performances of Cesium beam frequency standards"- Metrologia, 20 (1984).

[8] A.De Marchi: "Understanding environmental sensitivity and ageing of Cesium beam frequency standards"- Proc. 1st Europ. Time & Frequency Forum, Besancon (1986).

[9] A.De Marchi: "New insights into long term frequency instabilities (drift & long term noise) of Cesium beam frequency standards"- Proc. 41st FCS (1987).

[10] A.De Marchi: "Rabi pulling and long term stability in Cesium beam frequency standards" - IEEE Trans.on UFFC, 34 (1987).

Reproducibility of H Maser Standards

S.B. Crampton, J.J. Krupczak, and Phat Vu
Physics Department, Williams College, Williamstown, MA 01267, USA

Thermal noise limits the precision of H maser standards out to averaging times of order 10^4 seconds, beyond which it is limited by other factors despite attempts to stabilize everything that could cause a frequency shift. This paper examines the dependence of the hydrogen maser frequency on factors other than thermal noise, in order to provide a guide to improving stabilization efforts.

1. Introduction

Fig. 1 introduces the basic components of a hydrogen standard. Atoms in the c and d states are focused by a hexapole magnet into a Teflon-coated storage bulb centered in a microwave cavity tuned to the c-to-a ground state hyperfine transition. Magnetic shields reduce the ambient field to low values at which the frequency is only quadratically frequency dependent, and oscillation power of order 10^{-13} watts is coupled out of the cavity to a receiver.

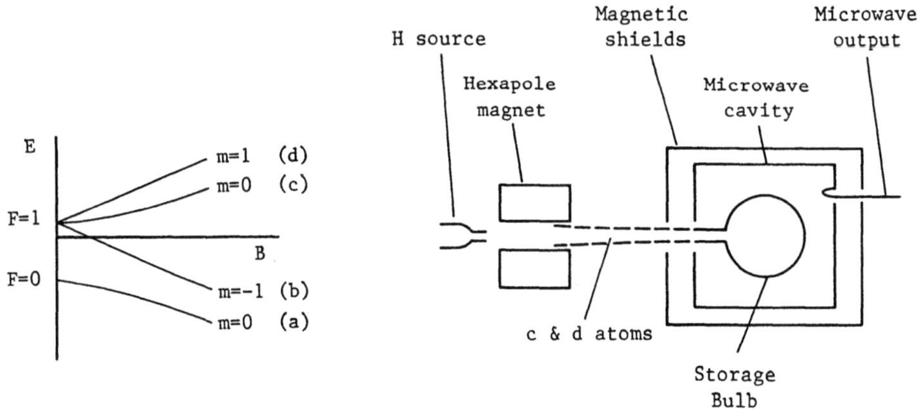

Fig. 1 Energy levels of ground state hydrogen atoms plotted against ambient magnetic field B (left) and schematic of hydrogen standards (right).

Frequency Standards and Metrology Editor: A. De Marchi

2. Measured Precision

Fig. 2 illustrates the performance of two Smithsonian VLG-11 hydrogen masers by plotting the Allan variance of comparisons of the two frequencies over successive measurement intervals against the length of the interval.[1] Precision of a part in 10^{15} or better is achieved by averaging out thermal noise for an hour or more. Degraded precision for much longer averaging times indicates slow drifting due to systematic effects. In this case cavity tuning instabilities due to temperature instabilities were thought to be sufficient to account for the slow drift. Recent data for two Sigma Tau Standards Corporation masers are very similar to this data out to 10^4 sec averaging times, the longest for which data are presented.[2] Evidently, hydrogen maser techniques have achieved the limit of precision set by the thermal noise and maximum oscillation power at conventional temperatures down to the level of parts in 10^{16}. Improvements depend on stabilizing parameters that affect the hydrogen maser frequency at the level.

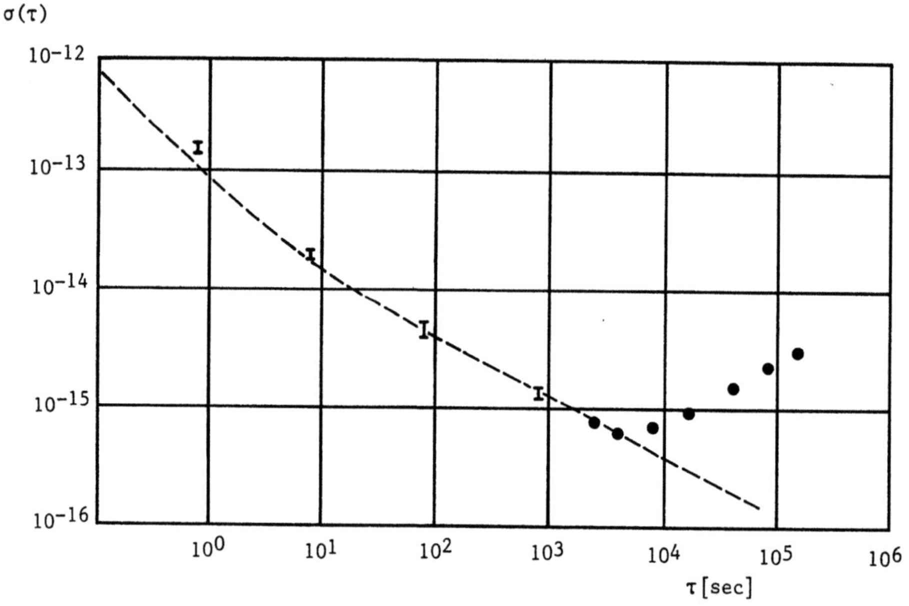

Fig. 2 Allan variance σ of two Smithsonian Astrophysical Observatory VLG-11 hydrogen masers, plotted against averaging time τ. The dashed line is the theoretical limit set by thermal noise.

3. Wall Shifts

Collisions with the inside wall of the storage bulb retard the phase of an atom's oscillating magnetic dipole moment by an average amount $\Delta\Phi$, shifting the oscillation frequency by $\delta\omega = \Delta\Phi/t_0$,[3] with t_0 the average time between wall collisions. These shifts are large, amounting to several parts in 10^{11} near room temperature. They vary exponentially with temperature and differ at a particular temperature by as much as a few percent from one coating to another. For example, the Orsay group reports that their new compact masers differ in frequency by as much as parts in 10^{12}; the Smithsonian and Sigma Tau groups report reproducibility from one coating to another to be typically a part in 10^{13}.[4] Comparing data obtained by the Smithsonian group for CF_4 wall coatings to data for Teflon suggests that the differences from one Teflon coating to another are due to differences in effective surface area and sticking coefficient rather than the energy with which atoms are bound to the surfaces.[5] There is no conclusive evidence for changes of wall shift once a coating is in place, although the Smithsonian group has reported degradation of the line Q due to wall coating changes,[6] and it would not be suprising to find some long term changes of wall shifts accompanying such changes.

4. Cavity Pulling

Cavity mistuning shifts the maser frequency by $\delta\omega = \Delta\ \Gamma$, with Δ twice the ratio of cavity mistuning to cavity resonance linewidth and Γ the aggregate relaxation rate, or oscillation frequency ω divided by twice the line Q.[3] Frequency pulling due to cavity mistuning is typically 2 parts in 10^{14} per Hz of cavity mistuning, corresponding to stability of the overall cavity dimensions to a part in 10^9. Even careful temperature control cannot guarantee that there will not be long term dimensional changes large enough to be significant. Either the apparent cavity frequency must itself be monitored and stabilized, or the effects of cavity mistuning on the maser frequency must be detected and corrected. An example of the latter approach is "spin exchange tuning"[7] whereby Γ is varied by varying the spin exchange collision rate, and the observed variation of oscillation frequency is used to correct the cavity tuning. The difficulty with this approach is that the precision with which the oscillation frequency can be corrected for cavity mistuning is not much better than the precision with which the oscillation frequency can be measured at all. Correcting for cavity mistuning to the parts in 10^{16} level would require measuring times of order 10^4 sec, during which the cavity tuning would drift appreciably. Moreover, spin exchange collisions introduce frequency shifts that are nonlinear in Γ at the parts in 10^{15} level.

5. Magnetic Field Effects

Changes of the oscillation frequency due to ambient magnetic field changes can be effectively eliminated by stabilizing the frequencies of the $d \to c$ and $c \to b$ transitions, provided that care is taken to eliminate magnetic field gradients.[8]

Frequency shifts due to gradients are proportional to the amplitude of any magnetic field gradients and to $\rho_{dd}-\rho_{bb}$, where ρ_{jj} is the relative population of the jth atomic state. These shifts decrease with the square of the ambient magnetic field at fields above 10 mOe. Because magnetic field gradients also produce damping effects that contribute to Γ an amount Γ_m that varies as the inverse fourth power of the field, the presence of magnetic field gradients can be detected by observing changes of oscillation power level or of Γ with ambient field. In order to achieve stabilities at the parts in 10^{16} level at fields below 10 mOe, Γ_m must be stable to about $10^{-3}\sqrt{\Gamma_m}$. For example, if Γ_m is of order 10^{-3}, Γ_m must be stable to a few percent. Magnetic field effects should not be troublesome at the parts in 10^{16} level provided that the maser shields are successfully demagnetized and the magnetic environment external to the maser does not change drastically. These effects can be suppressed by reducing $\rho_{dd}-\rho_{bb}$ and by using higher ambient fields, although the ambient field itself must be very stable then because of the quadratic field dependence of the oscillation frequency.

6. Collision Effects

Our recent analysis[9] of spin exchange collisions including hyperfine effects during the collisions themselves indicates that collisions shift the oscillation frequency by

$$\delta\omega_c = n_H \langle v\rangle \, [(\rho_{cc}-\rho_{aa})\bar{\lambda}_0 + (\rho_{cc}+\rho_{aa})\bar{\lambda}_1 + \bar{\lambda}_2], \tag{1}$$

where n_H is the atom density, $\langle v\rangle$ is the thermal average collision speed, and the $\bar{\lambda}_j$ are thermal average frequency shift cross sections. Collisions also contribute to Γ a term

$$\Gamma_c = n_H \langle v\rangle \, [(\rho_{cc}-\rho_{aa})\bar{\sigma}_0 + (\rho_{cc}+\rho_{aa})\bar{\sigma}_1 + \bar{\sigma}_2], \tag{2}$$

where the $\bar{\sigma}_j$ are thermal average broadening cross sections. The terms in $\delta\omega_c$ and Γ_c proportional to $\rho_{cc}+\rho_{aa}$ introduce oscillation frequency shifts that are nonlinear in Γ_c, as illustrated in Fig. 3 below for various values of cavity mistuning using typical conventional and cryogenic maser parameters. Note that there is no value of cavity mistuning for which the oscillation frequency is independent of Γ_c. Because it is difficult to maintain the stability of the maser input atomic beam intensity and therefore also the stability of Γ_c to better than a few percent, it is important to try to adjust the maser operating conditions so that the maser frequency is first order independent of Γ_c. Thermal frequency fluctuations preclude doing this by simply observing the variation of oscillation frequency near the operating point. To achieve stabilities of the order of parts in 10^{16} at conventional temperatures, it will be important to choose the operating point by precisely measuring the variation of oscillation frequency over the full range of oscillation for a variety of cavity tunings and and then fitting the data

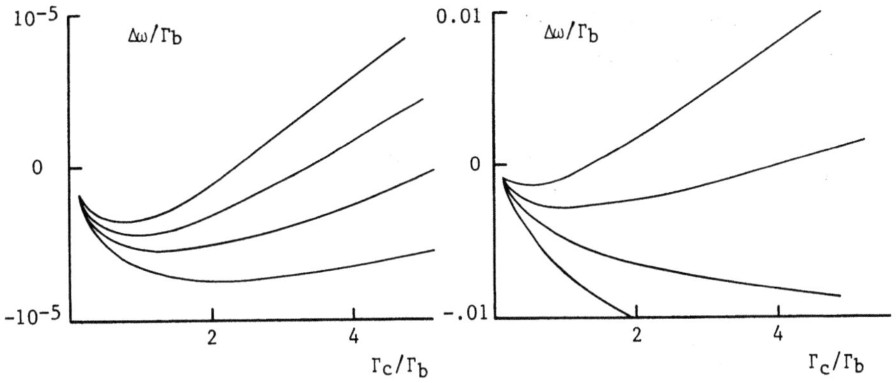

Fig. 3 Collisional frequency shifts $\Delta\omega$ for typical maser parameters at conventional temperatures (left) and at 4 K (right), plotted relative to values for $\Gamma_c=0$. Γ_b is the bulb escape rate.

to a rate equation model for the level populations. At cryogenic temperatures the hyperfine-induced frequency shift effects are three orders of magnitude larger, making it potentially very difficult to achieve the improved stabilities allowed by the lower thermal noise and smaller spin exchange broadening cross sections at liquid helium temperatures.

7. Conclusions

Careful attention to eliminating magnetic field gradient effects and hyperfine-induced collisional effects should make it possible to achieve hydrogen standard frequency stabilities at the parts in 10^{16} level for very long averaging times. It will be very much harder to achieve better stabilities or even comparable stabilities at liquid helium temperatures.

1. R. F. C. Vessot, M. W. Levine and E. M. Mattison, in Proc. of the 9th Annual Precise Time and Time Interval Conference, 1978 [NASA Tech. Memoeandum No. 78104, 1978 (unpublished)], p. 549.
2. H. E. Peters, Private Communication, 1988.
3. D. Kleppner, H. M. Goldenberg and N. F. Ramsey, Phys. Rev. 126, 603 (1962).
4. C. Audoin, H. E. Peters, and R. F. C. Vessot, Private Communications, 1988.
5. R. F. C. Vessot, Private Communication, 1988.
6. E. M. Mattison and R. F. C. Vessot, in Proc. of the 41st Annual Frequency Symposium, 1987 [IEEE Cat. No. 87CH2427-3, 1987 (unpublished)], p. 95.
7. D. Kleppner, H. C. Berg, S. B. Crampton, N. F. Ramsey, R. F. C. Vessot, H. E. Peters and J. Vanier, Phys. Rev. 138, A972 (1965).
8. S. B. Crampton, E. C. Fleri and H. T. M. Wang, Metrologia 13, 131 (1977).
9. J. M. V. A. Koelman, S. B. Crampton, H. T. C. Stoof, O. J. Luiten and B. J. Verhaar, submitted for publication, 1988.

Advances in the Theory of Gas-Cell Atomic Frequency Standards

J.C. Camparo

Chemistry and Physics Laboratory, The Aerospace Corporation,
P.O. Box 92957, Los Angeles, CA 90009, USA

I. Introduction

In general terms the operating principle of the optically pumped gas-cell atomic clock can be described quite simply: an optical field polarizes an atomic ensemble by optical pumping, and additionally monitors the degree of polarization that is attained; simultaneously, the atomic ensemble interacts with a microwave field, which reduces the degree of polarization in the atomic ensemble when the microwave field's frequency matches an appropriate atomic transition frequency. Specifically, the polarization corresponds to hyperfine polarization in the ground state of an alkali metal vapor (monitored by the level of optical transmission through the vapor, which is confined inside a microwave cavity), and the microwave field is resonant with the alkali's 0-0 hyperfine transition frequency. Since the measured atomic line Q in this case is quite high ($\sim 10^7$), the optically detected change in polarization can be used to discriminate against very small microwave field frequency changes, and recent results have shown that stabilities of one part in 10^{14} can be realized with this device [1]. However, behind this general simplicity is a high degree of complexity, so much so that until recently not even the shot noise limited performance of the device could be theoretically determined with accuracy.

The complexity in the gas-cell clock's atomic signal arises from two factors. First, the ground state of an alkali atom has a total of 2(2I+1) Zeeman sublevels, where I is the particular alkali isotope's nuclear spin. Consequently, the response of the alkali atom to the microwave field cannot be accurately described in terms of a simple two-level quantum system, even though the microwave field only couples two of the ground state Zeeman sublevels (those with $m_F=0$). The second factor that gives rise to complexity is associated with the non-uniformity of the optical and microwave fields over the gas-cell signal volume. In order to reduce the diffusional rate of depolarizing atom-wall collisions, the gas-cell is typically filled with several torr of an inert buffer gas. However, since the rate of diffusion throughout the gas-cell is much slower than the 0-0 hyperfine transition dephasing rate, the alkali atoms are unable to motionally average the optical, microwave and static magnetic fields that are present. In fact, for computational purposes it is reasonable to consider the atoms as "frozen in place" as a result of the buffer gas's presence. Not only does this immobility yield a potential for inhomogeneous broadening of the clock signal lineshape as a result of spatially varying AC Stark [2] and Quadratic Zeeman shifts [3], but even in the absence of these mechanisms the microwave field non-uniformity results in non-lorentzian lineshapes [4].

In the following sections of this paper, a theory describing the gas-cell clock's atomic signal will be reviewed. This theory differs from its predecessors in that the physical processes occurring in the gas-cell are considered on two different spatial scales. On the microscopic scale the alkali atoms interact with *locally* homogeneous fields, and the response of the atoms to the various fields is examined with

Frequency Standards and Metrology Editor: A. De Marchi

the density matrix. The results of this calculation are used to determine the degree of polarization in small regions of the signal volume, and hence the change in transmitted light intensity through these regions as a function of the microwave frequency. On the macroscopic scale the fields appearing in the density matrix equations vary, and the change in transmitted light intensity through the different regions of the gas-cell becomes a function of the various fields' spatial distributions. As will be shown below, these two scales of physics play equally important, but different, roles in determining the shape and behavior of the gas-cell frequency standard's atomic signal.

II. Microscopic Scale of Physics

On the microscopic scale the optical pumping/magnetic resonance evolution of all stable alkali isotopes can be described by what we have termed the generalized Vanier theory (GVT) [5,6]. In this theory all of the Zeeman sublevels of an alkali's ground state are considered. However, by making several reasonable assumptions, it can be shown that there are only five independent density matrix equations describing the evolution of the alkali ground state, regardless of the total alkali ground state degeneracy. These assumptions are: i) all Zeeman sublevels of a given hyperfine multiplet interact with the light equivalently (with A (B) defined as the photon absorption rate for the $F=a=I+1/2$ ($F=b=I-1/2$) hyperfine level), ii) all Zeeman sublevels of the ground state have an equal probability of being populated as a result of decay from the exicted electronic states, and iii) the microwaves only induce transitions between the ($F=a$, $m_F=0$) and ($F=b$, $m_F=0$) ground state Zeeman sublevels. Additionally, relaxation of the ground state Zeeman sublevels is included in the theory phenomenologically by considering "dark" longitudinal (γ_1) and transverse (γ_2) relaxation rates.

Since most gas-cell clocks operate near or above the saturation limit, where there is no change in the observed clock signal amplitude with increasing microwave power supplied to the cavity, an important aspect of the microscopic physics is the manner in which the 0-0 hyperfine transition exhibits saturation broadening. The GVT equations show that the linewidth $\Delta_{1/2}$ (HWHM) of this resonance is related to the Rabi frequency ω_1 by:

$$\Delta_{1/2} \simeq \omega_1 \sqrt{\Gamma_2/\Gamma_{1\beta}}, \tag{1}$$

where

$$\Gamma_2 = (A+B)/2 + \gamma_2 \tag{2}$$

and

$$\Gamma_{1\beta} = \frac{(A+\gamma_1)\,(B+\gamma_1)\,(g_a B+g_b A+g\gamma_1)}{g(A+B)^2/4 + A(B-A) + \gamma_1\{g[(A+B)/2+\gamma_1] + g_a B + g_b A\}} \tag{3}$$

Here the g_i's are hyperfine level degeneracy factors, with g the total ground state degeneracy factor equal to $2(2I+1) = g_a+g_b$. Equation (1) indicates that the ratio $\sqrt{\Gamma_2/\Gamma_{1\beta}}$ is inversely proportional to the atomic line Q, and hence influences the shot noise limited performance of the standard [7].

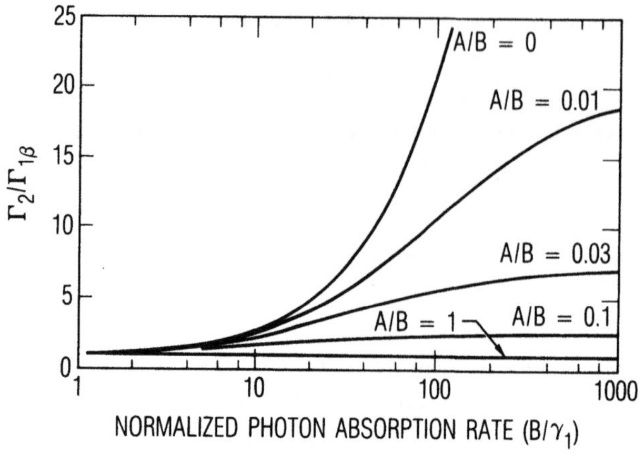

Figure 1.

The ratio $\Gamma_2/\Gamma_{1\beta}$ is illustrated in Fig. 1 as a function of B/γ_1 ($\gamma_1=\gamma_2$) for several values of the photon absorption rate ratio A/B. For efficient optical pumping (A/B → 0) and high light intensities, as might be achieved with a diode laser as the optical pumping light source [8], it is apparent that the ratio $\sqrt{\Gamma_2/\Gamma_{1\beta}}$ is very large. This implies that under these optical pumping conditions the atomic line Q may be degraded by more than an order of magnitude, compared to situations where $\sqrt{\Gamma_2/\Gamma_{1\beta}} \simeq 1$. Since a simple two-level quantum system model of the clock's signal yields $\Delta_{1/2} \simeq \omega_1$, it is clear that the complexity that arises from the microscopic scale of physics cannot be ignored if accurate predictions of the clock's performance are to be made under novel operating conditions.

III. Macroscopic Scale of Physics

In nearly all cases of gas-cell clock operation the density of alkali atoms in the vapor N is relatively high, so that it is not valid to assume that the vapor is optically thin to the optical pumping radiation. Thus, on the macroscopic scale of the gas-cell volume, the photon absorption rates A and B vary, and the equations describing the optical transmission through the vapor become transcendental. To illustrate this, consider a case where the photon absorption rate for the *a* hyperfine level is zero (A=0), then the optical attenuation only depends on the $\rho_{bb}(z)$ density matrix elements along the optical path at the positions z. These density matrix elements, however, are also a function of the various photon absorption rates B(z) (i.e., $\rho_{bb}(z) = \rho_{bb}(B,z)$). Assuming that the light attenuation can be described by a Bouger-Beer-Lambert type law we then have:

$$I(z) \sim B(z)/\sigma = (B_o/\sigma) \exp[-N\sigma z \, \rho_{bb}(B,z)], \tag{4}$$

where I(z) is the optical intensity at the position z, and σ is the photon absorption cross-section.

To handle this difficulty, we have employed iterative approximations. Initially $\rho_{bb}(B,z)$ in (4) is set equal to a constant, "global", density matrix element ς ($\rho_{aa}(z) \rightarrow 1-\varsigma$). ς is then evaluated through a self-consistency requirement, so that simple analytic functions are obtained as first approximations to A(z) and B(z) [9]. These functions are then employed in the microscopic density matrix equations, so as to determine $\rho_{bb}(z)$ and $\rho_{aa}(z)$, which in turn are used to numerically determine the value of I(z=L) (i.e., the optical intensity transmitted by the vapor).

A second aspect of the macroscopic scale of physics has to do with the spatial variation of the microwave field and hyperfine polarization over the signal volume. Since the atoms are localized by the buffer gas, atoms in different regions of the vapor experience different microwave field strengths, and hence reach saturation at different levels of the total microwave power supplied to the cavity. Additionally, as a result of diffusion to the gas-cell walls, where the atoms quickly depolarize, spatial modes of hyperfine polarization appear in the signal volume. To properly account for the effects of these spatial distributions requires a three dimensional description of the macroscopic physics.

In order to incorporate three dimensional field and polarization variations into the analysis, the cavity volume is divided into thousands of tubes [10]. The tube diameters are small, so that variations of field strength and hyperfine polarization across a tube face are negligible. In this way the optical power transmitted by each tube as a function of microwave frequency reduces to a problem in one spatial dimensional (z-axis). Each tube is associated with an appropriately normalized microwave field strength value, so that any particular transverse geometry of the field (xy plane) may be considered in the calculations. Additionally, to account for the transverse distribution of hyperfine polarization, an appropriate weighting function is superimposed onto the GVT solutions for each tube. The observed clock signal is computed by summing the optical power transmitted by all the tubes, and tabulating the sum as a function of microwave frequency.

To illustrate the influence of the macroscopic scale of physics on the clock's operation, Fig. 2 shows the behavior of the clock signal amplitude S and the "effective" clock signal linewidth $\delta\nu$ as a function of the microwave power supplied to a TE_{111} mode microwave cavity with a cavity Q of 100. $\delta\nu$ is the linewidth (HWHM) of a lorentzian that approximates the clock signal lineshape near line center, and is inversely proportional to the square root of the slope of the clock's error signal: as $\delta\nu$ increases, the clock's shot noise limited frequency stability degrades [7]. Figures 2a and 2b clearly show that for this particular calculation, saturation occurs near -40 dbm, that below this value the clock signal increases nearly linearly with microwave power, and that above this value the effective linewidth increases approximately like the fourth root of the microwave power.

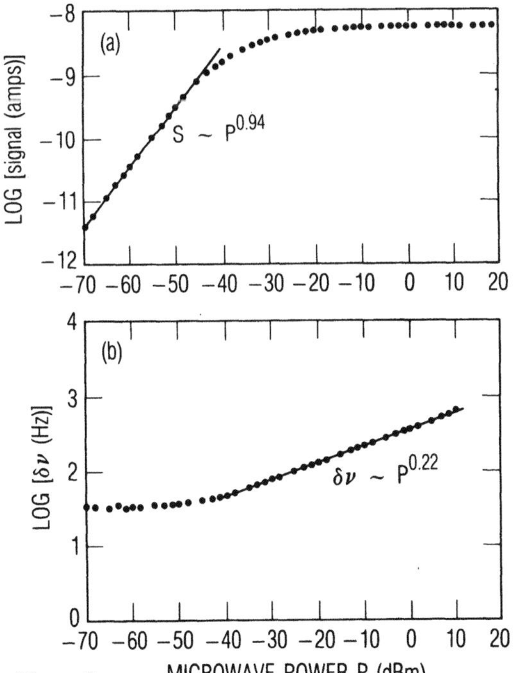

Figure 2

What is surprising about these results is that a simple two-level quantum system description of the clock's signal, as well as the generalized Vanier theory, predicts that $\delta\nu$ should increase like the square root of the microwave power in the saturation regime. The macroscopic scale of physics alters this behavior, because the observed lineshape near line center is dominated by those spatial regions of the atomic ensemble where the microwave field strength is below the saturation value, and because the volume containing this population shrinks in size as the microwave power supplied to the cavity increases [4]. With regard to clock operation, the implication of this finding is that high levels of microwave power can be supplied to the clock cavity, without seriously degrading the clock's shot noise limited performance [10].

IV. Conclusion

As a consequence of the atomic signal's inherent complexity, improvement in the gas-cell frequency standard's performance characteristics has been, to date, largely an empirical endeavor. However, with the advances in understanding of gas-cell atomic clock physics discussed above, as well as the recent clock modelling efforts of others [11,12], we are fast approaching a time when computer aided design of novel gas-cell frequency standards will be a common practice. In this regard it is worth noting that theoretical modelling has already indicated that shot noise limited performance of this standard could be improved by roughly three orders of magnitude simply by employing diode lasers as the optical pumping light source [13]. There are still, however, gaping holes in our understanding of gas-cell clock physics, and in our ability to model the clock's behavior. We do not yet know the origins of long term frequency instability in this standard, though ideas that merit further attention have been discussed [14,15]. Consequently, all modelling to date has focused on the short term, shot noise limited performance of the device. Additionally, none of the models has included all the major mechanisms that give rise to frequency shifts in the standard (i.e., buffer gas, spin exchange, AC Stark, Quadratic Zeeman, Bloch-Siegert, etc.), and as mentioned above several of these mechanisms could give rise to inhomogeneous broadening of the clock signal lineshape. Inclusion of these frequency shifts will most likely be the next major advance in the theory of the clock's atomic signal. Not only will this lead to more accurate descriptions of the clock's shot noise limited performance; it will also allow theoretical investigations into the origins of long term frequency instability. Moreover, with these next generation theories of the clock's atomic signal there will be an opportunity to calculate the total frequency shift of the standard, compare this prediction against experiment, and in this way critically test our detailed description of the gas-cell physics.

References

1. W. Riley and F. Danzy: Proc. 19th Annual Precise Time and Time Interval (PTTI) Applications and Planning Meeting (U.S. Naval Observatory, Washington D.C., 1987) pp. 267-274.
2. J. C. Camparo, R. P. Frueholz, and C. H. Volk: Phys. Rev. A 27, 1914 (1983).
3. A. Risley, S. Jarvis, and J. Vanier: J. Appl. Phys. 51, 4571 (1980).
4. J. C. Camparo, to be published.
5. G. Missout and J. Vanier: Can. J. Phys. 53, 1030 (1975).
6. J. C. Camparo and R. P. Frueholz: Phys. Rev. A 31, 1440 (1985); Phys. Rev. A 32, 1888 (1985).
7. J. Vanier and L-G. Bernier: IEEE Trans. Instrum. Meas. IM-30, 277 (1981).
8. J. C. Camparo: Contemp. Phys. 26, 443 (1985).
9. J. C. Camparo and R. P. Frueholz: J. Appl. Phys. 59, 301 (1986).

10. J. C. Camparo and R. P. Frueholz: to be publishd IEEE Trans. Ultrason., Ferroelec., and Freq. Control.
11. P. Tremblay, N. Cyr, and M. Tetu: Can. J. Phys. 63, 1563 (1985).
12. M. B. Gornyi, B. G. Matisov, G. M. Smirnova, and V. I. Khutorshchikov: Sov. Phys. Tech. Phys. 32, 448 (1987).
13. J. C. Camparo and R. P. Frueholz: J. Appl. Phys. 59, 3313 (1986); IEEE Trans. Ultrason., Ferroelec., and Freq. Control UFFC-34, 607 (1987).
14. C. H. Volk and R. P. Frueholz: J. Appl. Phys. 57, 980 (1985).
15. J. C. Camparo: Proc. 18th Annual Precise Time and Time Interval (PTTI) Applications and Planning Meeting (U. S. Naval Observatory, Washington, D. C., 1986) pp. 565-588.

Part IV

New Standards

Presentation of the Session Entitled "New Standards"

C. Audoin

Laboratoire de l'Horloge Atomique,
Equipe de Recherche du CNRS associée à l'Université Paris-Sud,
Bât. 221, Université Paris-Sud, F-91405 Orsay Cedex, France

The frequency and time standards to be described in this Session are micro-wave or sub-millimeter-wave frequency and time standards whose study has been started more recently than that of Rb, Cs and H frequency/time standards.

In the case of ion standards, the confinement method is specific to charged particles. In its present form, the Mg standard shows an original and very interesting method of preparing the atomic states.

Ion frequency standards have been the subject of a great deal of work, which followed the pionering work done in the Hans Dehmelt group at the University of Washington. This work led to the measurement of the hyperfine splitting of $^3He^+$ in the ground state /1,2/. An outstanding amount of work on ion frequency standards has been accomplished in the NBS group, in Boulder, under the leadership of Dave Wineland who will present the second talk of the Session. Use of $^{199}Hg^+$ for realizing an operational frequency standard has been proposed by Fouad Major, from the NASA Goddard Space Flight Center, in 1969 /3/. The first speaker, Len Cutler, and his associates, have developed three units showing a remarkable long term frequency stability and which are currently used to steer the USNO time scale.

The interest of using fine structure transitions in Mg and Ca atoms has been pointed out by Franco Strumia, from the University of Pisa, during the First Symposium held in 1971 near Quebec City /4/. The metrological aspect of the work has been taken over by the Istituto Elettrotecnico Nazionale Galileo Ferraris in Torino and the results achieved will be presented by the third speaker, Aldo Godone.

References

1. E.N. Fortson, F.G. Major and H.G. Dehmelt : Phys. Rev. Lett. 16, 221 (1966)
2. H.A. Schuessler, E.N. Fortson and H.G. Dehmelt : Phys. Rev. 187, 5 (1969)
3. F.G. Major : NASA Report X-521-69-167 (Goddard Space Flight Center, Greenbelt, Maryland, 1969)
4. F. Strumia : Proc. of the 1st Frequency Standards and Metrology Seminar, 193 (1971)

Frequency Standards and Metrology Editor: A. De Marchi

High Accuracy Spectroscopy of Stored Ions *

D.J. Wineland, W.M. Itano, J.C. Bergquist, J.J. Bollinger, F. Diedrich†*,*
and S.L. Gilbert

Time and Frequency Div.,
National Institute of Standards and Technology, Boulder, CO 80303, USA

ABSTRACT: Physical limitations to the achievement of accurate, high resolution spectroscopy on stored ions are briefly discussed. For experiments on ion clouds, a compromise between frequency stability and second order Doppler shift uncertainty must be made. Expected performance for specific examples using single ions and ion clouds is discussed.

1. Introduction

In this paper we discuss some of the basic physical limitations to the achievement of accurate, high resolution spectroscopy of electromagnetically confined ions. For brevity, we do not address many practical problems, such as local oscillator spectral purity. These problems, although important, will probably not cause the ultimate limit in accuracy and resolution.

This topic naturally divides between experiments on many ions and single trapped ions. We anticipate that for experiments using more than one stored ion, the uncertainty in the second order Doppler (time dilation) shift will be the accuracy-limiting systematic effect[1-3]. This has been the case in the past. For single ions, the uncertainty in the second order Doppler shift can be made extremely small. The limiting uncertainty for single ions may depend on the experiment[4-6], but inaccuracies should eventually be less than 1 part in 10^{18}. Use of single ions however might be precluded if the signal-to-noise ratio (S/N) is not high enough to achieve the desired measurement precision in a reasonable length of time. Therefore, we must consider a trade-off between second order Doppler shift and S/N, both of which increase as the ion number increases.

As in any spectroscopic experiment, we must account for the perturbations due to static and time varying multipole interactions for electric, magnetic, and gravitational fields. These include atom-trap field interactions, collisions, shifts due to uncontrolled electric and magnetic fields, gravitational red shifts, etc. (see refs 1-9 and other contributions in these proceedings).

†Present address: Max-Planck Inst. Quantenoptik, Garching.

Frequency Standards and Metrology Editor: A. De Marchi

2. Optical-Pumping, Double-Resonance Experiments on Stored Ions

We will assume ion spectra are observed by optical-pumping, double-resonance methods in which optical fluorescence changes are monitored. In Fig. 1, we assume that level 3 is at an optical energy above the ground state 1. The transition of interest for high resolution spectroscopy is between levels 1 and 2 (frequency ω_0) which we designate the "clock" transition. A_2 and A_3 are radiative decay rates and R is the excitation rate between levels 1 and 3. R' and R" denote pumping rates between the clock levels due to optical pumping sources (independent of the clock radiation) applied simultaneously with R.

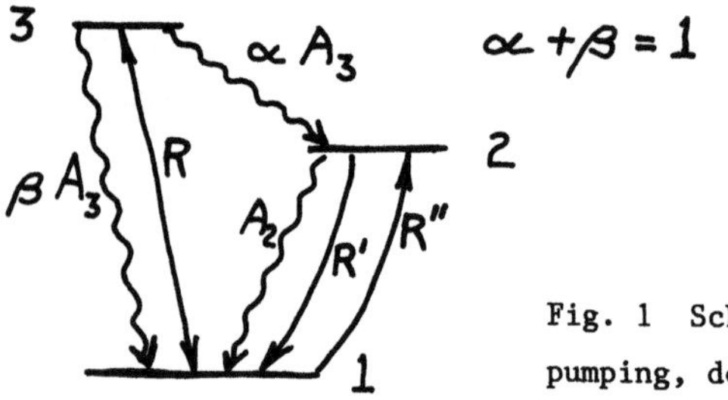

Fig. 1 Schematic diagram for optical pumping, double resonance experiments.

2.1. <u>Depletion Pumping</u> Assume A_2, R', R" « R < βA_3, and $\alpha \cong \beta$. When R is applied, atoms are pumped from level 1 to level 2 through level 3 (level 1 is depleted). If radiation (not shown in Fig. 1) is applied near the clock frequency $\omega_0 \equiv (E_2-E_1)/\hbar$, some of the atoms will be transferred back to level 1. This causes an increase in the detected fluorescence.[7,8]

2.2. <u>Electron Shelving</u> [4] We assume that the clock radiation and pumping radiation are applied sequentially to avoid light shifts on the clock transition. Assume $\alpha = 0$, R' » R", and $A_3 \gg (A_2+R'-R")$. In steady state with R, R' and R" applied, the ions spend most of the time in level 1. If ions are in level 2 after the clock radiation is applied and if $R \cong A_3$, we can observe the absence of about $N_d \cong \eta A_3(A_2+R'-R")^{-1}$ detected (3→1) photons for each photon absorbed on the clock transition (η is the net photon detection efficiency). When $N_d \gg 1$ and sources of technical noise are absent, the fluctuations in detected counts are caused not by the photon shot noise but by the fluctuations in the number of ions which have made the transition from level 1 to level 2 (Ref. 1). This is the fundamental noise limit. In Ref. 5, $\alpha \cong 10^{-7}$, R', R" = 0, $A_3 \cong 4.3\times10^8$ s^{-1}, $A_2 \cong 12$ s^{-1}, $\eta \cong 5\times10^{-4}$, $N_d \cong 20000$. In Ref. 9, $\alpha = 0$, $A_2 = 0$, $N_d \cong 1$, and R » R' » R" (R, R', and R" result from the same radiation source).

3. Frequency Stability

The signal is the response of the fluorescence detector to changes in frequency of the oscillator which drives the clock transition. As a specific example, we assume use of the Ramsey separated field method in time domain. That is, excitation of the clock transition is by two phase coherent pulses of radiation of duration ΔT_R separated by T_R.[9] For mathematical simplicity we assume $\Delta T_R \ll T_R \ll A_2^{-1}$ and $R'-R'' \gg A_2$ (In Ref. 5, R', R'' = 0). The electron shelving method of detection is assumed in the case where $N_d \gg 1$ in a time much less than $(A_2 + R'-R'')^{-1} \ll T_R$; that is, the detection time is less than the repumping time, which is less than T_R. If we probe the clock transition on each side of the central Ramsey peak, we can develop an error signal which steers the average frequency to ω_0. Under the above assumptions, the measurement precision as characterized by the two-sample Allan variance[10] for the locked oscillator is[1,11]

$$\sigma_y(\tau) = (\tau N T_R \omega_0{}^2)^{-\frac{1}{2}} \tag{1}$$

where τ is the measurement time and N is the ion number.

4. Second Order Doppler Shift in Ion Clouds

From Eq. 1, if ω_0 or T_R is too small, it may be necessary to use large N to obtain the desired measurement precision in a reasonable length of time. We anticipate that the largest systematic effect in an experiment on many stored ions will be the uncertainty in the second order Doppler shift, $\Delta\nu_{D2}$. Fractionally, this shift is equal to $\langle \Delta\nu_{D2}/\nu_0 \rangle \cong -\langle v^2 \rangle/2c^2$ where $\nu_0 = \omega_0/2\pi$, $\vec{v}$ is the ion velocity, c is the speed of light, and $\langle\ \rangle$ denotes an average over the N ions and over time. We would like to minimize the second order Doppler shift while simultaneously making the number of ions as large as possible to increase stability. These two requirements are contradictory and so a compromise between accuracy and stability must be made.

4.1. rf Trap

We assume that the pseudopotential approximation[12] is valid and the kinetic energy of the secular motion (the motion associated with the pseudopotential) is reduced to a value where the Debye length of the ion cloud is less than the cloud dimensions[13]. If the cloud is large enough, this can be accomplished by bringing the secular motion nearly into thermal equilibrium with a background gas (e.g., He) through ion-neutral collisions[7,14]. In principle it can also be accomplished by the method of laser cooling but so far this has been achieved with only a few ions in an rf trap because of rf heating.

Under the above conditions, the average kinetic energy per ion is dominated by the rf driven micromotion [3,14]. For a spheroidal cloud of radius r_{cl} and height $2z_{cl}$ we calculate $\langle\Delta\nu_{D2}/\nu_0\rangle = -2a(2z_{cl}^2 + r_{cl}^2)/5mc^2$ where $a \equiv q^2V_0{}^2/(m\Omega^2(r_0{}^2+2z_0{}^2)^2)$. Here, q and m are the ion mass and charge, V_0 and Ω are the applied rf potential and frequency and r_0 and z_0 are characteristic trap dimensions[12]. From this expression and the ion density[12,13], we have for a spherical cloud [3,14]

$$\langle\Delta\nu_{D2}/\nu_0\rangle = -3\omega_s{}^2r_{cl}^2/10c^2 = -3q^2N/10r_{cl}mc^2 = -3(\omega_sNq^2/m)^{2/3}/10c^2 \quad (2)$$

where ω_s is the secular frequency for a single ion in the trap. From Eq. 2, we derive a relation between N and $\langle\Delta\nu_{D2}/\nu_0\rangle$,

$$N = -2.16 \times 10^{16}r_{cl}M \langle\Delta\nu_{D2}/\nu_0\rangle/Z^2 \quad (3)$$

where Z is the ion charge (in units of the proton charge), M is the ion mass in atomic mass units, and r_{cl} is the radius of the cloud in centimeters. Equation 3 shows that for an assumed value of $\langle\Delta\nu_{D2}/\nu_0\rangle$, we would like to work with as large a cloud (low density) and ion mass as possible.

4.2. Penning Trap

We will assume that the internal (cyclotron and axial) temperature of the ions is reduced to a value where the second order Doppler shift is dominated by the velocity of ions due to cloud rotation at angular frequency ω[3,13,15]. This rotation is an essential part of ion storage in the Penning trap since it provides the inward Lorentz $(q(\vec{v}\times\vec{B})/c)$ force which overcomes the radially outward force due to the trapping fields and space charge. We have $\langle\Delta\nu_{D2}/\nu_0\rangle = -(\omega r_{cl})^2/5c^2$ where the cloud is a uniformly charged spheroid of height $2z_{cl}$ and radius r_{cl}[13]. From Ref. 13, the density of ions is given by $n_0 = m\omega(\Omega_c-\omega)/2\pi q^2$ where $\Omega_c = qB_0/mc$ is the ion cyclotron frequency. From these two expressions, the total number of ions in the cloud is [3]

$$N = 3.10 \times 10^{13}B(T)\langle-\Delta\nu_{D2}/\nu_0\rangle^{\frac{1}{2}}\ z_{cl}(r_{cl}-r_c')/Z \quad (4)$$

where B is in teslas, z_{cl} and r_{cl} in centimeters and $r_c' \equiv (5\ \langle-\Delta\nu_{D2}/\nu_0\rangle)^{\frac{1}{2}}\ c/\Omega_c$. For a given value of $\langle\Delta\nu_{D2}/\nu_0\rangle$, large B, z_{cl} and r_{cl} (low density) are desirable. N is independent of M (for the typical case $r_c' \ll r_{cl}$). We may choose B to make the clock transition field independent to first order[1,9].

5. Single Ions

From Eq. 1, if ω_0 and T_R are large enough, stability can be high enough for practical use, even with N = 1. For single ions, many systematic effects become negligible[1-6]. For example, from our recent experiments on single Hg^+ ions[5], we estimate uncertainties in the second order Doppler shift correction to be less than 1 part in 10^{20}. What will actually be the limiting systematic effect will depend on practical considerations and which ion is used. For the sake of definiteness we consider two cases of interest at Boulder and list what we think are the limiting systematic effects.

5.1 $^{199}Hg^+$, $5d^{10}6s\ ^2S_{\frac{1}{2}}(F=0,\ m_F=0) \leftrightarrow 5d^96s^2\ ^2D_{5/2}(F=2,\ m_F=0)$ optical quadrupole transition ($\nu_0 = 1.07 \times 10^{15}$ Hz).

(1) Magnetic field sensitivity. From the Breit-Rabi formula, we obtain $(\nu_0(B)-\nu_0(B=0))/\nu_0 = -1.8 \times 10^{-5}B^2$ where B is expressed in teslas. It should be possible to obtain fractional precision of better than 1 part in 10^{18}.

(2) $^2D_{5/2}$ state quadrupole interaction. The quadrupole moment of the $^2D_{5/2}$ state can couple to static electric field inhomogeneities to cause a shift of this level. In a suitable (Cartesian) coordinate system, the atomic quadrupole moment interacts with a potential of the form $\varphi = A(x^2+y^2-2z^2) + \delta A(x^2-y^2)$. In the specific case where $\delta A = 0$ and $\vec{B}$ (quantization axis) is parallel to z, then $\Delta\nu_Q \cong -1.8$ Hz for 1 V applied to the ring electrode for the trap of Ref. 5. The presence of contact potentials on the trap electrodes makes the direction of z (in φ) uncertain but $\Delta\nu_Q$ averages to zero for three measurements of ν_0 where the directions of $\vec{B}$ are mutually orthogonal.[16]

5.2 $^{199}Hg^+$ ground state $(F=0,\ m_F=0) \leftrightarrow (F=1,\ m_F=0)$ hyperfine transition. (ν_0 = 40.5 GHz)

(1) Magnetic field sensitivity. $(\nu_0(B) - \nu_0(B=0))/\nu_0 \cong 0.24\ B^2$ where B is expressed in teslas. For single ions, magnetic field homogeneity requirements are not stringent.

6. Examples

Table I lists some possibilities of interest at Boulder. Implicit in the Table is the use of laser cooling. Sympathetic cooling [9,17] may be necessary for N > 1. Reasonable short term stability is obtained for single $^{199}Hg^+$ ions using the ground state hyperfine transition. Systematic effects should be more easily controlled using single ions.[1-6] Fluctuations in cloud density and size (and therefore detected fluorescence) could degrade stability over that shown in the first two rows. For example, laser induced instabilities similar to those cited in Refs. 15 and 18 have been observed in all Penning traps at NBS; these instabilities must be suppressed to achieve $\sigma_y(\tau)$ given by Eq. 1.

Table I: Expected performance in particular systems. In rows 1 and 2, N is calculated from Eq. 4. Eq. 1 is used to determine $\sigma_y(1\ s)$. For N > 1, the ion cloud is assumed to be spherical (radius = r_{cl}). B is chosen to minimize field sensitivity. For $T_R > 1$ s, quoted stabilities are referred to 1 s.

SYSTEM	$\langle -\Delta\nu_{D2}/\nu_0 \rangle$	B(T)	r_{cl}(cm)	N	T_R(s)	$\sigma_y(1\ s)$
$^9Be^+$ hyperfine $\nu_0 \cong 303$ MHz (Penning trap)[9]	10^{-15}	0.819	0.5	2.0×10^5	100	1.2×10^{-13}
$^{201}Hg^+$ hyperfine $\nu_0 = 25.9$ GHz (Penning trap)[1]	10^{-16}	0.534	0.5	4.1×10^4	100	1.9×10^{-15}
$^{199}Hg^+$ hyperfine $\nu_0 = 40.5$ GHz (rf trap)	$<10^{-15}$	$\cong 0$	--	1	100	3.9×10^{-13}
$^{199}Hg^+$ optical quadrupole $\nu_0 = 1.07\times10^{15}$ Hz (rf trap)[5]	$<10^{-18}$	$\cong 0$	--	1	0.025	9.4×10^{-16}

7. Acknowledgements

We thank the U.S. Office of Naval Research and the U.S. Air Force Office of Scientific Research for support. We thank D. Heinzen and L. Hollberg for helpful comments on the manuscript.

1. D.J. Wineland, W.M.Itano, J.C. Bergquist, F.L. Walls, Proc. 35th Ann. Symp. Freq. Control (1981), p. 602 (copies available from Electronic Industries Association, 2001 Eye St., N.W., Washington, D.C. 20006)
2. D.J. Wineland, et. al., J. de Phys., Colloque C8, 42 (1981) p. C8-307
3. D.J. Wineland, in Prec. Meas. and Fundamental Constants II, B.N. Taylor and W.D. Phillips, eds., Nat. Bur. of Stand. (U.S.) Spec. Publ. 617 (1984), p. 83; D.J. Wineland, W.M. Itano, J.C. Bergquist, J.J. Bollinger and J.D. Prestage, Ann. Phys. (Fr.) 10, 737 (1985)
4. H.G. Dehmelt, IEEE Trans. Instrum. Meas. IM-31, 83 (1982); H.G. Dehmelt, in Advances in Laser Spectroscopy, ed. by F.T. Arecchi, F. Strumia, and H. Walther (Plenum, 1983) p. 153; H.G. Dehmelt, these proceedings

5. J.C. Bergquist, et. al., these proceedings; F. Diedrich et al., submitted for publication
6. D.J. Wineland, Science 226, 395 (1984) and references therein
7. L. Cutler, these proceedings
8. G. Werth, these proceedings
9. J.J. Bollinger, et. al., these proceedings
10. J.A. Barnes, et. al., IEEE Trans. Instrum. Meas. IM-20, (1971)
11. Ref. 1 is in error by a factor of $\sqrt{2}$
12. H.G. Dehmelt, Adv. At. Mol. Phys. 3, 53 (1967)
13. D.J. Wineland, J.J. Bollinger, W.M. Itano, and J.D. Prestage, J. Opt. Soc. Am. B2, 1721 (1985); L.R. Brewer, et al., Phys. Rev. A38, 859 (1988)
14. L.S. Cutler, R.P. Giffard, and M.D. McGuire, Appl. Phys. B36, 137 (1985)
15. W.M. Itano, et. al., these proceedings
16. W.M. Itano et. al., to be published
17. D.J. Larson, J.C. Bergquist, J.J. Bollinger, W.M. Itano, and D.J. Wineland, Phys. Rev. Lett. 57, 70 (1986)
18. R.C. Thompson, G.P. Barwood, and P. Gill, Opt. Acta 33, 535 (1986)

Mg Beam Frequency Standard

A. Godone, E. Bava, and C. Novero
Istituto Elettrotecnico Nazionale G. Ferraris, Torino, Italy

The description of the technical realization of a Mg beam frequency standard and the experimental results are here reported. Moreover the main sources of uncertainty are examined in order to provide the evaluation of the accuracy achievable with such a standard.

I) Introduction

Absence of hyperfine structure and existence of a natural selection process to provide population difference and efficient detection are the most attractive characteristics of a frequency standard based on the $^3P_1 - {}^3P_0$ $\Delta m_j = 0$ transition (λ = 498 μm) in the metastable triplet of Mg. Its frequency has been measured with respect to a Cs standard with an uncertainty of $\pm$ 3×10^{-11} [1]; in this experiment an atomic beam rich of metastable atoms is interrogated via a Ramsey-type technique, yielding a line Q-factor $Q_L = 5 \times 10^8$, and a 5 MHz quartz crystal oscillator is locked to the Mg transition via a low phase noise frequency multiplication chain.

In the following the experimental set-up of the Mg standard will be briefly described together with the main results achieved; moreover the accuracy budget will be reported after discussing the most important sources of frequency uncertainty for the practical apparatus

II) Experimental Set-up of the Mg Beam

The Mg frequency standard set-up is shown in fig. 1. The atoms effused by an oven at 520°C are excited into the metastable levels $^3P_{0,1,2}$ via electron collision; the technical details and the dynamic processes of excitation are reported in ref. [2]. After collimation the atomic flux is 8×10^{13} atoms/s, the total beam divergence is 10 mrad and the beam diameter in the first interaction zone is 16 mm.

The decay region (Z_1 = 1.15 m) allows a population difference before interaction of about 10% between the $^3P_1$ and $^3P_0$ levels, due to natural decay of $^3P_1$ to ground state with a time constant of some milliseconds.

The efficiency of production of metastable atoms has been measured to be 30% at a discharge current I = 300 mA; the velocity distribution in the metastable beam just outside the discharge region is given by [2]:

$$\rho_m(v,I) = A v^{\gamma} e^{-HI/v - v^2/\alpha^2} \operatorname{sh}(KI/v) \qquad (1)$$

where A is a normalization constant, γ an exponent higher than 3 which takes into account the elastic collision effects in the discharge region, H and K two constants related to the anelastic excitation and de-excitation processes and α the most probable atom velocity in the oven (α = 736 m/s at T = 520°C).

Frequency Standards and Metrology Editor: A. De Marchi

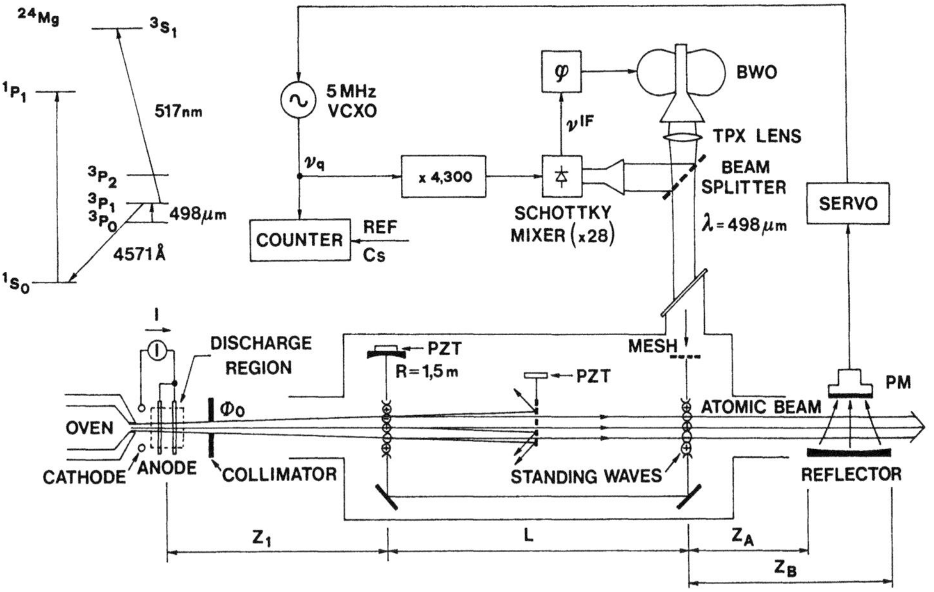

Fig. 1 - First excited Mg levels and experimental set-up (Z_I=1.15m, Z_A=0.3m, Z_B=0.5m).

The normalized time-of-flight distribution observed via the intercombination-line fluorescence 1.5 m downstream is shown in fig. 2 (triangles); in the same figure the continuous curve is the theoretical distribution corresponding to eq. 1 when the decay process along the beam is taken into account [2]. The experimental points are best fitted with γ = 4, H = 2200 and K = 1540.

The Ramsey-type interrogation is performed via a folded open resonator as indicated in fig. 1; with a coupling plane mirror made with a copper mesh, whose grid constant is 50 μm and reflectivity 98% at 600 GHz, the cavity loaded Q-factor is Q_c = 300,000.

A grid between the two interaction zones, separated by a distance L = 30 cm, acts as a spatial selector for different atomic passes in order to avoid the averaging out of the Ramsey interference fringes [3]; the beam is stopped for a length of $\lambda/4$ along its cross-section with a period of $\lambda/2$.

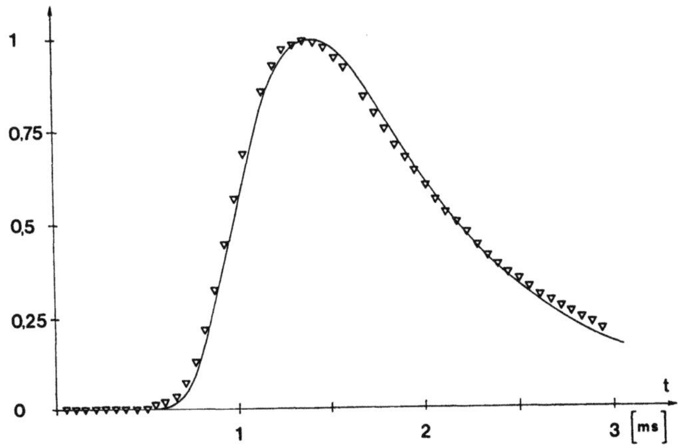

Fig. 2 - Time of flight distribution.

The clock transition is detected by a photomultiplier (PM) via the increase of the fluorescence light of the $^3P_1$ level. Both in the interaction and in the detection regions an axial-quantization magnetic field B is applied for the observation of the π-transition.

The radiation at 600 GHz is provided by a carcinotron (BWO) phase-locked to a 5 MHz quartz oscillator via a low phase-noise frequency multiplication chain; the closed-loop bandwidth is 450 kHz necessary to control the high flicker noise level of the free-running submillimetric source. Details of the RF electronic system are reported in [4,5].

III) Experimental Results

Experimental observations of the Rabi and Ramsey patterns are reported in [1] together with the dependence from the grid position; a typical record of Ramsey fringes is shown in fig. 3 with an equivalent video bandwidth of 0.1 Hz: the full width at half maximum (FWHM) of the central fringe is 1.2 kHz corresponding to a line Q-factor $Q_L = 5 \times 10^8$, the contrast is 45% with respect to the Rabi pattern and 2% with respect to the background fluorescence level. The signal-to-noise ratio in 1 Hz bandwidth turns out $(S/N)_{1Hz} \simeq 250$ with an RF power inside the cavity of about 50 uW (the optimum level is estimated 0.5 mW [6]).

In order to lock the 5 MHz quartz oscillator to the Mg resonance the well-known first-derivative technique is used: the BWO signal is sinusoidally frequency modulated at $f_m = 175$ Hz with a peak-to-peak frequency deviation $\Delta\nu_{pp} = 1.4$ kHz; the time constant of the control loop is 100 ms.

The short-term frequency stability has been evaluated through the two--sample standard deviation $\sigma_y(2,\tau)$; in the present experiments the following value has been achieved for counting times τ in the range $0.3 \lesssim \tau \lesssim 100$ s:

$$\sigma_y(2,\tau) = 8 \times 10^{-12} \tau^{-1/2} \qquad (2)$$

The frequency ν_0 of the $^3P_1 - {}^3P_0$ $\Delta m_j = 0$ transition of ^{24}Mg, at zero magnetic field, has been measured with respect to a Cs standard with an uncertainty of $\pm 3 \times 10^{-11}$; it turns out [1]:

$$\nu_0 = 601277157860 \pm 20 \text{ Hz}$$

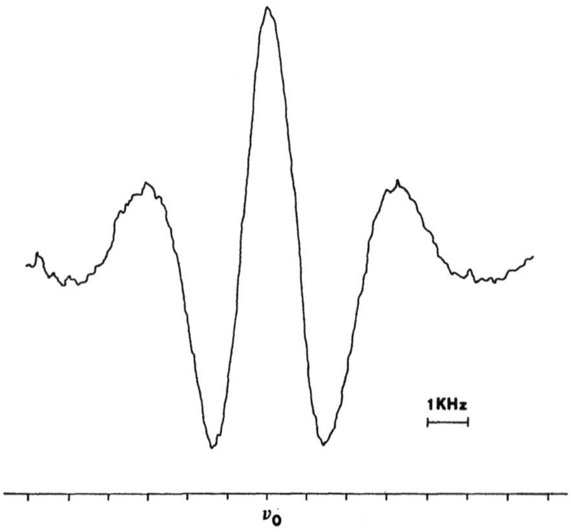

Fig. 3 - Observed Ramsey fringes.

On the basis of the experimental results reported and the comparison with the theoretical predictions [6] it seems possible to attain a short-term frequency stability of $\sigma_y(2,\tau) \simeq 1\times10^{-12}\,\tau^{-1/2}$ by optimizing the rf power and the detection efficiency; a stability of $\sigma_y(2,\tau) = 2\text{-}4\times10^{-14}\,\tau^{-1/2}$ could be reached via optical pumping on the $^3P_1 - {}^3S_1$ green transition in order to increase the population difference between the $^3P_1$ and $^3P_0$ levels and decrease the background noise.

IV) Accuracy Budget

The velocity-selection mechanism and the optical Ramsey scheme of the Mg standard are quite different with respect to the others microwave standards and this requires a complete re-examination of the main sources of uncertainty.

IVa) Relativistic Doppler Shift

Introducing the relative 2nd order Doppler shift $y = \Delta\nu/\nu_0$ the following expression may be written:

$$y = y_0 + \left.\frac{\partial y}{\partial\gamma/\gamma_0}\right|_{\gamma_0}\frac{\delta\gamma}{\gamma} + \left.\frac{\partial y}{\partial H/H_0}\right|_{H_0}\frac{\delta H}{H} + \left.\frac{\partial y}{\partial K/K_0}\right|_{K_0}\frac{\delta K}{K} + \left.\frac{\partial y}{\partial I/I_0}\right|_{I_0}\frac{\partial I}{I} + \left.\frac{\partial y}{\partial\chi/\chi_0}\right|_{\chi_0}\frac{\delta\chi}{\chi} \tag{3}$$

where the subscript "0" stands for "nominal value", y_0 is the bias term, the quantities γ, H, K and I define the velocity distribution (see eq. 1) and $\chi = \sqrt{2\pi}\,\omega_R\rho_0$, being ω_R the Rabi angular frequency and ρ_0 the e^{-1} radius of the gaussian field in the interaction regions.

The derivative terms of eq. 3 have been computed in [7] and for the experimental situation considered the following values have been found:

$$\left.\frac{\partial y}{\partial\gamma/\gamma_0}\right|_{\gamma_0} = -1.2\times10^{-12} \qquad \left.\frac{\partial y}{\partial H/H_0}\right|_{H_0} = -1.5\times10^{-12} \qquad \left.\frac{\partial y}{\partial K/K_0}\right|_{K_0} = 0.6\times10^{-12}$$

$$\left.\frac{\partial y}{\partial I/I_0}\right|_{I_0} = -2\times10^{-12} \qquad \left|\left.\frac{\partial y}{\partial\chi/\chi_0}\right|_{\chi_0}\right| = < 1\times10^{-12}$$

In order to reduce the 2nd order Doppler error to the 10^{-13} level the velocity distribution parameters , γ, H, K must be known within 5%; the same requirement stands for the discharge current I and rf field ($\alpha\ \chi$).

IVb) Residual 1st Order Doppler Shift

If the atomic beam axis is not perfectly perpendicular to the cavity arms, the net travelling wave always present in a real resonator produces a 1st order Doppler shift, which may be expressed in this case [7] by the relation:

$$\frac{\Delta\nu}{\nu_0} = A(\chi)\frac{2\varrho_0}{L}\frac{a}{c}\,\vartheta\,2\sqrt{2}\,r_m^{\perp} \tag{4}$$

where c is the speed of light, $A(\chi)$ is a function of the rf power and geometry, θ the small misalignment with respect to perpendicularity and $r_m^{\perp}$ the normalized surface resistance for normal incidence of the mirrors (skin-effect).

If we assume $L = 0.3$ m, $\theta \sim 0.4$ mrad, $r_m^{\perp} = 10^{-3}$ and $A(\chi) = 2$ the above effect amounts to 3×10^{-13}.

IVc) Cavity Phase-shift

If a collimated beam is considered, orthogonal to the resonator standing waves, the following expression is obtained for a displacement d of the grid from its ideal position or between the two standing waves:

$$\frac{\Delta\nu}{\nu_0} \simeq \frac{\sqrt{6}\left(1+\sqrt{2}\right) r_m^{\perp} \alpha d}{cL} f\left(\chi/\alpha\right) \tag{5}$$

The values assumed from $f(\chi/\alpha)$ depend on the case considered, anyway it is of the order of unity. For $d = 10$ µm in both cases it turns out $\Delta\nu/\nu_0 \simeq 4\times10^{-13}$ for optimum excitation power.

If a very small inclination θ ($\theta \ll \lambda/4L$) of the atomic beam is considered an approximate expression is:

$$\frac{\Delta\nu}{\nu_0} = \frac{\sqrt{3}\, r_m^{\perp} \alpha \,\mathrm{tg}\left(\beta L \vartheta/2\right)}{c\beta L/2} f\left(\chi/\alpha, \vartheta\right) \tag{6}$$

where $\beta = 2\pi/\lambda$. Again $f(\chi/\alpha, \theta)$ is of the order of unity, therefore for $\theta = 0.1$ mrad it turns out $\Delta\nu/\nu_0 \approx 4\times10^{-13}$.

For a divergent beam, the total error is an algebraic sum of contributions from beams with different inclinations; this situation cannot be evaluated directly from (5) and (6), but numerical integrations are required [8]. In fig. 4 the computed dependence of phase-shift with respect to the grid displacement from the ideal position is reported: an averaging effect, due to the beam divergence, takes place, reducing significantly the value obtained by eq. 5.

A similar reduction is observed also in the case of a net inclination of the divergent beam, so that the total phase-shift error is reduced below $\pm1\times10^{-13}$.

IVd) Other Effects

The uncertainties related to the rf electronics (synthesis chain and BWO phase-lock loop) have been considered in [4,5]: they may be reduced at a level lower than 1×10^{-13}.

The cavity-pulling may be evaluated with the well-known relation:

$$\frac{\Delta\nu}{\nu_0} = \left(\frac{Q_C}{Q_L}\right)^2 \frac{\Delta\nu_c}{\nu_0} \Lambda(\chi) \tag{7}$$

where $\Delta\nu$ is the cavity-detuning and $\Lambda(\chi)$ a factor depending mainly from the rf power [9]. Assuming $\Delta\nu_c \sim 100$ kHz and $\Lambda(\chi) \sim 0.2$ at optimum power, eq. 7 gives in our case $\Delta\nu/\nu_0 \approx 1\times10^{-14}$.

The 2nd order Zeeman shift (C-field) has been evaluated for the ^{24}Mg clock transition in [10]; the zero-field frequency ν_0 may be computed with the relation:

$$\nu = \nu_0 + KB^2, \quad K = 3.82706 \pm 0.00004 \ [\mathrm{Hz/Gauss}]^2 \tag{8}$$

Assuming a quantization field $B \lesssim 0.1$ Gauss (10^{-5}T), ν_0 may be evaluated with an uncertainty $\Delta\nu/\nu_0 \lesssim 1\times10^{-14}$.

Finally the black-body shift has been theoretically estimated [11] less than 1×10^{-14}.

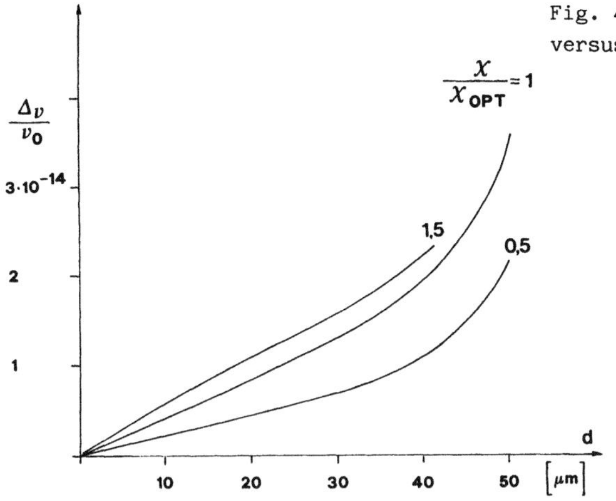

Fig. 4 - Phase-shift versus grid displacement.

V) Conclusions

The prototype of a frequency standard based on a Mg submillimetric transition has been described and the main results obtained have been reported.

A detailed analysis of the uncertainty sources has been sketched, leading to the following conclusions: a final accuracy level of $\pm 1\times10^{-13}$ may be achieved, mainly limited by the 2nd order Doppler effect; residual 1st order Doppler effect and cavity phase-shift may be controlled with a beam reversal technique (not difficult to realize) and reduced below 1×10^{-13}. A short-term frequency instability lower than 1×10^{-13} for τ = 1 s may be obtained with optical-pumping, which allows to reach the final accuracy with a counting time of a few seconds.

References

[1] A. Godone, E. Bava and C. Novero
Metrologia 24, 133 (1987).

[2] G. Giusfredi, A. Godone, E. Bava, C. Novero
Journ. Appl. Phys. (1988).

[3] A. Godone, E. Bava, G. Giusfredi, C. Novero, Y.Z. Wang
Opt. Comm. 59, 263 (1986).

[4] A. De Marchi, A. Godone, E. Bava
IEEE Trans. IM-30, 132 (1981).

[5] A. Godone, E. Bava, C. Novero
IEEE Trans. IM - to be published.

[6] E. Bava, A. Godone, G. Rietto
Appl. Phys. B41, 187 (1986).

[7] C. Novero, H.O. Di Rocco, E. Bava, A. Godone
in preparation.

[8] E. Bava, A. Godone, C. Novero
in preparation.

[9] C. Audoin, P. Lesage, A.G. Mungall
IEEE Trans. IM-23, 501 (1974).

[10] C. Novero, A. Godone, E. Bava, M. Causà
Lett. Nuovo Cimento - to be published.

[11] E. Bava, A. De Marchi, A. Godone
Lett. Nuovo Cimento 38, 107 (1983).

Part V

Cold H Masers

Introduction to Cold Hydrogen Masers

S.B. Crampton

Physics Department, Williams College, Williamstown, MA 01267, USA

Over ten years ago several laboratories started new programs to cool hydrogen atom gas to very low temperatures. The primary motivation was to produce the Bose-Einstein condensation of a weakly interacting gas.[1] At about the same time it was realized[2] that cold hydrogen atom gas would offer some advantages for hydrogen maser frequency standards. The principal advantage is that the spin exchange collision cross section, which limits the line Q and power radiated in conventional masers, decreases substantially at low temperatures. Other advantages include decreased thermal noise and the possibilities of improved magnetic shielding and improved cavity Q and stability. It is also hoped that the molecular solid or liquid helium surfaces used to store hydrogen at low temperatures can be made to be uniform and stable and so produce reproducible, stable wall shifts.

The frequency shifts at low temperatures due to the combined effects of collisions with the surfaces and collisions with surface material vapor are large, but they exhibit a minimum where they are first-order temperature independent, as illustrated in the figures. The left hand graph displays shifts due to hydrogen-helium collisions at the surfaces and in the saturated helium vapor over the liquid helium surfaces in a closed cell, as observed in Walter Hardy's laboratory.[3] In the right hand graph data[4] taken in the same laboratory using a closed cell containing solid neon surfaces has been extrapolated to lower temperatures using our measured value[5] of the binding energy of a hydrogen atom to a neon surface. At the low temperature side of each figure the frequency shifts decrease exponentially due to the exponential decrease of surface dwell time with increasing temperature. At the high temperature side the frequency shifts again increase exponentially due to the exponential increase with temperature of the density of surface material vapor over the surface. Both effects depend sensitively on any surface non-uniformities and on the surface thickness, unless the thickness is well over a hundred Å. It is clear that very good control of the surface structure and temperature will be needed in order to produce good cryogenic frequency standards.

The first cryogenic maser to oscillate below 10K used solid neon surfaces.[6] Subsequently, three laboratories have developed liquid helium surface hydrogen masers. The progress of two of these masers is reported at this session.

Frequency Standards and Metrology Editor: A. De Marchi

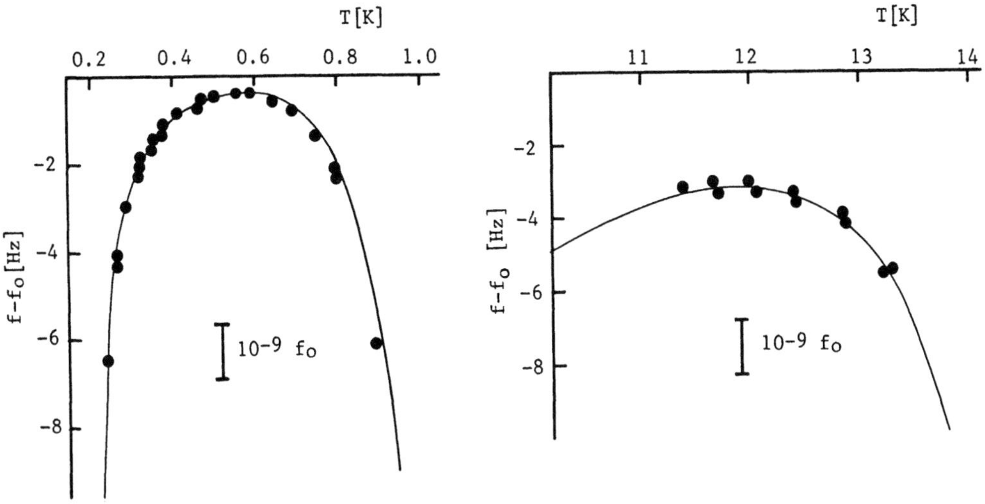

Fig. 1 Frequency shifts due to collisions with the surfaces and with the surface material vapor for liquid helium surfaces (left) and solid neon surfaces (right).

1. W. C. Swalley and L. H. Nosanow, Phys. Rev. Lett. 36, 910 (1976).

2. S. B. Crampton, W. D. Phillips and D. Kleppner, Bull. Am. Phys. Soc. 23, 86 (1978); R. F. C. Vessot, M. W. Levine and E. M. Mattison, in Proc. of the 9th Annual Precise Time and Time Interval Conference, 1978 [NASA Tech. Memorandum No. 78104, 1978 (unpublished)] p. 549.

3. M. Morrow, R. Jochemsen, A. J. Berlinsky and W. N. Hardy, Phys. Rev. Lett. 46, 195 (1980); Phys. Rev. Lett. 47, 455 (1981).

4. Stephen Steel and S. B. Crampton, unpublished data.

5. S. B. Crampton, K. M. Jones, G. Nunes and S. P. souza, in Proc. of the 16th Annual Precise Time and Time Interval Conference, 1985 [NASA Tech. Memorandum No. 8756, 1985 (unpublished)], p. 339.

6. H. F. Hess, G. P. Kochanski, J. M. Doyle, T. J. Greytak and D. Kleppner, Phys. Rev. Lett. A34, 1602 (1986); M. D. Hürlimann, W. N. Hardy, A. J. Berlinsky and R. W. Cline, Phys. Rev. Lett. A34, 1605 (1986); R. L. Walsworth, I. F. Silvera, H. P. Godfried, C. C. Agosta, R. F. C. Vessot and E. M. Mattison, Phys. Rev. Lett. A34, 2550 (1986).

The Cold Hydrogen Maser

R.F.C. Vessot[1], *E.M Mattison*[1], *R.L. Walsworth*[2], *and I.F. Silvera*[2]
[1]Smithsonian Astrophysical Observatory, Cambridge, MA 02138, USA
[2]Harvard University, Lyman Laboratory of Physics, Cambridge, MA 02138, USA

1. Abstract

Since the invention of the atomic hydrogen maser in 1968[1] ongoing developments have provided fractional frequency stability in the 10^{-16} region for averaging intervals of about 10^4 seconds[2,3]. Low temperature operation of the atomic hydrogen maser offers possibilities for frequency stability in the 10^{-18} domain.[4 -14] New low temperature H-H interactions have been predicted[15] that may obviate this possibility and these will be studied using a cryogenic hydrogen maser designed to provide the necessary variability of operating conditions. We describe the cryogenic hydrogen maser (CHM) and the frequency stability measuring system, and the expected frequency stability.

2. Benefits of Low Temperature Operation for H-Maser Frequency Stability

Temperature affects the maser's frequency stability through several mechanisms. The fundamental contribution, $\sigma_{yf}(\tau)$, to the Allan deviation is given by[1] $\sigma_{yf}(\tau) = (1/Q)\sqrt{kT/2P\tau}$, where τ is the averaging interval, k is the Boltzmann constant, T the absolute temperature, P the power delivered by the hydrogen atoms, and Q the maser's oscillation line Q. Q is given by $Q = 2\pi\nu/\gamma_2$, where ν is the oscillation frequency and γ_2 is the total transverse relaxation rate of the atoms in the storage volume:

$$\gamma_2 = \gamma_{bulb} + \gamma_{wall\ relaxation} + \gamma_{spin\ exchange} + \cdots \tag{1}$$

$$= \bar{v}\frac{A_a K}{4V_b} + \frac{\bar{v}}{\lambda}p(T) + \frac{\bar{v}n\sigma}{\sqrt{2}} + \ldots \tag{2}$$

$$= \left(\frac{8k}{\pi m}\right)^{1/2}\left[\frac{A_a K}{4V_b} + \frac{p(T)}{\lambda} + \frac{n\sigma}{\sqrt{2}}\right] T^{1/2} + \ldots \tag{3}$$

Here m is the mass of the H atom, A_aK is the storage bulb aperture area times the collimating factor, V_b is the bulb volume, $p(T)$ is the probability per wall collision of loss of phase, λ is the mean free distance between wall collisions, and $n\sigma$ is the atomic density times the spin exchange cross-section. If we include this representation of Q in the expression for $\sigma_{yf}(\tau)$, we see that $\sigma_{yf}(\tau)$ is directly proportional to temperature.

The instability due to additive white phase noise, given by[16] $\sigma_{ya}(\tau) = \frac{1}{2\pi\nu\tau}\left[\frac{k(T+T_n)B}{P_r}\right]$, also depends on T. Here T_n is the noise temperature of the maser's receiver, P_r is the power received, and B is the measurement system bandwidth.

Frequency Standards and Metrology Editor: A. De Marchi

The limitation on the maser power is proportional to $n\bar{v}\sigma/I$, the spin exchange collision rate divided by the flux, I, of atoms entering the storage volume.[17] For atomic hydrogen at low temperatures, T<1K, $\bar{v}\sigma$ is about 2000 times smaller than at 300K. The lower collision rate permits significantly more power output for a given loss of line Q.

In addition to stability improvements that result from the reduction of noise and collision rates, low temperature operation can yield substantial improvements in the mechanical and thermal stability of materials and in the control of magnetic fields by the use of superconducting shields.

3. The Harvard-Smithsonian Cryogenic Hydrogen Maser

The motivations behind the design of this maser are 1) to learn about H-H and H-He atomic collision interactions at very low temperatures, and 2) to produce an oscillator of unprecedentedly high frequency stability. The design of the maser follows the classical scheme for maser operation using atomic beam techniques and a hexapole magnet state selector, and affords us flexibility of operation. The maser cavity is cooled by a dilution refrigerator and operates in the 350-600 mK range.

Figure 1 is a schematic diagram of the CHM. Atomic hydrogen dissociated externally at room temperature is led through a Teflon tube into the lower region of the cryostat, where it is cooled to ~10K. The hydrogen beam then passes through a 1.3 cm long hexapole state selection magnet with bore diameter 3 mm and pole tip fields of .88T, that focuses atoms in the F=1, m_F=+1 and 0 states into the storage volume. A beam stopping disc at the exit of this magnet prevents undeflected atoms and molecules from reaching the storage volume. The cylindrical storage volume is defined by the inner surface of a thick-wall dielectrically loaded TE_{011}-mode cavity resonator made of C-cut single crystal sapphire,[18] silvered on its exterior surface. To optimize the filling factor, two Teflon septums keep the stored atoms from encountering the predominantly transverse r.f. magnetic fields near the ends of the resonator. The time constant for loss of atoms from the storage volume is made to be about 7.5 sec. by a Teflon entrance collimator tube (4 mm ID, 27mm long). The cavity is mounted in good thermal contact with a surrounding heavy-walled copper chamber that is suspended from the mixing chamber of the dilution refrigerator. A 3-section longitudinal solenoid outside of the copper chamber produces a uniform d.c. magnetic field in the storage volume. (It can also apply a field with known gradients so as to cause the atoms in the volume to undergo transverse relaxation in addition to the processes described in Eq. 1.) The earth's magnetic field is attenuated by a four-layer nested set of magnetic shields made of a material that provides effective shielding at low temperatures.

A high frequency coil wound transversely immediately outside the silver surface of the cavity can stimulate $\Delta m_F = \pm 1$ transitions in the F=1 manifold. This allows us to measure the internal magnetic field and to use Andresen's method[19] for determining the population among the hyperfine levels of the stored atoms. Liquid helium admitted into the chamber coats the interior of the storage volume and flows downward and outward through the entrance beam tube to a large capacity sorption pump shown at the lower part of Figure 1. The interior surface of the pump entrance tube ("neck") is lined with multi-layered copper foil to reduce the thickness of the flowing superfluid film below its saturation thickness and thus diminish the vapor pressure in the neck. A transverse high frequency coil is mounted inside the neck, and a longitudinal solenoid is placed outside. This combination allows us to alter the population distribution within the F=1 hyperfine manifold to implement Audoin's method[20] for determining T_1 and T_2,

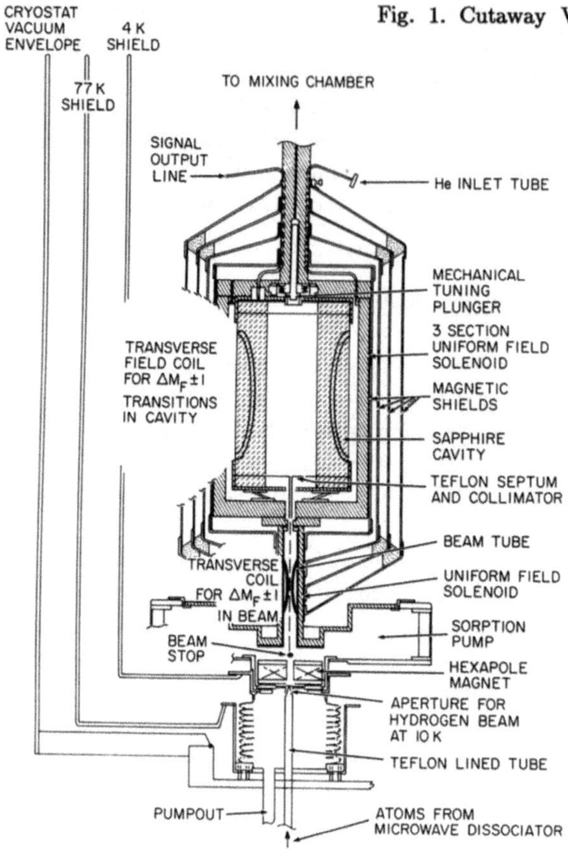

Fig. 1. Cutaway View of Cryogenic Hydrogen Maser

the longitudinal and transverse relaxation times of the stored atoms, as well as the relative flux of atoms in the ($F=1$, $m_F=0$) and ($F=0$, $m_F=0$) states.

In the temperature range at which the CHM will be operated, the frequency shift arising from collisions between hydrogen atoms and helium vapor atoms has a negative slope as a function of temperature, while the frequency shift from collision with the liquid helium storage surface has a positive slope. For a storage volume of the CHM's dimensions (4.44 cm in diameter and 13.22 cm long), the combined effects yield a frequency shift with a minimum value at approximately 0.53K. To control the output frequency at a level of one part in 10^{18} will require temperature control within 20μK. This is within the present state of the art.[6,21]

The cavity frequency is adjustable by an externally operated mechanical tuning plunger. The cavity is critically coupled and has a loaded Q of about 8×10^4 at temperatures below 4.2K.

4. Frequency Stability Measurement System

The system we have built for measuring the frequency stability, power output, and output frequency of the CHM is shown in Figure 2. It consists of a low noise phaselock receiver and a 3-way heterodyne frequency comparator, both mounted in temperature-controlled enclosures. The phaselock receiver uses a 100 MHz crystal oscillator of high spectral purity.

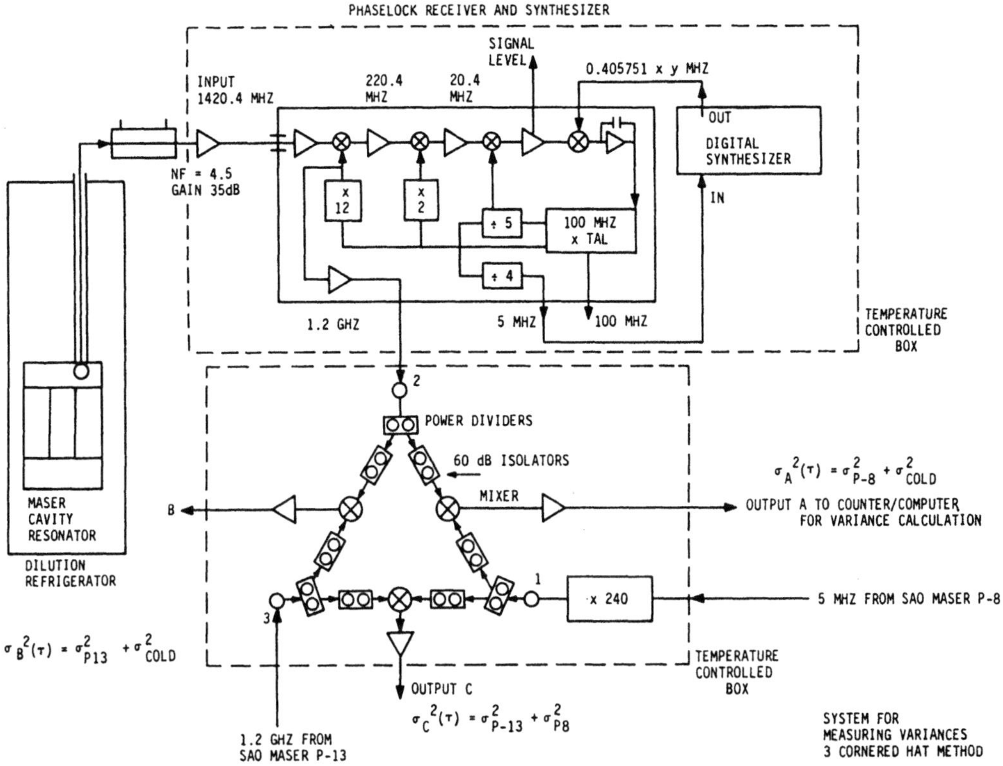

Fig. 2. Frequency Stability Measurement System

The expected Allan variance of the CHM with a 10K GASFET[22] preamplifier located on a low temperature station near the cavity resonator is shown in Figure 3, which also shows the stability of the SAO VLG-11 series masers at room temperatures. This plot indicates the two contributions of thermal noise, additive white phase noise of the receiver, and the fundamental noise within the linewidth of the oscillator. The conditions chosen here are for the optimum frequency stability, which is obtained in the CHM when the hydrogen-hydrogen spin exchange relaxation rate equals the escape rate from the bulb.[6]. In VLG-11 series-masers we observe that the Allan deviation turns up for long averaging intervals, going through a level of about 4×10^{-16} at 10,000 seconds. This behavior is due chiefly to frequency variations of the cavity resonator and of the circuitry involved in extracting the signal to the first mixer of the receiver. If we assume that the long-term frequency variations result from a combinaton of cavity pulling mechanisms, then the expected effect for the CHM is lower than that for the VLG-11 masers by the ratio of their line Qs. The typical line Q for the VLG-11 is about 1.8×10^9, while the line Q of the CHM with a storage time of 7.5 sec. is between 3.4×10^{10} and 1.7×10^{10}, depending on the flux level. It is reasonable to expect much better cavity resonator frequency stability at 0.5K with a solid sapphire resonator than that of the room temperature Cer-Vit cavities used in the VLG-11's. If we have ten times better cavity stability we can expect the results shown shaded at the bottom of Figure 3.

It is clear that the best way to measure the frequency stability of the CHM is to use oscillators of comparable stability as references. At present, such references are unavailable.

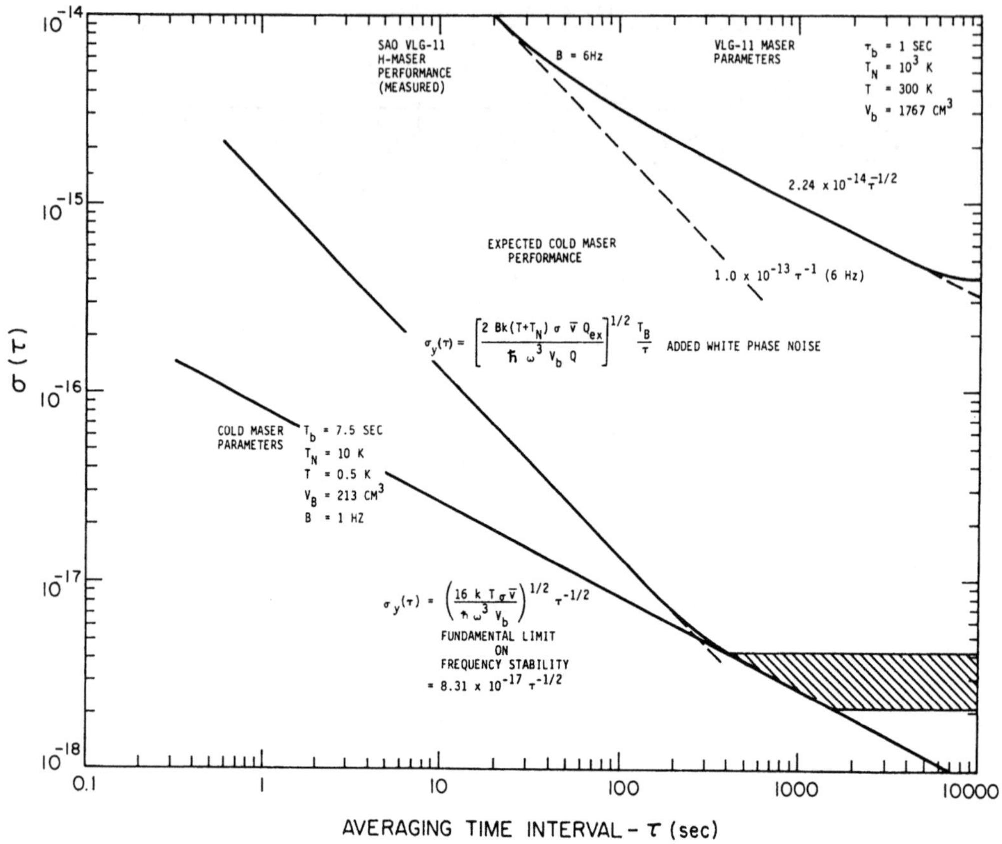

Fig. 3. Predicted Frequency Stability of CHM

However, an alternative measurement approach is to use the "3-cornered hat" technique, where three oscillators are intercompared simultaneously,[23] all three sets of frequency measurements being made at the same time. By differencing the data, the frequency stabilities of the individual oscillators can be extracted.

The effectiveness of this technique depends on there being no systematic effects that alter, simultaneously and equally, the frequencies of all three oscillators, and that there is sufficient isolation between inputs to prevent cross-talk. The designed isolation between channels for the system shown is 130 dB. We believe it is reasonable to expect a factor of 10 improvement in the determination of frequency stability beyond the stability of the two reference masers.

5. Conclusions

Whether we can realize the full predicted stability of the H maser depends on resolving a question that has been raised by Verhaar and his colleagues,[16] who have made a rigorous study of hyperfine interactions between ground state hydrogen atoms during spin exchange collisions and have predicted frequency shifts that are significant for low temperature hydrogen masers. The H-H collisional shifts observed in room temperature masers are proportional to the population difference between the two $m_F = 0$ states[24] and can be compensated for by tuning the maser resonator so there is no change of oscillation frequency with collision rate. In addition to these shifts, Verhaar et.al. predict shifts that are proportional to the part of the overall storage time that does not depend on collision rate. The predicted sensitivity of these shifts requires that the collision-independent storage time be stable to a few parts in 10^7 to achieve stability in the 10^{-18} domain.

The measurement of these shifts and their effect on the maser's frequency stability is a major objective of the Harvard-SAO Cryogenic Hydrogen Maser program.

6. Acknowledgements This work is supported by the Smithsonian Institution's Scholarly Studies program and by the U.S. Department of Energy under Grant Number DE-FG-02-85ER45190.

7. References

1. D. Kleppner, H.M. Goldenberg, N.F. Ramsey, Phys. Rev. **126**, 603 (1962).
2. R.F.C. Vessot, E.M. Mattison, R.L. Walsworth, I.F. Silvera, H.P. Godfried, C.C. Agosta, *IEEE, Transactions on Instrumentation and Measurement*, **IM-36**, p. 588 (1987).
3. E.M. Mattison, R.F.C. Vessot, W. Shen, *IEEE*, UFFC-34 No. 6, 622 (1987).
4. S.B. Crampton, H.T.M. Wang, *Phys. Rev. A* **12**, 1305 (1975).
5. R.F.C. Vessot, M.W. Levine, E.M. Mattison, *Proceedings of the 9th Annl. Precise Time and Time Interval meeting*, p. 549 (1977).
6. A.J. Berlinsky, W.N. Hardy, *Proceedings of the 13th Annl. Precise Time and Time Interval meeting*, p. 547 (1981).
7. R.F.C. Vessot, E.M. Mattison, E.L. Blomberg, *Proceedings of 33rd Annl. Symposium on Frequency Control*, p. 571 (1979).
8. K.E. Anderson, S.B. Crampton, K.M. Jones, G. Nunes, and S.P. Souza. In *Proceedings of LT-17, Low Temperature Conference*, Karlsruhe, F.R.G., Aug. 15-22, 1984.; S.B. Crampton, K.M. Jones, G. Nunes, S.D. Souza, *Proceedings of 16th Annl. Precise Time and Time Interval meeting*, p. 339 (1984).
9. I.F. Silvera and J.T.M. Walraven, *Phys. Rev. Lett.* **44**, 164 (1980).
10. M. Morrow, R. Jochensen, A.J. Berlinsky and W.N.Hardy, *Phys. Rev. Lett.* **46**, 145 (1981); Erratum **47**, 455 (1981).
11. W.N. Hardy and M. Morrow, *J. Phys. Colloq.* **42**, PC8/171 (1981).
12. G.P. Kochanski, J.M. Doyle, T.J. Greytak, and D. Kleppner, *Phys. Rev. A* **34**, 1602 (1986).
13. M.D. Hurliman, W.N. Hardy, A.J. Berlinsky, and R.W. Cline, *Phys. Rev. A* **34**, 1605 (1986).
14. R.L. Walsworth, I.F. Silvera, H.P. Godfried, C.C. Agosta, R.F.C. Vessot and E.M. Mattison, *Phys. Rev. A* **34**, 2550 (1986).
15. B.J. Verhaar, J.M.V.A. Koelman, H.T.C. Stoof, O.J. Luiten and S.B. Crampton, *Phys. Rev. A* **35**, 3825, (1987).
16. R.F.C. Vessot, L. Mueller, J. Vanier, *IEEE*, **54**, p. 199 (1966).
17. D. Kleppner, H.C. Berg, S.B. Crampton, N.F. Ramsey, R.F.C. Vessot, H.E. Peters and J. Vanier. *Phys. Rev. A* **138**, 972 (1965).
18. V.J. Folen, W.G. Maison, J.D. White, C.A. Bartholomew and A.J. Frank, *Hybrid Modes in Anisotropic Dielectrics*, NRL Memo. 4870, U.S. Naval Research Laboratory, Washington, DC.

19. H.G. Andresen, *Z. Physik*, **210**, 113 (1968)
20. M. Desaintfuscien and C. Audoin, *Journal dePhysique* **35**, 829 (1974).
21. J.A. Lipa, *Int. Conf. Low Temp. Phys.*, Physica **107** B+C, 343 (1981).
22. Berkshire Technologies Incorporated, Oakland CA 94609
23. J. Groslambert, D. Fest, M. Oliver and J.J. Gagnepain, *Proceedings of the 35th Annl. Frequency Control Symposium*, 458 (1981).

Performance of the UBC Cryogenic Hydrogen Maser

M.D. Hürlimann, W.N. Hardy, M.E. Hayden, and R.W. Cline

Department of Physics, University of British Columbia,
Vancouver, B.C. Canada V6T 2AG, Canada

1. Introduction

Hydrogen masers are presently the most stable atomic clocks for averaging times τ between 1 s and 1000 s. With the best conventional devices, frequency fluctuations of less than 10^{-15} for averaging times of about 1 hour are measured [1]. In 1978 it was realized by Crampton et al. [2] that cryogenic hydrogen masers (CHM) could have even better frequency stabilities, because of lower cavity thermal noise and the possibility of using cooled preamplifiers. In addition, it was noted that the spin exchange broadening at low temperatures is 1000 times smaller, which increases the optimal density by about the same factor and results in a much higher power output. Berlinsky and Hardy [3] concluded that the frequency stability of a cryogenic hydrogen maser, operated at 0.5 K, could be up to 1000 times better than the best state-of-the-art frequency sources, provided that the stability is still limited by thermal and receiver noise.

In the early 1980's several groups started to develop techniques for studying atomic hydrogen at low temperatures in their quest to observe Bose-Einstein condensation in spin-polarized hydrogen [4]. It was soon found that the only suitable wall coating below 1 K is liquid helium. One of the techniques used at UBC has been zero field magnetic resonance at the hyperfine transition and the results have been directly relevant to the design of the cryogenic hydrogen maser. It was observed that the temperature dependent frequency shift, a combination of wall shift and He "buffer gas" shift, has an extremum at $T_m \approx 0.5$ K. This makes it possible to design a maser, operated at T_m, where the requirement on the temperature regulation of the maser cavity is technically realistic.

2. Design of the UBC Prototype Cryogenic Hydrogen Maser

In a cryogenic hydrogen maser operated at 0.5 K, all the walls that are in contact with hydrogen atoms have to be covered by a saturated film of superfluid helium. In the conventional beam geometry, the helium film would rush towards the warm source and then reflux. In order to avoid this formidable technical difficulty, the UBC maser is all-cryogenic. The hydrogen atoms circulate back and forth between the state selector and the maser cavity. State selection is achieved by a 40 GHz microwave cavity, which is tuned to the $|b\rangle$ - $|c\rangle$ ESR transition and located inside a 14 kG superconducting magnet. The magnetic fields compress atoms in the $|a\rangle$ and $|b\rangle$ states and expel atoms in the upper two states ($|c\rangle$ and $|d\rangle$). Inside the microwave cavity, atoms in the

Frequency Standards and Metrology Editor: A. De Marchi

$|b\rangle$ state are pumped into the $|c\rangle$ state and then ejected from the high field state selector into the maser cavity. There the atoms radiate coherently and after an average time T_B leave the bulb. Atoms in the $|a\rangle$ state are attracted back into the state selector region. The $|a\rangle$ and $|b\rangle$ populations are equalized in the state selector by layers of magnetic foil. Hydrogen atoms moving across the foil with thermal velocity experience a fluctuating magnetic field. By choosing the size of the magnetic particles to be about 100 Å the spectrum of the field fluctuation is centered around the resonance frequency for the $|a\rangle$ to $|b\rangle$ nuclear spin flip. This closes the cycle. Since the recombination lifetime is of the order of hours, atoms go through many cycles before they eventually recombine to H_2 molecules and freeze out on the surfaces underneath the superfluid ^{4}He film.

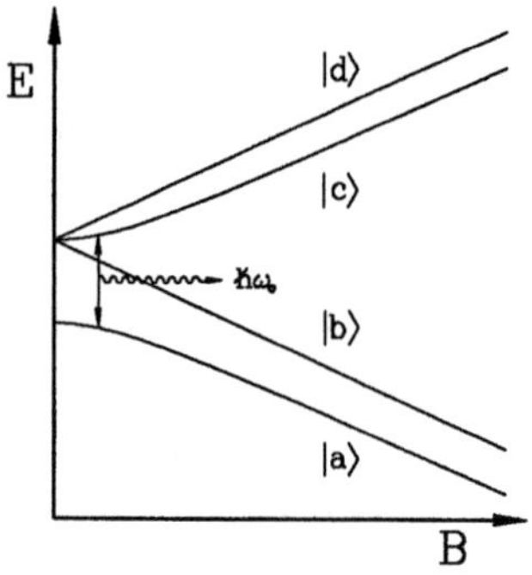

Fig. 1. Hyperfine energy level diagram for atomic hydrogen.

In order to simplify the cryogenic engineering we decided to build the first maser with a small split ring resonator, rather than with a full size high Q resonant cavity. We observed the first oscillations of our maser in 1986 and it was reported in [5],[6]. A more detailed description of our prototype apparatus and of the state selector can be found in Ref.[7].

3. Results

3.1. Output Power of Maser

The output power of the maser at the input of the preamplifier is shown in Fig.2 as a function of the total density of hydrogen atoms in the maser bulb n_H, for two different runs. In the first run the loaded cavity Q was 1700, whereas in the second run an added electronic tuning circuit lowered it to 1100. In both cases, the temperature of the maser bulb was 0.5 K. Within a simple theory the power emitted by the atoms is:

$$P = \frac{\hbar\omega_o}{2}\frac{V}{T_B}\left[n_H(\rho_{cc}-\rho_{aa})_o - n_o\frac{{T_B}^2}{T_1T_2}\right] \tag{1}$$

where $n_o = \frac{4}{\mu_o\hbar(\gamma_e+\gamma_n)^2\eta' QT_B}$, V is the volume of the maser bulb, $(\rho_{cc}-\rho_{aa})_o$ is the population inversion outside the bulb, η' is a filling factor, $1/T_1 = 1/T_B + \sigma_{ex}v\ n_H$, and $1/T_2 = 1/T_B + \sigma_{ex}v\ n_H/2$. For a constant population inversion and the range of densities n_H used in

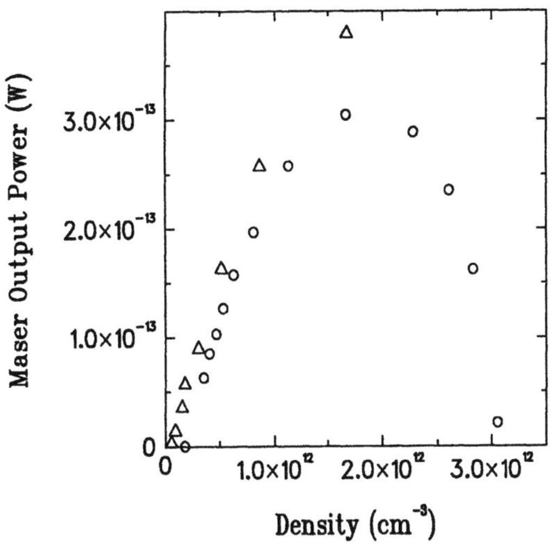

Fig. 2. Output power of the maser as a function of the total density of H atoms in the maser bulb for two runs. Triangles: May 1986; circles: May 1987.

our experiment, the power output should be essentially linear in density, because spin exchange effects give only a small contribution to the relaxation rates. Our data show that the state selector is not as efficient as expected and that the population inversion is density dependent. Based on the facts that the power output is insensitive to variations of the state selector microwave power ($\approx 50\ \mu$W) and that the power output from the two runs differed significantly, we believe that the relaxation rate of the magnetic foil is too low and forms a bottleneck. Presumably, the foil became further contaminated or oxidized in the time between the two runs. In order to substantiate this we have developed a computer program to simulate the operation of the maser.

3.2. Computer Simulation

The maser is modeled by dividing it into four distinct regions, representing: (1) maser cavity; (2) atom transport tube between the state selector magnet and the maser cavity; (3) 40 GHz microwave cavity; (4) region in high field with relaxing foil. The regions are connected by orifices and each of them is characterized by a temperature, a magnetic field and relaxation rates as discussed below. In the 16 rate equations (4 hyperfine states in 4 regions) we include contributions due to: (1) flow between the different regions, which is governed by effusion and can be calculated by knowing the size of the orifices, temperatures and magnetic fields; (2) microwave pumping in the state selector cavity; (3) spin exchange relaxation, as calculated by Verhaar and collaborators [8]; (4) one body relaxation caused by the magnetic foil and by magnetic impurities and field inhomogeneities in the other regions; (5) maser action in the maser cavity. The recombination rates at the temperatures and densities relevant to maser operation are much slower than the dominant relaxation rates. We therefore neglect recombination and solve the set of rate equations for quasi-stationary solutions. We also neglect the slow exchange of atoms between the state selector cavity and the waveguide which is connected to the discharge source of atomic H.

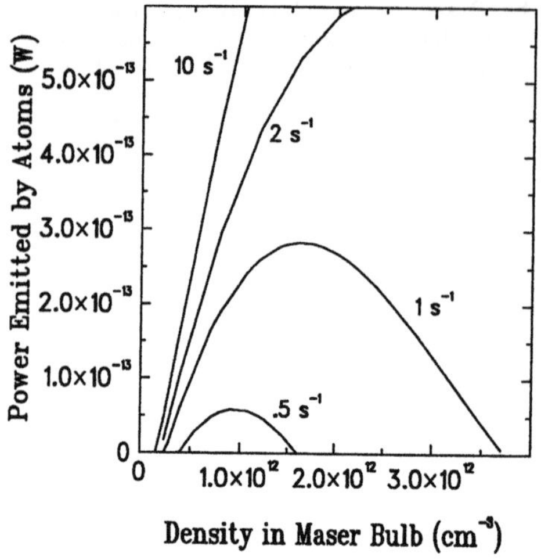

Fig. 3. Simulation results: power emitted by atoms versus total density of H atoms in the maser bulb for four different relaxation rates of the magnetic foil in the state selector.

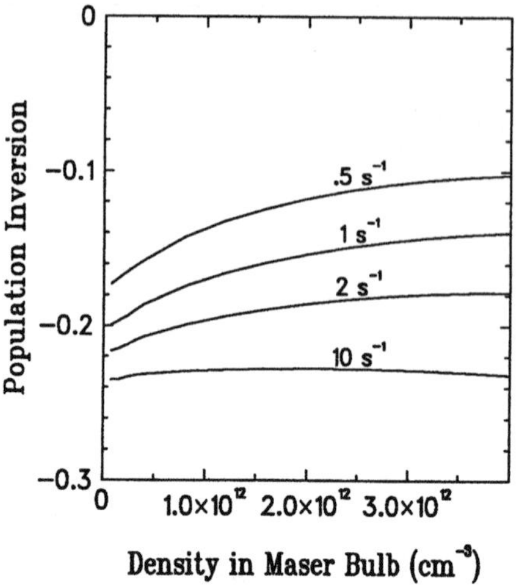

Fig. 4. Simulation results: population inversion outside the maser bulb versus total density of H atoms in the maser bulb for four different relaxation rates of the magnetic foil in the state selector.

Figures 3 and 4 show the results of a simulation where the foil relaxation rates are $10\,s^{-1}$, $2\,s^{-1}$, $1\,s^{-1}$, and $0.5\,s^{-1}$. Qualitatively, the calculated power emitted by the atoms agrees well with the observed behaviour and confirms our initial interpretation. The density dependence of the population inversion outside the maser bulb (Fig. 4) is strong for low relaxation rates, but disappears for higher relaxation rates.

We have not tried to improve the quantitative agreement between the measured and calculated power output, because there are too many relaxation rates which we are not able to measure in the present apparatus. Based on this simulation with "reasonable" relaxation rates, we are confident that the power output of the maser can be easily improved by replacing the present magnetic foil by a more efficient one. It is not an intrinsic limitation of our design. We have also verified that the population inversion can be further enhanced by reducing the excessive relaxation between the three upper states inside the maser cavity, which is due to some magnetic contamination.

3.3. Temperature Dependence of Maser Frequency

The temperature dependence of the maser frequency is dominated by the wall shift due to the changing fraction of atoms in the bound state on the the ^{4}He film, and by the helium "buffer gas" shift. The second shift is caused by collisions of H atoms with ^{4}He atoms in the vapour phase and is proportional to the helium vapour density n_{He}. We fit our data to:

$$f(T) - f_o = f_s \frac{\Lambda A}{V} \exp(E_B/kT) + n_{He}\delta \qquad (2)$$

where f_s is the surface hyperfine frequency shift, Λ is the thermal de Broglie wavelength, A/V is the surface area to volume ratio of the maser bulb, E_B is the binding energy of H to the l-^{4}He surface and δ is the pressure shift coefficient for the gaseous ^{4}He.

We were able to observe stable maser action for maser bulb temperatures between 230 mK and 660 mK. In Fig.5 the measured frequency shift is plotted together with a fit to (2). The obtained value $E_B = 1.011 \pm 0.010$ K is the most precise determination of the binding energy of H on ^{4}He.

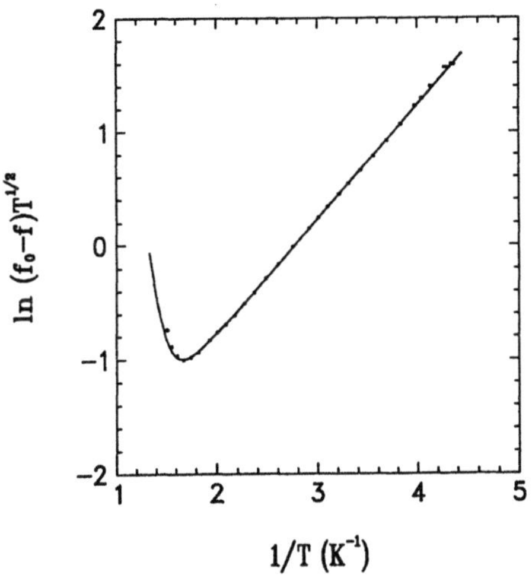

Fig. 5. Observed temperature dependence of the maser frequency shift (from the unperturbed value in the bias field). The solid line is a fit to (2), yielding $E_B = 1.011 \pm 0.010$ K and f_s= -71.5±3.0 kHz.

3.4. Frequency Stability

For short averaging times, the main limitation on frequency stability is due to broad band white phase noise added to the weak maser signal (typically 10^{-13} to 10^{-12} W) by the preamplifier. Using an amplifier with noise temperature T_N, a measuring bandwidth B and with a signal power P_A at the input of the amplifier, the resulting frequency fluctuations for an averaging time τ are:

$$(\frac{\Delta f}{f})_{Amp} = \left[\frac{k(T+T_N)B}{P_A}\right]^{\frac{1}{2}} \frac{1}{\omega_o \tau} \quad (3)$$

For intermediate averaging times, thermal noise in the cavity, modified by the maser action becomes the dominant noise source:

$$(\frac{\Delta f}{f})_{therm} = \left[\frac{kT}{2PQ_l^2\tau}\right]^{\frac{1}{2}} \quad (4)$$

where T is the temperature of the maser cavity, P is the power radiated by the atoms and $Q_l = \frac{1}{2}\omega_o T_2$ is the atomic line Q. The expected frequency stability of our cryogenic hydrogen maser is more than one order of magnitude worse than the prediction for a full size maser [3], mainly because of the small volume of the maser bulb (5.15 cm^3). Based on formulae (3) and (4) and the parameters of our maser: B = 6 Hz, T = 600 mK, $T_N \approx$ 25 K, $P_A = 1.2 \times 10^{-13}$ W, $P \approx 2P_A$, $T_2 = 0.25$ s the predicted stability is:

$$\frac{\Delta f}{f} = \left[\left(\frac{3.4\times10^{-15}}{\sqrt{\tau}}\right)^2 + \left(\frac{1.5\times10^{-14}}{\tau}\right)^2\right]^{\frac{1}{2}} \tag{5}$$

We were able to borrow two high quality crystal oscillators [9] and measure the frequency stability of the maser by the three cornered hat method for 1 s and 3 s (Fig.6, diamonds). For longer averaging times, the ratio of the CHM stability to the stability of the two reference clocks increases and the averaging required for this technique becomes impractical. [The open circles in Fig.6 are measurements of the frequency fluctuations between the CHM and a single crystal oscillator and thus are only upper limits.] The measured instabilities of the CHM are about 4 times higher than the predictions of (5). We have not been able to determine the source of this discrepancy. We have excluded flicker phase noise of the preamplifier [10], but there might be a contribution due to the fact that the 3 measurements required for the three cornered hat measurements were not performed simultaneously.

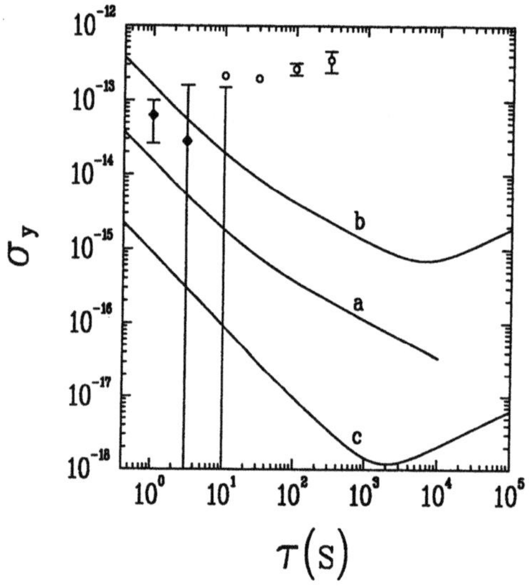

Fig. 6. Plot of Allan variance versus averaging time. Diamonds: frequency fluctuations of CHM measured with three cornered hat method; circles: comparison of CHM with the better of the two quartz crystal oscillators; a: predicted stability of CHM, see (5); b: measured stability of state-of-the-art hydrogen maser [1]; c: predicted stability of full size CHM [3].

4. Outlook

Recently, Verhaar and his collaborators analyzed the spin exchange broadening and the frequency shifts theoretically [11]. In their extensive calculations they found that the frequency shifts are of the form:

$$\delta\omega_{se} = [\lambda_0(\rho_{cc} - \rho_{aa}) + \lambda_1(\rho_{cc} + \rho_{aa}) + \lambda_2]\, v\, n_H \tag{6}$$

where $\lambda_0, \lambda_1, \lambda_2$ are thermally averaged spin exchange cross sections and v is the relative velocity of the H atoms. The first term, which has been the only term considered in previous calculations, can be compensated by cavity pulling. This is the well known "spin exchange tuning". When the additional two terms in (6) are included, the spin exchange tuning does not leave the maser frequency unaffected by changes in density and population. Measurements are currently in progress in our laboratory to verify these frequency shifts experimentally. Any instabilities in density or populations will directly result in frequency fluctuations, unless a new scheme to compensate for them can be found. It might be necessary to reduce the optimal density, derived for minimum thermal noise. A reduction by a factor of 100 will decrease the spin exchange induced fluctuations by the same factor, but will increase the thermal noise by only one order of magnitude.

In light of these frequency shifts it will be necessary for us to measure frequency stabilities over longer averaging times. We hope to accomplish this with the help of two conventional hydrogen masers, borrowed from S. Crampton. Furthermore it is essential to build a recirculating CHM with a larger cavity in order to evaluate the full potential of cryogenic hydrogen masers.

This work is supported by grants from the Natural Sciences and Engineering Research Council of Canada and in the earlier stages has been supported by a Precision Measurements Grant from the U.S. National Bureau of Standards. We would like to thank Stuart Crampton, Fred Walls, Dave Allan and Dick Davis for many useful discussions and loan of equipment. In addition, John Berlinsky's many contributions during the early phase of this project are gratefully acknowledged.

1. R.F.C. Vessot, M.W. Levine, and E.M. Mattison, *Proceedings of the 9th Annual Precise Time and Time Interval Conference*, 549 (1978).

2. S.B. Crampton, W.D. Phillips and D. Kleppner, *Bul. Am. Phys. Soc.* **23**, 86 (1978).

3. A.J. Berlinsky and W.N. Hardy, *Proceedings of the 13th Annual Precise Time and Time Interval Conference*, 547 (1982).

4. For reviews of the subject see T.J. Greytak and D. Kleppner, in *New Trends in Atomic Physics* , North Holland Pub. Co., Amsterdam (1984) and I.F. Silvera and J.T.M. Walraven in *Progress in Low Temperature Physics*, vol. X, chapter D, D. Brewer Ed.(1986).

5. M.D. Hürlimann, W.N. Hardy, A.J. Berlinsky and R.W. Cline, *Phys. Rev. A* **34**, 1605 (1986).

6. M.D. Hürlimann, A.J. Berlinsky, R.W. Cline, and W.N. Hardy, *IEEE Trans. on Instr. and Measurement IM* **36**, 584 (1987).

7. W.N. Hardy, M.D. Hürlimann, and R.W. Cline, *Jpn. J. Appl. Phys.* **26**, 2065 (1987). Supplement 26-3.

8. J.M.V.A. Koelman, S.B. Crampton, H.T.C. Stoof, O.J. Luiten, and B.J. Verhaar (preprint).

9. These were Oscilloquartz B5400 crystal oscillators on loan from F. Walls, NBS Boulder.

10. M.D. Hürlimann and W.N. Hardy, *Electronics Letters* **23**, 283 (1987).

11. B.J. Verhaar, J.M.V.A. Koelman, H.T.C. Stoof, and O.J. Luiten, *Phys. Rev. A* **35**, 3825 (1987).

Quantum Scattering of Atomic Hydrogen from the Surface of Liquid Helium

J.T.M. Walraven[+] *and J.J. Berkhout*

Natuurkundig Laboratorium, Universiteit van Amsterdam,
Valckenierstraat 65, NL-1018 XE Amsterdam, The Netherlands
[+]Present address: Van der Waals Laboratorium, Universiteit van Amsterdam,
Valckenierstraat 67, 1018 XE Amsterdam, The Netherlands

We studied the scattering of low-energetic neutral hydrogen atoms from the surface of liquid helium by a capillary flow method at sub-Kelvin temperatures. With decreasing temperature a decreasing flow impedance is observed, which may be interpreted as arising from an increasing probability of elastic surface scattering. This behavior is compared with a model for atom-surface scattering. Emphasis is put on the quantum nature of the scattering process.

I. Introduction

In many experimental situations surface effects influence measurements in a poorly known or hard to control manner and are as such a source of error. One may aim at minimizing these errors by using highly reproducible surfaces or eliminate the errors entirely by using a surface free design. Well-known examples are the beam design for the Cs clock and liquid helium coatings for the low-temperature hydrogen maser. For novel, hardly explored experimental conditions surfaces need not necessarily be harmful and could possibly be used even to advantage in precision measurements. For instance, if a gas of ultra-cold (laser cooled) atoms falls by gravity onto a horizontal surface, do the atoms simply bounce back or do they stick to the surface with high probability?

With our work we hope to contribute to the understanding of low-energy surface collision phenomena. We studied the low energy limit of surface scattering for an exceptionally well defined system: Hydrogen atoms (H) colliding with the surface of liquid helium. In our experiments we determine the sticking coefficient s, which is defined as the probability per wall collision that an atom enters a surface bound state. Knowledge of the value and temperature dependence of s is of interest in the present more general context of surface scattering but was in fact obtained in relation to experiments with spin-polarized hydrogen. This knowledge is of importance for the interpretation of compression experiments,[1][2][3][4] while it may be crucial to aim successfully at Bose-Einstein condensation (BEC) with spin-down polarized hydrogen (H↓) in open geometries[5] or in bubbles.[6] Furthermore, knowledge of the thermal accommodation coefficient α, for our system closely related to s,[7] is of importance for loading a minimum-B-field trap with spin-up-polarized H-atoms (H↑).[8][9] A review of the available experimental and theoretical information on s and α is given in ref.7.

In this contribution we give a brief discussion of the theory of atom surface scattering as it applies to hydrogen and the surface of liquid ^{4}He. We further discuss

Frequency Standards and Metrology Editor: A. De Marchi

the experimental method by which the value and temperature dependence of s was obtained and finally we speculate on future experiments.

II. Theoretical Considerations

For an atom colliding with a liquid or solid surface quite generally four scattering channels are open. The atom may scatter elastically or inelastically and may or may not stick to the surface, i.e. become adsorbed or absorbed. For liquid surfaces translational symmetry eliminates the possibility of surface diffraction. Therefore the momentum parallel to the surface is conserved in the scattering process and elastic scattering leads to specular reflection in the absence of surface roughness. For temperatures of experimental interest, $T < 0.5\,\mathrm{K}$, thermal excitations of the bulk and surface of liquid helium are essentially absent and a helium film covering a sufficiently flat substrate is an ideal scattering surface for testing scattering models.

Hydrogen is non-solvable in liquid helium but can be weakly bound to the liquid surface, having an adsorption energy $\epsilon_a \approx 1\,\mathrm{K}$ on pure ^{4}He and $\epsilon_a \approx 0.4\,\mathrm{K}$ on the surface of ^{3}He or an ^{3}He/^{4}He mixture. Therefore inelastic scattering processes may lead to surface adsorption (sticking) or to non-specular reflection but a bulk absorption channel is absent. According to Kagan, Shlyapnikov & Glukhov[10] and Castaing & Papoular[11] the inelastic non-sticking process is less efficient than the sticking process by a factor of order $(\epsilon_a/k_B T)^{-1.5}$, so this process becomes quickly unimportant for temperatures below 500 mK. Thus a very simple picture emerges. The atoms either scatter specularly or adsorb and desorb diffusively after a relatively long time (>10 ns) due to interaction with one of the rare thermal excitations.

The adsorption potential for the H-^{4}He system only supports a single bound state.[12] In directions parallel to the surface the atoms behave as a nearly free two-dimensional gas. The wave function of an adsorbed atom may therefore be written as

$$\Psi_{\mathbf{k}}(R,Z) = A^{-\frac{1}{2}}\, e^{i\mathbf{k}\cdot\mathbf{R}}\, \phi(Z), \tag{1}$$

in which $\boldsymbol{k}$ is the component of the wave vector parallel to the surface, $\boldsymbol{R}$ is the two-dimensional position vector and Z is the distance to the surface reference plane. It was shown by Kagan and Shlyapnikov[5] that the adsorption process is accompanied by the excitation of a single ripplon and not by phonons. Thus energy and momentum can only be conserved in the adsorption process if a substantial part of the binding energy is converted into kinetic energy of the atom in the adsorbed state. The ripplon dispersion relation for a ^{4}He film of thickness d is given by[13][16]

$$\omega_q^2 = \{(g+3\alpha/d^4)q + (\gamma/\rho_0)q^3\}\tanh(qd), \tag{2}$$

where ω_q is the angular frequency of a ripplon with wave vector $\boldsymbol{q}$, g the gravity acceleration, $\gamma = 3.78\times10^{-4}\,\mathrm{J/m^2}$ the surface tension,[14] $\rho_0 = 145\,\mathrm{kg/m^3}$ the bulk density and α the Van der Waals constant of the substrate. In the case of ^{4}He films of about 200 Å thickness the gravitational term is 7 orders of magnitude less than the Van der Waals term. Furthermore, in the region of interest for q ($\hbar\omega_q \approx \epsilon_a$) the Van der

Waals term and the thin film term $\tanh(qd)$ are not important, reducing eq.2 to $\omega_q^2 = (\gamma/\rho_0)q^3$. In the limit of zero incident energy a single ripplon with a wavelength of 42 Å is created while the atom picks up a kinetic energy equivalent to 400 mK; atom and ripplon have equal but opposite momentum. The ratio of double ripplon emission to single ripplon emission is estimated by Statt[15] to be about 0.2. After adsorption, two possible situations may occur. The atom may be evaporated by interaction with a ripplon while still in this energetic state or thermalize through ripplon interaction before it is evaporated at a later stage. Zimmerman and Berlinsky[16] calculated the mean time between ripplon interactions for an H-atom on the ^{4}He surface to range from 8.7×10^{-10} s at 400 mK to 2.4×10^{-8} s at 100 mK, much shorter than the mean residency time on the surface τ_a which ranges from 1.10^{-8} s at 400 mK to 3.10^{-4} s at 100 mK. The adsorption rate $R(E,\theta)$ per atom incident with energy E at an angle θ with the surface normal may be calculated using time dependent perturbation theory:[17]

$$R(E,\theta) = \frac{2\pi}{\hbar}\sum_{\mathbf{q}}\frac{\hbar q}{2\rho_0\omega_q}\left|\left\langle B\left|\frac{\partial U_{\mathbf{q}}(Z)}{\partial Z}\right|\sigma\right\rangle\right|^2 (1+n(\mathbf{q}))(1+N(\mathbf{k}-\mathbf{q}))\delta(E-E_f) \tag{3}$$

with $\langle Z|B\rangle$ the bound state wave function and $\langle Z|\sigma\rangle$ the free state wave function. Far from the surface σ is the component of the wave vector normal to the surface and $E\cos^2\theta = \hbar^2\sigma^2/2m$. Further, $U_{\mathbf{q}}(Z)$ is the $\mathbf{q}$ dependent adsorption potential, $n(\mathbf{q})$ is the occupation number for ripplons with momentum $\hbar\mathbf{q}$ ($n(\mathbf{q}) \ll 1$) and $N(\mathbf{k}-\mathbf{q})$ the analogous function for the H atoms. The sticking probability is obtained by normalizing the transition rate to the incident flux:

$$s(E,\theta) = \frac{V\,R(E,\theta)}{A\,(E/2m)^{\frac{1}{2}}\cos\theta}\ , \tag{4}$$

with V the volume of the sample cell. For low temperatures ($k_BT \ll \epsilon_a$) the matrix element is proportional to σ, so in this limit $s(E,\theta) = s_0E^{\frac{1}{2}}\cos\theta$ and the thermal average of s is proportional to $T^{\frac{1}{2}}$. As shown by Goldman[18] s is strongly dependent on the long-range part of the interaction potential with the surface, and it may be that the low temperature limit for some potentials is only reached for $T \ll 1\mu K$. The results of Goldman for a Morse potential and three potentials with a $1/z^3$ behavior are included in Fig.2.

III. The Experiment

We measure the sticking coefficient as a function of temperature using a capillary flow method.[19] The experimental cell is shown in Fig.1. A buffer volume V is filled with H↓ to a density n. The gas will flow through a capillary to a volume kept free of H↓ atoms by forced recombination on a pump plate.[20] All surfaces except that of the pump plate are covered with liquid helium. The flux ϕ at the end of the capillary depends on the H↓ flux at the entrance and the transmission coefficient (Clausing factor) K:

$$\frac{dN}{dt} = \phi = \tfrac{1}{4} n \bar{v} A K = \frac{N}{\tau} \tag{5}$$

with N the number of atoms in the buffer volume, and A the entrance orifice. By measuring τ the Clausing factor follows from $K = \tau_0/\tau$ where $\tau_0^{-1} = \frac{1}{4}\bar{v}A/V$. To determine s the relation between K and s must be known. In the Knudsen flow regime and for $s=1$ the Clausing factor is determined purely by geometrical factors and is thus temperature independent. In this limit we write $K = K_K$. For capillaries with large ratio of length

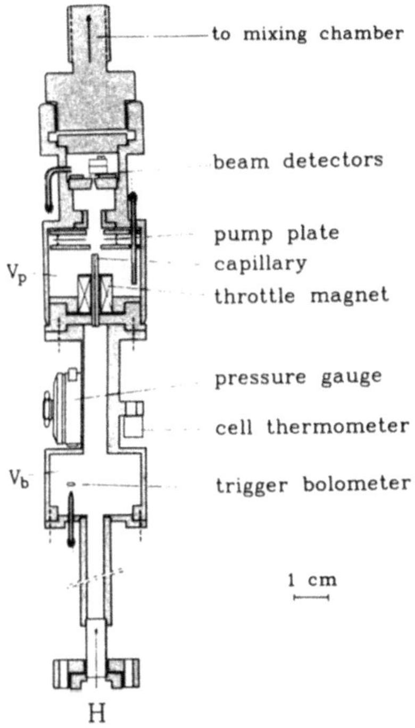

Fig. 1. The experimental cell. The H↓ is fed in from a room temperature source at the bottom into a buffer volume V_b. This volume is connected to a volume V_p by a capillary. V_p is kept clear of atoms by forced recombination on the pumping plate. The atoms are detected either by the pressure gauge or by monitoring the recombination energy dissipated in the pumping plate.

ℓ over radius a the relation $K_K = 8a/3\ell$ holds and the impedance Z of the tube is given by the Knudsen relation $1/Z = \frac{2}{3}\pi\bar{v}a^3/\ell$. For our geometry $K_K = 0.0284$. For $s=0$ the Clausing factor is 1, that is, every particle entering the capillary will scatter specularly until it leaves at the other end and the impedance of the tube equals that of the entrance orifice, $Z = Z_0$, where $1/Z_0 = \frac{1}{4}\bar{v}A$. Note the enormous enhancement of K for $s=0$ in comparison to K_K. The relation between s and K for several values of interest of s was determined by a Monte Carlo computer simulation, where in line with the theoretical model presented in section II the atoms were assumed to have a probability s to stick to the surface at each wall collision and $1-s$ to scatter specularly. Our results are given in Fig.2. The sticking coefficient turns out to be proportional to T and may be given within 10% experimental accuracy as $s = T/T_0$ with $T_0 = 3\,\mathrm{K}$ in the range $73\,\mathrm{mK} < T < 526\,\mathrm{mK}$. It is remarkable that within experimental error s is the same for ^4He surfaces and for surfaces of $^3He/^4He$ mixtures.

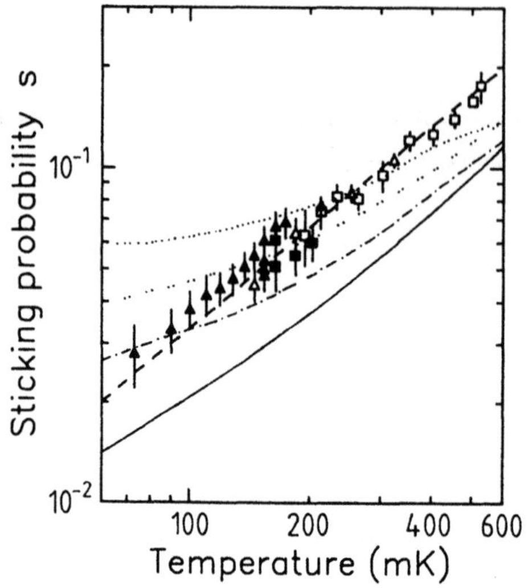

Fig.2. Experimental and theoretical values for the sticking coefficient. □: ^{4}He, ■: ^{3}He/^{4}He (pressure gauge data); Δ: ^{4}He, ▲: ^{3}He/^{4}He (pump plate data). Dashed line: $s = 0.33\,T$. The other curves are the theoretical results of Goldman.[18]

IV. Conclusion

Comparing our experimental results with the calculated values one finds reasonable agreement. Both have the same order of magnitude, but a difference in temperature dependence exists. The dependence of s on the angle of incidence and the predicted strong specular reflection have not yet been experimentally demonstrated. The surprisingly small difference between the results on ^{4}He and ^{3}He/^{4}He mixtures deserves attention. The recent experiments with trapped hydrogen in which temperatures down to 3 mK were obtained[8] open the perspective of temperatures down to micro-Kelvins. Our current understanding of the sticking process leads to the expectation that at these temperatures the equilibration time of the gas with the surface becomes of order minutes.

V. Acknowledgements

The experiments where done in collaboration with E.J. Wolters and R. van Roijen. Technical support was given by O.H. Höpfner. This work is part of the research program of the 'Stichting voor Fundamenteel Onderzoek der Materie (FOM)', which is financially supported by the 'Nederlandse Organisatie voor Wetenschappelijk Onderzoek (NWO)'.

References

1. R. Sprik and J.T.M. Walraven, Phys.Rev.B 32, 5668 (1985).
2. D.A. Bell, H.F. Hess, G.P. Kochanski, S. Buchman, L. Pollack, Y.M. Xiao, D. Kleppner, and T.J. Greytak, Phys.Rev.B 34, 7670 (1986).
3. T. Tommila, S. Jaakkola, M. Krusius, K. Salonen, and E. Tjukanov, in: Proc. 17th Int. Conf. on Low Temperature Physics LT-17, Karlsruhe, August 15-22, eds. U. Eckern, A. Schmid, W. Weber, and H. Wühl (North-Holland, Amsterdam) p. 543 (1984).
4. J.D. Gillaspy and I.F. Silvera, to be published.
5. Y. Kagan and G.V. Shlyapnikov, Phys.Lett. 95A, 309 (1983).
6. I.F.Silvera, J.D.Gillaspy, and J.G.Brisson, in "Proceedings of the Third Workshop on Spin-polarized Quantum Systems (SPOQS/3)", S. Stringari (Ed.), World Scientific Publ.(1988), to be published.
7. J.J.Berkhout and J.T.M. Walraven in "Proceedings of the Third Workshop on Spin-polarized Quantum Systems (SPOQS/3)", S. Stringari (Ed.), World Scientific Publ.(1988), to be published.
8. H.F. Hess, G.P. Kochanski, J.M. Doyle, N. Masuhara, D. Kleppner, and T.J. Greytak, Phys.Rev.Lett. 59, 672 (1987); N. Masuhara, J.M. Doyle, J.C. Sandberg, D.Kleppner, T.J. Greytak, H.F. Hess and G.P. Kochanski, Phys.Rev.Lett. 61, 935 (1988).
9. R. van Roijen, J.J. Berkhout, S. Jaakkola, and J.T.M. Walraven, Phys.Rev.Lett. 61, 931 (1988).
10. Y. Kagan, G.V. Shlyapnikov, and N.A. Glukhov, JETP Lett. 40, 1072 (1984).
11. B. Castaing and M. Papoular, J.Physique Lett. 44, L-537 (1983).
12. I.B. Mantz and D.O. Edwards, Phys.Rev.B. 20, 4518 (1979).
13. M.W. Cole, Phys.Rev.B 2, 4239 (1970).
14. H.M. Guo, D.O. Edwards, R.E. Sarwinski, and J.T. Tough, Phys.Rev.Lett. 27, 1259 (1971).
15. B.W. Statt, Phys.Rev.B 32, 7160 (1985).
16. D.S. Zimmerman and A.J. Berlinsky, Can.J.Phys. 62, 590 (1984).
17. D.S. Zimmerman and A.J. Berlinsky, Can.J.Phys. 61, 508 (1983).
18. V.V. Goldman, Phys.Rev.Lett. 56, 612 (1986).
19. J.J. Berkhout, E.J. Wolters, R. van Roijen, and J.T.M. Walraven, Phys.Rev.Lett. 57, 2387 (1986).
20. J.J. Berkhout, O.H. Höpfner, E.J. Wolters, and J.T.M. Walraven, Jap.J.Appl.Phys. 26, 231 (1987), suppl. 26-3 (LT18).

Part VI

Optically Pumped Cesium

Influence of Low Magnetic Field on Pumping Efficiency in an Optically Pumped Cesium Beam Resonator

G. Théobald, P. Cérez, N. Dimarcq, and V. Giordano

Laboratoire de l'Horloge Atomique,
Equipe de Recherche du CNRS associée à l'Université Paris-Sud,
Bât. 221, Université Paris-Sud, F-91405 Orsay Cedex, France

1. Introduction

In this paper, we report our investigation on the influence of the magnetic field on the optical pumping of a cesium beam using coherent excitation with a broadband laser. The generation of Zeeman coherences in the ground state level when optical transitions of the cesium D_2 line are excited by a linearly σ polarized laser may result in a decrease of the efficiency of the optical pumping and lead to an intensity minimum in the fluorescence $L_{\mathcal{F}}$ as the applied magnetic field decreases to zero value. Accordingly the population difference Δn and $L_{\mathcal{F}}$ are found to depend on the magnitude B of the magnetic field . Here we describe physical basis which indicate that the fluorescence intensity variation results from the coupling of the lower level $\Delta m = \pm 2$ coherences to the excited level population. Comparison of theoretical predictions with experimental records of the fluorescence are then made. Finally guidelines are given which may influence the magnetic field design in the light-atom interaction zones of an efficient optically pumped cesium beam frequency standard.

2. Light-atom interaction. Density matrix evolution

Figure 1 shows the energy levels of the $6^2S_{1/2}$ ground state and of the $6^2P_{3/2}$ excited state of the cesium atom. The typical arrangement for the light-atom interaction is indicated in Fig. 2.

A monokinetic atomic beam crosses the light beam at right angle to cancel first order Doppler effect . A dc magnetic field B directed along the z axis fixes the quantization axis and removes the Zeeman degeneracy. According to the selection rules F' - F = 0, ± 1, only six transitions are allowed between one of the two levels g or f of the ground state and a level e of the excited state. We can distinguish between two kinds of transitions :

- the pumping transitions such as the (F = 3 or 4 → F' = 3 or 4) transitions. In this case, the laser is tuned to the g → e transition. The atoms on the e level can decay either to g or f levels. After a few pumping cycles, atoms tend to accumulate in the f level.
- the cycling transitions which are the (F = 4 → F' = 5) and (F = 3 → F' = 2) transitions. The atoms transferred in the excited state by the laser tuned to the g → e transition can only decay to the absorbing level g.

Frequency Standards and Metrology Editor: A. De Marchi

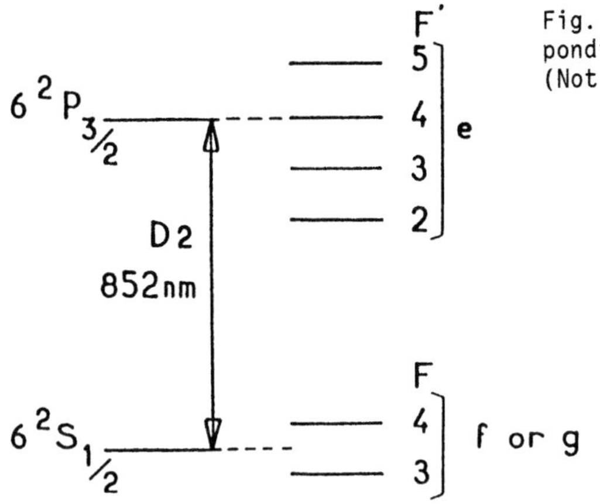

Fig. 1. Energy level diagram corresponding to the D_2 line of cesium atom (Not to scale)

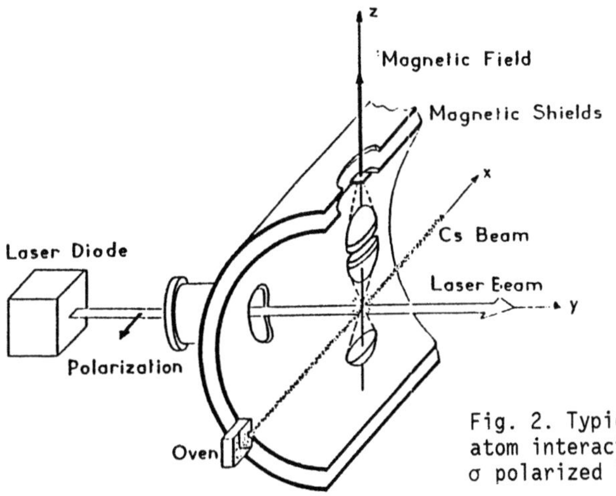

Fig. 2. Typical arrangement for the light-atom interaction. The arrow indicates a σ polarized light

Atoms are submitted both to the light beam interaction and to spontaneous emission. We have used a semi-classical formalism in which the laser field is considered classically and the atom is described by the density matrix σ . Spontaneous emission is taken into account by adding relaxation terms to the equations giving the rate of variation of the σ components /1/. When the laser is tuned to the g $\rightarrow$ e transition, the density matrix can be written in the basis $\{|F_i m_i>\}_{i = g, f, e}$ and takes the form :

σ^e	σ^{eg}	σ^{ef}
σ^{ge}	σ^g	0
σ^{fe}	0	σ^f

σ^e, σ^g and σ^f refer to the e, g, f levels respectively. The diagonal elements $\sigma^e_{m_e m_e}$, $\sigma^g_{m_g m_g}$ and $\sigma^f_{m_f m_f}$ are the populations of the e, g, f Zeeman sublevels. The corresponding off diagonal elements $\sigma^{e}_{m_e m'_e}$, $\sigma^{g}_{m_g m'_g}$ and $\sigma^{f}_{m_f m'_f}$ are called the Zeeman coherences. The σ^{eg} and σ^{ge} (resp. σ^{ef} and σ^{fe}) represent the optical coherences between the e and g (resp. f) levels which evolve at optical frequencies.

Under broadline excitation, the coherence time τ_c of the laser is much smaller than the lifetime τ of the excited state. Consequently, the optical coherences are not coupled to σ^{e}, σ^{g} and σ^{f} /2/.

Furthermore, we suppose that in the present study the hyperfine coherences σ^{fg} and σ^{gf} do not exist. Finally the time evolution of the components of σ is described by rate equations and we refer the reader to COHEN-TANNOUDJI /2/ for more details.

When the optical methods are applied to cesium beam clocks, we are mainly interested in two physical quantities /3/ which determine the clock signal amplitude :

* the fractional population difference Δn between the $m_F = 0$ ground state hyperfine sublevels

$$\Delta n = \frac{n(F = 4,\ m_F = 0) - n(F = 3,\ m_F = 0)}{n} \qquad (1)$$

where n is the total population of the cesium atom. Δn specifies the efficiency of the light to create a large population difference.

* the yield $L_{\mathcal{F}}$ of fluorescence photons emitted by a cesium atom under the action of the light beam.

Efficient state preparation for atomic clock purposes requires Δn values as high as possible. This is performed with a linearly σ or π polarized laser in which the polarization vector is perpendicular or parallel to the magnetic field.

The fluorescence yield $L_{\mathcal{F}}$ is mainly related to the populations of the excited state sublevels /4/. When a linearly σ-polarized light is used, successive absorption and stimulated emission of photons create $\Delta m = \pm 2$ Zeeman coherences in the ground state of Cs atoms in low magnetic fields. When the magnetic field B increases, degeneracy between the Zeeman sublevels is removed and the coherences are cancelled. Since $\Delta m = \pm 2$ coherences are coupled by the laser to the population of the excited state, it then results a decrease in the fluorescence intensity when the magnetic field goes to zero. Moreover the appearance of ground state coherences decreases the net number of transitions from the atomic ground level to the excited one. The atomic system behaves as it were a non absorbing medium /5/.

Rate equations have been numerically solved for any amplitude B of the field. Figure 3 reports predicted Δn values versus field amplitudes. The atom light interaction time was 12 µs and the laser intensity 2 mW x cm^{-2}. As it is readily seen on Fig. 3 the hyperfine pumping process is generally less efficient at low magnetic field and atoms are not completely transferred from one ground state to the other one.

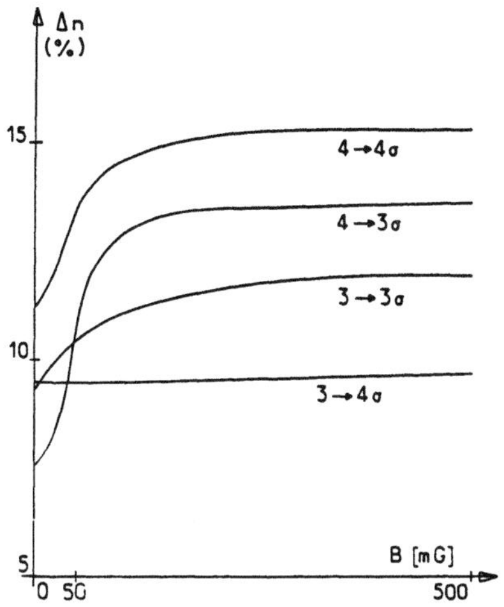

Fig. 3. Calculated population difference as a function of the magnetic field amplitude B

3. Comparison with Experiment

The experimental set-up for measuring the effect of $\Delta m = \pm 2$ coherences as the change of the fluorescence intensity was previously described by AVILA et al. /3/. Figure 4 shows the experimental records giving the fluorescence $L_{\mathcal{F}}$ versus the magnetic field for the $(3 \rightarrow 3)$, $(4 \rightarrow 4)$, $(4 \rightarrow 3)$ and $(3 \rightarrow 4)$ pumping transitions using σ -polarized excitation. Similar records are reported on Fig. 5 for the two cycling transitions $(3 \rightarrow 2)$ and $(4 \rightarrow 5)$. The experimental results agree well with the predicted ones. Some interesting features may be noted :

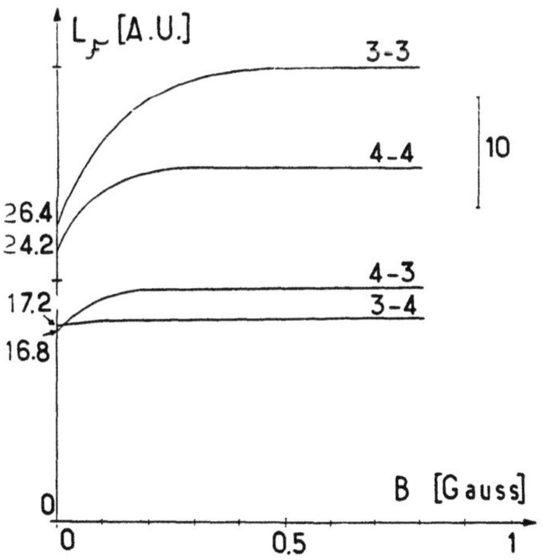

Fig. 4. Experimental variations of the fluorescence intensity versus B.
The laser light is σ- polarized and its intensity is 2 mW x cm^{-2} ; the interaction time is 12 µs

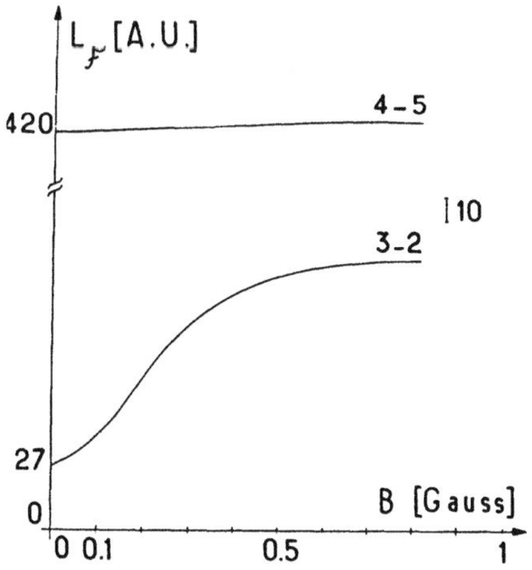

Fig. 5. The same as Fig. 4 for cycling transitions

i) the pumping transitions which exhibit a good pumping efficiency /3/ reflect a strong dependence on the magnetic field.

ii) the (3 → 2σ) transition is very sensitive to the magnetic field amplitude whereas the (4→ 5σ) transition does not exhibit a significant decrease in fluorescence when B goes to zero.

iii) in the case of the (3 → 2 σ) transition, the linewidth of the dip is considerably broadened by laser saturation effects.

4. Conclusion

We have pointed out the effects of $\Delta m = \pm 2$ coherences created in the Cs $6^2S_{1/2}$ ground state by coherent excitation with a laser diode tuned to the D_2 transition. They lead to a noticeable decrease of the population difference achieved by optical pumping and simultaneously to a reduction of the fluorescence when the magnetic field is lowered to values less than 200 milligauss. Two transitions, namely the (3 → 4 σ) and (4 → 5 σ) transitions, do not present variations versus B. Experimental results have been well interpreted by the rate equations model including Zeeman coherences.

These coherences may have important consequences on the clock signal amplitude in an optically pumped cesium beam resonator because it depends on :

i) Δn achieved in the first optical interaction region

ii) L_{fl} in the second optical interaction region.

The clock signal amplitude is then reduced at B values lower than ∿ 200 mGauss, which is detrimental to the short term frequency stability of the clock. In other respects, the microwave interrogation region needs a low B value ($\lesssim$ 60 mG) in order to minimize the frequency shifts related to the $\vec{C}$ field instabilities. Therefore it seems unrealistic to extend the low $\vec{C}$ field region of the cesium beam resonator to include the optical regions of state preparation and state detection so as to get a single uniform field in the three interaction regions. Contrary to what one would expect /6/, we think that the optically pumped basic device operating with a single laser diode /7/ may benefit from the magnetic field profile shown in Fig. 6.

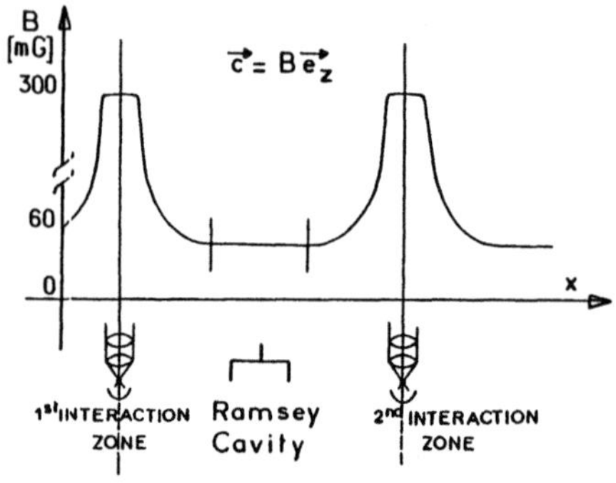

Fig. 6. Magnetic field profile along the beam axis

Acknowledgement

This work has been sponsored in part by the Bureau National de Métrologie, the Groupement d'Intérêt Public "Temps-Fréquence" and the Direction des Recherches, Etudes et Technique du Ministère de la Défense.

References

1. M. Ducloy : Phys. Rev. A8, 1844 (1973)
2. C. Cohen-Tannoudji : in Frontiers in Laser Spectroscopy, ed. by R. Balian, S. Haroche and S. Liberman, Vol. 1 (North Holland, Amsterdam, 1977) p. 1
3. G. Avila, V. Giordano, V. Candelier, E. de Clercq, G. Théobald, P. Cérez : Phys. Rev. A36, 3719 (1987)
4. A. Omont : Prog. Quantum Electronics 5, 69 (1977)
5. E. Arimondo, G. Orriols : Lett. Nuovo Cimento 17, 333 (1976)
6. R.E. Drullinger : Proc. of the IEEE 74, 140 (1986)
7. G. Théobald, V. Giordano, M. de Labachelerie, A. Hamel, N. Dimarcq, P. Cérez, C. Audoin : Proc. of the 42nd Frequency Control Symposium, Baltimore, USA (1988)

An Optically Pumped Primary Frequency Standard

R.E. Drullinger, J.H. Shirley, D.J. Glaze, and L. Hollberg

National Institute of Standards and Technology,
(formerly the National Bureau of Standards),
325 Broadway, Boulder, CO 80303, USA

An optically pumped primary frequency standard is being constructed at the National Institute for Science and Technology. To achieve the potential accuracy of this new technology a thorough reconsideration of all the potential systematic errors has been undertaken. Based on the needs of the optical pumping process and other improved technologies (ovens and cavities), a unique beam tube has been designed and fabricated. The laser performance necessary to allow full realization of the optically pumped clock's potential has been achieved by an optical feedback technique. The design goal is a short term performance of $\sigma_y(\tau) \leq 5\times10^{-13}/\sqrt{\tau}$ with a total frequency uncertainty of no more than 10^{-14}

Introduction

The concept of an optically pumped cesium frequency standard is now well established and a number of labs have programs to explore the technology[1-8]. These efforts indicate that an optically pumped primary standard can be constructed such that several of the accuracy-limiting systematic effects in conventional standards can be greatly reduced.

To build such a new standard that will realize its accuracy potential, careful consideration must be paid to all sources of systematic effects. Some of these effects, such as fluorescent light shift, are new to this type of standard. Others, such as those related to the velocity of the atoms in the beam, have been analyzed in conventional standards. However, substantial differences in the velocity profile between conventional standards and optically pumped standards, combined with increased accuracy expectations for the new standard, require that these effects be thoroughly reanalyzed. These studies are outlined in the following section on systematic effects.

The design of the atomic-beam tube must accommodate the optical pumping process as well as adequately address the various systematic effects. The essential features of the new standard are outlined in the section on the beam tube.

Systematic Effects

The systematic effects which have been reanalyzed include: fluorescent light shift[9]; velocity effects to include second-order Doppler shift, shifts vs. RF power and modulation parameters; Rabi pulling; cavity pulling; Majorana effects; distributed-cavity phase shift[10]; RF spectral purity and magnetic field uniformity.

Frequency Standards and Metrology Editor: A. De Marchi
© Springer-Verlag Berlin, Heidelberg 1989

Many of the shifts can be expressed as the quotient of two velocity integrals containing factors dependent on the microwave power, the modulation parameters and the particular shift mechanism. For very narrow velocity distributions, the velocity average can be ignored, and the power and modulation dependent factors cancel. The shifts then have little or no dependence on microwave power or modulation parameters. An optically pumped standard, however, will use almost all of the broad thermal distribution of velocities emerging from the oven. The shifts then acquire significant dependence on microwave power and modulation parameters. For example, the second-order Doppler shift and end-to-end cavity phase shift can change by 5-10% when the microwave power changes by 1 dB.

Second-order Doppler shifts are calculable if the effective velocity profile, effective microwave power, and modulation parameters are known to adequate accuracy. A numerical method for extracting both the velocity distribution and effective microwave power level from Ramsey lineshapes has been developed and tested with theoretical lineshape data. The agreement of the extracted distributions with the known input distributions is adequate to permit a 1% evaluation of the second-order Doppler shift. Details of the method will be published elsewhere.

Rabi pulling arises from the overlap of other Zeeman components of the hyperfine spectrum when sublevel populations or velocity distributions are asymmetric. When optical pumping is used for state selection, highly symmetric sublevel populations can be achieved. Optical excitation by linearly polarized light creates alignment (symmetric, non-uniform sublevel populations) but not orientation (asymmetric sublevel populations). Orientation can be produced only by pumping with elliptically polarized light traveling along the quantization axis. By using a quantization axis (C-field direction) along the atomic beam and exciting laser beams perpendicular to the atomic beam to eliminate the first-order Doppler effect, even a small elliptically polarized component of the pumping light will not create asymmetric populations. Hence, Rabi pulling should be reduced to insignificant levels.

Majorana transitions, or transitions among the Zeeman sublevels induced by changing magnetic fields, have long been suspected of causing population asymmetries and Rabi-pulling shifts in standards with magnetic state selection[11]. By eliminating the strong state-selecting magnets and performing optical pumping and detection within the shielded C-field region one expects no Majorana transitions. However, if they do occur, either by design or by accident, they should still cause no frequency shifts in an optically pumped standard. The evolution of Zeeman sublevels in a changing magnetic field can be described by the Hamiltonian $\hat{\mu}\cdot\vec{H}(T)$. The evolution operator for this Hamiltonian is a rotation operator. The angular momentum symmetry of rotation operators shows that initially symmetric sublevel populations remain symmetric, no matter what contortions the magnetic field goes through, even if its changes are resonant with the Zeeman frequency. Hence, even an intentionally excited Majorana transition should not shift the clock transition. This will be studied experimentally.

Theoretical studies of the magnetic field generated by a solenoidal winding closely surrounded by side and end shielding indicate that the inhomogeneity arises primarily from the gap between the last winding and the end shield. For a 5 mm gap at the end of a 17 cm diameter solenoid, the on-axis field one diameter from the end is only 0.2% less than that at the center. The inhomogeneity decreases exponentially with distance from the gap. This uniformity is sufficient to use the mean field as measured by the first-order, Zeeman-dependent transitions as a means of estimating the mean square field correction necessary for the clock transition.

Atomic Beam Tube

The design logic for the atomic beam tube and its realization have been described previously[12]. The beam tube is horizontal and of medium length. The heart of the system is a new Ramsey cavity designed to minimize effects of distributed cavity phase shift[10]. The length of the cavity is about 1.5 m and will result in a line Q of 10^8. The microwave field polarization is such as to require a longitudinal C-field. This was chosen to minimize Rabi-pulling effects and to produce a more uniform C-field.

The beam tube is symmetric end to end and designed to allow for simultaneous counter-propagating beam operation[12]. The optical detection zones are 15 cm from the ends of the Ramsey cavity, and the optical pumping zones are 15 cm farther out. This geometry should result in a fluorescent light shift in the 10^{-16} range. The fluorescence collection optics consist of two 5 cm radius mirrors arranged to image the fluorescence into a light pipe which conducts it out of the vacuum envelope to the detector and preamplifier. The system has been shown to collect 45% of the fluorescent light and the imaging properties provide the necessary isolation from scattered laser light.

The atomic beam cross section is 3 mm in diameter, established by the windows in the microwave cavity. The beam is generated in a hybrid recirculating oven[13] with a single 3 mm diameter opening. Operating at 110°C, this oven is expected to generate a beam flux at the detector of 2×10^9 atoms per second while emitting only 30 mg per year of cesium into the beam tube.

Summary

A new cesium atomic beam primary frequency standard which will use optical pumping for state preparation and detection is being built. A thorough reanalysis of systematic effects has led to a design with an expected line Q of 10^8, a short term stability of $\sigma_y(\tau) \leq 5 \times 10^{-13}/\sqrt{\tau}$ and a total frequency uncertainty of 10^{-14}. The laser technology necessary to support this level of performance has been demonstrated[14]. The line Q and accuracy goal place stringent requirements on the line-centering servo. The details of the servo remain an open area of research.

However, the design philosophy is to take advantage of the operational flexibility allowed in an optically pumped system to have a semi-automatic evaluation and to run the standard as a clock. It is expected that RF power, modulation parameters, the effective C-field strength and end-to-end cavity phase shift will be under active control.

References

1. A. Derbyshire, R. E. Drullinger, M. Feldman, D. J. Glaze, D. Hillard, D. A. Howe, L. L. Lewis, J. H. Shirley, I. Pascaru and D. Stanciulescu, Proc. 39th Annual Symposium on Frequency Control, Philadelphia, USA (1985), p.18

2 G. Avila, V. Giordano, V. Candelier, E. deClercq, G. Theobald, and P. Cerez, Phys. Rev. A, 36, 3719 (1987)

3 Shang Song-quan, Wu Xin-xin, Yao Shu-tong, Xie Lin-zhen and Wan Yi qui, Chinese Physics Letters, 2, 557 (1985)

4 S. Ohshima, Y. Nakadan and Y. Koga, Proc. 19th Annual Precise Time and Time Interval (PTTI) Applications and Planning Meeting, Redondo Beach, California, December (1987) p.245

5 J. Umezu, B. Komiyama, H. Saito and Y. Ohta, Proc. 2nd European Frequency and Time Forum, Neuchatel, Switzerland (1988) p.521

6 T. McClelland, I. Pascaru, J. Zacharski, N. H. Tran and M. Meirs, Proc. 41st Annual Symposium on Frequency Control, Philadelphia, USA (1987) p.59

7. A. Clairon, E. deClercq, B. Dahmani and A. H. Gerard, Proc. 2nd European Frequency and Time Forum, Neuchatel, Switzerland (1988) p.499

8. P. Thomann, H. Schweda and G. Busca, Proc. 2nd European Frequency and Time Forum, Neuchatel, Switzerland (1988) p.513

9. J. Shirley, Proc 39th Annual Symposium on Frequency Control, Philadelphia, USA (1985) p.22

10. Andrea DeMarchi, Jon Shirley, David J. Glaze and Robert Drullinger, IEEE Trans. Instrum. Meas. IM-37, 185 (1988)

11. G. Becker, IEEE Trans. Instrum. Meas. IM-12, 319 (1978), D. W. Allan, H. Hellwig, S. Jarvis, D. A. Howe and R. M. Garvey, Proc. 31st Annual Symposium on Frequency Control, Ft. Monmounth, N.J., (1977) p.555

12. R. E. Drullinger, Jon Shirley, D. J. Glaze, L. W. Hollberg and A. DeMarchi, Proc. 40th Annual Symposium on Frequency Control, Philadelphia, USA (1986) p.428

13. R. E. Drullinger, D. J. Glaze and D. B. Sullivan, Proc. 39th Annual Symposium on Frequency Control, Philadelphia, USA (1985) p.13

14. S. Ohshima, Y. Koga, Y. Nakadan, L. Hollberg and R. Drullinger, Proc. 2nd European Frequency and Time Forum, Neuchatel, Switzerland (1988) p.531

Design of an Optically Pumped Cs Laboratory Frequency Standard

E. de Clercq, A. Clairon, B. Dahmani, A. Gérard, and P. Aynié

Laboratoire Primaire du Temps et des Fréquences,
Bureau National de Métrologie-Observatoire de Paris,
61, av. de l'Observatoire, F-75014 Paris, France

I - INTRODUCTION

The use of optical pumping to replace state selecting magnets in a classical Rabi atomic-beam magnetic resonance apparatus was first proposed by KASTLER as early as 1950 [1]. A successful realization of this approach was achieved by ARDITI and CEREZ [2] in a Rb beam pumped by spectral lamps. The advent of laser sources in the near infrared led to the pioneering work of PICQUE [3] and ARDITI [4] on a Cs beam using the first available diode lasers and of LEWIS and FELDMAN with dye lasers [5]. Now many laboratories [6] are involved in the study or in the realization of optically pumped Cs beam frequency standards. In France, where, up to now, there was no laboratory primary frequency standard, the L.P.T.F. has started in 1982 the study of such an optically pumped standard. In this paper, we give the design of our Cs beam apparatus and some preliminary studies.

II - GENERAL DESIGN

In the design of our apparatus, we tried to take advantage of optical pumping characteristics [7] which are : symmetrical design and on axis alignment, larger atomic beam cross-section and extended static magnetic field region.

The uniformity and stability of the static magnetic field ("C field") is one of the most important source of uncertainty in the error budget of a Cs clock [8,9]. In order to avoid possible magnetic field uniformity problems that would arise with holes for laser beams in the case a solenoïd is used (axial magnetic field) and to have more space for the fluorescence collection systems, we chose a transverse magnetic field approach produced by means of four current-carrying aluminium rods as in the NRC design[10]. These rods are 2 m long and cover the cavity length plus the regions of interaction with laser beams (see Fig. 1).

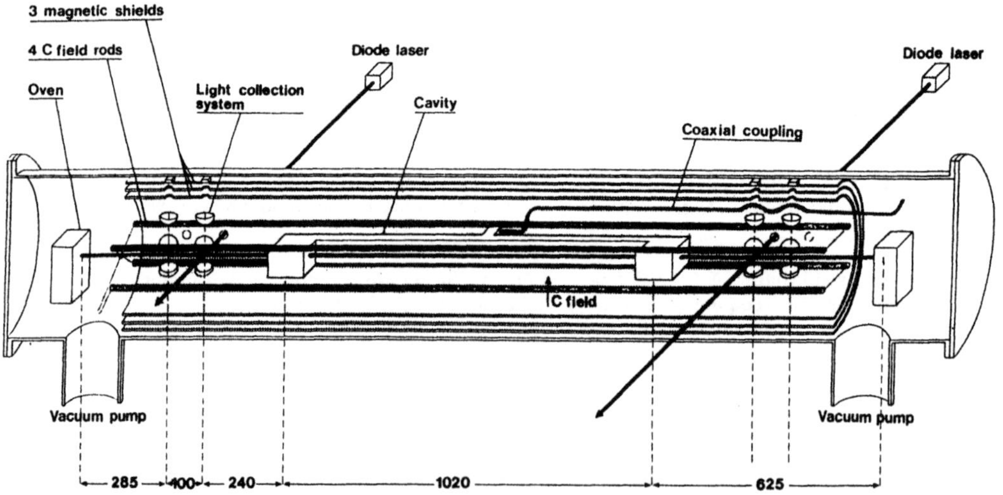

Fig. 1 - Artist's view of the laboratory standard

Frequency Standards and Metrology Editor: A. De Marchi

The Ramsey cavity is of the usual type for a transverse magnetic field. Its total electrical length is 48 $\lambda g/2$ for 1 m interaction length. The atomic beam passes at a distance of $\lambda g/2$ from the short circuits to allow greater dimensions of the slots (3 x 8 mm^2) and consequently a more intense atomic beam. It is known that this type of a cavity gives rise to Rabi pulling [11], but it is expected that this effect is not present with optical pumping [11,12]. The whole cavity is placed inside the magnetic shield and the RF power is fed in by a means of a coaxial cable from the outside in order to avoid holes in the magnetic shields.On each side of the cavity, there are two optical interaction regions in order to allow experimental investigation of light shifts [13], light pressure effects [14],the use of simultaneous counter-propagating beams [7], etc... The fluorescence collection system is described in Part IV.

All these parts (C fields rods, cavity, optics) fit inside an aluminium support cylinder (internal diameter 150 mm), which is not shown in Fig. 1. It is surrounded by three coaxial mumetal cylinders (2 mm wall thickness). An oven is mounted at each end of this cylinder for beam reversal purposes.The whole set-up is enclosed in a 25 cm diameter, 2.60 m long high vacuum enclosure.

III - OVEN AND ATOMIC BEAM

In an OP Cs standard, a high intensity and well collimated atomic beam is needed to insure low consumption and low residual Cs partial pressure. This can be obtained with a capillary array [15]. We have tested two large size multichannel collimators with high L/D ratio where L is the length of a channel and D its diameter. The first one is an array of long parallel cylindrical nickel tubes (15 mm long, 0.18 mm inside diameter, 0.30 mm outside diameter) which are laser welded together. The total transparency is estimated to be 30%. The second one is a stack of crinkled nickel foils (foil thickness = 0.05 mm, L = 15 mm, corrugation of 0.16 mm high and 0.6 mm wide and total transparency of 60%). In both cases, the overall collimator aperture is 5 x 11 mm^2. The beam intensity is monitored at 280 mm from the oven by moving a hot-wire detector. The intensity and the shape of the beam have been computed using the analysis of GIORDMAINE and WANG [16] and taking into account for each channel the angle between its axis and the detector. The calculation is performed in the free molecular flow regime, also called transparent mode, where the mean free path λ is greater than L : by this we mean that the molecular collisions in a channel are negligible. This assumption is valid up to an oven temperature of about 100°C.

Figure 2 shows the vertical beam profile obtained with a collimator made of circular tubes (oven temperature 67°C). The angular beam divergence (HWHM) is 11 mrad, in very good agreement with calculations. The beam divergence does not change with oven temperature as long as the atomic flow regime is the free molecule regime. This is observed on Fig. 3, the experimental beam divergence is constant below an oven temperature of 100°C, then increases rapidly. Above this temperature there is a transition flow regime followed at higher temperature by a viscous regime. In this case, there are collisions in the channels and in the immediate vicinity of the collimator, and slow atoms are preferentially scattered out of the beam axis. Consequently, a lack of slow atoms is expected in the velocity distribution [17,18].

We have studied the on axis velocity distribution using optical Doppler shift method : the laser beam is split into two arms, one crossing the atomic beam at 90° for frequency calibration and origin determination purposes and the other at 25° selecting a particular class of velocities according to the laser frequency. The fluorescence signals of each interaction region are simultaneously recorded. The Doppler free fluorescence spectrum is also used to calculate the contribution of the 4-3 and 4-4 pumping transitions to the Doppler shifted spectrum. This contribution is then numerically removed from the observed shifted spectrum in order to get only the signal due to the 4-5 cycling transitions which is much greater. The resulting signal is the convolution of the velocity distribution profile with the resonance width (laser linewidth and natural width about 30 m/s). Fig. 4 shows the curves obtained for two oven temperatures. The "+" in Fig. 4 are calculated using the convolution of a Lorentzian shape (atomic response) with a Maxwellian velocity distribution weighted by a 1/v factor since the interaction time and the number of photons increase as 1/v. We observed a Maxwellian velocity distribution up to an oven temperature of about 100°C ; above this temperature there is an apparent shift of the velocity distribution to higher velocities due to a loss of slow atoms as can be seen in Fig. 4 for an oven temperature of 147°C.

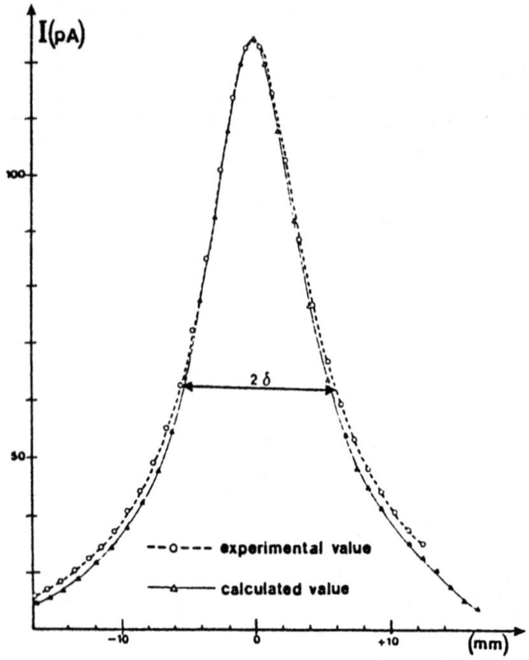

Fig. 2

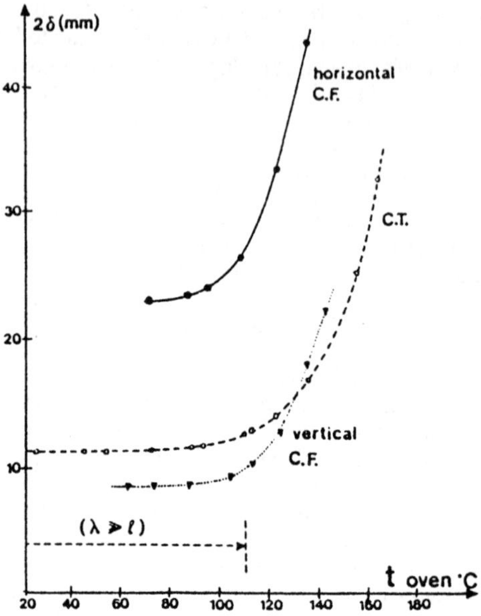

Fig. 3

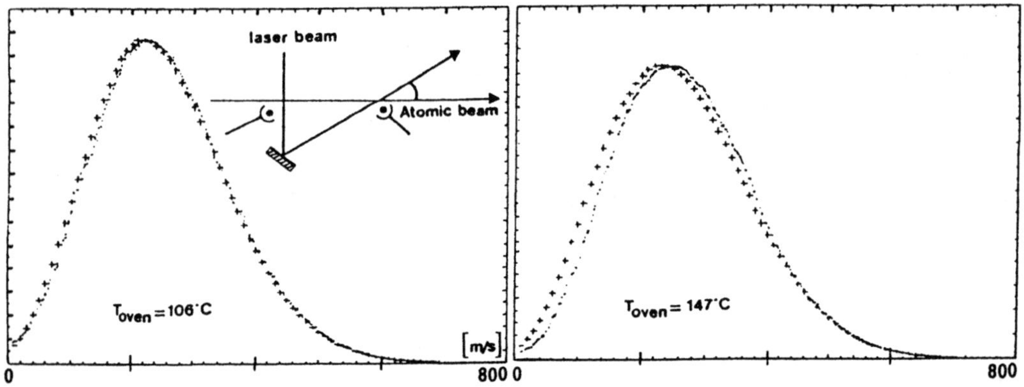

Fig.4 - Atomic beam velocity distribution

IV - LASER AND DETECTION OPTICS

In a Cs beam frequency standard with optical detection,minimization of laser frequency noise when a cycling transition is used [19] and efficiency of fluorescence light collection, are main goals in achieving a large signal to noise ratio .

The set-up used for pumping and detection is shown in Fig. 5. With the optical feedback technique from a Confocal Fabry-Perot (CFP) cavity [20,21,22], the laser linewidth can be easily reduced to 100 times less than the Cs natural linewidth. In this case, the FM noise contribution to the detection noise is negligible. However, for long term operations of the Cs standard, the frequency of

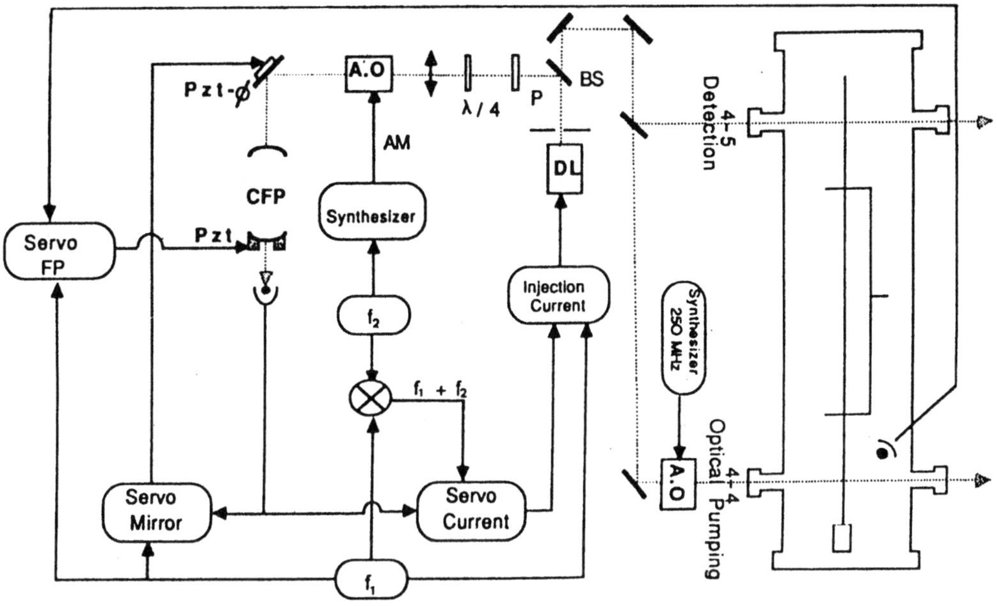

Fig. 5 - Laser stabilization scheme

the diode laser needs to be locked to the Cs resonance line. As explained in Ref [21], the frequency of a coupled cavity laser diode depends on the free running frequency diode, the distance between the diode and the CFP, and the CFP length. Consequently, all these three parameters must be controlled to obtain long term stabilization and tunability of the laser.

One way of obtaining signals for stabilizing the Coupled Cavity Laser (CCL) is through the use of the CFP itself as a frequency discriminator. The laser current or CFP-diode distance is modulated at f_1. This modulation induces FM modulation which is detected on the CFP transmission fringe and used, after synchronous detection and integration, to lock the distance CFP-diode (servo-miror). In this case, the coupled cavity frequency is equal to the CFP resonant frequency. It is important to note that the FM induced by current modulation is reduced by the feedback. In any case, the FM must be less than the fringe width (Free Spectral Range/Finesse) in order to get a stable operation. In this scheme, the CFP determines the CCL frequency and can be locked to the Cs frequency by using the first derivative of the fluorescence resonance line (Servo-FP). However, temperature fluctuations of the diode of the order of 10^{-2} K or ageing may set the CCL in an unstable state. Therefore, it is necessary to lock the free running frequency of the diode to the CFP resonance. In order to obtain a signal proportional to the difference between the diode laser frequency and the CFP frequency, the feedback level is modulated at f_2. After demodulation of the CFP transmitted power at $f_1 + f_2$, the signal obtained can be used, after integration, to control the diode current or temperature (Servo-current). The existence of this signal comes from the fact that the modulation of the feedback level induces an FM modulation of the CCL except if all eigen-frequencies of the elements of the CCL are equal. The detection at $f_1 + f_2$ is equivalent to the detection of the FM induced by the feedback modulation which is proportional to the detuning of the free running laser diode frequency with respect to the CFP eigen-frequency.So, when the three servo-loops are in operation the long term stabilization is only limited by possible mode hopping as for the free running diode laser.

In order to get a large S/N ratio the light collection device must satisfy simultaneously two contradictory requirements : a large efficiency and a low stray light level. Our design is shown in Fig. 6. In the bottom, a spherical mirror is centered on the atomic beam and then the Cs beam is imaged at the input of a light pipe by means of two fast AR coated aspheric lenses (Ø 52 mm, f/Ø = 0.63). The light pipe is simply used to locate the photodiode outside the magnetic shields. The stray light intensity on the detector is reduced by the imaging property of the device and by diaphragms located along the laser beam. With a preliminary device, a fluorescence light collection efficiency of about 20% was measured with a stray light power to laser power ratio of 1.10^{-7} to 5.10^{-7}.

V - SIGNAL TO NOISE RATIO AND SHORT TERM STABILITY

The amplitude signal to noise ratio of the clock signal is estimated taking into account all the processes involved in the generation of this signal. Let us consider a Cs thermal beam of flux I_a(at./s) in the detection region with a Maxwellian velocity distribution f(v) and a class of velocities centered on v. These atoms are optically pumped in the "good" quantum level with a probability ρ. After travelling across the Ramsey cavity, they can be found in the other clock level with a probability p(v) (at resonance $p(v) = \sin^2(2\,b\,l\,/\,v)$ [23]. The number of atoms entering in the detection region obeys Poisson statistics. The number of photons per atom follows also a Poissonian distribution whose mean is $\beta(v)$. The photons are collected with a probability $\gamma(\gamma = 0.2)$ and then detected with a quantum efficiency η (Si photodiode $\eta = 0.7$). Consequently, the amplitude signal to noise ratio of the clock signal is given by :

$$\frac{S}{N} = \frac{\rho\gamma\eta I_a \int \beta(v)\,p(v)\,f(v)\,dv}{\left[2\rho\gamma\eta I_a\left(\int \beta(v)\,p(v)\,f(v)dv + \gamma\eta\int \beta^2(v)\,p(v)\,f(v)\,dv\right) + 2\eta\, r\, P/h\nu + F_A\, 4kT/e^2 R_s\right]^{1/2} B^{1/2}} \qquad (1)$$

where B is the noise bandwidth. Two additional noise sources have been introduced :

- stray light (characterized by a rejection factor $r = 10^7$ and a photon flux in the incident beam of the detection laser $P/h\nu$; P = 3 mW)
- electronic noises (photodiode shunt resistance R_s = 100 MΩ and amplifier noise factor $F_A = 1.8$)

We have estimated the signal to noise ratio in two cases : (a) the one laser scheme and (b) the two laser scheme. Fig. 7 shows S/N versus the oven temperature; the beam intensity is calculated for a rectangular oven aperture (3x8 mm^2) and a collimator transparency of 30%. The two schemes are :

a)-pumping and detection with the 3-3 σ transition. This transition is the best one with a single laser [24]. In this case, $\rho = 0.12$ and $\beta(v)$ = constant = 3.8 photons per atom. The mean microwave transition probability in maximum for $2bl/\alpha = 0.6\pi$ [23] where α is the most probable velocity in the oven.

b)-pumping with 4-4 σ transition and detection with a monochromatic laser tuned to the 4-5 σ transition. This scheme is the best one when detection is performed with a cycling transition and pumping with a single laser frequency. In this cases $\rho = 0.15$ [24] and $\beta(v) = [s/(1+s)]\Gamma(w/v)/2$ where s is the saturation parameter (s = 3 in our calculation), Γ is the spontaneous emission rate (3.1 10^7/s), w is the width of the laser beam (w = 5 mm). The mean microwave transition probability is maximum for $2\,b\,l\,/\,\alpha = 1.63$.

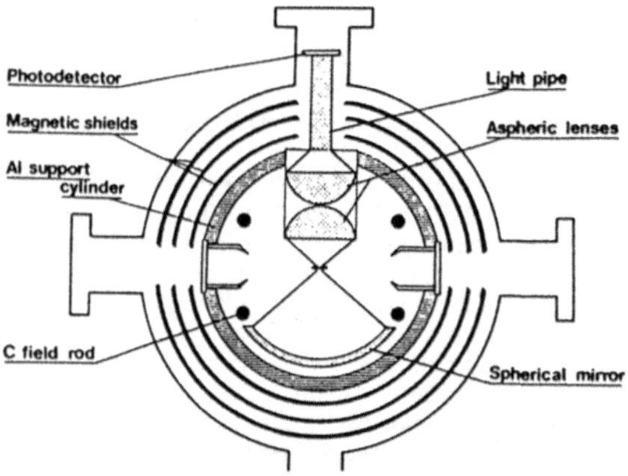

Fig. 6

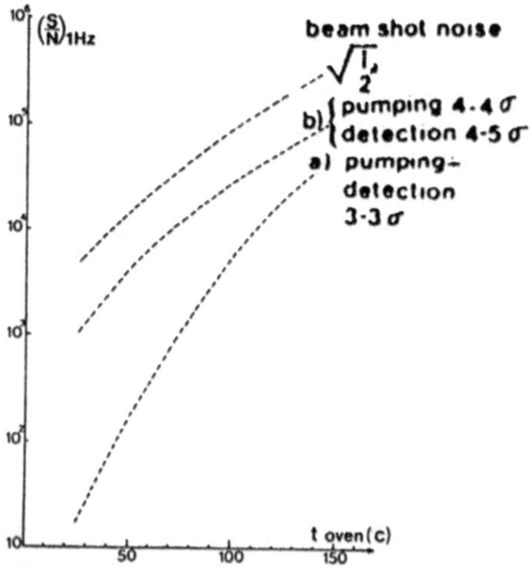

Fig. 7

Fig. 7

The short term stability can be estimated by :

$$\sigma_y(\tau) \sim (\Delta\nu/\nu)/(S/N)_{1Hz} \ , \qquad (2)$$

where $\Delta\nu$ is the linewidth of the central Ramsey fringe. The expected short term stability has been calculated in the two precedent cases for an oven temperature of 90°C, which gives $I_a = 9.10^9$ at/s in the detection region and $\alpha = 213$ m/s.
The results are shown in the following table :

	$(S/N)_{1Hz}$	velocity distribution	linewidth	$\sigma_y(\tau)$
(a) 1 laser	$3.1\ 10^3$	Maxwellian	120 Hz	$4\ 10^{-12}\,\tau^{-1/2}$
(b) 2 lasers	$2\ 10^4$	Maxwellian x1/v	100 Hz	$5.5\ 10^{-13}\tau^{-1/2}$

It is interesting to note that even in the 1 laser case, the expected stability compares favourably with results of the best present primary frequency standards. We hope that the stability forecast above will be paralleled by an improvement in accuracy relative to the conventional Cs beam standards. Finally, optically pumped standards lead the way to the next generation which, probably, will use laser cooling.

REFERENCES

1. A. Kastler, J. Phys. Rad. 11, 155 (1950), J. Opt. Soc. Am. 47, 460 (1957)
2. M. Arditi and P. Cerez, IEEE Trans. Instrum. Meas. IM21, 391 (1972)
3. J.L. Picqué, Metrologia 13, 115 (1977)
4. M. Arditi, J.L. Picqué , J. Phys. Lett. 41, L379 (1980)
5. L.L. Lewis and M. Feldman in Proc. 35th Ann. Freq. Control Symp. (1981)
6. See for example papers of P. Cerez, R. Drullinger, S. Ohshima, P. Thomann, P. Tremblay and C. Jacques, and Y. Wang in these proceedings.
7. L.L. Lewis, M. Feldman and J.C. Bergquist, J. Phys. Colloq. C8, 42, C8-271 (1981)
8. A.G. Mungall, H. Daams,, D. Morris and C. Costain, Metrologia 12, 129 (1976)
9. A. Bauch, H. de Boer, B. Fischer, T. Heindorff, R. Schröder, 2nd European Frequency and Time Forum, Neuchâtel, March 1988.
10. A.G. Mungall, R. Bailey, H. Daams, D. Morris and C. Costain, Metrologia, 9, 113 (1973)
11. A. De Marchi, G. D. Rovera and A. Premoli, Metrologia 20, 37 (1984)
12. M. Arditi, Metrologia 18, 59 (1982)
13. A. Brillet, Metrologia 17, 147 (1981)
14. A. Clairon, E. de Clercq, B. Dahmani, A.H. Gérard and C. Thomas, 2nd European Frequency and Time Forum, Neuchâtel, March 88
15. J.G. King and J.R. Zacharias, Adv. Electron. Electron Phys. 8, 1 (1956)
16. J.A. Giordmaine and T.C. Wang, J. Appl. Phys. 31, 463 (1960)
17. I. Estermann, O.C. Simpson and O. Stern, Phys. Rev. 71, 238 (1947)
18. H.C.W. Beijerinck and N.F. Verster, J. Appl. Phys. 46, 2083 (1975)
19. G. Avila, E. de Clercq, M. de Labachellerie and P. Cerez, IEEE Trans. Instrum. Meas. IM34, 139 (1985)
20. B. Dahmani, L. Hollberg, R. Drullinger, Opt. Lett. 12, 876 (1987)
21. P. Laurent, A. Clairon, C. Bréant, submitted to IEEE QE
22. See papers of L. Hollberg and C. Bréant in these proceedings
23. N.F. Ramsey, Molecular Beams, Oxford University Press, 1956
24. G. Avila, V. Giordano, V. Candelier, E. de Clercq, G. Theobald and P. Cerez, Phys. Rev. A36, 3719 (1987)

Preliminary Results and Some Considerations on Optically Pumped Cesium Beam Frequency Standard

Wang Yiqiu, Xie Linzhen, Wang Depei, and Yao Shutong

Department of Radio-Electronics, Peking University,
Beijing, 100871, People's Republic of China

I. Introduction - The advantages of optically pumped cesium beam frequency standards in comparison with the conventional one have become well recognized in the community of frequency metrology. Several laboratories [1-4] have been undertaking works on this new kind of apparatus either as primary standard or as practical one. Although big progress has been made, some problems remain yet to be solved. Among them the most important one is related to the noise and instabilities of characteristics of semiconductor laser diodes used for optical pumping and detection. Extensive work is being undertaken to solve these problems. In this report we will summarize our work with laser diodes and preliminary results of our optically pumped cesium beam standard. Some considerations on the effects of laser noises will be given at the end.

II. Work on Semiconductor Laser Diodes - Both CSP (model HLP-1400) and TJS (model ML-2701 and ML-2307) single-mode GaAlAs laser diodes have been used in different stages of our experiments. Work done for improving their properties to serve our purpose includes:

1. Extending the Wavelength Tuning Range In ordinary condition the lasing wavelength for some diodes could not be tuned exactly to the pumping transition because of mode hopping. This difficulty has been overcome to some extent by use of weak optical feedback (< 60 db) [5]. Figure 1 shows the experimental set up. A mirror is placed opposite the rear facet of diode by a small distance d, d ~ 5L-9L, L being the cavity length of diode laser is about 300 μm. This mirror is fixed on a 2.8 cm long temperature controlled copper rod. The distance d and thus the lasing frequency of diode can be monitored thermally. This approach allowed slightly shift the tuning range and force

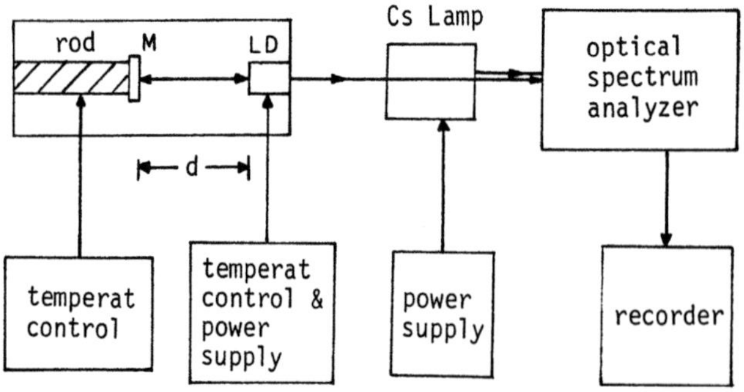

Fig. 1. Experimental set up for shifting and extending the tuning range of diode laser LD. A mirror M is fixed on a temperature controlled copper rod. Cs lamp is used as wavelength reference.

Frequency Standards and Metrology Editor: A. De Marchi

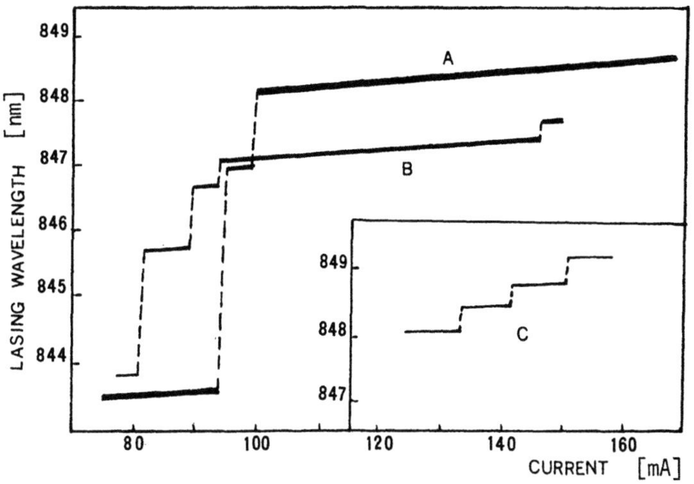

Fig. 2. Wavelength tuning characteristic of a diode laser with external cavity. A: d=100 μm, B: d=87 μm, C: without external cavity.

the wavelength of some CSP laserdiodes to coincide with the Cs-D_2 line. In addition, in the case of strong feedback when very small d (d~ L/3) was used and an external cavity formed, the continuous wavelength tuning range could be extended by about ten times, as shown in Fig. 2 [6]. This greatly improved the wavelength tunability of diode lasers.

2. Locking the Laser Frequency to the Pumping Transition The lasing frequency of diode is very sensitive to the change of temperature and injection current. For the TJS laser diodes the typical rates are -60 GHz/°C and -6GHz/mA, respectively. It is hard to keep the variation of temperature and current within 10^{-5}°C and 10^{-4}mA, so frequency locking the diode laser to the pumping transition of cesium beam is used for alternative[7] . The experimental set up is shown in Fig. 3 (left part), where the F=4-F'=4 transition is chosen as reference. For obtaining discriminator curve the laser frequency is modulated at 3KHz, which is limited by the response time of fluorescence detection circuit. The result shows (Fig.4), the frequency fluctuation at low Fourier frequencies is reduced significantly. The measured error signal, i. e. the voltage variation on the output of lock-in amplifier, converted into frequency fluctuation in Allan variance gives that the frequency instabilities for 1, 10, 100 seconds are 8×10^{-12}, 2.2×10^{-12} and 7.8×10^{-13}, respectively. This approach ensured the stable working of our optically pumped frequency standard for two days.

3. Suppressing the Frequency Noise. To get good signal-to-noise ratio of clock transition the high frequency noise of diode laser must be suppressed (cf. IV). The frequency locking method described above cannot reduce the noise higher than several tens hertz. So we have tested an electronical feedback scheme using an external Fabry-Perot cavity as reference to suppress the higher frequency fluctuation. One side of the transmission curve of Fabry-Perot cavity was used for frequency discrimination,

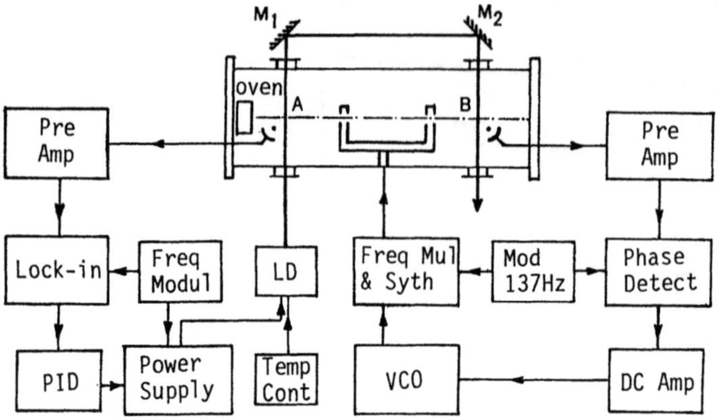

Fig. 3. Block diagram of an optically pumped cesium beam frequency standard

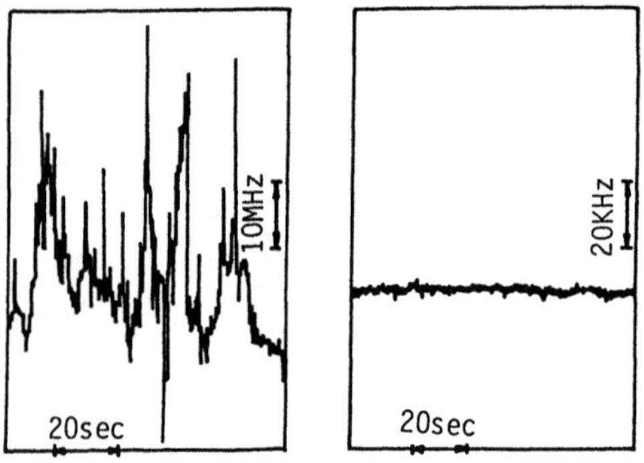

Fig. 4. Frequency noise of a laser diode before (A) and after (B) locking to the pumping transition of cesium beam

converting the frequency jitter into voltage fluctuation which feeds back negatively to the laser current supply through a wideband amplifier to adjust laser frequency. An avalanche photodiode along with the amplifier formed a fast loop with the bandwidth of 200MHz and a time delay of 10 ns. A second slow loop locks laser's center frequency either to the atomic line or to the transmission peak of second Fabry-Perot cavity. The noise spectrum was reduced in 30db up to 10 MHz, and a positive feedback peak appeared near 15 MHz. We believe that better result can be achieved with a more elaborated amplifier.

4. Frequency Offset Locking of One Diode Laser to Another In the two laser pumping scheme they have to be differed in frequency about 9 GHz, and for the scheme of F=4 → F'=4 pumping and F=4 – F'=5 cycling transition detection -- about 251 MHz. It is desirable to have similar frequency stability for both lasers. For this purpose one way is to frequency lock one laser with offset to another which is stabilized on a pumping transition of cesium. In this initial stage we have tried to lock one diode lasing at F=4 – F'=5 cycling transition to the another tuned on the pumping transition with offset frequency of 251 MHz. A double balanced mixer and a delay line served as a frequency discriminator. Because of the slow response of the loop only the slow frequency

variation of one laser could follow that of another. We have tested this method with success in our cycling transition detection experiment.

III. Preliminary Results of Optically Pumped Cesium Beam as Frequency Standard - The cesium beam tube with six optical windows for laser light with appropriate polarization is 0.8m long and o.25 m in diameter. The vacuum in it is about 6×10^{-6}Pa. The cesium oven with a collimator composed of about 120 copper capillaries 9 mm long and 0.08 mm in diameter is temperature controlled at about 130 °C. The effusive area is about 0.6 mm^2. The U-shape microwave cavity with drift region 168.7 mm long is integrated with C-field coil and 3 layers of magnetic shielding. The C-field direction is perpendicular to the atomic beam and its strength is 4.9 A/m. In this preliminary experiment only single laser pumping scheme was carried out. The diode laser (ML-2307, 2.8 mW) was tuned to the F=4→F'=4 transition with $\sigma^+ + \sigma^-$ circular polarizations together, which resulted from linear polarized laser beam parallel to the C field direction. This pumping scheme gives maximum relative population difference between clock sublevels, δ_{00} = -15.2%, the largest one for single laser pumping [8]. Laser frequency was locked to the fluorescence peak in this pumping redion A in Fig. 3. The microwave signal of 9192 631 771.6 Hz produced by frequency multiplication and synthesizing from a 5 MHz VCO was modulated at 137 Hz. Atoms, having undergone the mw transition were detected in region B by the light beam come from the same laser diode. To reduce the stray light the photodetector (HUV-4000) in this region is placed in a pipe coated with light absorption material inside.

Figure 5 shows the Ramsey pattern of mw clock transition. The linewidth W of the central peak is 623 Hz and the signal-to-noise ratio S/N in 1 Hz bandwidth at 137 Hz is about 280. The frequency stability of a 5 MHz VCO output after locking to the peak of Ramsey pattern was apparently improved for sampling time τ> 100 sec. while the loop time constant was 10 sec. The figures were 2.2×10^{-11}, 3.1×10^{-11}, 8.8×10^{-12}, 4.1×10^{-12}, 8.7×10^{-13}, and 6.3×10^{-13} for 1, 10, 100, 1200, 3600 and 10800 sec., respectively. These data are in agreement with those expected from the values of W and S/N.

The F=4→F'=5 cycling transition has been used also for detecting the clock signal. Though the signal increased greatly, the S/N was not improved because of laser noise.

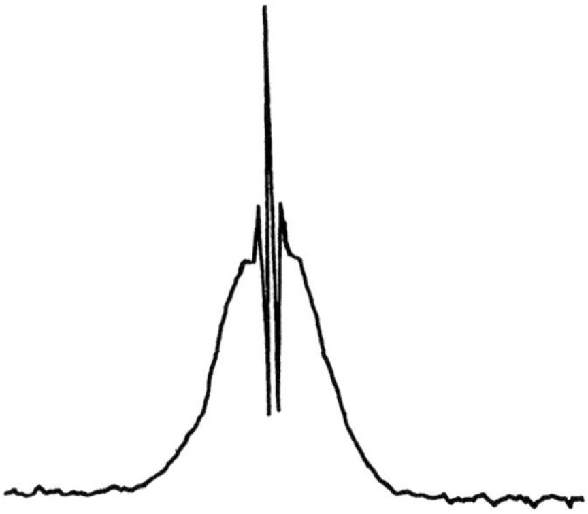

Fig. 5. The Ramsey pattern of microwave clock transition

Very good Ramsey pattern has been obtained in zero C-field, this illustrated the good magnetic field homogeneity in the mw interaction region.

IV. Considerations About the Effects of Laser Noise on the Performance of Frequency Standards - The frequency stability of an atomic standard for sampling time τ is given by

$$\sigma(\tau) = KW [f_0 (S/N)_{1Hz} \tau^{\frac{1}{2}}]^{-1}, \quad (1)$$

where K is a constant and f_0 is the microwave transition frequency. For a given apparatus W depends upon the oven temperature and $\sigma(\tau)$ is mainly determined by S/N. In the ideal condition this ratio is limited by the shot noise of atoms in beam, however, optical pumping and detection always introduce some additional noise come from laser diodes. Semiconductor lasers have similar noise spectrum both for amplitude and frequency fluctuation up to several GHz [9]. Typically, the relative intensity noise in flat region (1MHz $< f <$ 100MHz) is about $10^{-6}/\sqrt{Hz}$ and at low frequency--$10^{-5}/\sqrt{Hz}$, and the frequency noise spectral density is about 10^5 - 10^6 $Hz/\sqrt{Hz}$ and 10^7 $Hz/\sqrt{Hz}$,respectivly.

In the case of completed pumping the amplitude noise seems to have no serious effect on the clock signal except when cycling transition is used for detection, where signal amplitude is proportional to the pumping light intensity. Incident stray light may deteriorate S/N. Because narrow band amplifier is used so only the noise near modulation frequency is concerned.

On the other hand frequency noise will cause important influence on the performance of standard. The detected fluorescence power of clock signal is given by [10]

$$I_{fl} = FPSh\nu_0 \delta_{00} \xi, \quad (2)$$

where F is fluorescence receiving factor, P is mw transition probability, S is atomic beam intensity at detection region B, $h\nu_0$ is the photon energy and ξ is detection efficiency, i. e. average photon yield of an atom having undergone mw transition. Both δ_{00} and ξ are dependent on laser frequency ν through a line shape factor $g(\nu)$, which in our case takes the form

$$g(\nu) = [1 + 4(\frac{\nu-\nu_0}{\Delta\nu})^2]^{-1}, \quad (3)$$

where $\Delta\nu$ is natural line width of the pumping transition, being equal 16 MHz. The instantaneous frequency deviation of a free running diode laser may reach the same order of magnitude as $\Delta\nu$ and will be converted considerably into noise of clock signal. It is insufficient to suppress the frequency noise only near the modulation frequency of 137 Hz, because some nonlinear effects exist. These effects mix different higher Fourier components and produce the beat noise which can pass through narrow band amplifier and reduce the S/N.

Nonlinear effect comes, at first, from $g(\nu)$itself, cf. (3). Secondly, this effect arises from the fact that clock signal detection involves both pumping and detecting process at different time. Considering this we have

$$I_{fl} = FSh\nu_0\delta_{00}^{0}\xi_0 \int P(\upsilon)\int_0^{d/\upsilon} g[\nu(t) - \nu_0]\, g[\nu(t - \frac{l}{\upsilon}) - \nu_0]\, dt\, d\upsilon , \qquad (4)$$

where δ_{00}^{0} and ξ_0 are the maximum values of δ_{00} and ξ, d and l being the length of detection region and the distance between pumping and detection regions, respectively. This convolution gives rise to the nonlinear effect leading to mixing of different noise components. So the contribution of higher Fourier frequency noise must be taken into account. However, we may expect that noise frequency higher than 1/2t, here t is the lifetime of excited state (2×10^{-8} sec) will not affect the S/N because spontaneous emission cannot follow so fast variation of laser frequency. Therefore, it is sufficient for our purpose to suppress the frequency noise up to several tens MHz. The effort is being made in our laboratory.

V. Summary - Some methods to improve the properties of diode lasers are described. The preliminary results of an optically pumped cesium beam frequency standard are reported. It is shown that high frequency noise up to several tens MHz must be suppressed for getting good clock signal-to-noise ratio.

Acknowledgements - This work has been supported in part by the State Natural Science Foundation and the Ministry of Electronic Industry of P. R. China.

Reference

1. M. Arditi, J. L. Picque: J. Phys. Lett. 41, 379 (1980); M. Arditi: Metrologia 18, 59 (1982)
2. L. L. Lewis, M. Feldman: In Proc. 35th Ann. Freq. Control Symp. (1981) p. 612
3. R. E. Drullinger, J. Shirley, D. J. Glaze, L. W. Hollberg, A. DeMarchi: In Proc. 40th Ann. Freq. Control Symp. (1986)
   G. Avila, E. deClercq, M. de Labachellerie, P. Cerez: IEEE Trans. TM-34, 139 (1985)
4. S. Ohshima, Y. Nakadan, Y. Koga: To be published in IEEE Trans IM-37 (1988)
5. S. Shang, X. Wu, S. Yao, Y. Wang, L. Xie: In Proc. of SPIE Vol. 616 (1986) p.202
6. S. Yao, L. Xie: In Proc. of SPIE Vol. 740 (1987) p.8
7. D. Wang, L. Xie, Y. Wang: To be published in Optics Letters, (1988)
8. X. Wu, S. Yao, S. Shang, J. Xue, L. Xie: Chinese Phys. Lett. 2, 9 (1985)
9. Y. Yamamoto, S. Saito, T. Mukai: IEEE J. QE-19, 47 (1983)
10. X. Wu, Y. Xu, K. Tian, S.Yao, L. Xie, Y. Wang: In Proc. of Intern. Symp. on Time and Frequency (Hangzhou, China, 1983); Acta Metrologia Sinica 6 15 (1985)

Experimental Results on an Optically Pumped Cs Frequency Standard

S. Ohshima, Y. Nakadan, T. Ikegami, and Y. Koga

National Research Laboratory of Metrology,
1-1-4, Umezono, Tsukuba, Ibaraki 305, Japan

Abstract - Frequency stability and microwave power dependent frequency shift of an optically pumped Cs frequency standard are reported. The stability was better than $8 \times 10^{-12}/\sqrt{\tau}$ when a fast electrical servo control loop was used for laser diode frequency stabilization. The symmetry of the microwave power dependent frequency shifts in beam reversal is quite good, and frequency uncertainty of correction to the microwave power shift was estimated to be about 2×10^{-14} in case of good experimental condition.

1. Introduction

An optically pumped Cs frequency standard using a laser diode was proposed in 1977 [1], and many experiments were carried out to investigate the feasibility and the performance [2, 3]. At the National Research Laboratory of Metrology (NRLM), a laboratory type optically pumped Cs frequency standard was fabricated [4], and the experiments to investigate its performance is being continued.

Frequency stabilization of laser diodes is important in optically pumped Cs standards. Although it is known that an optical feedback technique is an effective method to stabilize a laser diode frequency [5, 6], we tried to use a somewhat fast electrical feedback method and the result was rather good.

It is expected that the beam reversal symmetry will be improved in the optically pumped standard because the beam trajectory is simple compared with that of conventional type standards. Microwave power shifts in both beam directions were measured to confirm it, and a quite symmetrical characteristic was obtained.

2. System of the Optically Pumped Cs Frequency Standard

The system of the optically pumped Cs frequency standard is shown in Fig. 1. The cavity length is 0.96m which is the same as the NRLM-II [7], and an individual laser was used for pumping and detection respectively.

We adopted a computer controlled single loop system to lock the microwave frequency to Cs resonance frequency, which consist of a 5.000MHz VCXO, a commercial frequency synthesizer and a Gunn diode oscillator. The frequency of the Gunn diode is con-

Frequency Standards and Metrology Editor: A. De Marchi

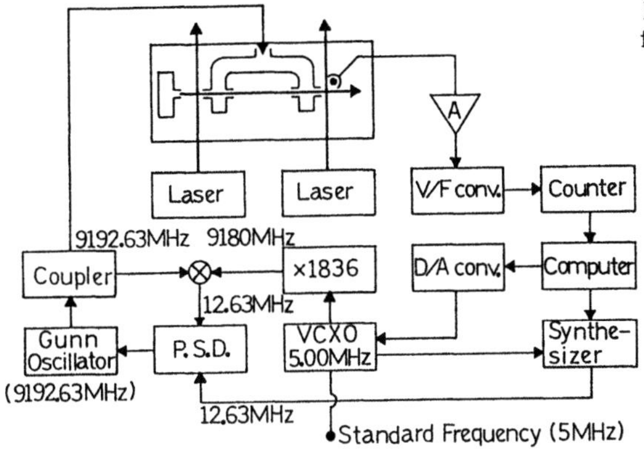

Fig.1 System of the optically pumped Cs frequency standard

tinuously phase locked to the sum of the multiplied frequency (9180MHz) of the VCXO and the frequency of the synthesizer (12.63MHz). Frequency modulation is 0.5Hz square wave in this experiment. This system is capable of selecting the control parameters such as modulation width, modulation frequency, frequency offset, servo loop gain, etc.

3. Optical Pumping Efficiency

Laser frequency is stabilized to a saturated absorption spectrum of a Cs gas cell. Small frequency modulation and phase sensitive detection are used for the stabilization. Both the pumping and detection laser light are σ polarized, and detection laser's frequency is locked to the cycling transition spectrum $6^2S_{1/2}(F=4)-6^2P_{3/2}(F'=5)$.

When the pumping laser's frequency is swept, the fluorescence intensity at the detection region varies like Fig.2. Pumping laser power is about 10mW and its beam cross section is about 2mm×4mm. F=4-F'=3 and F=4-F'=4 spectra are overlapped and the pumping efficiency of F=4-F'=4 is almost 100% because of the strong intensity of the pumping laser. The pumping efficiency of F=4-F'=3 is saturated and about 4% is remained. As this residual atoms produce an excess noise in the detection, we used F=4-F'=4 pumping. We can see the spectrum of F=4-F'=5, but it is not due to the optical pumping but that the atoms are blown off by the laser radiation pressure. We can

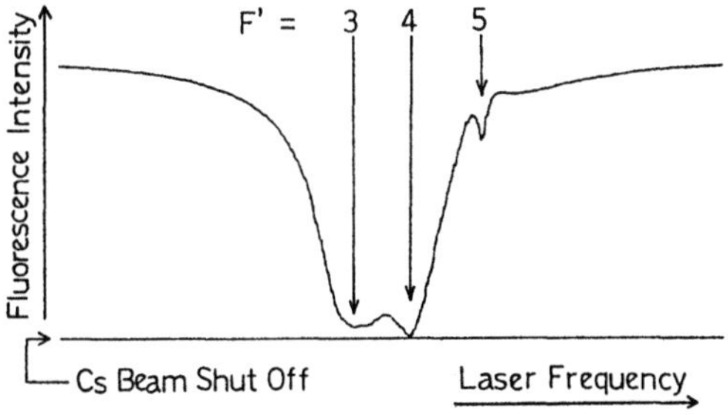

Fig.2 Optical pumping efficiency

also see interesting structure in both side of F=4-F'=5 spectrum, which will be also due to the mechanical effects of laser light.

4. Frequency Stability

In order to reduce the frequency noise of laser diodes, we changed modulation frequency from 2kHz to 100kHz and expanded the band width of control loop from roughly 100Hz to 10kHz. A signal-to-noise ratio of the Ramsey resonance was improved several times and was about $2100/\sqrt{Hz}$. Although the signal-to-noise ratio was a couple of times less than that of the optical feedback stabilization yet, there is room which we can make faster the electrical feedback loop without particular technique.

The frequency stability of the standard was measured by using a commercial atomic clock as reference. The result is shown in Fig. 3 and is represented as $8\times 10^{-12}/\sqrt{\tau}$. The data in longer τ are scattered and it is due to the lack of the number of data. It is considered that the above stability is limited by the reference atomic clock [8].

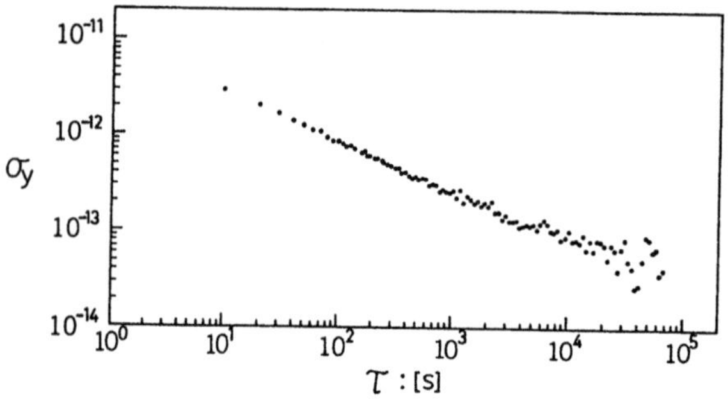

Fig. 3 Frequency stability

5. Beam Reversal Symmetry of Power Dependent Frequency Shifts

Figure 4 shows measured power dependent frequency shifts. The frequency was compared with that of the atomic clock which was used for keeping of UTC(NRLM). We measured in both directions of the Cs beam and of the C-field. Where, for example, WE-SN means that the beam direction is from the west to the east and C-field direction is from the south to the north. The frequency shift due to the Zeeman effect is corrected. Measuring time of every point was three hours, and measurement uncertainty is estimated to be on the order of $\pm 8\times 10^{-14}$ judging from the stability of Fig. 3. The agreement of EW-SN and EW-NS is consistent with that very well. On the other hand, there is a slight discrepancy between WE-SN and WE-NS, and it may be due to that the experimental condition of WE-NS was not good. (The laser diode temperature controller was not in normal condition when we took that data, and C-field might be fluctuated a little.)

The averaged points of WE-SN and EW-SN are exhibited by ×. The points between the relative powers of 0.2 and 1.0 are almost on the linear line, and the standard deviation from the line which is obtained by least square method is 2×10^{-14}. As their total measuring time was amount to 48 hours, that standard deviation is consistent with the extrapolation of Fig. 3. The line is almost horizontal, and it means that the beam

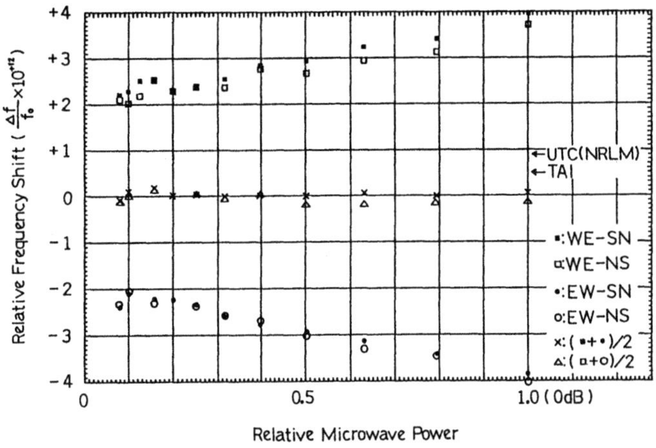

Fig. 4 Microwave power dependent frequency shift

reversal symmetry is quite good. The frequency at the microwave powers of 0 which is obtained by the extrapolation of that line is 8.9×10^{-13} lower than the frequency of UTC(NRLM). When we extrapolate the average of WE-NS and EW-NS (△) to the microwave powers of 0 by the same way, the frequency is 8.8×10^{-13} below the UTC(NRLM) and it is consistent with the former value.

The following problems are remained. Although the frequency at the powers of 0.2 should be higher than the frequency at the powers of 1.0 in the degree of 1×10^{-13} as a result of the second order Doppler effect, we did not confirm it yet. Secondly, the characteristic of the frequency shift is unnatural at the lower microwave power than 0.2, and the reason is not clear yet.

We estimated the frequency difference between UTC(NRLM) and TAI using BIPM circular T and exhibited it in the figure for reference. As that difference was 3.7×10^{-13}, the optically pumped standard frequency is 5.2×10^{-13} below that of TAI. The second order Doppler shift is not corrected yet, but it is difficult to explain the frequency difference between the optically pumped standard and TAI only by the second order Doppler shift and estimation error of the frequency difference of UTC(NRLM) and TAI. There may be other unexpected frequency shifts or systematic errors.

Acknowledgments

The authors would like to thank Mr. T. Inouye for his cooperation to this experiment.

References

1. J.L.Picqué: Metrologia 13, 115 (1977)
2. G.Avila, E.de Clelrcq, M.de Labachellerie, P.Cérez: IEEE Trans. IM-34, 139 (1985)
3. A.Derbyshire, R.E.Drullinger, M.Feldman, D.J.Glaze, D.Hilliard, D.A.Howe, L.L.Lewis, J.H.Shirley, I.Pascaru, D.Stanciulescu: Proc. of 39th Ann. Freq. Control Symp., 18 (1985)

4. S.Ohshima, Y.Nakadan, Y.Koga: IEEE Trans. IM-37, 409 (1988)
5. B.Dahmani, L.Hollberg, R.Drullinger: Opt. Lett. 12, 876 (1987)
6. S.Ohshima, Y.Koga, Y.Nakadan, L.Hollberg, R.Drullinger: Proc. of 2nd Europ. Freq. Time Forum, 531 (1988)
7. Y.Nakadan, Y.Koga: IEEE Trans. IM-34, 133 (1985)
8. S.Ohshima, Y.Nakadan, T.Ikegami, Y.Koga, R.Drullinger, L.Hollberg: submitted to IEEE Trans. Instrum. Meas.

Temperature Measurements of Laser Cooled Cesium Atoms

C. Salomon and J. Dalibard

Laboratoire de Spectroscopie Hertzienne de l'E.N.S.,
24, rue Lhomond, F-75231 Paris Cedex 05, France

Cesium atoms in optical molasses have been laser cooled to a temperature of $70\mu K \pm 40\mu K$. The observed temperature and its variation with laser power and detuning are in strong disagreement with the conventional laser cooling theory. We present the experimental results and we give a tentative explanation for these anomalously low temperatures, which is based on the existence of a light polarization gradient.

INTRODUCTION

While the first laser deceleration and laser cooling of atomic beams have been performed with Sodium atoms [1–3], it is clear that the laser manipulation of Cesium atoms is also of interest for at least two reasons : first, the resonance line at $852nm$ can be easily reached with simple and unexpensive laser diodes [3]. This enables one to perform experiments where 3, 4, 5, ... different lasers are needed without loss of simplicity. Second, the prospect of improving the Cesium time and frequency standard is attractive.

Using the laser cooling technique first proposed by T. Hänsch and A. Schawlow for free atoms, extremely low temperatures have been achieved : $240\mu K$ for Sodium atoms [6] and $120\mu K$ for Cesium atoms [7]. This sample of cold atoms illuminated by three orthogonal standing waves of weak intensity and detuned slightly red from the atomic resonance has been called "optical molasses" because of the very strong velocity damping coefficient [6]. Up to very recently, these experiments were in good agreement with the commonly accepted theory of Doppler cooling. However, the NBS Washington group recently published temperature measurements of Sodium atoms which were nearly one order of magnitude lower than the temperature predicted by the usual Doppler cooling theory for a two level system [8]. Here we will present similar measurements performed with Cesium atoms, which are also in strong disagreement with theory. We first recall the basic predictions of the Doppler cooling theory. We then describe the experiment and we finally give a tentative explanation for the observed anomalously low temperatures.

1) DOPPLER COOLING of a 2-LEVEL SYSTEM

Consider first the case of a 1 dimension experiment along $0z$. The idea of the initial proposal by Hänsch and Schawlow [5] is to illuminate an atom or a collection of atoms by two counterpropagating laser waves along $0z$ detuned slightly below an atomic resonance transition. Because of the Doppler effect, an atom moving along $0z$ with a velocity v_z will preferentially absorb photons from the wave counterpropagating to its motion rather than from the copropagating one. This creates a net radiation pressure force opposed to its motion, which damps its velocity along the laser beam axis.

Frequency Standards and Metrology Editor: A. De Marchi

In the low intensity limit, the total force is the sum of the two radiation pressure forces f_+ and f_- :

$$f_\pm = \pm \hbar k \vec{u}_z \frac{\Gamma}{4} \frac{\omega_1^2}{(\Delta_\mp k v_z)^2 + \Gamma^2/4} \tag{1}$$

where k is the wave vector, $\Delta = \omega_L - \omega_0$ is the laser detuning, Γ the excited state natural linewidth and ω_1 the Rabi frequency of each laser traveling wave. For small velocity atoms $kv_z \ll \Gamma$, $\vec{f} = \vec{f}_+ + \vec{f}_-$ is linear in velocity :

$$\vec{f} = -\alpha v_z \vec{u}_z \tag{2}$$

with

$$\alpha = \hbar k^2 \frac{(-\Delta)\Gamma\omega_1^2}{(\Delta^2 + \Gamma^2/4)^2} \tag{3}$$

Clearly α is a friction coefficient ($\alpha > 0$) if $\Delta < 0$. Note that in the low intensity limit $\omega_1 \ll \Gamma$, α is proportional to the laser power.

In the three dimensional case we apply three pairs of orthogonal standing waves having equal intensity and the same detuning. In the weak intensity limit, eq. (2) is readily generalized to :

$$\vec{f} = -\alpha \vec{v}$$

with α being still given by (3). The stationary temperature of Doppler cooled atoms is the result of a balance between this friction coefficient and the heating mecanism due to the randomness of the absorption and emission processes [9]. This heating is described by a diffusion coefficient :

$$D = (\hbar k)^2 \frac{\Gamma}{2} \frac{\omega_1^2}{\Delta^2 + \Gamma^2/4} \tag{4}$$

The equilibrium temperature is then written as :

$$k_B T = \frac{D}{\alpha} = \hbar \frac{\Gamma}{4} \left| \frac{2\Delta}{\Gamma} + \frac{\Gamma}{2\Delta} \right| \tag{5}$$

The minimum temperature is then reached for $\Delta = -\Gamma/2$

$$k_B T_{min} = \frac{\hbar\Gamma}{2} \tag{6}$$

We note that this temperature is independent of the laser power. As an example, the minimum temperature for Cesium atoms laser cooled on the D_2 line at $852nm$ ($\Gamma^{-1} = 32ns$) is $120\mu K$ for a detuning of $\Delta = -\Gamma/2$.

2) EXPERIMENTAL SET–UP AND TEMPERATURE MEASUREMENTS

The experimental set–up is sketched in Fig. 1. A Cesium atomic beam is slowed by two laser diodes whose frequencies are chirped in order to maintain the atom–laser resonance condition while the atoms are decelerating [3,4,10]. The nearly stopped atoms ($| v_x | \sim | v_y | \sim | v_z | \leq 3m/s$) are then catched into the molasses region formed by the intersection of three orthogonal, $7mm$ diameter laser standing waves. The polarization of these three waves are linear and mutually orthogonal. These standing waves are derived from a frequency

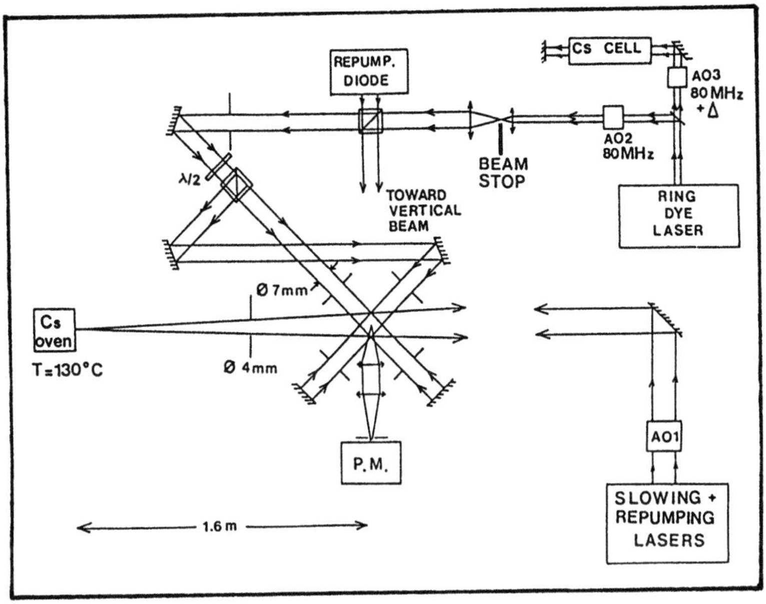

Fig. 1 : *Experimental set–up for the observation of Cs optical molasses. Laser decelerated Cs atoms are injected into the intersection of 3 orthogonal laser standing waves. The fluorescence of atoms inside the molasses is imaged and detected on a photomultiplier.*

stabilized ring dye laser (Coherent 699) whose frequency can be precisely tuned below the Cesium resonance transition $6s_{1/2}, F = 4 \longrightarrow 6p_{3/2}, F = 5$ using acousto–optic modulators. The $A - 0$ 2 modulator acts also as a fast beam shutter. A repumping diode laser beam tuned to the $6s_{1/2}, F = 3 \longrightarrow 6p_{3/2}, F = 3$ transition is mixed to the 3 molasses beams and prevents from optical pumping into the $F = 3$ ground state.

All three laser diodes are only long–term locked to the Cesium line as opposed to Ref. 7 which uses more sophisticated optical feedback techniques. The Earth magnetic field in the molasses region is compensated for by three pairs of Helmholtz coils to a precision of about $10 mGauss$. The information about atomic velocities and molasses storage time is obtained by monitoring the fluorescence light emitted by a well defined volume of atoms inside the molasses region. A typical fluorescence *vs* time sequence is shown in Fig. 2. In the first 0.7 second, the slowing laser diodes are "on" and they fill the molasses volume with slow atoms (48 sweeps of $15ms$). The slowing lasers are then blocked using $A - 0's$ and the slow decrease of the fluorescence ($1/e$ time of $0.4s$) corresponds to atoms slowly escaping from the observation volume. The longest $1/e$ time that we observed with $7mm$ diameter beams was $\tau = 0.7$ second for a detuning $\Delta = -2\Gamma$ and a Rabi frequency $\omega_1 = 0.6\Gamma$ for the strongest transition $| g, F = 4, m_F = 4 > \longrightarrow | e, F = 5, m_F = 5 >$. From the amount of fluorescence, we estimate the density of atoms in the molasses to be $5 \times 10^6 At/cm^3$ for an oven temperature of $127° C$. This density corresponds to about 20 times the unperturbed thermal atomic beam density.

In order to measure the temperature of the cooled atoms, we have used the release and recapture method [6–8]. This consists in switching off the molasses lasers for a given amount of time off (t_{off}), so that the atoms of the molasses move ballistically in the dark and are only experiencing the gravity acceleration. The lasers are then turned on again and we infer the fraction of atoms remaining in the observation volume from the fluorescence emitted at this time (see Fig. 3). This fraction as a function of t_{off} is compared to a numerical calculation for which we assume the initial velocity distribution to be Gaussian. This calculation takes into account the size of the molasses region, the geometry of the detection system and the gravity acceleration. A least square fit adjustement between experimental and numerical results then leads to a determination of the average quadratic velocity. As an example, Fig. 3 presents such a fit for $\Delta = -3\Gamma$, $\omega_1 = 3.4\Gamma$, giving :

$$v_{r.m.s.} = 7cm/s \pm 2cm/s$$

so that the corresponding temperature is :

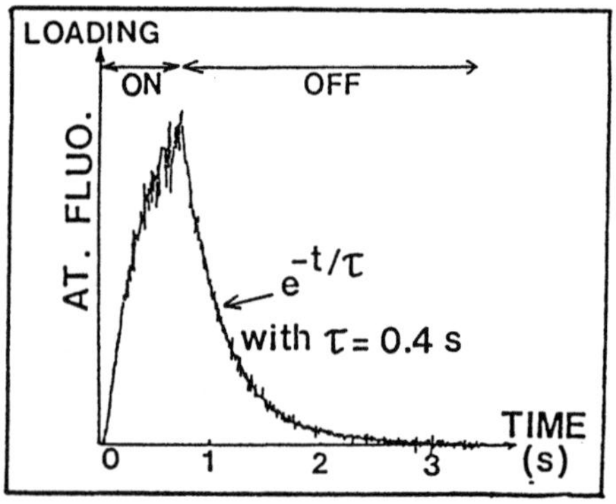

Fig. 2 : Time evolution of the fluorescence of atoms inside the molasses, when the loading lasers are blocked.

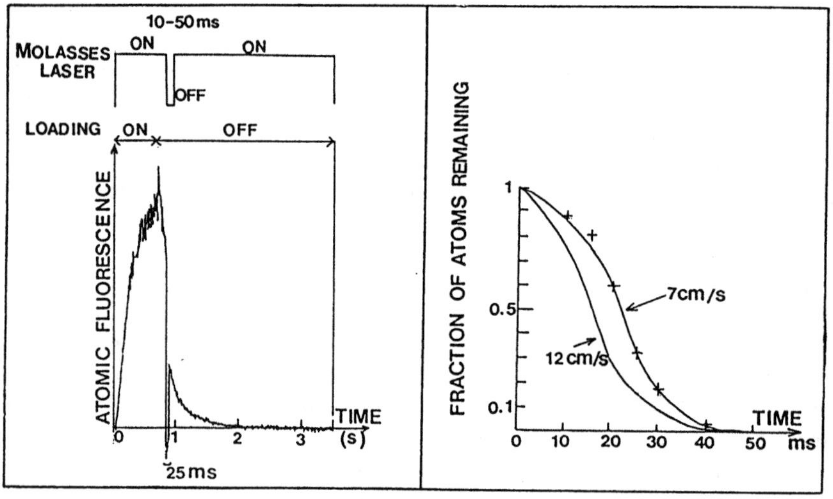

Fig. 3 : Measurement of the average quadratic velocity of atoms inside the molasses.

a. Typical fluorescence signal in a Release & Recapture sequence.

b. Fit of the fraction of atoms remaining after a given time–off ofthe molasses laser, with two theoretical results obtained with average quadratic velocities of 7 cm/s and 12 cm/s.

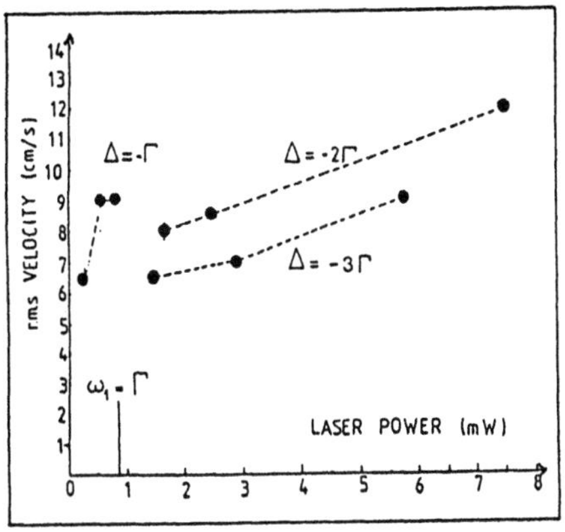

Fig. 4 : Average quadratic velocity measurements for various powers and detunings.

$$T = 70 \pm 40 \mu K$$

This temperature violates eq. 5 by more than a factor 5 at this laser detuning.

The dependence of the measured average quadratic velocity upon laser power and laser detuning is presented in Fig. 4. At a given detuning, the temperature decreases when the laser intensity decreases. At a given laser power, the temperature decreases when increasing the laser detuning. This, again, is in strong disagreement with eq. 5. Finally we note that for our lowest measured velocities around $6cm/s$, the release and recapture method becomes very unprecise. Indeed the fraction of atoms remaining in the molasses region after $10ms$ is mostly determined by the gravity acceleration and no longer by the randomly distributed initial velocities.

TENTATIVE EXPLANATION FOR THESE VERY LOW TEMPERATURES : THE POLARIZATION GRADIENT FORCE

We briefly present here the tentative explanation proposed by our group and also by the Stanford group at the ELICAP conference [11,12].

First we note that, for a 1 dimensional molasses, it has been possible to check the conventional 2–level system Doppler cooling theory [13,14] : at $1D$ one can indeed realize a 2–level system by a judicious choice of laser polarization and optical pumping. On the contrary, in the $3D$ case the Zeeman sublevels cannot be ignored anymore since the light polarization rotates from point to point on the optical wavelength scale. The 2–level approximation is then much too crude to describe the atomic dynamics. For instance, the existence of several Zeeman ground sublevels may lead to pumping times among these levels much longer than the usual Γ^{-1} 2–level relaxation time. This long pumping time, in turn may lead to temperatures much lower than $\hbar\Gamma$. In the mechanism proposed in [11] cooling occurs via non adiabatic transitions between various Zeeman ground state levels. The key point is that these magnetic sublevels are light–shifted differently because of the various oscillator strengths for transitions with the excited states. For an atom at rest and red detuning ($\Delta < 0$), the most light–shifted level is the lowest energy level and the most populated level. We take now a moving atom. If the atomic velocity is not infinitely small, the atom will not follow adiabatically this most light–shifted level. Some "non adiabatic transitions" will occur from the lowest energy state to other states. Such a transition increases the atomic potential energy while decreasing the atomic kinetic energy. After this non adiabatic transition, the atom returns to the most light–shifted (lowest energy) level by optical pumping : atomic potential energy is then converted into field energy. Globally, such a cycle "non adiabatic transition–optical pumping relaxation" leads to a transfer of atomic kinetic energy into field energy. Quantitative treatments of simplified models are presently under way in our laboratory. At present they are in qualitative agreement with the experimental observations.

CONCLUSION

We have produced a cold sample of Cesium atoms with a density of $5 \times 10^6 At/cm^3$ at a temperature of $70\mu K \pm 40\mu K$. The free falling atoms escaping from such molasses when the laser beams are turned off are ideal for observing the $9GHz$ microwave resonance with an unprecedented precision. A reasonable observation time of $100ms$ limited only by gravity would produce a resonance width of about $1.5Hz$. Using optical detection of the microwave transition on a cycling transition one may expect a 10^3 signal to noise ratio as well as an excellent control of systematic effects. Note that a micro gravity experiment would at this point provide even lower linewidths ! We hope that a better understanding of the polarization gradient force will allow us to produce Cesium atoms with a temperature approaching the recoil limit. This limit is about $100nKelvins$ and provides an exciting experimental challenge.

AKNOWLEDGMENTS

We aknowledge very fruitful discussions with A. Aspect, C. Cohen–Tannoudji, E. Arimondo, R. Kaiser, N. Vansteenkiste ; we also thank W.D. Phillips, H. Metcalf and coworkers for numerous and stimulating discussions. This work has been supported in part by D.R.E.T. and Collège de France.

[1] V. Balykin, V.S. Lethokov and V.I. Mushin, JETP Lett. 29, 560 (1979).
[2] J.V. Prodan, W.D. Phillips and H. Metcalf, Phys. Rev. Lett. 48, 596 (1982).
[3] W. Ertmer, R. Blatt, J. Hall and M. Zhu, Phys. Rev. Lett. 54, 996 (1985).
[4] R. Watts and C. Wieman, Opt. Lett. 11, 291 (1986).
[5] T.W. Hänsch and A.W. Schawlow, Opt. Comm. 13, 68 (1975).
[6] S. Chu, L. Hollberg, J.E. Bjorkholm, A. Cable and A. Ashkin, Phys. Rev. Lett. 55, 48 (1985).
[7] D. Sesko, C. Fan and C. Wieman, JOSA B 5, 1225 (1988).
[8] P. Lett, R. Watts, C. Westbrook, W. Phillips, P. Gould and H. Metcalf, Phys. Rev. Lett. 61, 169 (1988).
[9] D. Wineland and W. Itano, Phys. Rev. A20, 1521 (1979).
[10] C. Salomon and J. Dalibard, C. R. Acad. Sci. Paris 306, II 1319 (1988).
[11] J. Dalibard, C. Salomon, A. Aspect, E. Arimondo, R. Kaiser, N. Vansteenkiste and C. Cohen–Tannoudji, Eleventh Int. Conf. on At. Physics, Paris (1988). Ed. S. Haroche, World Scientific (1989).
[12] S. Chu, D.S. Weiss, Y. Shevy, P.J. Ungar, Eleventh Int. Conf. on At. Physics, Paris (1988). Ed. S. Haroche, World Scientific (1989).
[13] R. Kaiser, N. Vansteenkiste, E. Arimondo, A. Aspect, C. Cohen–Tannoudji, M. Karrais and H. Haberland, Eleventh Int. Conf. on At. Physics, Paris, Poster III 4 (1988).
[14] E. de Clercq, C. Thomas, A. Gerard, A. Clairon and J.L. Picqué, Eleventh Int. Conf. on At. Physics, Paris, Poster III 1 (1988).

Part VII

Local Oscillators

Local Oscillators – Introductory Remarks

V.F. Kroupa

Institute of Radio Engineering and Electronics,
Czechoslovak Academy of Sciences, 182 51 Praha, Czechoslovakia

The heart of any stable frequency source of the classic type is a crystal oscillator, the desired properties of which are slow aging rates, low close-to-the-carrier noise (Kroupa, Gagnepain) and insensibility to the environmental changes. These requirements are met with careful manufacture of resonators and proper circuit design (Besson) with modern double-rotated cuts, and by choosing the resonant frequencies in the range of 5 to 10 MHz.

For generation of other frequencies we, nowadays, use frequency synthesis based on a variety of techniques (IC digital circuits, phase-locked-loop, etc.). One particular task in frequency standards and metrology is to link the atomic transition frequency to the fly wheel crystal oscillator with the lowest possible additional noise (Walls).

The complete exploitation of the electromagnetic spectrum at radio and microwave frequencies with the simultaneous progress in generation of coherent visible frequencies do compel researchers to open this vast reservoir of information channels. And here, we witness, as we all believe, the beginning of a new era in communications and frequency synthesis. In the latter case both arts of low noise generators and narrow line lasers shake hands (Hils and Hall, Maleki).

Kroupa, V.F., Frequency stability: Fundamentals and Measurement. IEEE Press, 1983

Gagnepain, J.J., Besson, R., Walls, F., Hils, D. and Hall, J.L., Maleki, L. in this section of the 4th Symposium on Frequency Standards and Metrology.

Frequency Standards and Metrology Editor: A. De Marchi

Stability of Frequency Locked Loops

F.L. Walls

National Institute of Standards and Technology,
Boulder, CO 80303, USA

1. Introduction

Passive frequency standards are characterized by the use of a reference resonance to stabilize the frequency of an external probe oscillator. A common configuration is shown in Fig. 1 [1-6]. The probe oscillator generally has phase modulation imposed on the carrier in order to interrogate the resonance with a minimum of offset. The resulting amplitude modulation is demodulated to yield an error curve that is essentially the derivative of the resonance. Although Fig. 1 shows a transmission system, similar schemes are sometimes possible in reflection [5]. The error signal from the demodulator is used to steer the probe signal toward the center of the resonance line [1-9]. For analysis times longer than one period of the modulation cycle, and under the condition that the probe oscillator wanders less than the half width of the error curve in the loop attack time, we can treat this curve as approximately static[1-4]. Near line center the loop error voltage V_D at the synchronous detector, is approximately $V_D = k(\nu_o^o - \nu_R) + V_n$ where k is the slope of the error curve, ν_R is the resonance frequency of the reference, and ν_o^o is the open loop frequency of the probe oscillator and V_n is the detector noise. If we now close the loop with gain G(f), it can be shown that the spectral density of fractional frequency fluctuations $S_y(f)$ for the probe source becomes

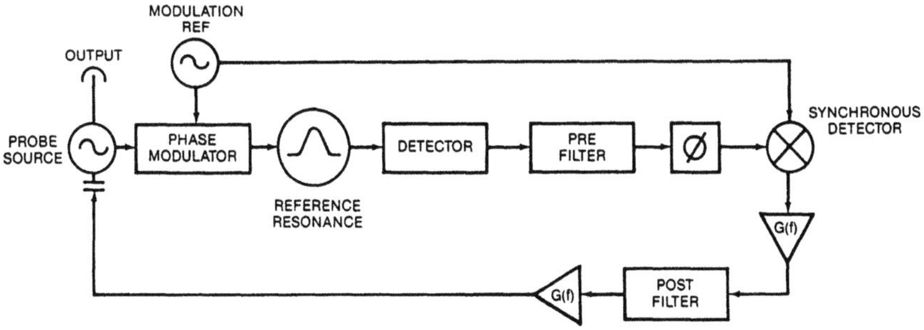

Figure 1. Generalized block diagram of a probe source locked to a reference resonance.

Contribution of the U.S. Goverment, not subject to U.S. copyright.

Frequency Standards and Metrology Editor: A. De Marchi

$$S_y(f) = \left(\frac{G(f)}{G(f)+1}\right)^2 \left(S_y{}^R + \frac{S_y{}^o(f)}{G^2(f)}\right) + S_y{}^N(f) \qquad (1)$$

where $Sy^o(f)$ is the open-loop spectral density of fractional frequency fluctuations of the probe source, $Sy^R(f)$ is that of the reference, and $Sy^N(f)$ is that of the detector and interrogation noise referred to the demodulator output. Cutler [10] has pointed out that noise in the local oscillator at the 2nd harmonic of the modulation frequency, which is usually ignored, causes a time varying frequency offset that is undistinguishable from reference noise. This sets the lower limit to the interrogation noise and often sets the lower limit of the noise performance of the local oscillator necessary not to degrade the overall performance. The magnitude of the 2nd harmonic noise modulation in radians/s is estimated in appendix B of [1] to be $k_3 = 2\pi\nu(S_y(\Omega/\pi))^{1/2}$, where $\Omega/(2\pi)$ is the modulation frequency. This leads to an interrogation noise term which is of order [1,4,10] $S_y{}^N(f) = 1/(16\pi)\ S_y{}^o(\Omega/\pi)$. For large values of G(f), $S_y(f)$ of the probe source reflects that of the reference plus the added noise of the detection system [1-6].

The primary goal of this paper is to investigate the effect of various realistic forms of G(f) on the spectral density of frequency and fractional-frequency stability. It will be shown that mod $\sigma_y(\tau)$ [11,12] is better suited than the traditional two-sample or Allan Variance $\sigma_y(\tau)$ [8], for evaluating the locked performance when the frequency stability of the reference is much greater than that of the local oscillator [7,8,11,12].

2. The Effect of Different Forms of Servo Gain

Figure 2 shows the effect of locking a probe source with $S_y{}^o(f) = 2 \times 10^{-28}/f^2 + 1 \times 10^{-24}/f + 2 \times 10^{-30}f^2$ (which roughly corresponds to that of a low-noise 5 MHz quartz oscillator) to a reference resonance with $S_y{}^R(f)$ of 2×10^{-30} and $Sy^N(f) = 5.6 \times 10^{-28}$. $S_y{}^N(f)$ is the estimated interrogation noise for a modulation frequency of 47 Hz [1,4,10]. Curves A, B, and C show the effect of using a first-, second-, or third- order loop each having a loop bandwidth of approximately 0.1 Hz or an attack time of 1.6 s[1]. The solid Curves A, B, and C of Fig. 3 show $\sigma_y(\tau)$ calculated for curves A, B, and C from Fig. 2 with a noise bandwidth of 3 Hz. The improvement in stability of the local oscillator due to the servo scales roughly as τ/τ_o where τ_o is the attack time and τ is the measurement time. Considerable insight into the effect of various gain stages on the frequency stability can be obtained by considering

$$\frac{\Delta\nu}{\nu_0} \sim \sqrt{\Delta\varphi^2/(2\pi\nu_R\tau)^2} \qquad (2)$$

as a measure of fractional time or frequency stability in the region of measurement times from about 1 to 10^5 s where the effects of low frequency

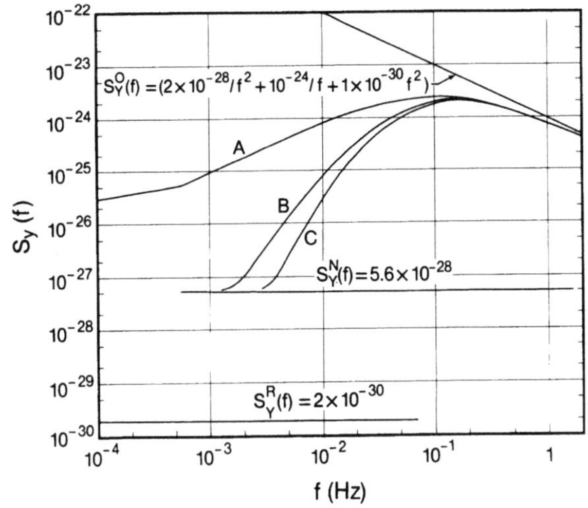

Figure 2. $S_y(f)$ of the probe source locked to a reference resonance. For A $G(f) = (1/(10f)$, for B $G(f) = (1/(10f))(1 + 1/(40f))$, for C $G(f) = 1/(10f))$ $(1 + 1/(40f))(1 + 1/(160f))$.

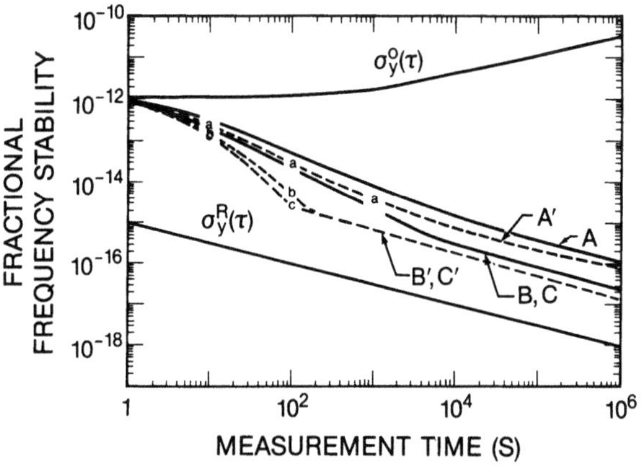

Figure 3. The solid curves show $\sigma_y(\tau)$ and the dashed curves show mod $\sigma_y(\tau)$ versus τ for curves A, B, and C of Fig. 2. The a, b, and c points show estimations of mod $\sigma_y(\tau)$ using Eqs. 2 and 3 with $f_h = f_o$. Also shown is $\sigma_y^o(\tau)$ and $\sigma_y^r(\tau)$.

divergence can often be ignored [7,8]. The squared phase deviation of the zero crossings is given approximately by[8]

$$\Delta\varphi^2 = \nu_R^2 \int_{1/2\pi^2\tau}^{f_h} (S_y\,(f))/f^2\, df \tag{3}$$

for a sample time τ and noise bandwidth f_h. For a noise spectrum which varies as $Sy(f) = K' f^n$ where $n > 2$, the integral is dominated by the high frequency bandwidth even for very long measurement times. The fractional-frequency (or time) stability given by Eq. 2 decreases as τ^{-1} just as does $\sigma_y(\tau)$, due to the increase in measurement time and not to a decrease in the value of $\Delta\varphi^2$. The integral in Eq. 3 can be integrated by parts for the various segments of Sy(f). This makes it easier to evaluate and optimize the performance of the overall system than by using a process based on $\sigma_y(\tau)$.

The contribution of the high frequency noise can be greatly reduced by phase averaging the data points[9]. The data at measurement time $\tau = n\tau_o$ (where τ_o is the data interval) is obtained by averaging the n adjacent phase points. This is equivalent to using mod $\sigma_y(\tau)$ to analyze the data[9,11,12]. The standard expression for mod $\sigma_y(\tau)$ contains an enormous number of terms and is quite laborious to compute[11-12]. Gary [13] has pointed out that mod $\sigma_y(\tau)$ can be reduced to

$$\text{mod } \sigma_y^2(\tau) = \frac{2}{n^4\pi^2\tau_o^2} \int_0^{f_h} \frac{S_y(f)\sin^6(\pi\tau_o nf)}{f^2\sin^2(\pi\tau_o f)}\, df. \tag{4}$$

which is much more manageable. Never the less it still requires numerical calculations to determine which segment of the phase noise dominates the integral. We can estimate mod $\sigma_y(\tau)$ from Eqs. 2 and 3 by using $f_h = f_{ho}/n$, where f_{ho} is the hardware bandwidth of the measurement system and $\tau = n\tau_o$ is the measurement time. The integral for $\Delta\varphi^2$ in Eq. 3 is now substantially reduced for measurement times large enough that f_{ho}/n is less than the bandwidth of the servo system. This is in contrast to the original calculation for Eq. 3 where the integral must always increase with τ. This integral can also be divided into parts and integrated analytically. The fractional frequency stability computed for A, B, and C of Fig. 2 using mod $\sigma_y(\tau)$ are shown as the dashed curves A', B', and C' of Fig. 3. The points labeled a, b, c show the results of estimating mod $\sigma_y(\tau)$ using Eqs. 2 and 3 with $f_h = f_{ho}/n$. The agreement with the dashed curves is very good. The calculations for $\sigma_y(\tau)$ and mod $\sigma_y(\tau)$ are virtually independent of using an upper cutoff frequency for the integration or a simple low pass filter of the same bandwidth. The use of mod $\sigma_y(\tau)$ reduces by a factor of about 100 the time necessary to reach the performance of the reference plus the interrogation noise, which in this case limits the performance for times longer than 100 s.

3. Discussion

The primary utility of these results is the insight into the origin of the major contributions to $\sigma_y(\tau)$ and mod $\sigma_y(\tau)$ as a function of the servo gain and the analysis bandwidth, and the effect of narrowing the bandwidth with longer measurement times. The specific examples illustrate that it is generally necessary to use a second-order loop to lock the probe source to the reference resonance, but that a third-order loop offers little additional improvement when drift in the probe is not serious. In addition, if the reference resonance is substantially more stable than the probe source, it can be very useful to use mod $\sigma_y(\tau)$ to analyze the output of the probe source. We have introduced a simple measure of frequency stability that is easy to use for optimizing the servo and the analysis system. With the techniques introduced here it is possible to realize fractional frequency stabilities and time prediction of the local oscillator which are characteristic of the reference resonance in a way that is orders of magnitude faster than those using more traditional approaches.

4. Acknowledgements

It is a pleasure to acknowledge the many enlightening discussions on this topic with L. S. Cutler, A. DeMarchi, R. E. Drullinger, and D. W. Allan. M. Cline made some of the initial calculations, A. O'Gallagher, J. Gary, and R. Sweet wrote the routines to calculate $\sigma_y(\tau)$ and mod $\sigma_y(\tau)$ from the specified phase noise.

5. References

1. F. L. Walls and S. R. Stein, NBS Tech Note 692, U.S. Government Printing Office, Washington, D.C., SD Cat. # C1346:692
2. C. Audoin and V. Candelier, IEEE Trans. on UFFC, UFFC-34, 573-581 (1987)
3. A. DeMarchi, G. D. Rovera, and A. Premoli, IEEE Trans. on UFFC, UFFC-34, 582-591 (1987)
4. F. W. Walls, IEEE Trans. on UFFC, UFFC-34, 592-597 (1987)
5. S. R. Stein, Proc. of 29th Symposium on Frequency Control, 321-327 (1975)
6. J. Vanier, M. Tetu, and L. G. Berniez, IEEE Trans. on I&M IM-28, 188-193
7. D. W. Allan, Proc. IEEE, 54, 221-230 (1966)
8. J. A. Barnes et al., IEEE Trans. I&M, IM-20, 120 (1971)
9. J. J. Snyder, Appl. Opt., 19, 1223-1225 (1980)
10. L. S. Cutler, Hewlett Packard, Private communication, (Sept. 1988).
11. D. W. Allan and J. A. Barnes, Proc. 35th Annu. Symp. Frequency Control, 470-474 (1981)
12. P. Lesage and T. Ayi, IEEE Trans. I&M, IM-33, 332-336 (1984)
13. J. Gary, National Bureau of Standards, Boulder, CO, private communication.

1/f Noise in Oscillators: Theoretical and Experimental Progresses

J.-J. Gagnepain

Laboratoire de Physique et Métrologie des Oscillateurs
du C.N.R.S. associé à l'Université
de Franche-Comté-Besançon, 32 avenue de l'Observatoire,
F-25000 Besançon, France

1. Introduction

Flicker noise or 1/f noise is still being intensively studied in many physical systems and devices, even after almost 60 years of investigations. Its ubiquitous nature is such that it can be observed in resistors, semiconductors, light, biological systems, etc. But its origin is not so well known and its mechanisms are not really understood.Each year brings at least one new theory, sometimes a controversial one. Basic questions (fluctuations of the number of particles, or of a mobility, or of other parameters ? ... [1]) remain to be answered but a guide to theoretical interpretation is found in correlations between the noise and some specific parameters of the system, when experimentally observed.

Local oscillators also experience 1/f noise, in both the electronic components and the resonators.These different contributions induce phase and frequency fluctuations and in the time domain a corresponding flicker floor. During recent years, a large number of data on resonators or oscillators has become available and it is now possible to show a correlation between the noise levels and the resonator Q-factors, over a wide frequency range. These results, from different authors, will be presented and discussed.

2. Measurement Methods

Noise can be measured by operating a resonator in a passive phase bridge, or in an oscillator. The first method was proposed by F.L. Walls in 1975 [2]. Its advantage is the direct measurement of the resonator with a good rejection of electronic noise ; but, for the best crystals, this rejection requires a pair of identical crystals in order to balance the two arms of the bridge.The second method consists of using the resonator in an oscillator and in evaluating the resonator noise from the measured phase or frequency fluctuations of the output signal.

The power spectrum of a resonator measured with the passive phase bridge method shows 1/f noise and, for the lowest Fourier frequencies, $1/f^2$ noise.The latter was attributed to temperature fluctuations [3]. Recently frequency stability measurements have been performed by F.L. Walls on oscillators placed in a closed environment, with controls of temperature, humidity, and pressure [4]; and it was clearly demonstrated that the $1/f^2$ (frequency random walk) was reduced. It was possible on one oscillator to observe the

Frequency Standards and Metrology Editor: A. De Marchi

flicker floor (after removal of drift) over seven decades of averaging time or of Fouri. frequency.

3. 1/f Noise in Various Resonators and Local Oscillators

Data on 1/f noise are available for various resonators and oscillators including bulk acoustic wave and surface acoustic wave quartz resonators, dielectric resonators, microwave cavities, etc., with resonance frequencies extending over a wide frequency range (1 MHz - 1 GHz).

- Noise in AT and BT quartz resonators was measured by using the passive phase bridge method [2,3,5]. Measurements performed in 1982 are summarized in Fig. 1. These results show a clear correlation between noise level and Q-factor. A $1/Q_o^4$ dependence law was found between the power spectrum of the fractional frequency fluctuations S_y(1Hz) and the unloaded Q-factor Q_o.

To explain this Q dependence, consider the contribution of phonon interactions in the crystal lattice to the noise mechanisms. It is known that the interaction between an acoustic wave and thermal phonons is temperature dependent ; and, for a frequency of 5 MHz, a maximum of phonon interaction is obtained at 20 K (Fig. 2). In addition, a second peak, with even higher intensity, can be observed at 50 K. This second peak is due to the relaxation of interstitial sodium atoms trapped in aluminum defects center.The 1/f noise level of a 5 MHz AT-cut crystal was measured by Planat in 1985 [6] over a temperature range corresponding to the phonon peak, i.e. from 4.2 K to 24 K. The noise level strongly depends on the acoustic attenuation.This dependence, illustrated in Fig. 3, is close to $1/Q_o^4$. Only one measurement of noise at low temperature has been made previously but only at 4.2 K and at 1.5 K [7].

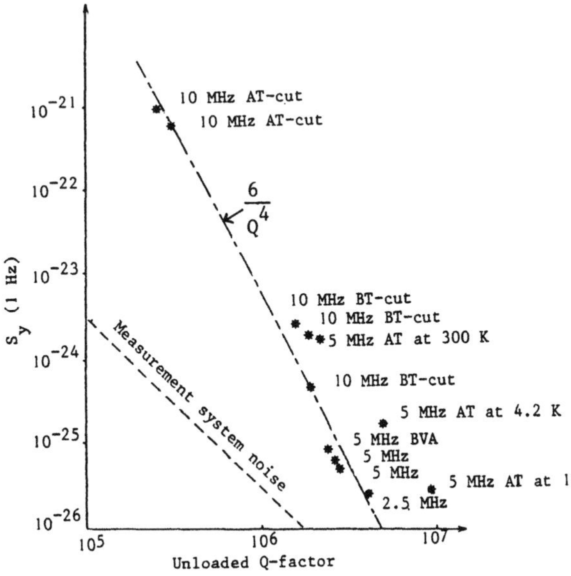

Fig. 1 : 1/f noise level S_y(1 Hz) of AT and BT cut resonators as a function of the unloaded Q-factors (from Gagnepain et al., ref. 5)

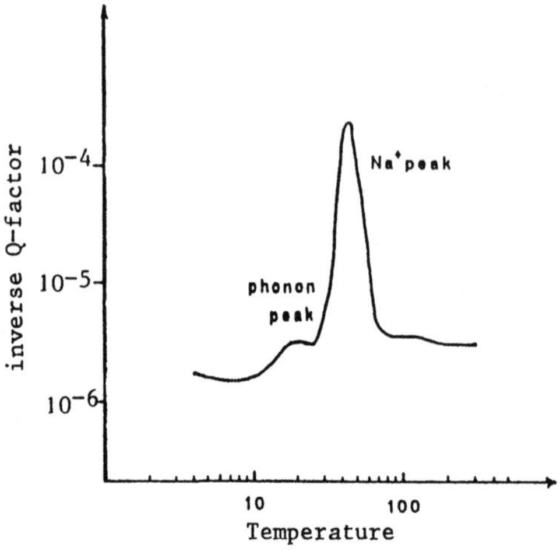

Fig. 2 : Acoustic attenuation (or inverse Q-factor) of a 5 MHz AT-cut crystal as a function of temperature (from Planat & Gagnepain, ref. 6)

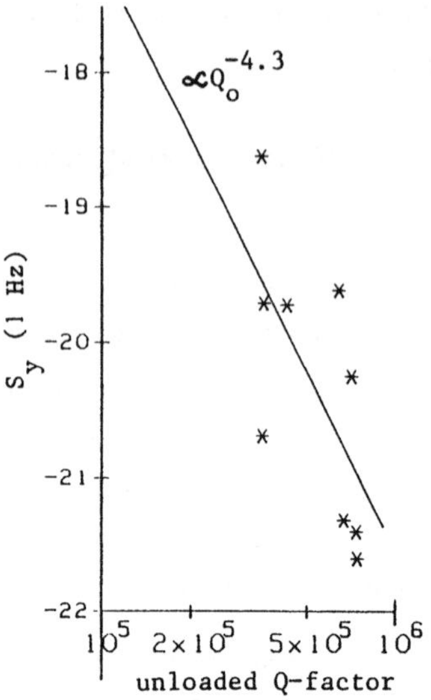

Fig. 3 : 1/f noise level versus the unloaded Q-factor of a 5 MHz AT-cut crystal measure between 4.2 K and 24 K on the phonon peak (from Planat & Gagnepain, ref. 6)

- Measurements of 1/f noise in surface acoustic quartz resonators by Parker [8,9,10] show again a clear dependence of the noise level versus the unloaded Q-factor.But by contrast with the previous results, the $1/Q_0^4$ dependence law applies in that case to the power spectrum of the frequency fluctuations, rather than to the "fractional" frequency fluctuations. Data on resonators at different frequencies from $\simeq$ 300 MHz to $\simeq$ 1 GHz are shown in Fig. 4.

Experimental points corresponding to the previously mentioned BAW and SAW resonators, and some data on microwave oscillators, were put on the same graph (Fig. 5) by Parker [10], and a surprising but clear dependence $S_{\Delta F}(f) \propto 1/Q_0^4$ appeared. The same $S_{\Delta F}(f)$ data points were also plotted by Kroupa as a function of frequency [11].

If the relation between the Q-factor and the frequency

$$Q_o . F_o = k = constant \tag{1}$$

is applied, the $1/Q_0^4$ dependence of $S_{\Delta F}(f)$ should in principle correspond to $1/Q_0^2$ for $S_y(f)$ [10,11]. This is not in agreement with the results presented in Fig. 1 and Fig. 3. For BAW resonators, $S_y(f) \propto 1/Q_0^4$ is supported, in particular, by the data corresponding to AT and BT 10 MHz quartz resonators (for which $k_{AT} \neq k_{BT}$) and by measurements at low temperatures (for which $k = k(T)$). Therefore equation (1) can be used only for resonators having strictly the same material factor k ; this is not so often the case because for many devices acoustic attenuation includes some additional effects. However, it is obvious from Fig. 5 that, over a wide frequency range, the $1/Q_0^4$ dependence is true only for $S_{\Delta F}(f)$ and not for $S_y(f)$. But there is no explanation why the noise of resonators as different as BAW, SAW, dielectric, metal cavities, etc. can follow a same fundamental law.

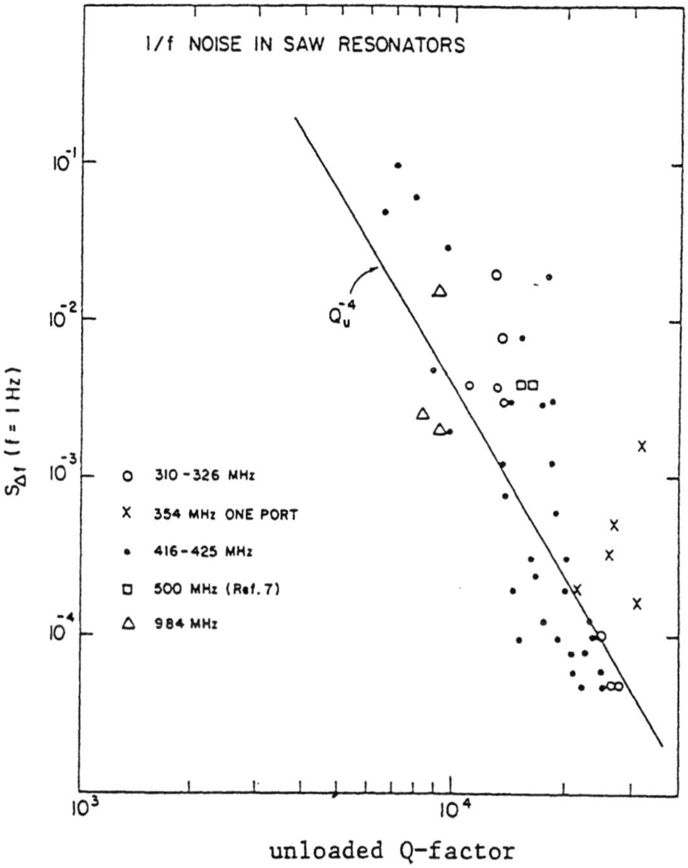

Fig. 4 : 1/f noise level of surface acoustic resonators in the 300 MHz-1 GHz frequency range (from Parker, ref. 9) as a function of the unloaded Q-factors

4. Theoretical Analysis

An interpretation of $S_y(f) \propto Q_o^{-4}$ in acoustic resonators was given in 1981 [7]. Based on the theory of phonons the calculation leads to the relation

$$S_y(f) = \frac{1}{Q_o^4}\left(\frac{E_o}{\Delta E}\right)^2 S_{\delta\tau/\tau}(f) \qquad (2)$$

where E_0 is Young's modulus and $\Delta E = CT\langle\gamma_{eff}^2\rangle$, C being the specific heat, T the absolute temperature and $\langle\gamma_{eff}^2\rangle$ the mean square of the effective Grüneisen constant γ_{eff}. The quantity τ in the subscript is the mean relaxation time of the thermal phonons that is assumed to have fluctuations with a 1/f spectrum. This relation has the expected Q_o^{-4} dependence of the fractional frequency fluctuations, but the assumption of 1/f fluctuations of the phonon lifetime τ has yet to be validated.

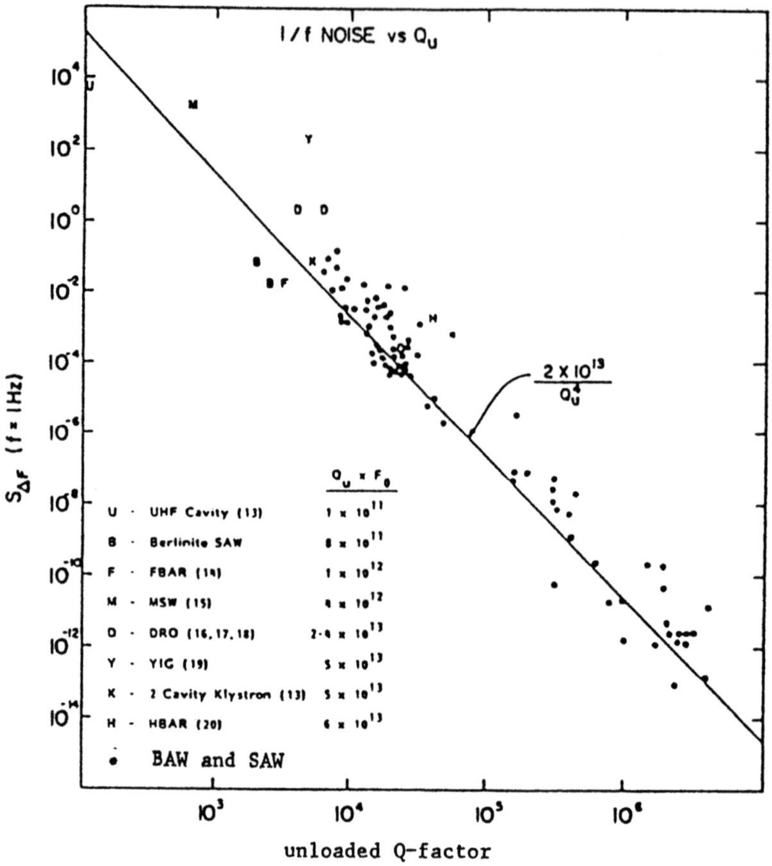

Fig. 5 : 1/f noise data of BAW, SAW, and various microwave resonators and oscillators versus unloaded Q-factors (from Parker, ref. 10)

On the other hand, a new approach to 1/f noise has been proposed by Planat [12]. Considering a dispersive system with wavenumber k and frequency $\omega = \omega(k)$, the general response V(x,t) of the system can be written

$$V(x,t) = \int_k a(k)\, e^{i[\omega(k)t - kx]}\, dk \tag{3}$$

The main contribution in this integral correspond to the value of k for which the phase is stationary, i.e. for $d\omega/dk = \omega'(k) = v_g = x/t$ where v_g is the group velocity.

Using this condition, the integration can be performed for large values of time, i.e. $t \gg 2\pi/\omega(k_o)$. This yields a $t^{-1/2}$ dependence of v(x,t) versus time. Considering a series of random impulses with white spectrum at the system input, it is then shown that for large observation times the response power spectrum is

$$S_V(f) \simeq \frac{v\,\pi^2 a^2(k_o)}{2\,|\omega''(k_o)|}\;\frac{1}{\omega(k_o) - 2\pi f} \tag{4}$$

corresponding to a 1/f spectrum. (v is the mean frequency of the random impulses, $\omega''(k_0)$ is the 2nd derivative of ω).

The use of dispersion to explain the transformation of white noise into 1/f noise has been implicitely included in more specific models (Musha [13], Keshner [14]). Planat in his theory extends this concept to a variety of physical systems and, in particular to plane waves. This approach is important because dispersion is a property of all the physical systems, at some degree, as 1/f noise is.

5. Conclusion

Results published on resonators and oscillators during these last years give new insights into 1/f noise, even though they are sometimes controversial and inconsistent. The fact that the noise level depends on the unloaded Q-factor is demonstrated without any doubt, but the observed dependence raises many unanswered questions. For example : is the $1/Q_o^{-4}$ law valid for frequency or for fractional frequency fluctuations ? Data plotted over a wide frequency range for resonators of different types are clearly in favor of frequency fluctuations. But why is the dependence law the same for different types of resonators ? Data measured at low temperatures or with crystals at the same frequency (10 MHz) but with different material constants (AT and BT crystals), show a $1/Q_o^4$ dependence for the fractional frequency fluctuations. This is consistent with (2), but not with relation (1). However, this relation must be used with caution since for many resonators the QoF product is not a constant.

A complete theory of the phenomenon does not yet exist. The phonon interaction model links frequency and Q but assumes, as yet unproven, 1/f fluctuations. The dispersion model has the great advantage of being very general and leading to 1/f spectrum, but a connection with frequency and Q remains still to be rigorously established.

Similar comments can be made about 1/f noise in other fields of physics. There also, refinement of both theory and experiment will continue in the future.

References

1. F.N. Hooge : "Comparison of 1/f noise theories and experiments", Noise in Physical Systems and 1/f noise 1985. Elsevier Sci. Publ. 27-33 (1986)
2. F.L. Walls, A.E. Wainwright : "Measurement of the short term stability of quartz crystal resonators and the implications for crystal oscillator design and applications", IEEE Trans. Inst. Meas., IM-24, n° 1, 15 (1975)
3. J.J. Gagnepain : "Fundamental noise studies of quartz crystal resonators", 30th Ann. Symp. on Freq. Cont., Atlantic-City (1976)
4. F.L. Walls : "The influence of pressure and humidity on the medium and long-term frequency stability of quartz oscillators", 42nd Ann. Symp. on Freq. Cont., Baltimore (1988)
5. J.J. Gagnepain, M. Olivier, F.L. Walls : "Excess noise in quartz crystal resonators", 37th Ann. Symp. on Freq.Cont., Philadelphia (1983)

6. M. Planat, J.J. Gagnepain : "Temperature dependence of 1/f noise in quartz resonators in relation with acoustic attenuation", Noise in Physical Systems and 1/f noise 1985. Elsevier Science Publi. B.U. (1986)
7. J.J. Gagnepain, J. Uebersfeld, G. Goujon, P. Haendel : "Relation between 1/f noise and Q-factor in quartz resonators at room and low temperatures.First theoretical interpretation", 35th Ann. Freq. Cont. Symp., Philadelphia (1981)
8. T.E. Parker : "1/f phase noise in quartz SAW devices", Electronics Letters, vol. 15, n° 10, (1979)
9. T.E. Parker : "1/f frequency fluctuations in quartz acoustic resonators", Appl. Phys. Lett. 46, n° 3 (1985)
10. T.E. Parker : "1/f frequency fluctuations in acoustic and other stable oscillators", 39th Ann. Freq. Cont. Symp. (1985)
11. V.F. Kroupa : "The state of the art of the flicker frequency noise in BAW and SAW quartz resonators", IEEE Trans. Inst. Meas. 35, n° 3 (1988)
12. M. Planat : "A dispersive wave approach for 1/f noise", 9th Conference on Noise in Physical Systems, World Scientific (1987)
13. T. Musha : "1/f resonance frequency fluctuations of a quartz crystal", 29th Ann. Freq. Cont. Symp. (1975)
14. M.S. Keshner : "1/f noise", Proc. IEEE, vol. 70, n° 3 (1982).

RF Local Oscillators for Future Frequency Standards

R.J. Besson

Ecole Nationale Supérieure de Mécanique et des Microtechniques,
Route de Gray, La Bouloie, F-25030 Besançon Cedex, France

1. Introduction

For some years, there has been a need for much better oscillators in frequency standards applications. Recent progress in resonator and oscillator understanding and design leads to overcome most previous limitations. As a result, a single crystal oscillator can now provide, at the same time, excellent short term stability and excellent long term drift. Fast warm-up can be obtained together with other excellent characteristics. Also impressive (and necessary) is the achievement of very small g sensitivity (some 10^{-11}/g).

In this paper, a presentation given at the Fourth Symposium on Frequency Standards and Metrology in Ancona is summarized. In particular recent progress of resonators and new advances of quartz oscillators devoted to atomic standards are discussed. Late improvements of resonators are largely due either to better theoretical understanding or to improvements of boundary conditions. Both lead to new concepts and new designs. Quartz oscillators electronics have also been improved and environmental conditions influence greatly reduced.

In conclusion, the most recent advances are be tabulated and indications are given on future possible trends.

2. General Presentation of Quartz Oscillators and Resonators

R.F. local oscillators finally identify to quartz oscillators since no other candidate can really compete to date. Quartz oscillators are present in almost any frequency control equipment so being the real workhorses of time and frequency systems, including atomic clocks. In fact, an atomic clock is a Xtal oscillator whose frequency is locked to an atomic reference resonator through appropriate circuitry.

Frequency Standards and Metrology Editor: A. De Marchi

Principles of quartz oscillators are rapidly discussed as a particular case of resonant circuit oscillators (whose resonator simply has a very large Q factor). Of importance is the fact that it is now possible to separate influences on stability due to crystal and to electronic circuitry through passive tests of resonators /1/.

After a brief historical review on piezoelectricity and its applications, quartz resonators are introduced as cavities for acoustic waves. In fact, piezoelectric resonators appear as solids (plates, bars, ...) of given configuration shape and dimensions prepared from high quality crystal under precise control of orientation. Geometric shape is fundamental since frequency depends on dimensions of the cavity. Principles are largely discussed stressing the influence of boundary conditions. It is shown that more effort should be devoted to the resonant cavity with probable success.

3. Recent Progress of Resonators

Resonator is the heart of piezoelectric oscillator. In particular long term drift, g and vibrational sensitivities of a well designed oscillator only depend on resonator. Tremendous improvements recently obtained are largely due to efforts in several directions /2/ :

a) Resonator theoretical understanding together with experimental verifications

A resonator is now analysed from a 3 dimentional point of view eventually including non linear aspects. As a consequence, theoretical models are much more realistic and yield almost any characteristic of a resonator, given its design parameters. Several examples are given including solutions to the problem of coupling between modes (this coupling may cause so said "activity dips" which prevent normal use of resonator at given temperatures). Very fundamental choices in the theory may also be important. For instance, the choice of so called "effective elastic coefficients" will automatically take into account the fact that a quartz sphere expands into an ellipsoïd. Similarly, a new three-dimensional model for heat transfer in quartz resonators may now be used which largely overcomes the old model of the infinite plane plate. It is also rapidly shown that experimental verifications feed back theory yielding more and more realistic models.

b) Boundary conditions

Much more realistic boundary conditions can now be introduced in theoretical models. Also, much better achievement of surface limits, fixation and mounting structures are now obtained. This point is largely emphasized using the example of computerized complex ultrasonic machining systems or the example of chemical etching (both may lead to monolithic structures).

c) Improvement in material's quality (still needed !)

Recent needs in hardened materials have strongly boosted this research which in principle may now come up with ultra high purity, (almost) dislocation free quartz material. However, better reproductibility is still needed in industrial supplies and nobody really knows exactly what ultimate quartz material quality could be obtained by appropriate effort /3/.

d) Non linear phenomena

Harmonic generation, amplitude frequency effects and intermodulation correspond to propagation of a finite amplitude wave in a non linear medium. Sensitivities to external or internal perturbations (temperature, force, pressure, acceleration, ...) are generally due to the coupling between the high frequency wave and the non linear strained medium, they also deserve a non linear treatment. Several examples are given.

e) Evolution of resonator design

Some new techniques like the use of doubly rotated cuts of crystals have recently been introduced, corresponding improvements including lower thermal transients and better g sensitivity are rapidly discussed. Several new resonators like ceramic flat pack /4/ or BVA resonators /5/ recently introduced are rapidly presented. BVA oscillators seem to be successfully used in connection with atomic references. BVA SC cut resonators recently designed allow a $5\ 10^{-11}$/g sensitivity together with a $5\ 10^{-11}$/bar pressure sensitivity, an aging rate of less than 10^{-9}/year, excellent $\sigma_y(\tau)$ and very low retrace.

4. Recent Progress of Quartz Oscillator

In this section only precision quartz controlled oscillators (i.e. oscillators devoted to atomic standards) will be considered. Improvements recently obtained are largely due to efforts in several directions :

a) The quartz crystal can be checked in passive test beds separetely from the oscillator electronics (which may in certain cases also be tested separately).

b) New understanding of the limiting mechanisms have been proposed and subsequently new electronic circuits have been introduced with better short-term and medium-term frequency stabilities. This includes improved match between resonator and electronic circuitry due to introduction by several authors of computer aided simulation of the oscillator circuit.

c) Long-term or medium-term frequency changes due to pressure and humidity (basically through electronic circuitry) is now much better understood **/6/**.

d) The influence of electronic circuitry in the phase loop shift condition (active devices, voltage controlled capacitor, load impedance, connection and cables capacitances ...) and the influence of oven controls are much better understood.

It is clear, however, that some more effort is necessary in aging processes and noise processes understanding.

5. Future Possible Trend and Conclusions

Last two decades have yielded impressive progress in resonator and oscillator field as well. Reduced aging better short term stability, low thermal transient and small environmental dependance have been obtained. This leads to interesting advantages in atomic standards design. Following table compares performances obtained 10 years ago with performances available to day with commercial precision units. From recent laboratory results possible near future trends are foreseen.

	10 years ago	now	less than 10 years from now
$\sigma_y(\tau)$ 1s, 10s	10^{-12}	2 to 4 10^{-13}	5 to 10 x 10^{-14}
Aging/year	2 to 3 10^{-8}	3 to 6 10^{-9}	5 x 10^{-10} predictable
g sens. max	2 10^{-9}/g	3 10^{-10}/g	4 to 7 10^{-11}/g (SC cut Xtals)

References

1. F.L. Walls, A.E. Wainwright, I.E.E.E. Trans. Instrum. Meas., IM-24, 15-20 (1975)

2. R.J. Besson, 1rst European Frequency and Time Forum, 20-30 (1987)

3. R.A. Laudise, R.L. Barns, I.E.E.E. Trans. on Ultrasonics, Ferroel. and Freq. Contr., Vol. 35, N° 3, 227-287 (1988)

4. J.R. Vig et al, Proc. 29th Ann. Freq. Contr. Symp., 220-229 (1975)

5. R.J. Besson, Proc. 31rst Ann. Freq. Cont. Symp., 147-152 (1977)

6. F.L. Walls, Proc. 42nd Ann. Symp. on Freq. Contr. (1988).

Ultra-Stable Cavity-Stabilized Lasers with Subhertz Linewidth

*D. Hils and J.L. Hall**

Joint Institute for Laboratory Astrophysics,
National Institute of Standards and Technology,
Boulder, CO 80303, USA, and
University of Colorado, Boulder CO 80309, USA

Many of the most interesting new applications for optical sources will depend essentially on the well-advertised properties of the laser, namely coherence and short wavelength relative to laboratory dimensions. Attractive measurement possibilities include improved tests of cherished principles in physics such as Lorentz invariance [1] and the absence of preferred-frame or "one-way" effects in the propagation of light [2]. The implications of any observed non-null result would be so profound that it is essential to explore each new avenue when the advance of technology allows a possibility of dramatic improvement.

A second class of applications for stable optical lasers envisions them as the coherent source for antennas designed to be sensitive to weak gravitational radiation [3]. One cannot seriously doubt the existence of such radiation, particularly in view of the remarkable agreement of the change of period of the neutron star binary PSR1913+16, serving as an orbital clock, with the prediction of General Relativity [4]. Still, the gravitational waves, due to presently known sources, produce at the earth a weak enough signal that only with the highest degree of invention and diligence will detection be possible.

The third area of application for ultrastable lasers is as the "local oscillator" for optical frequency standards based on the absorption spectrum of a single laser-cooled ion in an ion trap or a million atoms in some "atomic fountain" prepared by laser manipulation of the trajectories of essentially free atoms.

We begin our discussion of ultrastable lasers by remarking that the usual and real lasers which we can buy and use in the lab do not in fact emit coherent light. The critical issue here is <u>phase stability</u>, required for any meaningful claim of coherence. We use lasers to allow us to excite transitions in the optical domain, since the higher frequency can lead to a higher line Q. However it also immediately brings us the expectation of phase <u>in</u>coherence within the few millisecond domain due to vibrations imposed on the resonator structure. Our position is that many of the most interesting physical measurements using lasers will only be possible if the full potential coherence time is available -- limited only by inescapable fundamental processes. Thus it is interesting work to identify and eliminate non-essential noise sources so that one may begin to appreciate what the real limits may be. This report discusses development of techniques which have allowed demonstration of laser coherence times of 10 seconds with corresponding fractional frequency instability in the $\approx 10^{-17}$ domain. Further, the way is apparently still open for significant additional progress.

The first question is whether, for high-stability experiments, one must use a laser with special properties, for example massive invar construction. The answer from both theory and experiment is no, absolutely no. What is required is a laser source with a smoothly evolving phase and a frequency-control access with speed appropriate for the time-scale and level of instability present in the laser. It is not particularly important whether the laser has rather good intrinsic stability (for example, a low power hard-sealed HeNe laser in a hermetically-sealed environment, $\tau_c \lesssim 1$ ms) or is instead a rather rough source (for example, a cw visible dye laser using a liquid jet stream inside its resonator, $\tau_c \sim 1$ µs) or an exceptionally fast-changing optical source (such as a laser diode, $\tau_c \approx$ ns). Here τ_c is the mean time for the phase to diffuse by one radian. The only consequence of the choice of

*Staff Member, Quantum Physics Division, National Institute of Standards and Technology.

Frequency Standards and Metrology Editor: A. De Marchi

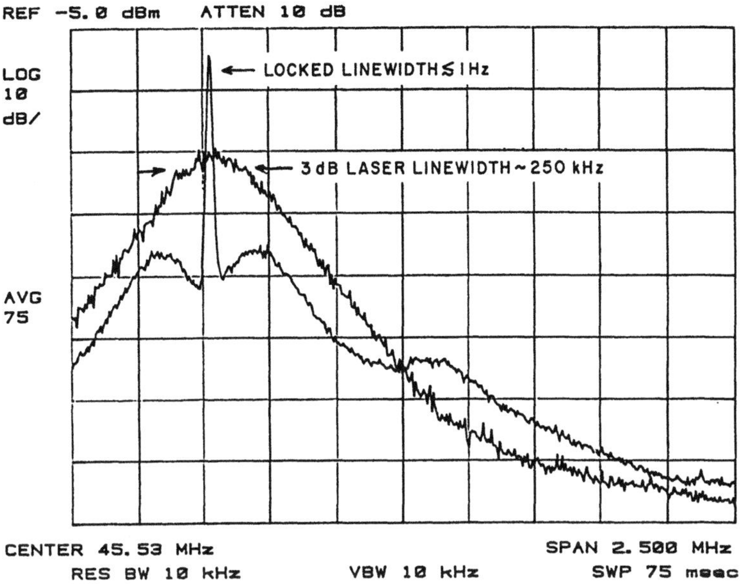

Fig. 1. Phase lock of commercial dye laser to HeNe reference laser by use of external phase/frequency transducer. Approximately 95% of the output is contained in the sharp line of subHertz linewidth.

a fast-evolving source is that the control-loop design and "actuator" must be of significantly greater bandwidth and lower time delay relative to the intrinsic phase-changing process so that the loop may guide the emission phase to be well within 1 radian of the desired sinusoidal pattern. Up to the present time we have been able to phase-lock a semiconductor laser only by operating it within an external resonator. The other two sources may be routinely controlled by phase-lock techniques.

As an example, in Fig. 1 we show the spectrum of a commercial dye laser subject to phase lock. The control is based on heterodyne detection against a reference HeNe laser which can be stabilized on the I_2 line at 612 nm. The figure shows that approximately 95% of the emitted power is captured in the phase-locked carrier. An interesting and attractive aspect of this experiment is that, while the error signal source (ECL phase/frequency-lock chip) is conventional, the "actuator" for the servo control is definitely not conventional. It is in fact located totally outside the cavity of the commercial ring dye laser and does not really influence its intrinsic performance.

With the aid of Fig. 2, we can indicate the functioning of this external stabilizer [5], the so-called "laser-man's magic box". In this figure we show the configuration for locking the laser onto a reference cavity, which provides the phase-reference field by "leakage" of the stored intracavity field [6]. When a laser phase error occurs relative to the guiding or reference optical sinewave, a fast error amplifier reduces this to near zero by means of the external electro-optic phase modulator. By inclusion of cascaded integrators in this loop we can provide a very tight lock indeed, with ≈zero phase "position" and phase "velocity" errors. If the input error were a frequency error instead, this fast loop would need to suppress a constant rate of change of phase and the servo would accordingly need to provide a constant rate of rise of the voltage supplied to the EO phase modulator. Eventually saturation would be a problem. Since the time response of any direct piezoelectric mechanical control of the laser's frequency is in the 10 or 100 μs domain -- far too slow to avoid saturation of the EO amplifier -- it is necessary to provide also a rather fast external frequency-shifting element. We use an acousto-optic modulator which provides an optical output frequency equal to the sum (or difference) of the input optical and radio frequencies. This AO "trans-

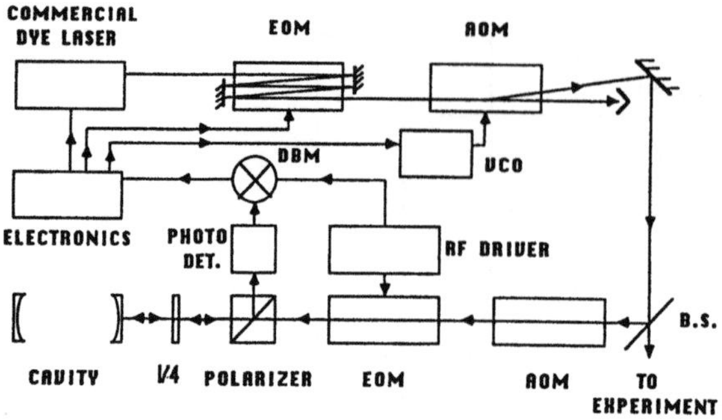

Fig. 2. Schematic view of External Frequency Stabilization. Using an Acousto-Optic Modulator to shift the laser beam's frequency (after the laser has produced it!) gives much quicker control than is possible with the laser electronics alone. Similarly the Electro Optic Phase Modulator allows still faster response: the frequency servo is locked so tightly that no phase error as large as a radian occurs until at some long time (>100 s) where the S/N is finally insufficient.

ducer" needs to be fed by a variable rf frequency which will be guided by the servo system to minimize the voltage requirements of the EO phase modulator driver (see Fig. 2). A problem for the servo design arises from the 300-500 μs delay of response of the AO while the fresh frequency change information is propagating as a sound wave from the rf/compression-wave transducer to the optical interaction region. As explained in Ref. [5], it is useful to introduce in the EO electrical loop a fast parallel channel: a delay-line driven by an inverting amplifier. By this delayed destructive interference, effectively the error input information to the EO actuator is reduced to zero just as the fresh frequency-correction signal reaches the interaction point in the AO. Thus at longer times the full responsibility for correction of an assumed step input error has now been passed over to the AO loop. A subsidiary feedback path slowly recenters the voltage on the EO amplifier so that it will be ready for the next challenge. Similarly, the long duration presence of a correction in the frequency-controlled oscillator loop gives indication that the laser's basic output frequency is not quite right and a correction via the laser's piezo (and ultimately, galvo-tuned Brewster-plate tuner) is undertaken to restore the AO to its nominal operating frequency. With the usual parameters of the commercially stabilized dye lasers this correction is rapid enough that the AO does not need to be double-passed to suppress the angular deviations implicit in tuning the frequency. Thus the optical input/output power efficiency can be >75%.

This external «Magic Box» technique of laser stabilization is interesting because it orthogonalizes the tasks of designing and producing a tunable and powerful laser source from the task of providing the desired level of frequency control. We can also strongly suppress laser amplitude noise by using a variable magnitude of rf drive to control the effective Bragg reflection efficiency of the AO modulator [7]. With thoughtful design one can reach amplitude noise levels of ~10 ppm (in 1 kHz BW) and presumably even better results for longer integrations, although other complications such as spatial inhomogeneity of the noise begin to intervene.

So we have described a successful approach to the "actuator" part of our general laser frequency (and intensity) control system and the focus now appropriately turns to consideration about the source of the frequency-error control information.

For the best frequency stability, it might be initially supposed desirable to lock the laser immediately onto a suitable atomic molecular reference transition. This is not correct, in our opinion, for a number of reasons, and we instead advocate locking the laser onto a high finesse passive interferometer. Now the stability of this physical system does matter. In comparing a passive, stable interferometer to atoms/molecules for this locking purpose, the interferometer approach is vastly superior [8]:

1) The resonance system is linear so that the signal/noise ratio may always be improved by using more power, in contrast to the power broadening of quantum resonances.

2) The resonance signal contrast of an appropriately designed resonator is very high, for example a fringe contrast of 25% or more is easy to obtain in reflection.

3) The interferometer offers a vast number of possible lock points separated by convenient rf intervals ($c/2L \sim 0.1$ to 5 GHz) covering a significant fractional frequency width, about ±10% [9]. These rf intervals can be measured very precisely [10].

4) The linewidth of the cavity resonances can be very sharp, easily less than 100 kHz; mirrors to give < 10 kHz linewidths are commercially produced.

5) The overwhelming disadvantage of a reference cavity as opposed to a reference quantum absorption is that the cavity spacer has no special length: each spacer is different one from another. It even is different from itself after some time due to thermal expansion, body-forces which produce distortions, and long-time creep. Environmental factors could be impossibly serious.

It is our assertion that these environmental problems in fact can be vastly reduced by thoughtful design: in the following we report cavity-stabilized laser oscillators at 5×10^{14} Hz, of dramatically subHertz linewidth, with a frequency drift below 1/4 Hz/s, with a frequency predictability of ±0.1 Hz for several minutes and ±100 Hz for a day [11]. A flicker floor below 10^{-17} was observed for $100 < \tau < 1000$ s.

The projected performance based on shot-noise limitations have been considered in a recent publication [12]. An operational summary is that at present shot noise does not represent a serious limit for the laser stabilization industry, with projected linewidths in the milliHertz domain or below. Still, it is useful to discuss the physical basis for this estimate. There are two distinct ideas in this calculation of the effective laser coherence time (or linewidth): discriminator slope [13] and signal-to-noise ratio leading to a noise spectral density Δ in Hertz/$\sqrt{\text{Hz}}$ and, secondly, conversion of the noise density into a linewidth.

We suppose the reference cavity will use mirrors fabricated with contemporary coating technology. For example, mirrors to give finesse, $\mathcal{F} = 10{,}000$ and resonant efficiency $D = 10\%$ are available from several suppliers. With an interferometer length of 30 cm, the free spectral range would be 500 MHz and the resonance linewidth FWHM would be $2\Delta\nu_c = 50$ kHz. The resonant dip contrast viewed in reflection would be about 50% if the cavity were excited by a single frequency. In fact, one will want to use phase modulation (ϕM) sidebands to allow detection of the resonance information as an rf signal, thereby allowing signal processing at a frequency high enough that the detection process is no longer troubled by laser amplitude modulation noise of a technical origin [14]. For this purpose a phase modulation index β of about 1 radian would be chosen, leading to a spectral distribution of carrier power $[J_0(\beta)]^2 = 59\%$, and each $[J_1(\beta)]^2$ first order sideband of 19%, with the second order sidebands of about 1.5% each. We suppose the cavity resolves the sideband structure. The amplitude of the "leakage" of the cavity buildup field toward the detector will have the form

$$E_{\text{leakage}} = \sqrt{D} \cdot \frac{J_0(\beta)}{1-j\left(\frac{\nu-\nu_c}{\Delta\nu}\right)} = J_0(\beta)\sqrt{D}\left[\frac{1}{1+\left(\frac{\nu-\nu_c}{\Delta\nu}\right)^2} + \frac{j\,\frac{\nu-\nu_c}{\Delta\nu}}{1+\left(\frac{\nu-\nu_c}{\Delta\nu}\right)^2}\right]$$

keeping only the desired term with the carrier frequency, ν, nearly matching the cavity resonance frequency, ν_c. The optical intensity presented to the detector is

$$I = (-RJ_0(\beta) + \text{Re}\,E_{\text{leakage}})^2 + 2R^2J_1^2(\beta) + 2R^2J_2^2(\beta)$$

$$+\ 2R\ \text{Im}(E_{\text{leakage}}) \times J_1(\beta) \times 2\sin\omega t$$

where the first line represents the dc light level which results in shot noise. The $\sin\omega t$ term will be amplified and phase-sensitively detected relative to the modulation sinusoid at ω, and thus represents the discriminator signal. For a laser power P, a photodetector quantum yield η, and a detector bandwidth B, we obtain a signal to shot noise ratio of

$$S/N \simeq \frac{P\eta\, J_0(\beta)\, J_1(\beta)\, \sqrt{D}\, \left(\frac{\nu-\nu_c}{\Delta\nu}\right) \times 4}{\left[\left(1-J_0^2\,(2\sqrt{D}-D)\right) P\eta e 2B\right]^{1/2}} \cdot \frac{1}{\sqrt{2}}$$

where the $\sqrt{2}$ loss of S/N arises from shot noise in down-mixed image sidebands.

If the servo loop is very tightly locked to the center of the resonance as we have supposed, the finite shot noise of the discriminator when input at the error point will induce a frequency error sufficient to cancel this noise with the voltage produced by the discriminator output. This leads to the following rms frequency excursion under feedback:

$$\delta\nu_{rms} = \Delta\nu_c \frac{\left(1-J_o^2(2\sqrt{\bar{D}}-D)\right)^{1/2}}{\sqrt{D}} \frac{\sqrt{4P\eta eB}}{4P\eta\ J_0 J_1}$$

and the corresponding frequency noise density

$$\Delta^2 \equiv \frac{(\delta\nu_{rms})^2}{B} = (\Delta\nu_c)^2 \cdot \frac{\left(1-J_o^2(2\sqrt{\bar{D}}-D)\right)}{D} \frac{e}{4J_0^2\ J_1^2\ P\eta}$$

For the numerical values $P = 0.2$ mW, $\eta = 0.5$ mA/mW, $D = 0.1$, $2\Delta\nu = 50$ kHz, $\beta = 1$, we have $\Delta = 3.2$ milliHertz/$\sqrt{\text{Hz}}$, with a white noise spectrum.

This represents the maximum performance possible since no noise beyond shot noise has been considered. If we suppose that a particular laser may need a bandwidth $B = 10$ kHz for the servo to be locked tightly enough, this noise density will amount to a frequency excursion of 0.32 Hz rms.

However, it is important to distinguish between this rms frequency excursion and the (Lorentzian) laser linewidth which would appear in the spectroscopy of a sharp atomic transition. For the slow optical phase diffusion associated with this wide band, low amplitude FM noise, the atomic Bloch vector will respond effectively to the average phase of the applied radiation. Thus we can see that small frequency excursions are not too serious if they occur at a high frequency because there will not be enough time to integrate a radian phase error before the fluctuation has reversed sign and begins to integrate the phase error back to zero. When we reach ~1 radian phase error the phase modulation sidebands do become prominent (19%), so it is reasonable to take $\beta = 1$ as the division point between phase coherence and incoherence. OWENS and WEISS [15] quote a relationship between the noise spectral density and linewidth, namely $\delta\nu_{1/2} = \pi\Delta^2$. Later ELLIOTT et al. [16] studied this noise modulation situation in more detail (including non-white spectral distributions) with the result that in the limit of interest here, small fast excursions ($B \gg \delta\nu_{rms}$), one obtains a Lorentzian line shape of the width given by Owens and Weiss. In our numerical example, we will have a laser linewidth due to shot noise of

$$\delta\nu_{1/2} = \pi \times \Delta^2 = \pi \cdot \left(\frac{3.2\ \text{mHz}}{\sqrt{\text{Hz}}}\right)^2 = 0.033\ \text{milliHertz!}$$

Thus, this estimation shows that under feedback control the laser would be locked to the cavity resonance with a precision of 1/30 milliHertz and would generate an optical wavetrain with a coherence time of more than 4800 s.

As these values seem rather attractive and perhaps surprising, it has been interesting to explore these performance possibilities experimentally. A first question is: Can we really lock to a cavity this well? Let's lock two independent lasers to one cavity and study their beat. This will focus attention on the locking quality and essentially eliminate the cavity drift problems. A second question is clearly: Even if the locking works as expected, can one really isolate the reference cavity from the environment well enough to approach these values? Our provisional answers are almost and perhaps, respectively.

Experiments of the first type have been described in our recent publication [12] in which two HeNe lasers were locked to a single resonant cavity. As noted, it really doesn't matter what kind of lasers were used as long as the servo loops are fast enough. To remove any ambiguity about injection-locking effects, the lasers were locked on adjacent axial interference orders in the reference cavity. The resulting 500 MHz difference in the optical frequencies is recovered when the two lasers are mode matched on a beam splitter and heterodyned in a fast photodetector.

This resulting 500 MHz rf signal contains the full phase and spectral information of the optical fields. For convenience in further analysis we then heterodyned this rf frequency down to the 30 kHz range using an rf frequency synthesizer referenced to a contemporary 5 MHz stable quartz oscillator. The resulting beat waveform passed through a bandpass filter 1 kHz $< f <$ 100 kHz and was counted with a high resolution frequency counter. During a quiet evening a set of 9000 consecutive 1 s

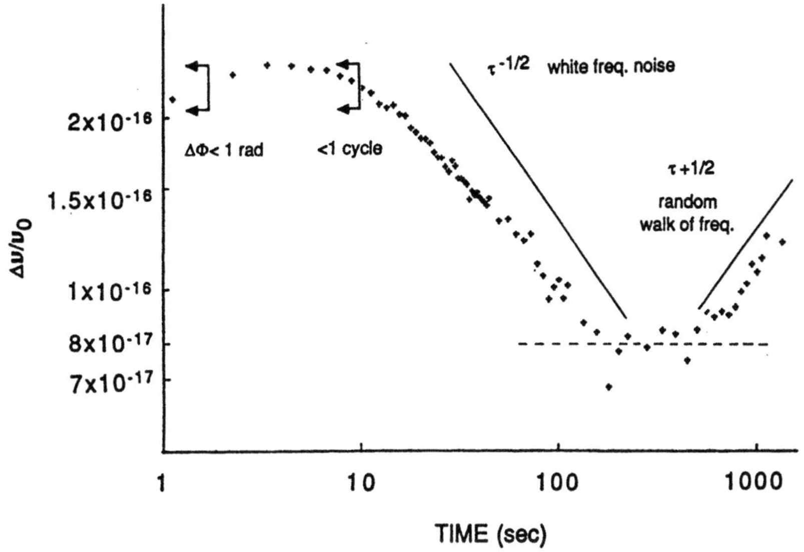

Fig. 3. Allan variance of beat of two HeNe lasers locked to same cavity. Averaged over the entire record length (9000 samples of 1 s), the coherence time is only a few seconds. A number of large rather abrupt "events" may contribute to the rise ~10 s. See text and Figs. 4 and 5.

measurements were obtained and stored on disk. Figure 3 shows the Allan variance for this 2.5-hr run. A drift rate of the beat below 10^{-4} Hz/sec was observed. It is clear that these optical sources are rather well locked! Still, we are quantitatively far from the 300-s coherence time estimated for the shot noise limit of this particular cavity and setup. Instead, by 300 s we have about 75 radians phase error averaged over this 9000-s run. Further, the shot noise model leads to a $\tau^{-1/2}$ prediction and we see experimentally a clear flicker region (τ^0) and $\tau^{+1/2}$ region, random walk of frequency.

To study the coherence time question in more detail, it was useful to reset the frequency synthesizer value to more nearly cancel the 500 MHz optical beat frequency. If the beat frequency were set to be above a Hertz or two, the down-converted optical beat wave simply looks like a sinewave, as seen in Fig. 4. To give an impression of the Fourier frequencies of importance in the variations of long times, we show in Fig. 5A, on an expanded time scale, a representative portion of the long data record leading to Fig. 3. This plot is of directly counted frequency data without any narrow filtering. It is obvious that the white noise, calculated to be a dozen milliHertz, is not the dominant process. Rather some larger amplitude excursions are seen with Fourier frequencies of 0.1 Hz or less. To emphasize this low frequency domain and perhaps jog our imagination about its possible cause, it was useful to average 10 adjacent points as shown in Fig. 5B. Now it is obvious that different physical processes are at work. For example, from $150 < t < 300$ s and from $500 < t < 730$ s there are two zones of rather smooth evolution, separated and surrounded by times of extreme and abrupt change. Perhaps one should be somewhat more cautious with labeling such ~0.5 Hz jumps at 5×10^{14} Hz as "extreme," but it is clear that the underlying physics is different. We might first conjecture that these "glitches" arise from environmental factors such as abrupt tilt of the optical table under fine hysteresis-controlled motion. However, both optical cavity modes are so close in frequency that they would be similarly affected. A better picture may be that the stress-induced birefringence of the mirrors may be changing due to symmetry changes in the optical contact area surrounding the mirror surface: unfortunately these measurements were made with crossed polarizations.

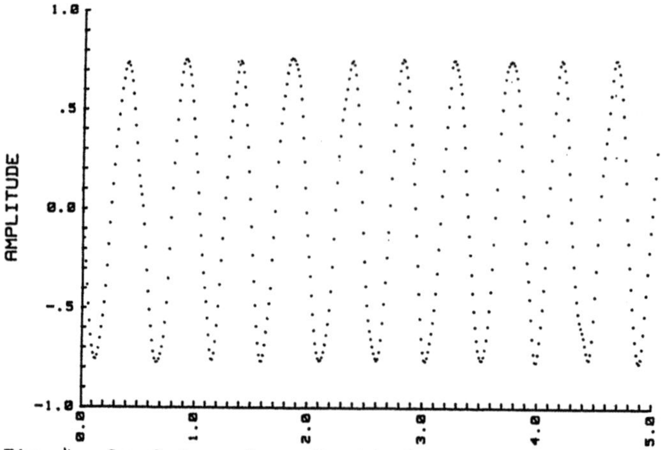

Fig. 4. Sampled waveform of optical beat between two lasers independently stabilized to a cavity. Optical beat frequencies were separated by one free spectral range (~500 MHz). Resulting 500 MHz beat heterodyned to 2 Hz with a phase-stable frequency synthesizer locked to an excellent quartz reference. Fitting this waveform gives a beat linewidth ~50 milliHertz.

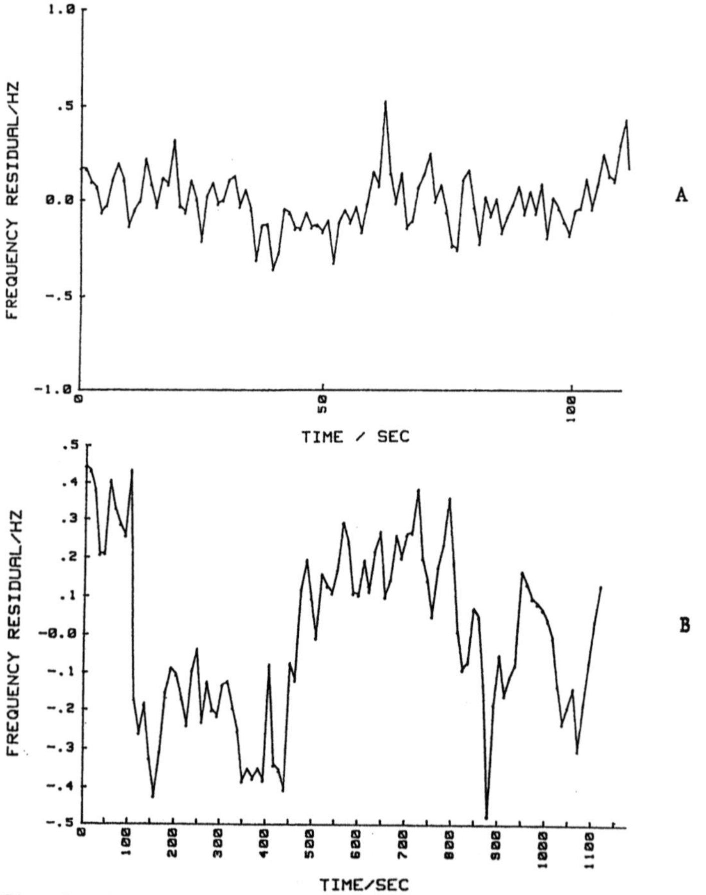

Fig. 5. Frequency noise in heterodyne of two lasers independently locked to a cavity. A) Data as recorded, 1 s measurements. B) 10 point averages. Zones of nearly phase-coherent operation (180<t<330 and 530<t<730 s) are accented by abrupt transitions perhaps associated with the cavity mirrors. See text.

The other clear feature in the two aforementioned time intervals is the presence of rather pronounced frequency excursions of about 40-s period. This frequency may be connected to an awkward choice of poles in the servo-amplifier design. The "oscillation" magnitude is about 1.3 radians peak. Absent this feature the 180-330 s interval would correspond to a coherence time of the two-laser beat of ≈150 s. The second interval shows a small frequency drift but is qualitatively similar. Thus we feel that -- away from the moments of large "glitches" -- we are within a few dB of reaching coherence time of the beat which are approaching those expected on the basis of the basic shot noise limitation. Thus we can conclude that the locking ideas work OK.

We note, however, that looking at the beat suppresses cavity drift and noise by a factor of 10^6. One must wonder: What about the cavity noise itself? The ideal way to investigate this point would be to have two similar cavity-locked systems and use heterodyne techniques to study the beat. Because of fabrication delays and our concern about the lowered sensitivity to equivalent long-term drifts in the two systems, we have made our first long term stability measurements using a $^{127}I_2$-stabilized HeNe laser as the reference. At short times (≈1 s) its finite signal/noise produces a stability limitation of about 3 kHz, which integrates at 100 s to about 500 Hz. A flicker floor at ~80 Hz instability is reached at about 1 hr for the best of these devices, and appears to be caused by drifts in the systematic offsets, perhaps associated with saturation-dependent gas lens effects, and drift of the temperature set point controlling the I_2 pressure and hence the pressure shift [17]. Perhaps electronic gain and/or zero drifts contribute as well. On the other hand, the cavity-stabilized laser can be expected to be best at the short times (~1 s) and to degrade toward longer times. Temperature drifts of the reference cavity of 10 μK would shift the cavity resonance frequency by 100 Hz, as would a tilt by 8-10 μrad which would change the gravity body-force distortion of the interferometer. We feel these are unreasonably large changes considering our present environmental control [18] with the "quiet house" and precision thermal servos. So it would be difficult to check the ultimate limits of the cavity-stabilized system using a customary I_2-stabilized laser as the reference.

Of course most materials, Zerodur [19,20] included, have internal (thermally-driven?) processes which tend to make the material "creep" in length. This phenomenon is manifestly demonstrated in Fig. 6 which shows a record for nearly 900 days of the beat frequency of our cavity-stabilized laser relative to the 633 nm I_2 "d"-peak. Over this interval the original slope, +2.1 Hz/s ($\Delta\ell/\ell = -3.8 \times 10^{-10}$/day), has moderated to the present 1 Hz/s (80 kHz/d; $\Delta\ell/\ell = -1.8 \times 10^{-10}$/d). These values are in substantial accord with the results of BAYER-HELMS et al. [21] who followed some Zerodur gauge blocks for 10 years during which an initial $\Delta\ell/\ell \simeq -7 \times 10^{-7}$/y ($-1.8 \times 10^{-9}$/d) creep rate aged down to $\Delta\ell/\ell \sim -3 \times 10^{-8}$/y (-2.9

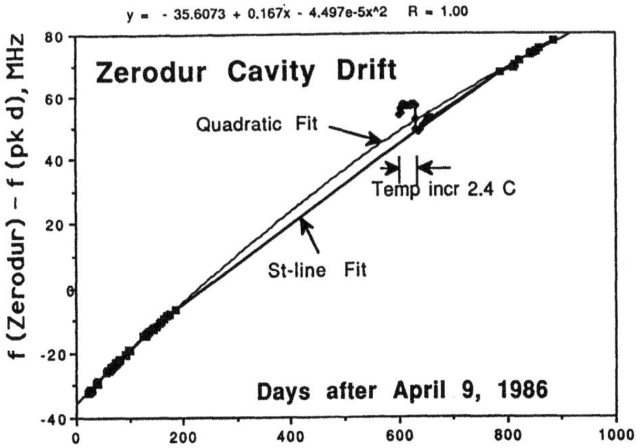

Fig. 6. Drift of Zerodur cavity. The 15 cmϕ × 30 cm Zerodur cavity has been hanging under vacuum for about 900 days. Initial drift rate of 167 kHz/day has decreased ~2 fold. Thermal set point was increased by 2.4 K on day 600 from 26.6° C. Note drastic effect on drift rate.

$\times 10^{-11}$/d). As our material was aged about 1.5 y after heat treatment before our first measurements, we regard these two measurements as being in good accord. An interesting anomaly is the report by JACOBS [22] of low drift ($<10^{-9}$/y), but the material of his sample displayed an anomalous temperature dependence relative to the Zerodur norm. It is possible that this particular sample was subjected to a non-standard thermal anneal cycle.

An interesting side experiment is associated with the data for days 580-650. We observed that in a few hours the drift rate could be measured to about 10^{-3} of its value. If the creep were indeed thermally activated as we had supposed, change of temperature by 2 or 3 K should provide an easily-observed change of the creep rate. So on day 600 the temperature set point was changed from 26.6° to 29°C. As may be seen in Fig. 6, the primary effect is a frequency shift by some +10 MHz, corresponding to a temperature coefficient $(\Delta\ell/\ell)/\Delta T \simeq -9 \times 10^{-9}$/K. This value is in agreement with the manufacturer's specification for this material, $|(\Delta\ell/\ell)/\Delta T| < 2 \times 10^{-8}$.

We were surprised initially by the large change of the drift rate, but from inspection of Fig. 6 it is clear that complicated physics is occurring. After day 612 a reasonably smooth drift rate is re-established, but the slope appears to be much lower than the creep rate at the original, lower temperature.

In Fig. 6, another interesting aspect of this temperature-change experiment can be seen in its long-term consequence for the creep rate: The 30 day period during which the temperature was elevated is clearly visible. So also is the "memory" of this rather gentle stress, lasting perhaps 100 days until the original creep rate was resumed. For conserving frequency information by using a stable interferometer fabricated with Zerodur [20], the lesson from Fig. 6 is clear: temperature variations even at the levels characteristic of room temperature variations will have long-term consequences for the frequency predictability. The Zerodur [20] remembers.

One may regard the property of frequency predictability to be one of the key attributes for our optical local oscillators stabilized by locking to a cavity. In general for an atomic clock it seems to be advantageous to make the maximum use of the LO intrinsic stability: corrections made less frequently correspond to reduced bandwidth and hence lower noise from the atomic resonance. Furthermore in some cases such as spectroscopy of a single trapped ion, one may need 100 lifetimes to develop an adequate control signal.

Thus it is interesting to take the I_2 vs. cavity frequency data for a reasonable period, say one day. One will then follow the frequency up to some moment and try to predict the frequency in the future. Figure 7A shows 24 hr's data, while Fig. 7B shows the residuals of these data after suppression of the average drift. A 24-h signal of ~75 Hz amplitude is determined via FFT. Working from the 10 MHz induced by our 2.4 K shift of the thermal controller, the effective temperature coefficient of the cavity is 4.2 MHz/K. So the apparent temperature variation of the cavity is 18 μK pk. This can be compared with the observed 24-h component of 2.5 mK of the air in the quiet house near the cavity housing. From this information one could deduce that the cavity temperature controller has a thermal stability gain of 140-fold at 24 h. As this number is vastly less than the gain deduced from electrical measurements, we surmise that other influences, perhaps thermally-driven tilt of the optical table top, are of more significance. A new single-point mounting should provide clarification of this point.

As to the predictability of the optical frequency from knowledge of its recent past, we show in Table I our estimate of the growth of frequency instability with prediction time, τ. These values are obtained from the data of Fig. 7A by fitting to a quadratic the data from $t = t' - 3\tau$ to $t = t'$, calculating the predicted value at $t = t' + \tau$, comparing with the observed frequency at that time, and averaging over the possible choices of t'.

As a final topic we discuss some very preliminary experiments using a different low expansion material, ULE [20,23]. In contrast to Zerodur which is a mixed-phase ceramic-glass material, ULE is a true glass formed of SiO_2 and TiO_2. Proprietary heat treatments are able to smoothly adjust thru zero the thermal expansion coefficient on the 10^{-9}/deg K range [24]. Unfortunately, growth limitations controlled the bar size available: our finished bar turned out to be $5.76 \times 7.16\ cm^2 \times 27.7$ cm long. The idea of directly replacing the Zerodur rod (15 cmϕ × 30 cm long) was also affected by the rectangular cross section of the available ULE specimen, necessi-

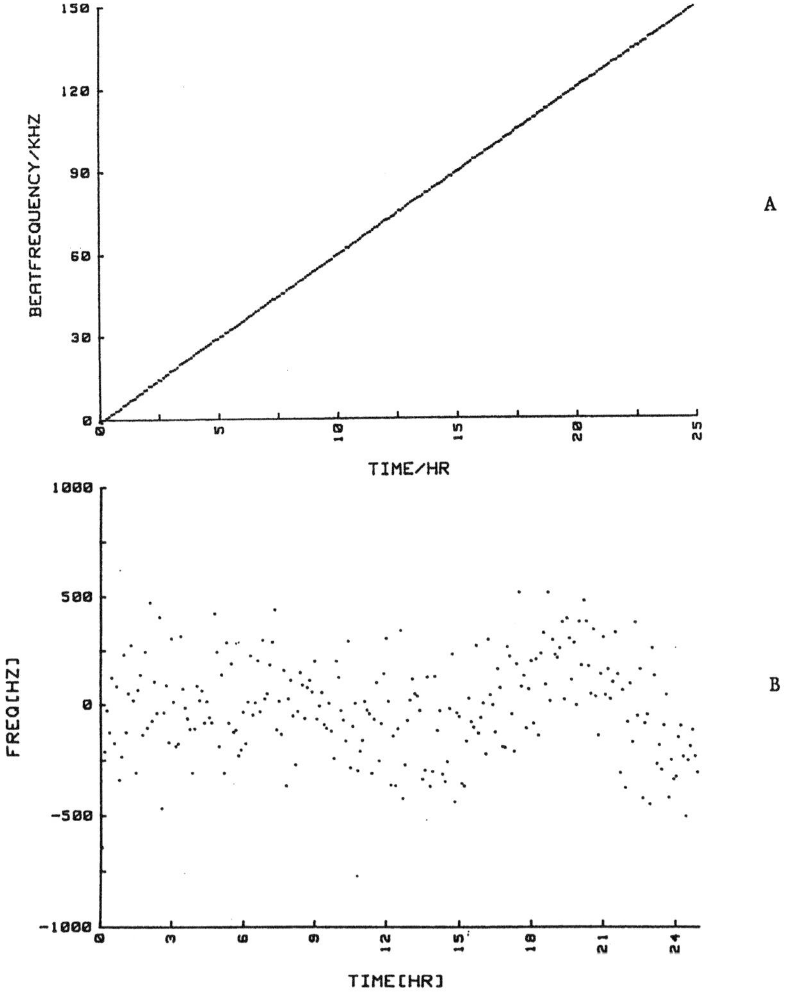

Fig. 7. Beat of Zerodur- vs. I_2-stabilized lasers. A) Data averaged for 6 minutes per point. Slope is ~150 kHz/d. B) Same data st-line slope removed. Some variations at 1-3 cycles/day are visible, but mainly we are at the S/N limit of the 127 I_2-stabilized reference laser. See text and Table I.

Table I. Estimated frequency unpredictability, based on extrapolation from 3 previous values. $\sigma(\tau)$, derived from beat with I_2, mainly shows I_2 laser noise.

time τ (s)	$\sigma(\tau)$(Hz)	total linear drift (Hz)	unpredictability (Hz)
1	3000	1.7	0.05
10	1200	17	0.1
100	300	170	0.1
1000	110	1,700	1.0
10^4	80	17,000	25

tating a change in the mounting details. Further we decided to investigate a single-strap suspension rather than the two-strap system previously employed. Thus at this particular point in time there are still some mixtures of effects to be clarified. One point however, is clearly evident and worth mentioning: the drift rate of the ULE [20] interferometer was about 20 Hz/s in the first week after optically-contacting its mirrors. After 20 days the drift rate is below 1/4 Hz/s. This is 4-fold less than the Zerodur [20] rate, even after several years.

To summarize this report, we find that it is readily possible to lock to an interferometer resonance with a noise level of ~ 50 milliHertz. Other experiments reported elsewhere [12] show that we were locking onto the 80 kHz linewidth cavity resonances with an accuracy of 1 or 2 Hz. The long-term stability of these suspended reference interferometers is surprisingly good. The observed length change of $\sim -2 \times 10^{-10}$/day appears to be pure uniform drift which can be predicted with 1% accuracy or better for a day. There is preliminary evidence that, as far as aging is concerned, ULE [20] may be a more attractive interferometer material than Zerodur [20]. Finally, we must admit some disappointment in the first experiments directly heterodyning two such cavity stabilized lasers. We observed that the beat linewidth is very narrow, well below 1 Hz, but Doppler shifts and mechanical stresses associated with the presently under-damped swinging motion lead to a frequency modulation of ~ 200 Hz pp. We are considering an isolated platform with cavity, EO phase modulator, AO isolators, Faraday isolators, and detector built as a single unit. Fiber optic illumination will be used along with active damping of the pendulum modes. Then we will be ready for the Hertz-level spectroscopy of the usual trapped ions of interest, e.g., Hg^+, Ba^+, Yb^+ and/or of free falling Ba, Ca, Mg atoms in an atomic fountain.

Acknowledgments. We are pleased to acknowledge and thank Christophe Salomon and Jean-Marie Chartier who participated in these experiments at an earlier stage. We thank Miao Zhu for his help with the dye laser stabilization. This research has been supported in part by the National Institute of Standards and Technology and in part by NASA, the Office of Naval Research, and the National Science Foundation.

References

1. D. Hils, J. L. Hall: "A New Kennedy-Thorndike Experiment," in preparation.
2. E. Riis, L-U A. Andersen, N. Bjerre, O. Poulsen, S. A. Lee, J. L. Hall: Phys. Rev. Lett. 60, 81 (1988).
3. J. E. Faller, P. L. Bender, J. L. Hall, D. Hils, and M. A. Vincent: "Space Antenna for Gravitational Wave Astronomy," in Proc., Colloq. on Kilometric Optical Arrays in Space, 23-25 Oct. 1984 Corgesa, Corsica (ESA SP-226, April 1985) pp. 157-163.
4. See e.g. J. H. Taylor, in these proceedings.
5. J. L. Hall, T. W. Hänsch: Opt. Lett. 9, 502 (1984).
6. R. W. P. Drever, J. L. Hall, F. V. Kowalski, J. Hough, G. M. Ford, A. J. Munley, M. Ward: Appl. Phys. B31, 97 (1983).
7. J. L. Hall, H. P. Layer, R. D. Deslattes: Proc. of 1977 IEEE/OSA Conf. on Laser Engineering and Applications (Opt. Soc. Am., Washington, DC, 1977), p. 45.
8. Some additional disucssion of these points may be found in J. Helmcke, S. A. Lee, J. L. Hall: Appl. Opt. 21, 1686 (1982).
9. For example with our latest "gyro-quality" mirrors we have a measured finesse of 60,000 at the 560 nm design wavelength, 38,000 finesse at 633 nm and still >10,000 finesse at 514 nm.
10. R. G. DeVoe, R. G. Brewer: Phys. Rev. A 30, 2827 (1984); R. G. DeVoe, C. Fabre, K. Jungmann, J. Hoffnagle, R. G. Brewer: Phys. Rev. A 37, 1802 (1988).
11. Earlier results have been presented by D. Hils, J. L. Hall: In XV International Conference in Quantum Electronics 1987 Technical Digest V21 (Optical Soc. of Am., Washington DC, 1987), p. 102.
12. Ch. Salomon, D. Hils, J. L. Hall: J. Opt. Soc. Am. B 5, 1576 (1988).
13. A general analysis of the line shapes of a cavity excited by FM light has been performed. Transmission line shapes are considered in D. Hils, J. L. Hall: Rev. Sci. Instrum. 58, 1406 (1987). Reflection-mode signal shapes for arbitrary modulation index and frequency will be published shortly.

14. See e.g. J. L. Hall, L. Hollberg, L-S. Ma, T. Baer, H. G. Robinson: "Progress toward phase-stable optical frequency standards," in 3rd Symposium on Frequency Standards and Metrology, J. Physique Colloq. 42, Suppl. 12, C8 59-71 (1981).
15. D. K. Owens, R. Weiss: Rev. Sci. Instrum. 45, 1060 (1974).
16. D. S. Elliott, R. Roy, S. J. Smith: Phys. Rev. A 26, 12 (1982).
17. M. P. Winters and J.-M. Chartier, private communication. Also see P. Cerez and A. Brillet, Metrologia 13, 29 (1977).
18. D. Hils, J. E. Faller, J. L. Hall: Rev. Sci. Instrum. 57, 2532 (1986).
19. Schott Optical Glass Inc., Duryea, Penn. See Ref. 20.
20. Trade names are used for purposes of technical communication only and have no other significance herein.
21. F. Bayer-Helms, M. Darnedde, G. Exner: Metrologia 21, 49 (1985).
22. S. F. Jacobs: Optica Acta (GB): V33, 1377 (1986);
J. W. Berthold III, S. F. Jacobs, M. A. Norton: Metrologia 13, 9 (1977).
23. ULE is code 7971 materials, Corning Glass Works, Corning NY. See Ref. 20.
24. H. E. Hagy, paper WS6, annual meeting of Optical Society of America, Seattle, WA, 1986, p. 101.

Low-Noise and Stable LO's for Deep Space Applications

L. Maleki

Jet Propulsion Laboratory, California Institute of Technology,
Pasadena, CA91109, USA

1. INTRODUCTION

The traditional role of local oscillators in communications systems is much expanded in the NASA's Deep Space Network (DSN) which supports communications, tracking, and radio science experiments associated with interplanetary missions such as Voyager, Ulysses and Galileo. In this world-wide network of three complexes, local oscillators provide spectrally pure signals in receivers which track spacecrafts as far away as the edge of the solar system. Spectrally pure signals are also crucial to the success of a variety of radio science investigations studying interplanetary particles, and planetary rings and atmospheres. Finally, local oscillators function as flywheels in advanced frequency standards currently under development for the DSN, and influence the ultimate stabilities obtainable with such devices.

In this paper the requirements of local oscillators for each class of applications in the DSN are first identified. The work for the development of local oscillators currently underway at the Jet Propulsion Laboratory (JPL) which manages the DSN for NASA will be then presented, followed by a projection of the future local oscillator capabilities in the DSN.

2. LOCAL OSCILLATOR REQUIREMENTS

The applications of local oscillators (LO's) in the DSN are navigation, radio sciences, and frequency standards. Each type of application requires particular levels of phase coherence performance, specified by the power spectral density of phase around the carrier.

Before the communications frequency range in the DSN was extended from 2.4 GHz (S-band) to 8.4 GHz (X-band), quartz oscillators were adequate for nearly all local oscillator requirements. This was because the noise characteristics remained within the acceptable margins as signals from commercial 5 MHz quartz oscillators were multiplied up to S-band. This condition was greatly altered however as 5 or 100 MHz signals from LO's were multiplied up to 8.4 GHz and higher. The increase of the multiplicative noise of the oscillator, on the one hand, and the progression of ever increasing requirements of various applications, have made even the state of the art commercial oscillators unsuitable for most applications in the DSN. The situation is expected to worsen as the DSN moves towards higher frequencies above X-band, and towards 32 GHz, a frequency currently under consideration for certain future missions.

The navigation requirements of local oscillators stem from the application of voltage controlled or programmable oscillators in phase locked loops within the ground receivers [1]. Phase locked loops are essential to Doppler tracking systems, and to systems where signals are weak and data rates are low. These are precisely the

Frequency Standards and Metrology Editor: A. De Marchi

conditions that characterize nearly all signals received from the interplanetary spacecrafts. The function of the phase locked loops in the ground receivers is to generate a reference signal that is coherent in phase with the signal arriving from the spacecraft. Local oscillators then represent the most critical component for the generation of phase coherent signals.

Despite the stringent requirements for spectral purity, the requirements in phase locked loops are greatly aided by the relatively large loop bandwidth, typically greater than 3 Hz. This renders the power spectral density of phase close to the carrier of lesser concern. Nevertheless only state of the art quartz oscillators are employed in the current designs of the Mark IV receiver in order to meet specifications for spectral purity. In the case of the Advanced Receiver currently under development at JPL, loop bandwidth of 1 mHz is envisioned. Thus spectral purity requirements will be much more stringent. Specific requirements for the noise of the VCO's to be used in the Advanced Receiver are currently being developed.

The above considerations for the case of phase locked loop receivers is significantly different from the case of the open loop receivers employed for radio science experiments. In this class of experiments signals from the spacecraft in a planetary occultation are monitored for variations of amplitude, phase, and frequency [2]. These parameters permit the determination of absorption and refraction of signals received from the spacecraft by the intervening medium. Thus physical properties of media scattering the spacecraft signals my be obtained by careful measurements of the received signals. In this way information regarding the atmospheric constituents of planets, the size of particles in planetary rings, and the density profile of plasma about the solar system or in an ionosphere is obtained.

Since in these investigations the parameters of interest are the variations in the amplitude, phase, or frequency of the received signals, the use of open loop receivers is required. Table 1 lists the current radio science requirements for the VCO expected to be used for the encounter of Voyager 2 spacecraft with Neptune. Note that the table includes requirements for acceleration sensitivity of the VCO, since there is a need for the deployment of a "clean up" loop in the antenna cone where vibrations due to antenna motion exist.

Only specialized, high performance crystal oscillators are capable of meeting the requirements stated in Table 1. These requirements are significantly increased for future missions such as Mars Observer, to a degree that the application of best crystal oscillators may be inadequate. The situation will undoubtedly worsen if currently stated radio science requirements at 8.4 GHz are expected to be met or exceeded at higher frequencies.

Table 1. Local oscillator requirements for the radio service receiver Voyager – Neptune encounter.

Frequency From Carrier (Hz)	Power Spectral Density of Phase, (Double Sided), dBC/Hz			
	100 MHz	20 MHz	10 MHz	5 MHz
1	−92	−108	−114	−120
10	−106	−118	−124	−130
100	−120	−130	−136	−142
1000	−130	−140	−146	−152
10000	−150	−140	−146	−152
Floor	−150	−140	−146	−152

The acceleration sensitivity will be less than $3 * 10^{-10}$ per G, static and dynamic (up to 100 Hz) for each axis.

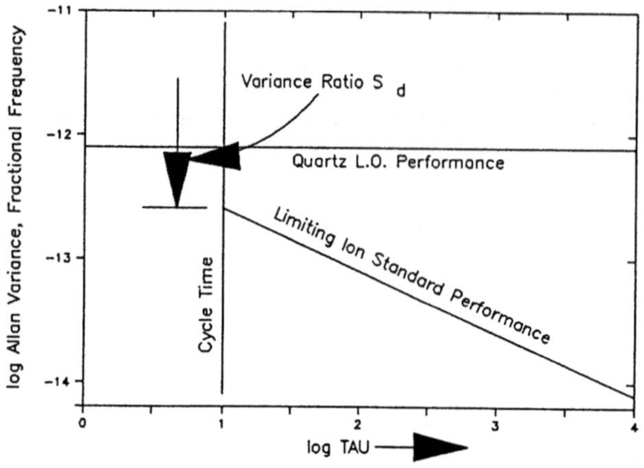

Fig. 1. Trapped Ion L.O. Requirements

The third class of applications of local oscillator in the DSN pertains to their use as flywheels of advanced frequency standards. Here again, radio science applications are the primary drivers of the requirements for stabilities higher than that currently achievable with hydrogen masers. The frequency stability requirements of spacecraft gravitational wave experiments currently are at the level of parts in 10^{17} at 1000 second averaging times.

These and other considerations have led to a directed effort for the development of advanced frequency sources, such as the trapped mercury ion standard, at JPL [3]. While the physics of the trapped mercury ion standard supports the projection of stabilities exceeding that of the hydrogen maser, the performance of the device is strongly dependent on the stability of the flywheel. Recent work in our group [4] has produced numerical techniques to calculate the effect of the flywheel oscillator noise on the long-term stability of the trapped ion standard. One of the significant outcomes of this work is the identification of approaches which minimize the effect of local oscillator noise on the stability of the standard. Nevertheless it is clear from the analysis that to exceed the performance of the hydrogen maser the trapped ion standard requires flywheels with a level of stability that is unlikely to be achieved with quartz oscillators. Figure 1 summarizes the results of calculations depicting the required performance of the flywheel oscillator used in an optimal configuration with the mercury ion standard.

The three cavity design further enhances the performance of the SCMO by affording reduced thermal expansion coefficient for all electronic components and increased temperature stability. As a result recent measurements with the first operating SCMO have hinted at the possibility of operation around 1.8 K temperature, with a sustained high stability extending to 1000 second averaging intervals. We expect that a demonstration of a stability performance at a level of parts in 10^{15} from 1 to 10^3 seconds will be made in 1989. The ultimate stability goal of this device is achieving stability of one part in 10^{17} for averaging intervals less than 10^3 seconds. Since the current operating frequency of this device is at 2.8 GHz and requires less multiplication steps to achieve higher frequencies, SCMO is considered the prime candidate for providing the required spectral purity at 8.4 GHz and higher for the future DSN missions. Improvements in the performance of the SCMO may be obtained by simply modifying the design to achieve even higher operating frequencies.

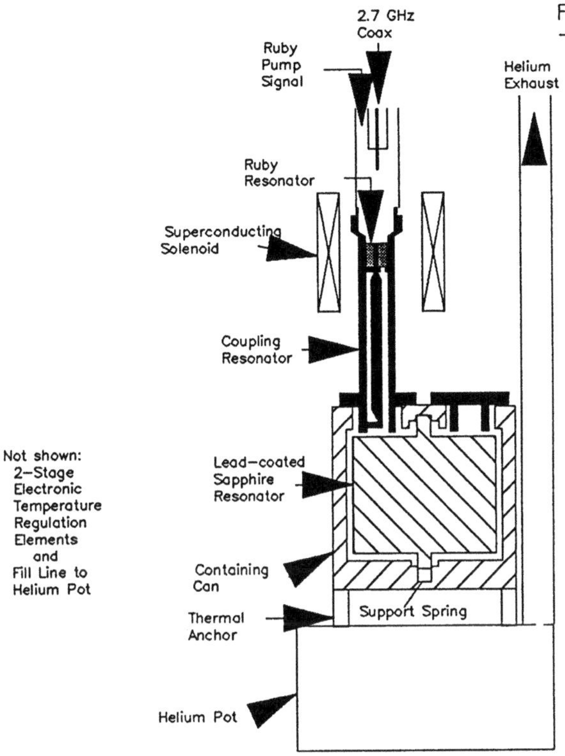

Fig. 2. Schematic drawing of SCMO.

A second novel oscillator is also under development at JPL and is based on the high Q of sapphire resonators. The resonator consists of a sapphire cylindrical ring in a metallic shield. Analysis of the performance of this dielectric resonator [6] indicates that it equals Q's of quartz resonators at temperatures achievable with thermoelectric coolers, and that at 10 to 20 Kelvin it approaches the performance of the SCMO. Measurements in our laboratory have produced Q's in the range of 3×10^7 to 10^9 at temperatures of 77 to 4.2 Kelvin, respectively.

The high Q's are achieved both as a result of reduction of internal losses of sapphire at low temperatures and a design which isolates the resonator energy from loss to the metal shield. The basis of the design is in the selection and use of the so-called "whispering gallery" modes with properties similar to waves internally reflecting in optical waveguides. A first version of this oscillator operating at room temperature is currently under development at JPL. Advanced versions of the oscillator utilizing the low noise ruby maser as the active element are also under investigation, and are expected to produce about 30 dB better phase noise compared to the best 5 MHz quartz oscillator 1 Hz from the carrier at 10 GHz.

4. CONCLUSION

The need for stable local oscillators in the DSN is to support the unique requirements of communications, navigation, and radio sciences associated with various present and planned interplanetary missions. Efforts at JPL to develop the superconducting cavity stabilized ruby maser oscillators is expected to provide the DSN with the highest phase coherent source to support certain radio science investigations. The development of cryogenic dielectric resonator oscillator will support the needs of the trapped ion frequency standard for a spectrally pure flywheel to achieve stabilities exceeding the hydrogen maser stability at time intervals beyond a few thousand seconds. Other versions of the dielectric resonator oscillator operating with thermoelectric coolers are expected to provide the phase coherent references for the DSN receivers operating at frequencies in the 32 GHz range.

Finally, in the decade of 1990's and beyond it is expected that some missions will carry optical communications packages. Heterodyne detection of optical signals require the use of optical local oscillators. This need is also recognized at JPL and current efforts directed towards stabilization of diode lasers to atomic resonances using optogalvanic spectroscopy are placing the ground work for the development of phase stable optical local oscillators.

This work represents the results of one phase of research carried out at the Jet Propulsion Laboratory, California Institute of Technology, under contract sponsored by the National Aeronautics and Space Administration.

REFERENCES

1. M. K. Simon and J. H. Yuen: In Deep Space Telecommunications Systems Engineering, ed. by J. H. Yuen,

2. V. R. Eshleman: Space Sc. Rev. 21, 207 (1977)

3. J. D. Prestage, G. J. Dick and L. Maleki: Proc. 19th Ann. PTTI Meeting (1987) p.285

4. G. J. Dick: Proc. 19th Ann. PTTI Meeting (1987) p. 133

5. D. M. Strayer and G. J. Dick: Proc. 18th Ann. PTTI Meeting (1986) p. 601

6. G. J. Dick and D. M. Strayer: Proc. 41st Ann. Conf. Freq. Control Symp. (1987) p.487

Part VIII

Frequency Synthesis

Progress in Infrared Frequency Synthesis

F. Bertinetto

Istituto di Metrologia G. Colonnetti, Torino, Italy

The progress in frequency synthesis to near IR and to the visible region was essential for the formulation of the new definition of the metre approved by the XVII CGPM (Conférence Generale des Poids et Mêsures) in Nov.1983 /1/.

It would seem that since then the initial stimulus has been lost for performing "true" frequency measurements of visible radiations in addition to or for comparison with the only two existing at present /2,3/. However mention has to be made for the improvement in accuracy (to $10^{-12}\cdot\nu$) achieved in the measurement of the frequency of a OsO_4 transition at 30 THz with the use of a coherent frequency chain /4/. The advisability of phase locking of all intermediate oscillators was suggested in Aussois /5,6/; the extensive use of this technique allowed new frequency measurements of the CH_4 transition at 88 THz to be made with 10^{-13} uncertainty and the synthesis of a 130 THz to be applied for the frequency measurement of a visible radiation, as was decribed in the papers presented at the symposium of Ancona.

To confine ourselves to the comparison between visible and microwave radiations, the advantage of eliminating intermediate steps is obvious. However, the accuracy of a recently proposed method, combining interferometry and double frequency modulation /7/, is intrinsically limited by diffraction; till now no attempt has been made to measure the frequency of the syncrothron radiation from an electron whose orbit is governed by laser light /8/.

1 - "Documents Concerning the New Definition of the Metre", Metrologia 19, 163-177 (1984)

2 - C.R.Pollock et al., Opt.Lett. 8, 133-135 (1983)

3 - D.A.Jennings et al., Opt.Lett. 8, 136-138 (1983)

4 - A.Clairon et al., IEEE Trans.Instr.Meas. IM-34, 265-268 (1986)

5 - E.Bava et al., J.de Physique 42, 463-472 (1981)

6 - V.P.Chebotayev, J.de Physique 42, 505-512 (1981)

7 - R.G. De Voe et al., Laser Spectr.VII, Springer Verlag (1986)

8 - D.J. Wineland, J.Appl.Phys. 50, (1979)

Frequency Standards and Metrology Editor: A. De Marchi

Coherent Frequency Measurements of the Hfs-Resolved Methane Line

G. Kramer, C.O. Weiss, and B. Lipphardt

Physikalisch-Technische Bundesanstalt, Bundesallee 100,
D-3300 Braunschweig, Fed. Rep. of Germany

We have set up a coherent frequency measurement chain to measure the frequency of the central component of the ν_3 (7) HFS multiplet of methane at 88 THz, which was resolved by a linear optical Ramsey resonance experiment. A frequency of

88 376 181 599.5 kHz $\pm$ 1 kHz

was found. The complete experiment worked continuously for many hours as an "Optical Clock" with a stability comparable to a commercial cesium clock. The uncertainty of $\pm$ 1 kHz is preliminary, improvement by an order of magnitude is expected for the near future.

As the layout of the chain as well as details of the first measurements can be found in a paper /1/ currently in press, only one additional detail will be treated here.

Tracking Oscillators

Coherent measurement of a frequency means counting the zero-crossings of a signal, as opposed to the determination of the center of a spectral line. For the generation of unambiguously defined zero-crossings, we make use of (phase-locked) electronic tracking oscillators, which process the beat signals. They are not only useful for the final data taking by means of a frequency counter, but also work as signal conditioners for the prescalers and digital phase-lock loops /2/, which reliably and coherently lock the various members of the chain. While piezo driven mirrors severely limit the loop band width in phase-controlling a laser, the electronic tracking oscillator can track the signal with an optimized band width which may exceed several MHz, over a tuning range of more than 100 MHz.

Coherence Margin

Counting errors (cycle slipping) may occur, if the phase excursions at the phase detector amount to more than about a radian. These phase excursions are caused by the frequency fluctuations of the oscillators involved, and by additive noise from the harmonic mixers. The phase detector cannot distinguish the noise from the frequency fluctuations (which constitute the signal to be tracked!), and the control loop tries to balance the noise by additional fluctuations introduced into the VCO frequency.

If the noise spectra are known, the total rms phase excursion can be calculated as the sum of the additive noise S_n picked up by the control loop and the portion of the

Frequency Standards and Metrology Editor: A. De Marchi

oscillator phase noise S_φ not being tracked out, i.e. falling outside of the control band width f_c:

$$\langle\varphi^2\rangle \approx \int_0^{f_c} S_n df + \int_{f_c}^{\infty} S_\varphi df.$$

With the loop transfer functions and for white frequency noise:

$$\langle\varphi^2\rangle = \pi/2 \cdot f_c \cdot [S_n + S_\varphi(f_c)].$$

The phase noise diagram illustrates the situation for measurement of the 4 THz laser in the chain. If the frequency axis is interpreted as the selected control band width, then the dashed line indicates the 1 radian reference. The intercept point of the cavity Gunn phase noise with the mixer noise level is below this line by a "coherence margin" of 20 dB. That such a margin is sufficient, was tested by locking two tracking oscillators with widely different band widths (1 MHz and 10 kHz) and comparing their counts. For this purpose their outputs were each frequency divided (by 10) and the phase difference of the divider outputs was recorded. Cycle slipping rates below 1 per 600 s were observed, which, at 4 THz, corresponds to a frequency measurement error of less than a part in 10^{15}.

Fig. 1: Schematic of a phase-locked tracking oscillator

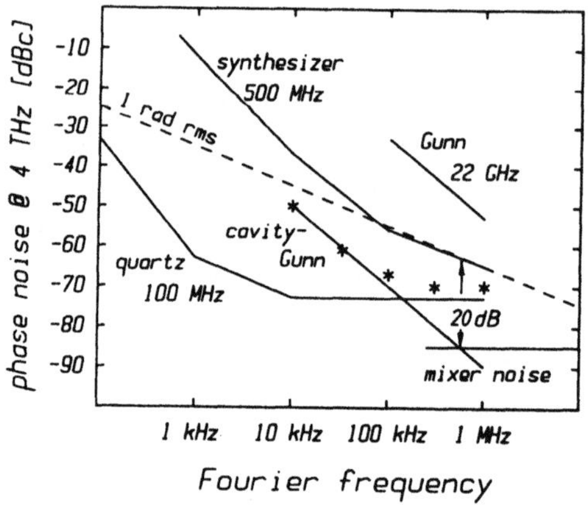

Fig. 2: Spectral density of phase noise of various oscillators referred to 4 THz

1 C.O. Weiss, G. Kramer, B. Lipphardt, E. Garcia: IEEE J. QE, Oct. 1988
2 J.L. Hall, M. Long-Sheng, G. Kramer: IEEE J. QE-23, Apr. 1987

Accuracy of HFS Resolved CH_4 Beam Frequency Standard

(G. Kramer)

In this second part of the paper we shall deal with progress made in developing a type of optical frequency standard that I proposed in 1973 already /3/. First experiments were reported ten years ago /4/, so it seems to be the time to say something about the level of accuracy achieved with the present apparatus and what the future potential of the method may be.

The method of producing "Linear Optical Ramsey Resonances" by means of a "Spatially Modulated Molecular Beam" is illustrated in figures 3 and 4. An uncollimated molecular beam traverses twice the standing wave field of a folded laser beam. In the middle between the two interaction regions a transparent grating with a grating constant equal to one half the wavelength is inserted, so as to obstruct certain trajectories of the molecular beam. For each trajectory we find the transition probability to take the familiar line shape of a microwave Ramsey resonance. As the phase of a standing wave field changes sign at every node plane, there are two classes of trajectories, those which cross the laser field in zones of equal phases and others with opposite phases. The corresponding Ramsey lines will have a maximum at line center or a minimum respectively. This means that as long as both classes of trajectories contribute equally to the total absorption, no narrow Ramsey interference fringes will be observed. The grating breaks this symmetry by preferably passing one of the two classes, which one, being determined by the position of the grating with respect to the node planes. By changing this position, positive or inverted Ramsey lines can be produced, which gives a means of completely separating the Ramsey fringes from other contributions to the absorption line shape.

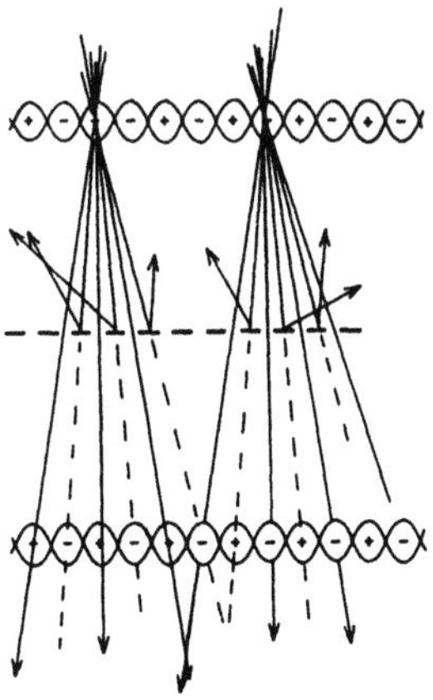

Fig. 3: Sample trajectories of molecules in an optical Ramsey arrangement

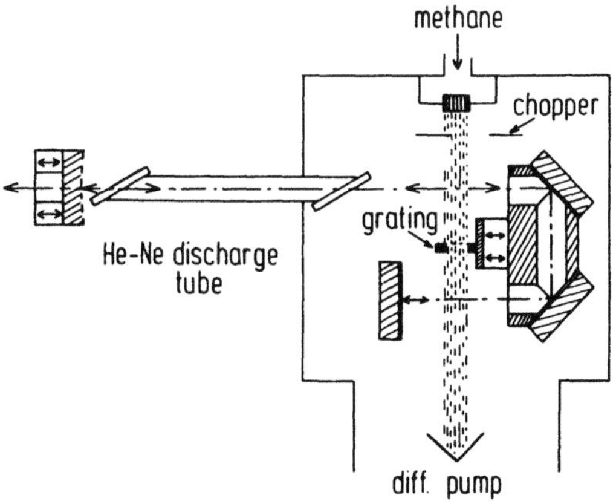

Fig. 4: Molecular beam optical frequency standard

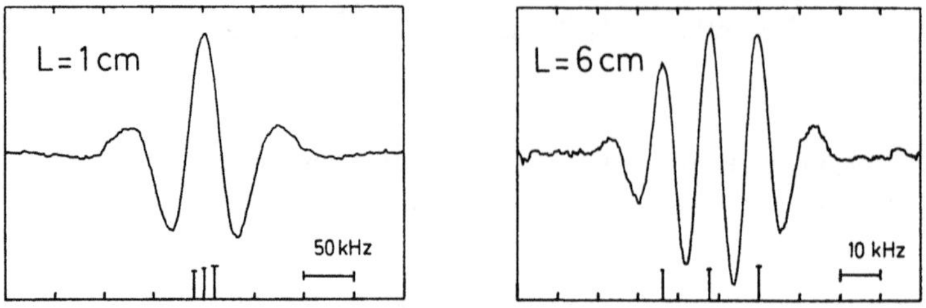

Fig. 5: Line shapes observed at two different settings of the field separation L. At L=6 cm, the HFS is resolved, linewidth is 6 kHz at 88 THz

1. Application to Methane at 3.4 μm

Figure 5 shows the lineshapes observed at two different resolutions, i.e. spacings of the interaction regions. At 6 cm spacing the hyperfine structure of the line is resolved. The linewidth of 6 kHz corresponds to a Q-factor of almost $2 \cdot 10^{10}$. Obviously such narrow lines cannot be observed by simply scanning the laser resonator, but require an offset lock spectrometer with a number of analog and digital frequency control loops, impossible to describe here in any detail. Thoroughly made, they should not contribute to the biases and uncertainties, which are the subject of this report.

a) Short-Term Stability

Short term stability is not a high-light of the molecular beam method, and much better results have been achieved by using saturated absorption (or dispersion) in telescopically expanded laser modes. But the S/N-ratio is sufficient to give $s=1.2 \cdot 10^{-12}$ for 1 second averaging with the hfs unresolved and $4 \cdot 10^{-12}$ for the 6 kHz linewidth; which is still somewhat better than a high performance commercial cesium beam tube. It is a favorable feature of the Ramsey excitation that the signal goes down only as the first power of the line-width, so with a singular line, short-term stability would be independent of the selected resolution. (In saturated absorption, the signal per unit volume varies as the 3rd power of the line-width.) With the present apparatus, signal strength is determined by the available size of the grating (2 x 10 mm) and by the intensity of the molecular beam. This beam is formed by a nozzle made from a microchannel plate with 12 μm pore size. By employing new technology 2 μm channels and/or assembling more than one grating, signal enhancement by a factor of 5 is expected.

b) Resonator Phase Difference

In an ideal, infinite Q-factor cavity resonator, the phase of the standing-wave field is uniform over the whole volume (except for the sign inversion at the node planes). Losses in the mirrors give rise to some running-wave component, which may cause a residual 1st order Doppler shift. Magnitude and sign of this shift depend on the relative position of the grating with respect to the node planes in the two interaction regions. The shift vanishes if the grating is aligned for maximum signal strength. For the lin-

ear limit, a straightforward analysis yields: $df = 3\cdot(1-R)\varphi/2\pi B$
with mirror reflectivity R, line width B, and φ the misalignment of the grating phase. At a reflectivity R=99%, an alignment precision of 0.1 rad is sufficient to keep the error below $3\cdot10^{-14}$.

c) 2nd Order Doppler Shift

While with the residual 1st order Doppler shift we gain from the high resolution possible in an optical frequency standard, this is not so with the 2nd order or relativistic Doppler shift, which causes a relative frequency bias, depending on the molecular velocity only. Due to the low molecular mass, for methane the shift is rather large: $1.5\cdot10^{-12}$ at room temperature. What counts, though, is not the magnitude of the bias, but the accuracy of its determination from other measurements. And here we can take advantage of a feature unique to the Ramsey excitation: the fact, that each velocity sub-group in the beam contributes a cosine function to the line shape. So a simple Fourier transform of the line shape produces the velocity distribution (precisely: the time-of-flight spectrum), already properly weighted with the corresponding transition probability. From this the shift can be precisely calculated, a technique I already applied when evaluating the PTB cesium clock many years ago.

With the methane F-line the situation is complicated by the hyper-fine structure, but as the splitting is known, it can to some degree be removed by de-convolution. Analysis of line-shapes taken at various temperatures and power settings show the expected dependencies, so an uncertainty of 20% of the bias seems to be a conservative estimate. This would contribute $3\cdot10^{-13}$ to the error budget.

Using the unstructured E-line of methane at LN temperature should reduce the uncertainty by at least a factor of 10. A heavy molecule like OsO_4 would also help.

d) Magnetic Field

The Zeeman effect, like most other features of the methane line, was first measured by J.L. HALL /5/. To first order it gives a symmetrical splitting of 235 Hz/G or less than $(3\cdot10^{-12})$/G. By avoiding circular polarisation and/or moderate shielding, the effect can always be kept sufficiently small.

e) Pulling by Neighbouring Lines

As we can not completely separate the hfs components, the central maximum of the central component appears shifted by a small amount due to asymmetric pulling from the neighbouring lines. Using the de-convolution technique, as for the Doppler shift, we can quite well determine the "slit function" of the apparatus and estimate the net pulling to be $(-3 \pm 1.5)\cdot10^{-13}$.

The bias could be reduced by better matching the Ramsey spacing to the hfs splitting. Obviously the pulling would not occur, if one used a solitary line.

f) Diffraction or "Photon Recoil"

What is usually called the photon recoil splitting, cannot be resolved with the present Ramsey apparatus. Compared to the accuracy level in question here, it is still a large effect and requires consideration. The common explanation of the effect describes the interaction of a molecule with the electromagnetic field as a collision process, in which a photon with the quantum mechanical momentum $h \cdot f/c$ is absorbed or emitted. Conservation of energy and momentum then requires that besides the Doppler shift there must be another shift, which causes the emission line to be on the red side of the absorption line by $2 \cdot df = (h \cdot f^2)/(m \cdot c^2)$. For methane at 88THz: df = 1.1 kHz.

For the linear optical Ramsey scheme, the photon picture cannot be used. As long as spontaneous emission is not relevant, the field can be assumed classical, but, at very high resolution, the molecular motion has to be treated quantum mechanically. The straight line trajectories, used to explain the method, describe the propagation only to the limit of Heisenberg's uncertainty, which means that we have to consider diffraction of the molecular beam at the grating. This can be done as follows: We start considering a molecule, which undergoes its first interaction at a precisely specified location z_1, say the center of an antinode. Then its momentum has to be completely uncertain, it propagates as a spherical wave. Using Kirchhoff diffraction calculus with the deBroglie wavelength of the molecule, we calculate the probability of finding the same molecule at a specified location z_2 along the laser beam in the second interaction region. With the field at z_2 and after averaging over z_2, we get the transition probability for our molecule under test, and summing over the velocity distribution yields the shape of the Ramsey line. For linear absorption, the result is that the line simply splits up symmetrically in two components separated by the well known amount of $2 \cdot df = (h \cdot f^2)/(2 \cdot m \cdot c^2)$. Relaxation was not in the model; what asymmetry this may cause, remains to be investigated.

Conclusion

I have considered a number of error sources, which for the present stage add up to an uncertainty of $\pm\, 4 \cdot 10^{-13}$. With a little effort quite some improvement still seems possible. There may be other biases not yet considered, but for simplicity of evaluation, a simple linear absorption experiment on free molecules should have its merits.

3 G. Kramer: PTB Jahresbericht 1973 (Braunschweig Feb.1974) p. 134

4 G. Kramer: J. Opt. Soc. Am. 68, 1634 (1978)

5 see e.g. J.L. Hall: In Atomic Physics, Vol. 3, Plenum Publ. Corp., New York (1973) p. 615

Phase-Locked Frequency Chains to 130 THz at NRC

B.G. Whitford

Division of Physics, National Research Council of Canada,
Ottawa, Ontario K1A 0R6, Canada

1. Introduction

In the seventies several schemes were designed at NRC for relating CO_2 laser frequencies near 30 THz to the Cs standard [1]; two of these were implemented, and were used in our measurement of the speed of light [2] and of the HeNe/CH_4 laser frequency [3,4].

None of our early chains contained CO_2 frequencies suitable for realizing our current goal, which is the measurement of the frequency of a 260 THz HeNe laser that has been locked to the transition in $^{127}I_2$ by the use of intra-cavity second-harmonic generation [5]. A new chain had to be developed. However, in our case, this was not difficult. Only a different set of CO_2 laser frequencies had to be chosen.

2. Choosing a New Chain

For our first chains, the number of well known CO_2 lines from which one could choose was limited to about 400 regular band lines from four isotopes. In 1985, when we were choosing a new chain for extension of the measurement range to 520 THz, 1053 lines were considered. Under the constraints that the total number of lasers in the chain be limited to 4 (previous chains had 5), that W-Ni diode output frequencies be less than 6 GHz and that 3 or fewer frequencies be involved at each mixing stage, 434 four-laser chains were found. Of these, only two were suitable for reaching both the 130 THz of a colour centre laser needed in the 520 THz measurement, and the 88 THz HeNe/CH_4 laser frequency. However, the addition of a fifth laser offset-locked to one of the basic chain frequencies increased this to six possible chains.

2.1 New NRC Chain

Details of the chosen, phase-locked chain are shown in Fig. 1. A 5th laser, E (not shown) must be offset locked to B in the extension to 520 THz, or, a 5th laser G, to C, in a possible extension to 474 THz. The transitions used for lasers A to G respectively are $R_{II}(29)$ of $^{12}C^{17}O_2$, $P_{II}(14)$ of $^{12}C^{18}O_2$, $R_{II}(28)$ of $^{13}C^{16}O_2$, P(12) of the hot D band of $^{13}C^{16}O_2$, $R_{II}(18)$ of $^{12}C^{16}O_2$, and $R_{II}(24)$ of $^{12}C^{17}O_2$. At each stage, a combination frequency and phase-lock is used. The frequency relations at each mixer are given on the right. The relations to be used for offset locking of the auxilliary lasers are shown on the extreme right. The beat signals generated by the

Frequency Standards and Metrology Editor: A. De Marchi

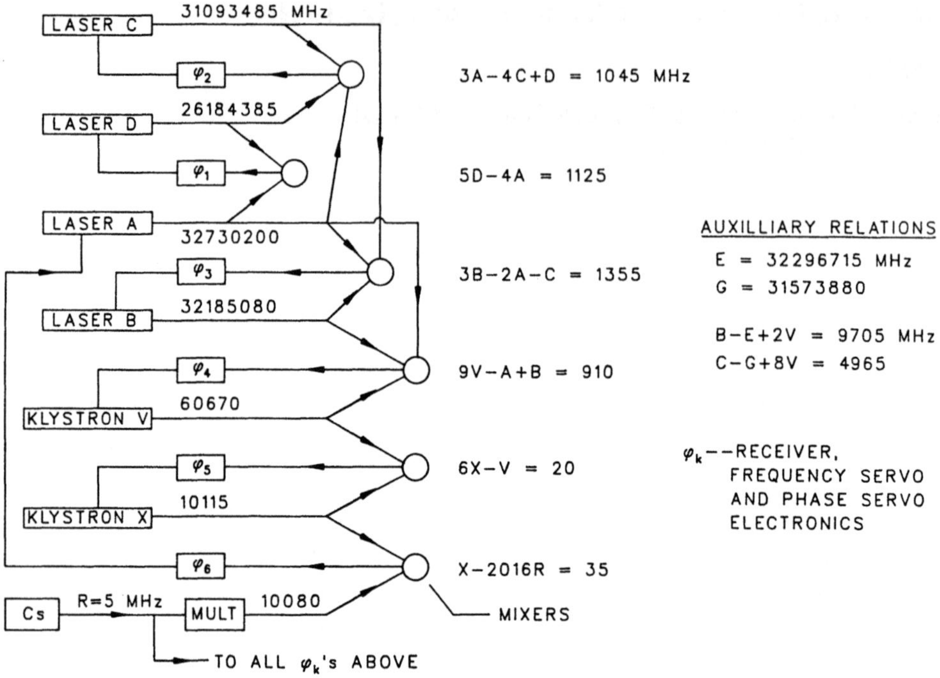

Fig. 1. Basic elements of the four-laser phase-locked frequency chain

mixers are further downconverted to frequencies of 20 or 35 MHz for amplification, limiting, frequency detection and phase-detection.

3. Chain Utilization

The scheme being realized at NRC for measuring the 520 THz frequency is shown in Fig. 2. The 130 THz colour centre laser frequency will be phase locked to the HeNe laser; its frequency, at one-quarter of the iodine frequency, will then determined by mixing with frequencies A and E from the chain.

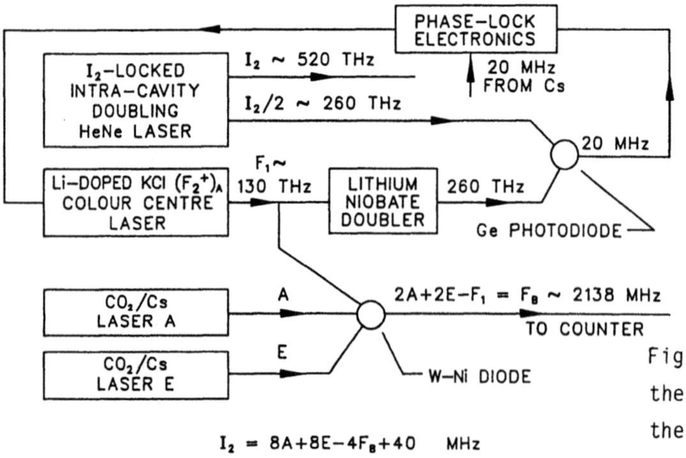

Fig. 2. Scheme for measuring the 260/520 THz frequencies of the HeNe/I_2 laser

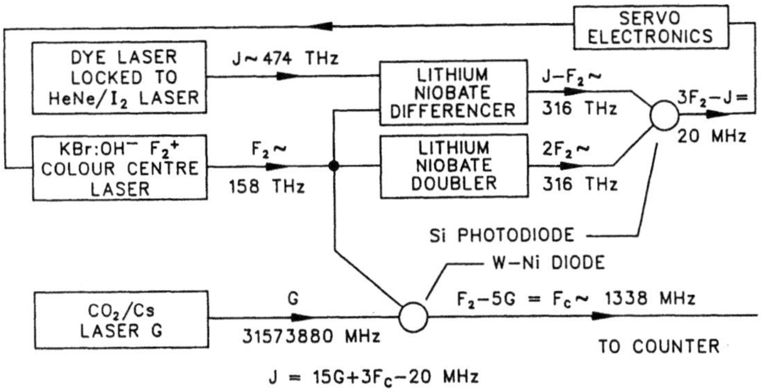

Fig. 3. Proposed scheme for measuring the 474 THz frequency of the $HeNe/I_2$ laser

This chain is particularly good for measuring the $HeNe/CH_4$ frequency M. This can be done by making use of the relation M-2C-D = 4826.6 MHz.

A third use which we are considering is the measurement of the 474 THz frequency of the $HeNe/I_2$ laser. The proposed technique is shown in Fig. 3. The mixing and doubling in the lithium niobate crystals and in the silicon photodiode effectively compare the dye laser frequency against the the 3rd harmonic of the colour centre laser frequency to generate a signal for locking the colour-centre laser frequency at one-third of the dye laser frequency. The 158 THz frequency is then measured against the 5th harmonic of G by mixing in a W-Ni diode. The major problems anticipated in the realization of this scheme are attaining the 6th order mixing, and operating the differencer crystal at 630 ^{0}C.

In addition, the basic frequencies A, B, C, and D, with the auxilliary frequency E can be used to generate many other Cs based reference frequencies spanning the range 1 GHz to 160 THz (near W-Ni diode limit) by harmonic generation and mixing in W-Ni diodes. The mixing can be represented by

$$iA + jB + kC + mD + nE = f_r,$$

in which i, j, k, m, and n, are integers, each ranging from -5 to +5, including 0, and f_r is the generated reference frequency. With mixing order limited to 7, there are 2860 such frequencies in this range. Each of our former phase-locked chains of 5 lasers provides another unique set of 3000 values in this range. Since at least 400 frequency chains of this type are feasible, one can conclude that about one million Cs based reference frequencies could be synthesized. This suggests that virtually any frequency below 160 THz could be measured against Cs by this technique. Incorporating continuously tunable colour center lasers, which now span the range 80 to 370 THz, and harmonic generation and mixing in nonlinear crystals, one could extend this coverage by Cs-referenced frequencies to the near IR and the visible region. The technique described above for the measurement of the 474 THz frequency might be used for general frequency measurement in the visible region.

Acknowledgement
The author wishes to thank G.R. Hanes for making the computer-aided searches for the new chains, and while doing so, originating the concept of four-laser chains.

References

1. B.G. Whitford: IEEE Trans. IM-29, 3 (1980)
2. K.M. Baird, D.S.Smith and B.G.Whitford: Opt. Comm. 31, 3 (1979)
3. B.G. Whitford and G.R. Hanes: CPEM 1986 Digest, p.277
4. B.G. Whitford and G.R. Hanes: IEEE Trans. IM-37, 2 (1988)
5. G.R. Hanes: Appl. Opt. 18, 23 (1979)

Direct Comparison of the Characteristics of Optical and Hydrogen Frequency Standards

S.N. Bagayev, V.P. Chebotayev, V.M. Klementyev, B.A. Timchenko, and V.F. Zakharyash

Institute of Thermophysics, Siberian Branch of the USSR Academy of Sciences, SU-630090 Novosibirsk, USSR

The first optical clock (OC) /1, 2/ comprised a reference frequency stable He-Ne/CH_4 laser at 3.39 μm stabilized to the optical resonance on the $F_2^{(2)}$ methane line with width of about 30 kHz (magnetic hyperfine structure (MHS) was not resolved) and a rather complicated system of division of the optical frequency to the frequencies of the radio-frequency range. The latter consisted of the chain of oscillators successively synchronized to the stable oscillations of the reference laser.

Progress in creation of more perfect OC can be achieved by using as an optical reference (the resonance) on an unperturbed transition of the central (7-6) component of the magnetic hyperfine structure at the $F_2^{(2)}$ absorption line of methane under the conditions where the influence of the second-order Doppler effect is negligible. Improvement of the OC is also bound up with simplification of the system of division of an optical frequency by using the high order harmonics for frequency synthesis.

Unperturbed optical resonances can be obtained by using cold particles, which allows substantial reduction of the influence of a second-order Doppler effect. We make experiments on optical selection of cold particles with an effective temperature of 10^{-1} - 10^{-2} K in a low-pressure gas to obtain saturated absorption resonances of such particles with width of 10-100 Hz, which will allow making on this basis optical frequency standards with frequency reproducibility of about 10^{-16} /3,6/.

Figure 1 presents the block-diagram of the new system of the OC which is now under development at our laboratory. As an optical frequency standard we used the He-Ne/CH_4 laser stabilized to the resonance at the 7-6 transition of the $F_2^{(2)}$ line of methane. The system of division comprises a CO_2 laser and an optically pumped CH_2F_2 laser (λ = 184.3 μm). The 18th harmonics of its generation frequency was used for synthesis. From the low-frequency side there are Gunn diode oscillators and a backward-wave tube (BWT) in the system of division. The frequency of the CH_2F_2 laser (ν = 1 626 602.95±1.5 MHz)

Frequency Standards and Metrology Editor: A. De Marchi

was synthesized and measured. The described optical clock also comprises a measuring-computing complex (not shown in Figure). A spectrometer with a frequency scale and resolution of about 100 Hz was made on the basis of phase locking of the BWT to the oscillations of the submillimeter laser. As nonlinear elements we used point-contact MOM-diodes, Schottky and metal-silicon diodes.

An important experiment with the optical time scale is the comparison of frequency characteristics of the optical and UHF frequency standards. When comparing the frequency characteristics of two standards in the system of optical clock in different ranges, one should observe simultaneously three processes: beatings between the oscillations of two UHF standards, two optical standards (He-Ne/CH_4 lasers) and the signal of beatings between an UHF and optical standards. For the first time the experiments on direct comparison of the frequency characteristics of the optical and UHF standards were made by us in 1983 /2/. At that time we compared the characteristics of a He-Ne/CH_4 laser and a rubidium standard. The frequency instability of beats between the standards was determined by the frequency stability of the Rb standard. Of great interest is the comparison of the characteristics of the most stable UHF standard, e.g. a hydrogen standard, and the high stable He-Ne/CH_4 laser, whose frequency characteristics are close at averaging times of about 100 s.

Figure 2 is the block-diagram of comparison of the characteristics of the He-Ne/CH_4 laser and the H-standard. The system of division comprises two comparable optical standards and two H-standards. The beat signal of the optical and H standards and that of two H-standards

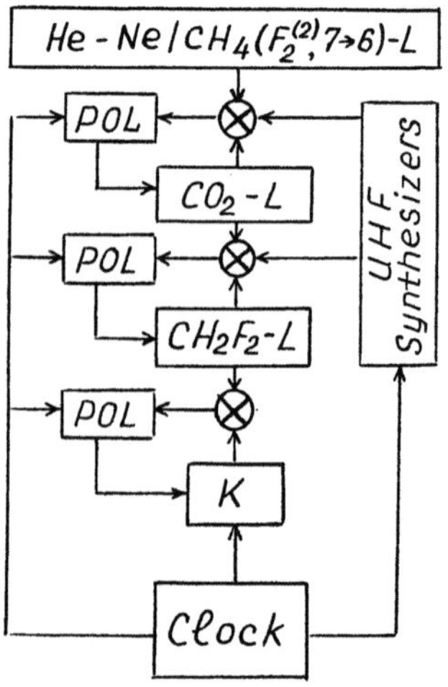

Fig. 1

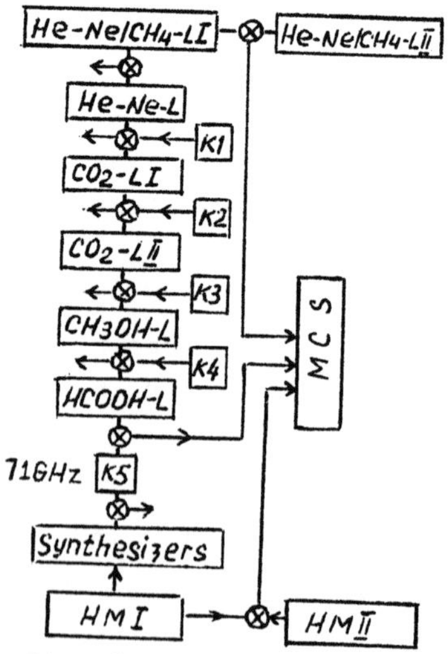

Fig. 2

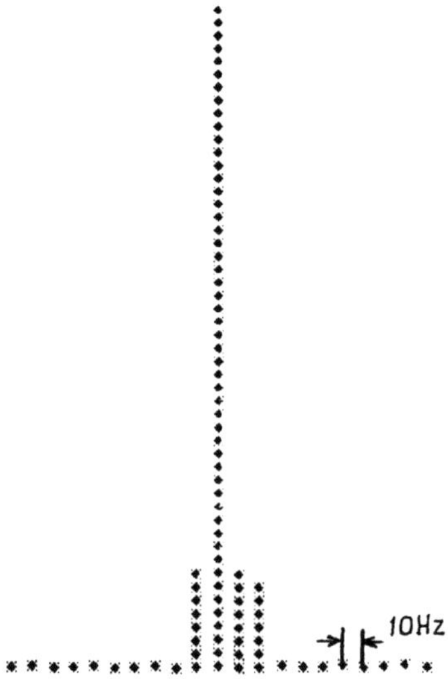

Fig. 3

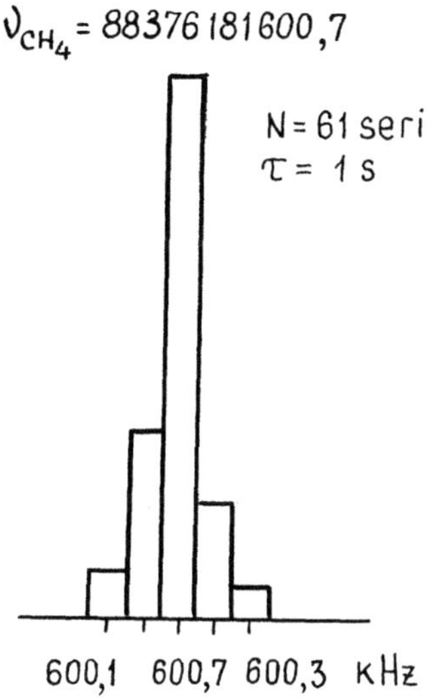

Fig. 4

are fed to the measuring-computing complex (MCC) from the output of which we obtain the results of comparison in the form of histogram and other statistic charactersitics.

Figure 3 shows the histogram of measurements for one series from 100 measurements at τ = 0.1 s. As can be seen, a root-mean square error for this series is about 10 Hz and, hence, the system of optical clock allows measurements at an accuracy not worse than 10^{-13}. Fig. 4 presents the histogram for 61 series of measurements, each series includes 10 measurements at an averaging time of each measurement of 1 s (the measurement results are reduced to the optical frequency). From the histogram one can see that the measurement accuracy is lower than 0,5 kHz. The similar result was obtained at simultaneous measurement of the difference frequency instability of two H-standards, i.e., there is an evident time correlation between the variations of beat frequencies of two H-standards and of the H and He-Ne/CH_4 standards.

Thus, the direct comparison of the characteristics of the optical and hydrogen standards have indicated that in its frequency characteristics the He-Ne/CH_4 laser exceeds the investigated H-standards.

Over the last years various research groups performed absolute frequency measurements of the He-Ne/CH_4 laser stabilized to the $F_2^{(2)}$ line of the P(7) transition of the ν_3 band in methane. The results are presented in /4/. From an analysis of the results of the most precise measurements it follows that from 1980 till 1987 an accuracy of relative measurements was gradually growing and reached the value of about 10^{-11} and the results of measurements of different research groups became closer. By data of /4/ the average value of an absolute frequency of a He-Ne/CH_4 laser is 88 376 181 602.3±0.8 kHz. However these measurements are related to the He-Ne/CH_4 lasers whose frequencies are stabilized to an optical saturated absorption resonance when the MHS components are not resolved. Ref. /5/ reports on comparison of the frequency of a He-Ne/CH_4 laser stabilized to the central MHS component with a known frequency of a portable He-Ne/CH_4 laser. The result is ν_{CH_4} = 88 376 181 599 900±90 Hz.

Recently we made the direct absolute frequency measurements of the MHS 7-6 component of the $F_2^{(2)}$ line of methane (λ = 3.39 μm) by using the optical clock and the hydrogen standard. Optical selection of cold particles in a low-pressure gas enabled us to obtain absorption resonances of cold CH_4 molecules with a homogeneous width of about 1 kHz in a 2 mm light beam in the He-Ne/CH_4 laser with an internal absorption cell and to observe a recoil doublet at the 7-6 transition of the hyperfine structure /3/. An effective temperature of cold molecules was about 0.1 K. The according second-order Doppler resonance shift bound up with cold particles is $\Delta\nu \simeq$ 0.06 Hz at a gas temperature of about 300 K.

Under these conditions the measured frequency value ν_{CH_4} = 88 376 181 600.7±0.4 kHz corresponds to that of the unperturbed frequency of the $F_2^{(2)}$ transition in methane. A great measurement error was due to an accuracy of estimation of an absolute H-standard frequency.

References

1. V.P.Chebotayev: J.de Phys. 42, Colloq. C8, Suppl., 5505 (1981)
2. V.P.Chebotayev, V.M.Klementyev, M.V.Nikitin, B.A. Timchenko, V.F.Zakharyash: Appl.Phys. B36, 59 (1985)
3. S.N.Bagayev, A.E.Baklanov, A.S.Dychkov, P.V.Pokasov, V.P.Chebotayev: ZhETF Pisma Red. 45, 372 (1987); Rev. Roum. Phys. 33, 361 (1988)
4. D.J.E.Knight: Metrologia 22, 251 (1986)
5. Yu.S.Domnin, A.N.Malimon, V.M.Tatarenkov, P.S.Shumyatsky: Izmeritelnaya Tekhnika 6, 5 (1986)
6. S.N.Bagaev, V.P.Chebotyev, A.K.Dmitriev, A.E.Om, Yu.V.Nekrasov, B.N.Skvortsov. Obtaining of Supernarrow Saturated Absorption Resonances on Cold Particles and Optical Frequency Standards.: this Proceed.

Part IX

Near IR Frequency Standards

The Physics of Optical Frequency Standards Using Saturation Methods

C.J. Bordé

Laboratoire de Gravitation et Cosmologie Relativistes,
Paris, France, and
Laboratoire de Physique des Lasers-Villetaneuse, France

This paper presents a survey of our present theoretical description of the physical phenomena which play an important role in optical frequency standards based on saturation methods. Such systems consist of a set of laser beams (usually counter-propagating along the same axis but sometimes separated in space) which propagate in a TEM mode and interact with 2 or 3 level systems (atoms or molecules in gas phase in a cell or in a molecular beam). The interaction physics is dominated by the relative motion of these two systems which gives rise to residual first-order Doppler effects usually taking the form of transit effects (wavefront curvature shift...) and to the important second-order Doppler effect.

As we shall see, the interaction is strongly velocity selective in the transit regime and this has dramatic consequences on the second-order Doppler shift. More generally, even if we know perfectly well the initial phase space distributions (e.g. Maxwell-Boltzmann for the molecules, well-defined boundary conditions for the laser beams) the two systems perturb each other in a non-local way : the medium may be optically thick but more seriously it will behave as a gas lens because of saturated dispersion and this will strongly affect the laser beam propagation and hence the signal power (first part of this paper) ; similarly, the laser beam may be strong enough to affect importantly the atomic motion in momentum space (recoil physics) or to induce large light shifts. The two parts of the system (atoms and light) should therefore be treated in a self-consistent manner. In practice, one is obliged to separate the two problems (and then iterate) : in a first step, one studies the light propagation with a given model of medium polarization and then one studies the molecular response to a given laser field geometry.

Let us start with gas lens effects on the laser propagation. These originate from the existence of a saturated dispersion χ'_{NL} inevitably associated with the saturated absorption χ''_{NL}. The existence of this χ'_{NL} has been demonstrated from the very beginning of saturation methods and often used in stabilization schemes: In heterodyne beat frequency experiments [1-3], in direct interferometry [4] and even beam deflection [5] outside the laser resonator. The transverse Gaussian dependence of the laser field is directly imaged into a transverse dependence of the non-linear dispersion χ'_{NL} and the medium behaves as a converging/diverging lens depending on which side of the resonance the laser frequency is tuned. The beam diameter increase or decrease leads to a locally frequency-dependent power and to frequency-dependent losses for any diaphragm along the path, with a dispersive signature. This phenomenon was first recognized by B.K. GARSIDE for He-Ne lasers [6]. Many further studies were conducted on the Lamb dip or on the CH_4 saturation peak asymmetries [7-11] inside He-Ne laser resonators, culminating with the proposals of critical geometries [12-14]. Here, for simplicity, we shall only consider the case of an external cell in which the asymmetry is directly related to the fact that the laser beam may be partly blocked before detection [15,16].

A theoretical approach to these gas-lens effects, which has been widely used, is the ABCD formalism of geometrical optics, in which the position and angle of a beam in the output plane are related to corresponding quantities in the input plane by a ray transfer matrix :

$$\begin{pmatrix} x_2 \\ y_2 \end{pmatrix} = \begin{pmatrix} A & B \\ C & D \end{pmatrix} \begin{pmatrix} x_1 \\ y_1 \end{pmatrix} \tag{1}$$

The use of these matrices for Gaussian laser beams was made possible by the work of H. KOGELNIK [17] who showed that the two corresponding complex curvatures ($1/q = 1/R - i\,\lambda / \pi w^2$), are related through the ABCD law $q_2 = (Aq_1 + B) / (Cq_1 + D)$ and gave the expression of the matrix for a medium whose refractive index n varies quadratically with the distance r from the optical axis : $n = n_0 - n_2 r^2 / 2$. A

Frequency Standards and Metrology Editor: A. De Marchi

possible extension to complex indices was then considered but the physical significance of complex x and y was not found. In what follows we show that such a generalization is indeed possible and more elegant that the usual approach. If the electric laser field is written as $(1/2)\, \hat{e} E_0(z,t) U_0(x,y,z) \exp i(\omega t - kz + \phi)$ + c.c. the departure from a plane wave can be written as :

$$U_0 = (1/X) \exp[-i(Y/X) k_0 r^2/2] \quad \text{with } k_0 = \omega/c, \tag{2}$$

and where $X = i\, w \exp(-i\,\Phi)$ and Y are respectively a complex beam waist radius and a complex divergence such that X / Y is the complex curvature q. With a quadratic law for k : $k^2 = k^2(o) - k(o)\, k_2\, r^2$ the wave equation gives, in the paraxial approximation :

$$i(d/dz) U_0 = [1/(2k(0))] \Delta_t U_0 - (k_2/2) r^2 U_0 \quad \text{with } \Delta_t = \Delta - \partial^2/\partial z^2, \tag{3}$$

which is completely analogous to the Schroedinger equation for a particle moving in a quadratic potential and from which a first-order differential equation can be written in matrix form for the unnormalized "spinor" $\begin{bmatrix} k_0 X \\ Y \end{bmatrix}$:

$$\frac{d}{dz} \begin{pmatrix} k_0 X \\ Y \end{pmatrix} = \begin{pmatrix} 0 & k_0/n_0 \\ -n_2/k_0 & 0 \end{pmatrix} \begin{pmatrix} k_0 X \\ Y \end{pmatrix} = \frac{i\,\sigma.\mathbf{N}}{2} \begin{pmatrix} k_0 X \\ Y \end{pmatrix} \tag{4}$$

where σ are the Pauli matrices and **N** is the vector $\{-ik_0(1-n_0 n_2/k_0^2)/n_0 \;,\; k_0(1+n_0 n_2/k_0^2)/n_0 \;,\; 0\}$. Alike the case of the atomic spinor equation, the integration of this equation requires generally a "chronological operator" T but becomes trivial if **N** is independent of z :

$$\begin{pmatrix} k_0 X_1 \\ Y_1 \end{pmatrix} = T \exp(i/2)\, \sigma.\!\int \mathbf{N}\, dz \begin{pmatrix} k_0 X_0 \\ Y_0 \end{pmatrix} = \exp\{ i(\sigma.\mathbf{n})\, L(n_2/n_0)^{1/2} \} \begin{pmatrix} k_0 X_0 \\ Y_0 \end{pmatrix} \tag{5}$$

n is the vector such that $\mathbf{n}^2 = 1$ of components:

$\{-ik_0(1-n_0 n_2/k_0^2)/2(n_0 n_2)^{1/2},\; k_0(1+n_0 n_2/k_0^2)/2(n_0 n_2)^{1/2},\; 0\}$ and L the length of the medium. Using the properties of Pauli matrices (parametrization formula of the SL(2C) group) we obtain directly the transformation matrix for $\begin{bmatrix} k_0 X \\ Y \end{bmatrix}$:

$$\exp\{ i(\sigma.\mathbf{n})\, L(n_2/n_0)^{1/2} \} = \cos L(n_2/n_0)^{1/2} + i(\mathbf{n}.\sigma) \sin L(n_2/n_0)^{1/2} \tag{6}$$

which corresponds exactly to the ABCD matrix of Kogelnik for $\begin{bmatrix} X \\ Y \end{bmatrix}$ and is now valid for complex index in an unambiguous formalism.

There are many other elegant features brought by this formalism. From equation (4) one can see that the second component $R_2 = i\, k_0(XY^* - X^*Y)$ of the 4-vector associated to the spinor is invariant when the index is real. This corresponds to the conservation of "étendue" of the beam. Using the covariant components of the spinor, the "Wigner function"

$$f(k_0 r, s) = [(XX^*)^{1/2}/2\pi] \int dR \exp(i k_0 s R)\, U_0(r+R/2)\, U_0^*(r-R/2)$$

for each transverse coordinate, is proportional to :

$$\exp -2\left\{ \widehat{k_0 r \;\; s} \begin{pmatrix} Y \\ -k_0 X \end{pmatrix} \widehat{Y^* \;\; -k_0 X^*} \begin{pmatrix} k_0 r \\ s \end{pmatrix} / (-R_2) \right\} \tag{7}$$

and defines an ellipse of constant area in phase space for constant R_2, which tilts and squeezes along the $k_0 r$ axis for free-space propagation. Finally, the formalism has been extended to higher order modes using Laguerre polynomials.

From (6) we get :

$$X=(1/2)\,[X_0 - iY_0/(n_0n_2)^{1/2}]\exp[i\,L\,(n_2/n_0)^{1/2}] + (1/2)\,[X_0 + iY_0/(n_0n_2)^{1/2}]\exp[-\,iL\,(n_2/n_0)^{1/2}] \text{ and}$$

$$Y=(1/2)\,[Y_0 + iX_0(n_0n_2)^{1/2}]\exp[i\,L\,(n_2/n_0)^{1/2}] + (1/2)\,[Y_0 - iX_0(n_0n_2)^{1/2}]\exp[-\,i\,L\,(n_2/n_0)^{1/2}] \qquad (8)$$

For $X_0 = i\,Y_0/(\,n_0\,n_2\,)^{1/2}$ we get the eigenvector of equation (4) which corresponds to fields propagating with a constant geometrical profile in a converging medium (analog of the wavefunctions of the harmonic oscillator).

Otherwise, if $n_2 \to 0$, we recover the usual laws for free-space propagation

$$\begin{pmatrix} X \\ Y \end{pmatrix} = \begin{pmatrix} i\,w\exp(-\,i\Phi) \\ \theta \end{pmatrix} = \begin{pmatrix} 1 & z/n_0 \\ 0 & 1 \end{pmatrix}\begin{pmatrix} i\,w_0 \\ 2w_0\,n_0/b \end{pmatrix} = \begin{pmatrix} i\,w_0(1-2iz/b) \\ 2w_0\,n_0/b \end{pmatrix} \quad \text{with } b = kw_0^2 \qquad (9)$$

and U_0 has the form of a two-dimensional minimum uncertainty wave packet. If $n_2 \neq 0$,

$$w^2/w_0^2 = XX^*/X_0X_0^* = |\cos[L(n_2/n_0)^{1/2}]|^2 + [(Y_0/X_0)(\cos L(n_2/n_0)^{1/2})^*\,\sin L(n_2/n_0)^{1/2}/(n_0n_2)^{1/2} + c.c.] + (Y_0Y_0^*/X_0X_0^*)\,|\sin L(n_2/n_0)^{1/2}|^2/|(n_0n_2)^{1/2}|^2 \qquad (10)$$

which can be approximated by $1-\mathrm{Re}\{n_2\}\,L^2 + \mathrm{Im}\{n_2\}\,L^2.(2L/b)$ and a similar law for the radius of curvature induced by the medium :

$$1/R = \mathrm{Re}\,\{\,Y/X\,\} = \mathrm{Re}\,\{\,(Y_0/X_0 - (n_0n_2)^{1/2}\,\mathrm{tg}\,L(n_2/n_0)^{1/2})\,/\,(1+(Y_0/X_0)\,\mathrm{tg}\,L(n_2/n_0)^{1/2}/(n_2/n_0)^{1/2})\,\}$$

$$\text{and } \Delta(1/R) \approx -\mathrm{Re}\{n_2\}L - \mathrm{Im}\{n_2\}L(2L/b) \qquad (11)$$

The main effect of interest to us is the change in beam waist radius which requires an evaluation of the quantity n_2L^2. The complex quadratic index is related to the non-linear susceptibility by $\chi_{NL}(E(r)) - \chi_{NL}(E(0)) = -\,n_2r^2$ and can be obtained from various models such as Lamb's 3rd-order formula [18] or others in strong laser fields [19].

For the purpose of giving an estimate of the effects we can simply remember that χ' and χ'' will have the same order of magnitude (they satisfy the Kramers-Kronig relations in weak fields [20]) and that χ'' is directly related to the absorption coefficient ($\chi''=\Delta\alpha\lambda/2\pi$) and hence to the signal power contrast

$C = \Delta P_{SIG}/P \approx \Delta\alpha L$ by $\chi'' = C\,\lambda/2\,\pi\,L$

Since $n_2w^2 \approx \chi$ (rigorously $2(S\,d\chi/dS)_{r=0}$ where S is the saturation parameter)

$$n_2\,L^2 \approx C\,\lambda/2\,\pi\,w_0^2 = C\,L/b.$$

Diaphragming the beam gives a contribution of gas lens effects to the signal which is given by the relative change in w^2 : $\Delta P_{GAS\text{-}LENS}/P \approx \Delta w^2/w^2 \approx n_2L^2 \approx CL/b$
so that finally $\Delta P_{GAS\text{-}LENS}/\Delta P_{SIGNAL} \approx L/b$ whatever the signal and the regime (transit or collision-dominated). For typical experiments this quantity may be of the order of a few percents resulting in an asymmetry which can be of the same order if there is any significant truncation of the laser beam.

Let us now turn to the rigorous calculation of the molecular response. Such a calculation requires to solve the molecular density matrix equations with the actual spatial dependence of the laser fields. For counter-propagating waves along the same z axis, a serious complication arises already in the classical motion approach, owing to the fact that the molecules generate all Fourier harmonics of exp $(\pm\,ikz)$: even ones for the populations $n_\alpha = \Sigma\,n_{\alpha,2p}\exp(2ipkz)$ and odd ones for off-diagonal density matrix elements (coherences) $\rho_{ba} = \Sigma\,\rho_{ba,2p+1}\exp[(2p+1)ikz]\exp[-\,i\omega t]$ It is therefore necessary to write an equation for each Fourier component. Fortunately these can be reduced to an infinite set of coupled first-order ordinary differential equations by a suitable rotation of the axes along the molecular trajectory [20]

$$v_x\,(d/dx)\,n_{\alpha,2p} = -\,(2ipkv_z+\gamma_\alpha)(n_{\alpha,2p} - n_\alpha^{(0)}\delta_{p0}) - i\varepsilon_\alpha\Omega^+\exp(i\phi^+)\,U_0^+\,\rho_{ba,2p+1} + \ldots \text{with } \varepsilon_b = 1,\ \varepsilon_a = -1$$

$$v_x\,(d/dx)\,\rho_{ba,2p+1} = -\{\,i\,(\omega_0-\omega+(2p+1)kv_z+\gamma_{ba}\}\,\rho_{ba,2p+1} - i\Omega^+\exp[-\,i\phi^+]\,U_0^{+*}(n_{b,2p} - n_{a,2p})] + \ldots (12)$$

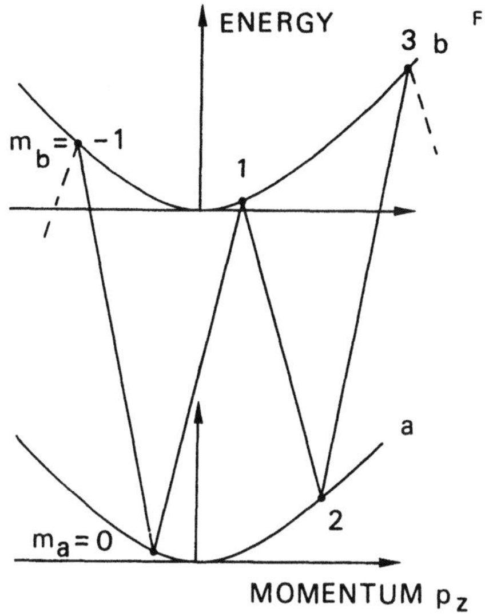

Figure 1

where $\Omega^{\pm} = (\pi\mu E_0^{\pm}/h)$ and $U_0^{\pm}(x,y) = L^{\pm}(Z)\exp[-L^{\pm}(Z)(r/w_0)^2]$; $L^{\pm}(Z) = (1 \pm iZ)/(1+Z^2)$; $Z = 2z/b$ which apart from a coefficient (iw_0) are the functions given before for Gaussian laser beams.

From the solution of these equations one can then compute either the power absorbed :
$\mathcal{P}^{\pm}_{abs}/h\nu = 2\Omega^{\pm}\,\mathrm{Im}\{\int d^3r\, d^3v\ \rho_{ba,\pm 1}\, U_0^{\pm}\exp(i\phi^{\pm})\}$.
or the flux of excited atoms reaching a detector $\Phi_\alpha(X) = \int dydz \int d^3v\, v_x\ n_\alpha(X,y,z,\mathbf{v},t)$
or even generate a map of susceptibilities in space.

An important piece of the physics of optical standards is missing in this classical motion approach namely the recoil physics associated with momentum exchanges between the light and the molecules along the optical z-axis. It is essential to consider independently the momentum states $\mathbf{p}_0 \pm m\, h\, \mathbf{k}/2\pi$ generated from an initial $\mathbf{p}_0$ through the interaction with the (±) waves(Fig. 1). These states have a width $h\Delta k/2\pi \approx h/2\pi w_0$ which is much less than their separation $hk/2\pi$. One can therefore build isolated wave packets characterized by an integer number m and write the density matrix elements as a sum of density matrix elements between energy momentum states :

$$<\mathbf{r}\,|\,\rho_{\alpha\beta}(\alpha_0\,,\mathbf{p}_0\,,t)\,|\,\mathbf{r}> = \Sigma_{m,m'}\int d^3p\, d^3p' <\mathbf{r}\,|\,\mathbf{p}><\mathbf{p}\,|\,\rho_{\alpha\beta}\,|\,\mathbf{p}'><\mathbf{p}'\,|\,\mathbf{r}> = \Sigma_{m,m'} <m\,|\,\rho_{\alpha\beta}(\mathbf{r})\,|\,m'>/h^3$$

Each of these terms varies proportionally to exp [i(m-m') kz] and therefore contributes to a given Fourier component of the expansion of the classical density matrix element. As an example the calculation of the absorbed power $\mathcal{P}^+$ requires the following sum over m_b of coherences between energy momentum states $\Sigma <m_b\,|\,\rho_{ba}\,|\,m_b - 1>/h^3 = \rho_{ba+1}\exp[-i(\omega t - kz)]$

Each wave packet can be treated classically with a suitable recoil correction to the energy.It leads this time to a double infinity of coupled differential equations [21].

$$v_x\,(d/dx) <m_b|\rho_{ba}|m_b - (2p+1)> = -\{i(\omega_0-\omega+(2p+1)kv_z+[m_b^2-(m_b-(2p+1))^2]\delta)+\gamma_{ba}\}<m_b|\rho_{ba}|m_b-(2p+1)>$$
$$+\,i\,\Omega^+U_0^{+*}\exp[-i(\omega t - kz)]<m_b-1|\,\rho_{aa}\,|\,m_b-(2p+1)>+\ldots \qquad (13)$$

where $\delta = h\nu/2Mc^2$ enters the recoil corrections.

Similar equations are obtained for the Fourier components of the Wigner functions $f_{\alpha\beta}{}^{(n)}(\mathbf{r},\mathbf{p},t,\alpha_0,\mathbf{p}_0)$ when summed over isolated wave packets in $\mathbf{p}$ space.
If the recoil shift is very small compared to the line-width (case of OsO_4 for example) then by summation over all energy momentum states (collapsing the zig-zag picture) one recovers the equations (12).

On the other hand if the recoil shift is large (case of CH_4 or of Ca) molecules get quickly out of resonance and only a few energy-momentum states need to be considered.

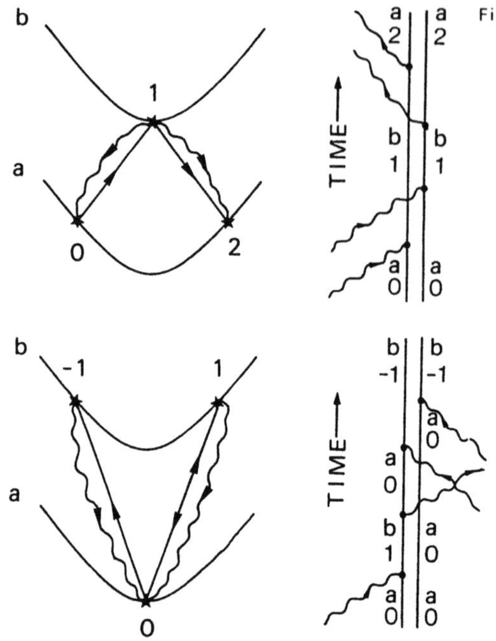

Figure 2

Such a truncation is also valid in the weak field situation which we shall first summarize before any strong field consideration. In this case, to 4th-order, only three energy-momentum states need to be taken into account either in the Λ configuration (lower-frequency recoil peak) or in the V configuration (higher-frequency recoil peak) (Fig. 2). Each of these two processes can also be represented by double-sided Feynman diagrams embedded in space-time, as the interference between two quantum processes. For the lower-frequency recoil peak these two processes are a photon absorption from the saturation wave followed by a photon emission in the probe wave. For the higher-frequency recoil peak a three-photon process (absorption followed by emission of a photon from the saturation wave followed by the absorption of a photon from the probe wave) interferes with a single photon absorption from the probe wave.

A set of diagrammatic rules is associated with these space-time diagrams, from which the signal power or any other physical quantity can be automatically calculated [21]. As a reminder of the physics known to us, that plays a role in the weak field domain, we shall step through the various physical effects which can be included in these rules in a unified way :

1 -The relaxation processes (collisions and radiative decay) are introduced in the propagators $\exp[-(\gamma_{\alpha\beta} + i\omega_{\alpha\beta})\,\tau]$ for each density matrix element $\rho_{\alpha\beta}$ through $\gamma_{\alpha\beta}$ and a pressure shift correction to $\omega_{\alpha\beta}$.

Then, we have all the effects connected with the atomic motion:
2 -Weak elastic collisions which represent another relaxation process directly affecting this motion, are included in a modified propagator given in Ref [22] in terms of Fourier transforms of the collision kernels. These collisions are responsible for a non-linear dependence of the width and shift of saturation peaks with pressure

3/4 - The first-order Doppler effect and transit effects come in through the v_z and v_r dependences of the argument of the laser field at each interaction vertex.When U_0 is complex this accounts very well for the wave-front curvature induced shift [20,23].

5 - One should emphasize that these double-sided Feynman diagrams and the corresponding rules can be made fully covariant with respect to Special Relativity thanks to relativistic quantum field theory restricted to the one-particle sector [21]. Hence the time which appears between two vertices is a proper time of the molecule and the space-time coordinates of each interaction vertex are corrected by the usual factor $\gamma = (1-v^2/c^2)^{-1/2}$ which automatically brings in the second-order Doppler effect.

6 -Recoil shifts appear as $(m_\alpha^2 - m_\beta^2)\,\delta$ corrections to the frequencies in the propagators and yield the recoil splitting.

7 -Space-time dependent light shifts can be introduced also via a modified propagator [21] which, to the lowest order, is equivalent to the computation of sixth-order diagrams [24].This approach accounts for the light shift of the recoil peaks [25] (recoil splitting contraction) as well as for the light shifts in three-level systems [24].

8 -Accelerations (e.g. gravitation or gravitational waves), which add an extra $\mathbf{a}.\nabla_v$ operator in the Liouville equation, can be treated thanks to frame transformations [26].

Finally : 9- Any discrete simple Fourier content of the laser spectrum such as a frequency modulation requires a simple addition of diagrams [26,27].

A diagram starts with the unperturbed initial phase space distribution. In this way both statistical mechanics and QED are incorporated. The product of all factors appearing on the diagram is summed over phase space and over all proper times between vertices. In a number of cases, all these integrals can be performed analytically. An example of interest to this conference is the lineshape in the transit dominated regime for which the ratio of transit time to lifetime $\eta = \gamma w_0/u$ is small ($u = (2kT/M)^{1/2}$).

Quite generally, we can write the signal power as a product of dimensionless quantities [28]

$$\Delta \mathcal{P}^{\pm} = -2\pi^{1/2}\,[h\nu/(w_0/u)]\,[N_0 L \int dx\,dy\, U_0^{+}U_0^{+*}U_0^{-}U_0^{-*}]\,[(u/w_0)/ku]\,(\Omega^{+}w_0/u)^2\,(\Omega^{-}w_0/u)^2\,J(\omega-\omega_0)$$
$$\sim -h\nu N_0 u L/k \qquad (14)$$

where the first factor is the maximum energy that a single molecule can provide during its interaction time (the transit time w_0/u in the present case), the second factor is the total number of molecules, the third one is a partition function in v_z space (ratio of hole burnt to velocity distribution widths). The two efficiency factors involving the Rabi pulsations $\Omega^{\pm}$ are proportional to Rabi nutation angles ($\pi^{1/2}/4$ for a $\pi/2$ pulse) and J is a lineshape of the order unity at line center. In the weak field limit, an analytic expression for the lineshape J is given in Appendix C of Ref [20]. Simplified expressions are :
$J = -G + (\pi/4)\ln 2 - (\pi/2)\Gamma - (\pi/2)\ln\eta$ at line center and
$J = G - (\pi/2)\Gamma - (\pi/2)\ln\xi$ off-resonance for $\xi = (\omega-\omega_0)w_0/u \gg \eta$ which gives the half-width at half-maximum as a combination of Euler's and Catalan's constants Γ and G:

$$\Delta\xi = 2^{-1/4}[\exp(3G/\pi - \Gamma/2)]\,\eta^{1/2} = 1.511\,\eta^{1/2} \qquad (15)$$

formula which has been verified experimentally in Novossibirsk [29]. The line gets very peaked at the center and very narrow owing to the enhanced contribution of slow molecules which have a much longer interaction time (up to γ^{-1}) than those having the velocity u which may interact only during w_0/u. This selective effect of the interaction has the other major advantage to reduce dramatically the second-order Doppler shift but it works only in an unsaturated regime : the Rabi factor has to be at most ~1 for the transverse velocity class v_r effectively selected $\Omega w_0/v_r \sim 1$. So that the efficiency factor $\Omega w_0/u$, which comes to the 4th power in the signal expression, is reduced by v_r/u and even should not exceed $\eta \ll 1$ for a full use of the effect. Otherwise the maximum of the effective contribution of various transverse velocity classes will shift towards higher values with the value of Ω as illustrated by Fig. 3.

A realistic estimate of the second-order Doppler shift has been a first motivation to develop a strong field theory of the saturation signal around 1973 [20, 30, 31]. A second reason to put again the emphasis on the strong field regime is the new phenomenon reported in [32] which occurs as a consequence of a small amount of wavefront curvature and which is illustrated in Fig. 4. For strong fields the line blows up into a well-centered central core with a dispersion shape and a broad pedestal of reversed asymmetry. This behaviour requires both fields to be very strong, a situation which had never been explored theoretically before, given the large number of coupled differential equations to be solved. The 1973 program solver of equations (12) used a Hamming's predictor-corrector method to integrate the spatial dependence of the coupled differential equations. For convenience we also integrated in this step the two following differential equations representing the local energy exchanges :

$$dP^{\pm}/dx = \mathrm{Im}\{\rho_{ba\pm1}\,U_0^{\pm}\exp(i\phi^{\pm})\} \qquad (16)$$

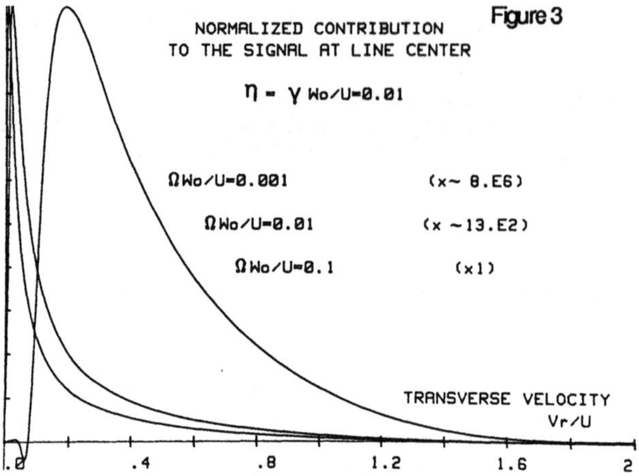

Figure 3

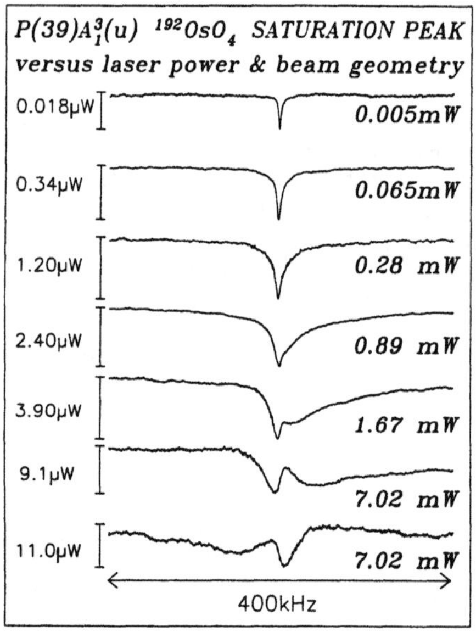

Figure 4

The saturated absorption signal is obtained as the numerical difference between solutions corresponding to modulating one of the beams "on" and "off".

In the present microcomputer program, the multistep predictor-corrector method has been replaced by a single-step Runge-Kutta method with the same success. A remarkable result of the Runge-Kutta algorithm for a linear differential equation such as the Liouville equation is the fact that the solution may be written in closed-form as a simple product of matrices.

The axial velocity integral, calculated with cubic splines, then gives graphs of the relative contribution of different transverse velocity classes for several detuning points of the line (Fig. 3 for $\omega=\omega_0$). A final integration over transverse velocities gives the saturation signal. The main difficulty is to define at which order the Fourier series should be truncated. A practical solution is to increase the number of harmonics until a satisfactorily reproducible result is obtained. Another clue to this problem is brought by the known analytic solution for $\omega=\omega_0$ and $\gamma = 0$ in terms of Bessel functions of integer order [28,32,33] :

$$n_b(+\infty) = n_a^{(0)} (1 - \cos\Theta)/2, \quad n_a(+\infty) = n_a^{(0)} (1 + \cos\Theta)/2, \quad \rho_{ba}(+\infty) = (1/2)\, n_a^{(0)} \sin\Theta \tag{17}$$

with $\cos\Theta = J_0(\zeta) + 2\sum (-1)^j J_{2j}(\zeta) \cos[2jkz\,(1 - v_z^2/2v_x^2)]$

and $\sin\Theta = 2\sum (-1)^j J_{2j+1}(\zeta) \cos[(2j+1)\,kz\,(1 - v_z^2/2v_x^2)]$

where $\zeta = 4\pi^{1/2}\,\Omega\,(w_0/v_x)\,|L^{1/2}|\,\exp[-(kv_z/2)^2\, w_0^2/v_x^2]$ is the critical parameter for fast convergence of the series (the curvature comes in through the complex Lorentzian L (Z)).

Excellent agreement with this solution, which is valid for matched curved wavefronts, gave us great confidence in the numerical computation program which also connected smoothly with the low field analytical results. The program was then used to explore the lineshape for equal intensities in the case of equal transit times and lifetimes (η = 1) at z = - b/4 (diverging probe beam matched with a converging saturation beam) first in the rate-equation approximation (REA) in which all harmonics with p≥2 are neglected (Fig. 5) and then retaining all contributing harmonics up to 2p = 26 (81 coupled equations) (Fig. 6).

In the rate equation approximation a central narrow dispersion feature appears on top of a symmetric pedestal (side fringes develop for lower values of η). This feature reverses if the curvature of both beams is reversed and disappears for beams with opposite curvature. In the rigorous calculation a slightly broadened dispersive contribution subsists on top of an asymmetric pedestal.

To try to understand this behaviour, the simpler case of a single well-resolved recoil component was investigated (three coupled energy-momentum states). In the rate-equation approximation, only a broad and slightly shifted resonance is obtained, whereas with the exp(±i2kz) coherences the line splits in an asymmetric way (Fig. 7). The splitting, which is symmetric in the absence of curvature, increases in proportion with the Rabi frequency and can be interpreted as a dynamic Stark effect (Autler-Townes effect) which gets dissymetric in the presence of curvature. Therefore, this explanation works for the pedestal but not for the central dispersion feature. Since this resonance is present in the REA approach to the 2-level system and since, with a negligible recoil shift, the resummed folded multilevel system of energy momentum states is equivalent to a 2-level system, we may infer that the single recoil peak case does not allow for enough states in momentum space. By working out the lineshapes with an increasing number of energy-momentum states, from 3 to 49 (Fig. 8), the lineshape evolves from the broad line of the single recoil peak to that of Fig. 5. This proves that the central resonance comes from molecules which drift in one direction in momentum (m) space absorbing one wave and emitting in the other.

The dispersion shape of this resonance may be briefly explained along the same lines as the curvature shift in the low field case. Because of curvature, molecules see linear frequency chirps of opposite signs for both waves as a function of the transverse coordinate ; a red shift for the converging saturation (+) wave followed by a smaller blue shift at some later time for the diverging (-) wave (Fig. 9). If this process is repeated in many alternate successive interactions, there will be a net phase shift accumulated which leads to either constructive or destructive interference with the detuning. The sign of the phase shift changes when the order or the roles of both beams are reversed. To have a direct confirmation of this hypothesis in phase space, we have plotted the populations of the various energy-momentum states in the ground level as a function of the transverse coordinate for a given detuning $(\omega_0-\omega)\,(w_0/u) = 1$ and starting in $v_z = 0$ (m = 0). Figure 10 also shows the laser field Gaussian profile (same conditions as Fig.8). The (m = 0) population decreases, rings and comes back to equilibrium. The (m = ± 2) populations build

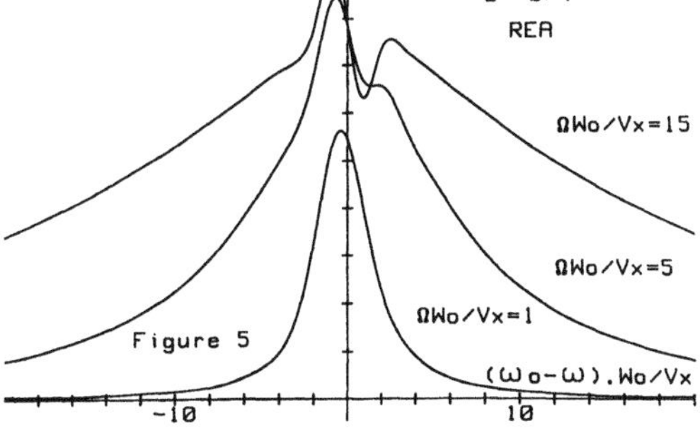

Figure 5

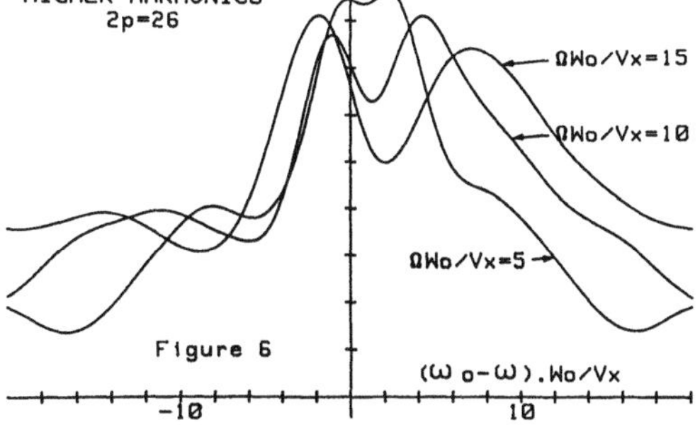

Figure 6

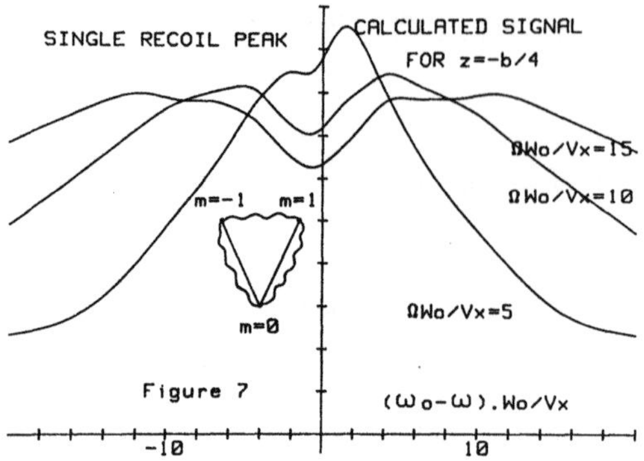

Figure 7

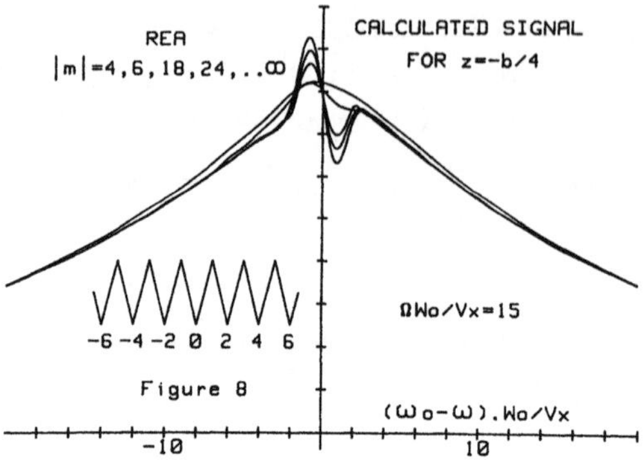

Figure 8

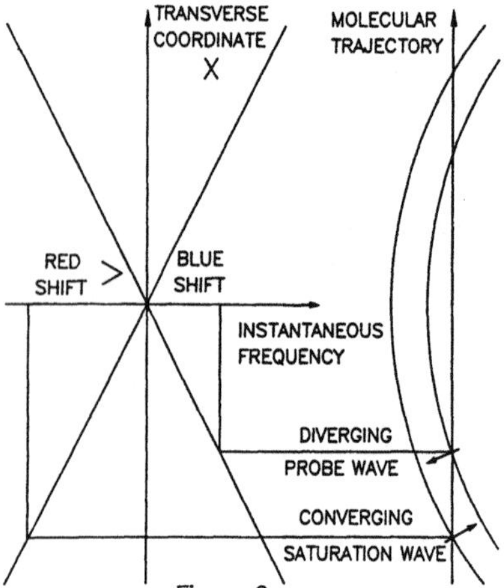

Figure 9

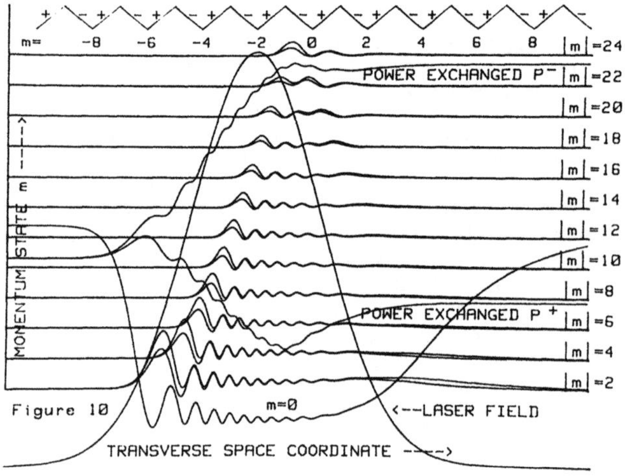

Figure 10

up preferentially in the (m = - 2) state, ring and decay. This preferred direction is also manifest on the m = ± 4...± 24 cases. This corresponds to more absorption of the (-)wave and more emission in the (+) wave and the effect is cumulative. This is clearly illustrated by a plot of the power exchanged for both waves. The (-) wave is continuously absorbed whereas the (+) wave, after a short initial absorption region, gets amplified. The opposite situation is obtained when the detuning has the opposite sign. This new resonance may have a number of applications. For example, we have demonstrated that it provides a way to adjust the laser geometry by first cancelling this signal in a strong laser field and then restoring the weak field conditions [32]. This point is the first one of a larger set of conclusions which are some possible answers to the various problems that we have raised in saturation spectroscopy. To continue with the curvature shift one should recommend reflection symmetry of the experiment and high beam quality (curvature effects disappear slowly with higher pressures).

The gas lens effects require a small L/b, an external absorbing system (in a laser the power level is intrinsically determined by diffraction) no diaphragm and a good collecting optics for detection. The recoil structure can be reduced to a single peak by various tricks [24,28,33] most of which use traveling waves (alternating the propagation direction, to achieve time reversal symmetry, is then recommended to avoid phase shifts). Traveling waves (either separated saturation and probe waves or 4-zones Ramsey fringes) are also a good way to avoid light shifts of the recoil peaks. More generally, one should avoid any off-resonant field and use isolated pairs of levels (no hyperfine structure). The second-order Doppler shift asks for heavy molecules, molecular beams with controlled (cooled or monocinetic) or measured velocity (via Ramsey fringes or first order Doppler shift). In a cell one can use either the η <<1 regime with very weak fields or the η > 5 regime in which the Maxwell-Boltzmann distribution is restored. But in this last case, one meets the difficulty of estimating the pressure shift which needs to be measured and in any case one should prefer transitions of symmetric molecules (e.g. vibration-rotation transitions of spherical tops) with very small pressure shifts. Finally molecular beams are a good way out of this last problem. Most present systems (CH_4, Ca, I_2, OsO_4, CO_2....) meet one or several of the previous difficulties to various extents. This has resulted in the present 10^{-12} level of reproducibility or accuracy. Taking all these factors seriously into account to design systems aimed at accuracy should lead us very soon at the 10^{-13}-10^{-14} level. Any step further will clearly require much more skill and/or a severe loss of signal.

Work supported by DRET and CNRS

References

1. R.L. Barger and J.L. Hall, Phys. Rev. Lett. 22, 4 (1969)
2. G. Kramer , C.O. Weiss and J. Helmcke, Z. Naturforsch. 30a, 1128 (1975)
3. N.G. Basov et al, Sov. J. Quantum Electron. 17, 545-547 (1987) and references therein
4. Ch. J. Bordé et al., C.R. Acad. Sc. Paris 277B, 381 (1973) ; J. de Physique France, 42 1393 (1981)
5. B. Couillaud and A. Ducasse, Phys. Rev. Lett. 35, 1276 (1975)
6. B.K. Garside, IEEE J. of Qu. El., QE4, 940 (1968)
7. H. Maeda and K. Shimoda, J. Appl. Phys. 46, 1235 (1975)
8. A. Le Floch et al., Phys. Rev. Lett. 45, 544 (1980) ; IEEE J. of Qu. El. QE-19, 1474 (1983)
9. G. Stephan et al., J. Physique, France 42, 1623 (1981), 43, 255 (1982), Phys. Rev. A 28, 2344 (1983), A 28, 3450 (1983), A 30, 1925 (1984), J. Physique Lett. 45, L653 (1984)
10. R. Felder, Metrologia 23, 101 (1986)
11. A. Titov, Kvant. Elektronika. 8, 2040 (1981), Opt. Comm. 43, 419 (1982)
12. P. Cerez and R. Felder, Appl. Opt. 22, 1251 (1983), 22, 3313 (1983)
13. A. Le Floch et al., J. Physique Lett. 43, L 493 (1982)
14. G. Stephan, J. Physique Lett. 44, L 361 (1983)
15. A. Le Floch et al., Opt. Lett. 6, 48 (1981)
16. E.N. Bazarov et al., Sov. J. Quantum Electron. 15, 1036 (1985), 17, 1421 (1987)
17. H. Kogelnik, Appl. Opt. 4, 1562 (1965)
18. W.E. Lamb, Phys. Rev. A134, 1429 (1964)
19. H. Greenstein, Phys. Rev. 175, 438 (1968)
20. Ch. J. Bordé, J.L. Hall, C.V. Kunasz and D.G. Hummer, Phys. Rev. A14, 236 (1976)
21. Ch. J. Bordé, Density matrix equations and diagrams for high resolution non-linear laser spectroscopy, Advances in laser spectroscopy, Plenum Press, 1-70 (1982)
22. Ch. J. Bordé, S. Avrillier and M. Gorlicki, J. Phys. Lett. 38, L249-252 (1977), 40, L35 (1979)
23. J.L. Hall and Ch. J. Bordé, Appl. Phys. Lett. 29, 788 (1976)
24. Ch. J. Bordé, G. Camy and B. Decomps, Phys. Rev. 20A, 254 (1979)
25. J.L. Hall, Ch. J. Bordé and K. Uehara, Phys. Rev. Lett. 37, 1339 (1976)Ch. J. Bordé, C.R. Acad. Sc. Paris, 283B, 181 (1976)
26. Ch. J. Bordé, J. Sharma, Ph. Tourrenc and Th. Damour, J. Phys. Lett. 44, L983 (1983)
27. G. Camy, Ch. J. Bordé and M. Ducloy, Opt. Comm. 41, 325 (1982)
28. Ch. J. Bordé, Revue du CETHEDEC, Ondes et Signal, NS83-1, 1-118 (1983)
29. E.V. Baklanov and B. Ya. Dubetsky, J. Phys. France 49, 1307-1313 (1988) and references therein
30. Ch. J. Bordé and J.L. Hall, VIIIth International Quantum Electronics Conference, San Francisco (1974)
31. J.L. Hall, Ch. J. Bordé and C.V. Kunasz, Bull. Am. Phys. Soc. 19, 448 (1974)
32. Ch. J. Bordé, Ch. Chardonnet and D. Mayou, in Laser Spectroscopy VIII, Springer,381-385 (1987)
33. Ch. J. Bordé et al., Phys. Rev. A30, 1836 (1984)

Double-Mode He-Ne/CH_4 Laser Stabilized by Methane Resonances with Relative Width of 10^{-11}

M.A. Gubin, A.V. Nikulchin, and D.A. Tyurikov

P.N. Lebedev Physical Institute, USSR Academy of Science,
Leninsky pr. 53, SU-117924 Moscow, USSR

Presently He-Ne laser (λ =3.39 μm) with an internal methane absorption cell ($F_2^{(2)}$line, P(7) branch, ν_3 band) is the most developed system among optical frequency standards stabilized over nonlinear resonances of saturation absorption or dispersion /1-4/. Most accurate standards of such a type operate under methane pressure of 10^{-4}-10^{-5}Torr, when collisional width of the line (Γ) is comparable or less than the flight-time width ($1/\tau$). In this case, as shown in /5/, the contribution of slow molecules to the resonance is increased and the line shape is essentially changed. At the centre of the nonlinear resonance, characterized by homogeneous width γ, a more narrow structure of the width close to the collisional one (Γ) appears.

When "wide" resonances of methane ($\Gamma > 10$ kHz) are used, the main reason for frequency shifts is the magnetic hyperfine structure (MHPS) of $F_2^{(2)}$line /6/, and the accuracy of He-Ne/CH_4 references is limited at the level $\sim 10^{-11}$, and the reproducibility is at the level about 10^{-12}-10^{13} /7/. Further increasing of the reproducibility and the accuracy is possible only with resolved MHPS, i.e. with the width $\Gamma < 10$ kHz, or, in relative units, $\Gamma/\omega < 10^{-10}$. As the resonance narrowing is accompanied by a dramatic decrease of the signal, the high sensitivity of the registration is needed.

In the present work a double-mode method of subdoppler spectroscopy is used to register narrow resonances of the saturation dispersion /8,9/. The laser containing a methane cell operates in the regime of two axial modes, ω_1 and ω_2, with linear orthogonal polarizations. The beat frequency is much greater than the homogeneous width of the methane line : $|\omega_1 - \omega_2| \gg \gamma$, and narrow resonances of the saturation dispersion appear in the beat frequency when one of the modes is pulled to the Doppler-broadened dispersion line. The presence of the second mode is important, as it plays the role of an "ideal" internal heterodyne, and technical fluctuations are eliminated from the beat frequency.

Frequency Standards and Metrology Editor: A. De Marchi

The registration sensitivity is determined by natural frequency fluctuations of each of the modes. High sensitivity of the method and the use of telescopic expanders of the laser beam inside the absorption cell allow one to achieve spectral resolution $\gamma/\omega \simeq 10^{-11}$ with good signal-to-noise ratio in small-size lasers.

1. Experimental Results

Frequency measurements were performed with three different double-mode He-Ne/CH_4 lasers (DML), containing telescopic expanders. Parameters of the lasers were as follows: DML-30, is a stationary laser, the resonator length is L=1.8 m, the cell length is l=0.9 m, the beam diameter inside the cell is $2w_0$ =30 mm; DML-60 is a stationary laser, L=2.2 m, l=0.6 m, $2w_0$ =60 mm; DML/T-30 is a transportable laser , L=1.8 m, l=0.6 m, $2w_0$ =30 mm. The methane cells were usually cooled down to T=77 K. Detailed schemes of the lasers are given in /4/.

The most narrow resonances of the saturation dispersion ($\gamma \simeq 600$ Hz, $\gamma/\omega \simeq 8.10^{-12}$) were obtained with the laser DML-60 (Fig. 1). The clearly resolved recoil doublet exists in each MHPS component. For frequency stabilization all the three lasers were tuned to lower spectral resolution (the region of homogeneous width was γ =5-8 kHz), when S/N ratio is substantially higher than the one in Fig.1, and MHPS is well resolved (Fig.2).

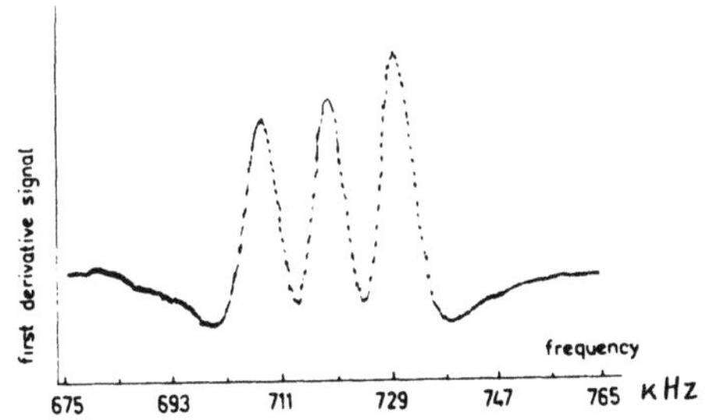

Fig. 1.
Recoil doublet at $F_2^{(2)}$ methane line, observed in DML-60 laser.

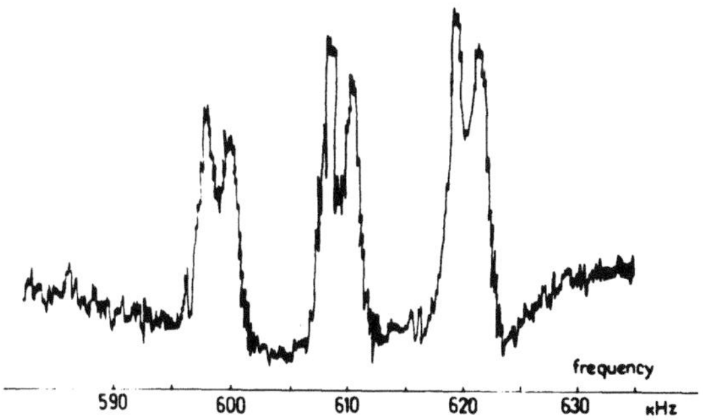

Fig. 2.
MHPS spectrogram showing spectral resolution at which frequency stabilization has been made of all the three compared lasers.

The frequency was stabilized over the central component of the MHPS and similar shape of the MHPS triplet served as the criterion of the laser parameters (methane pressure and field intensity). The results of frequency measurements are as follows :

a) frequency stability (Allan parameter of average time 100s) is $(1-2) \cdot 10^{-14}$ for DML-30 and DML/T-30, less than $1 \cdot 10^{-14}$ for DML-60;

b) for all the three lasers the domain of the parameters, corres - ponding to the linewidth γ =5-8 kHz, methane pressure, field intensity, modulation amplitude was found where the frequency shifts were minimal. The shift values under ± 10 % changes of the above listed parameters were equal to less than 5 Hz in the regime of the 2nd harmonic of modulation, and less than ± 1 Hz for the 4th harmonic. For DML-30 laser the dependences of stabilized frequency vs parameters are given in /4/, and for the transportable one are given in /10/.

c) the most important result is that the frequencies of all the three lasers, which were tuned on approximately equal level of spectral resolution, have coincided with an accuracy of $\pm 3 \cdot 10^{-13}$. For example, Fig. 3 show the temporal dependence of frequency difference between the lasers DML-60 and DML/T-30. The parameters of DML-60 laser during the experiment were stable, but the DML/T-30 resonator was readjusted.

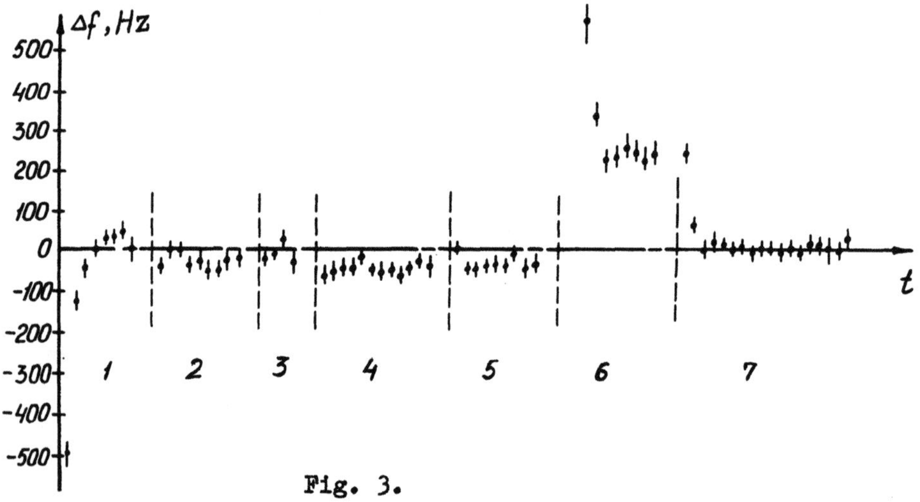

Fig. 3.

Temporal dependence of frequency difference between DML-60 and DML/T-30 lasers under readjustments of the DML/T-30 laser resonator.

It was repeatedly tuned at the power maximum (intervals 1-5,7), and detuned down to one half of the total power (interval 6). Measuring time at each point is 10 s, except interval 3, where it was 100 s.

Duration of the experiment was about 1.5 h (readjustment time is taken into account). Fig. 3. presents also the points corresponding to the intermediate processes appearing in the servo system after the resonator readjustment. The average laser frequency difference in this experiment was: $\omega_{DML-30} - \omega_{DML-60} = -18 \pm 4$ Hz (the points corresponding to the detuned resonator and intermediate pro cesses were excluded.

d) to determine absolute frequency of the central component of MHPS we have compared th e transportable laser DML/T-30 with a one-mode transportable laser developed at VNIIFTRI (Moscow). The frequency of the last one was measured before and after the comparison at the radio-optical chain of the VNIIFTRI /11/ with the accuracy better than 1.10^{-12}. It was determined that absolute frequency of the central component, reproduced by the transportable double-mode laser, is : $\omega_{7\rightarrow 6}$= 88 376 181 599 903 ± 90 Hz. The measurement error is determined mainly by inaccuracies of the radio-optical chain and one-mode transportable laser. (It was stabilized by a wide, γ = 150 kHz, resonance of saturation absorption with nonresolved MHPS). The value of $\omega_{7\rightarrow 6}$ is close enough to the result of /12/, but differs from the result of /13/. For more accurate determination of the frequency and checking the radio-optical chains with a systematic error less than 1.10^{-12} one needs a transportable laser with the reproducibility $\sim 10^{-13}$. The laser DML/T-30 can do this.

2. Discussion

The main reasons for frequency shifts in such frequency standards are : MHPS, 2nd-order Doppler effect (SDE), recoil effect that are theoretically investigated for many times /14-21/. In particular, in /22/ are given the results from the calculation in the third order of the field perturbation theory for the shift of the centre of the first derivative of the saturation dispersion resonance, appearing from a simultaneous action of flight-time and collisional broadening, MHPS, SDE and the line splitting due to recoil effect. The paper covers the domain of the values of Γ and $1/\tau$ parameters close to the experimental ones. Unfortunately, at the present time there is no theory reliably describing the Gaussian beam shape and the field saturation for flight-time domain of the linewidth ($\Gamma\tau \lessapprox 1$). Therefore for the comparison with the experiment it seems reasonable to take into account the field saturation by substitution of for Γ' : $\Gamma' = \sqrt{\Gamma^2 + V^2/2}$, where $V = dE/\hbar$ is the saturation parameter. The calculation shows that for resolved line ($\Gamma \lesssim 2$ kHz) the shift due to SDE is the main one , but it is significantly reduced by flight-time effects. In the region before splitting ($\Gamma >$ 2 kHz) the shift is also determined by the SDE plus slight influence of the MHPS of the line. In this region the saturation dispersion derivative shift very weakly depends on the collisional width. To change the value of Γ from 2500 Hz to 5000Hz the frequency shift is not greater than 5 Hz. The presence of this domain is in good agreement with the experimental data. Further increasing of the level of reproducibility and the accuracy may be expected with the well resolved recoil doublet and a rather narrow light beam. As follows from the calculations performed for the temperature T=77 K, $1/\tau$ = 1100 Hz (the beam diameter $2w_0$ = 90 mm), Γ = 500 Hz the residual SDE shift makes $\sim$4 Hz ($4 \cdot 10^{-14}$). The preserved accuracy of the flight-time and collisional widths being 10 %, the frequency reproducibility will be $1 \cdot 10^{-14}$. Note that here the characteristic parameter $\Gamma\tau$, which determines the resonance value /23/, slightly differs from unit ($\Gamma\tau \sim 0.5$). This allows to eliminate the difficulties in registering supersmall signals within the region $\Gamma\tau \ll 1$.

In conclusion note that the performed measurements of frequency shifts, as well as direct comparison of the frequencies from three different types of double-mode He-Ne/CH_4 lasers (the lasers were stabilized over the central MHPS F_2 methane line component),have shown that the frequency reproducibility was not more than $\pm 3 \cdot 10^{-13}$.

This is in agreement with the calculation within the range of the resonance width γ =5-7 kHZ, which corresponds to the resolved MHPS line. Realization of the resonances within the region of a splitted recoil doublet (γ =400-500 Hz), close to the achieved ones, but with better S/N ratio may allow one to reach the reproducibility and accuracy of double-mode He-Ne/CH_4 optical standards at the level of $\sim 1 \cdot 10^{-14}$.

References

1. Hall J.L., Science, v.202, 147 (1978).
2. Bagaev S.N., Chebotaev V.P. Usp.fiz.nauk, 148, N 1, 143 (1986).
3. Domnin Yu.S. et al. "Measuring techniques", 6, N 3, p.7 (in Rus.).
4. Basov N.G. et al. Kvantovaya Elektronika, 10, 702 (1983)(in Rus.).

5. Rautian S.G., Shalagin A.M. ZhETP, 58, N 3, p.962 (1970).
6. Hall J.L., Borde C., Phys.Rev.Lett., v.30, N 22, p.1101
7. Brillet A., Cerez P., Journal de Physique, N 12, v.42, p.c8-73 (1981) .
8. Basov N.G. et al. Journal de Physique, N12, v.42, p.c8-89.
9. Basov N.G. et al. Kvantovaya Elektronika, 11, N 6, p.1084 (1984).
10. Basov N.G. et al. Kvantovaya Elektronika, 14, p.866 (1987).
11. Domnin Yu.S. et al. "Measuring techniques", 6, N 3, p.5 (1986).
12. Kramer G. Report at the IX Vavilov Conference on Nonlinear Optics, 16-18 June, 1987, Novosibirsk.
13. Zakhar'yash V.F., Klementiev V.M. Report at IX Vavilov Conf. on Nonlinear Optics, 16-18 June, 1987, Novosobirsk.
14. Baklanov E.V., Dubetski B.Ya. Kvantovaya Elektronika, 2, 2041 (1975).
15. Krylova D.D. Kvantovaya Elektronika, 12, N 9, 1962, (1985).
16. Borde C.J. et al. Phys.Rev., ser. A, 14, 236 (1976).
17. Vorob'ev F.G. et al. Optics and Spectroscopy, 27, v.5, 728 (1969) (in Russian) .
18. Semibalamut V.M.,et al. In: , Kvantovaya Elektronika, 5, N 7, 1485 (1978).
19. Semibalamut V.M. et al. "Opticheskie standarty chastoty i vremeni", (ITF, Novosibirsk, 1985), p.98.
20. Dubetski B.Ya., Semibalamut V.M. Kvantovaya Elekttonika, 14, 2349 (1987).
21. Danileiko M.V., Nechiporenko V.N., Yashchenko L.P. Kvantovaya Elektronika , 15, N 10, (1988).
22. Krylova D.D., Kiselev A.E. Kvantovaya Elektronika (in press).
23. Alekseev V.A. et al. ZhETP, 84, 1980 (1983).

State-of-the-Art for High Accuracy Frequency Standards in the 28 THz Range Using Saturated Absorption Resonances of OsO_4 and CO_2

A. Clairon[1], *O. Acef*[1], *C. Chardonnet*[2], *and C.J. Bordé*[2]

[1]Laboratoire Primaire du Temps et des Fréquences, Observatoire de Paris, 61 av. de l'Observatoire, F-75014 Paris, France
[2]Laboratoire de Physique des Lasers (LPL), Université Paris-Nord, F-93430 Villetaneuse, France

I- INTRODUCTION

CO_2 lasers play an important role in optical frequency metrology. In fact, most infrared frequency synthesis chains implemented up to now, include at least one CO_2 laser. In practice, it is possible to measure any frequency from a few GHz to 150THz by means of only two stabilized CO_2 lasers and a microwave source. Furthermore, reproducibilities and stabilities now obtained with OsO_4 stabilized CO_2 lasers allow us to use them as secondary frequency standards for frequency measurements of visible radiation with accuracies approaching 10^{-12}.

The Lamb dip was first used for frequency stabilization of CO_2 lasers [1]. The saturated fluorescence technique developed later by FREED and JAVAN [2] brought significant improvements. It made possible the first accurate absolute frequency measurements of CO_2 lasers by EVENSON. Then PETERSEN, FREED et al. were able to develop a CO_2 frequency reference grid limited to 10^{-10} accuracy by the CO_2 saturation resonance width (10^6 Hz) and pressure shift.

The use of OsO_4 molecules to stabilize CO_2 lasers was proposed and demonstrated by LETOKHOV in 1976 [3]. By using narrow OsO_4 resonances we have been able to improve the accuracy and the stability of CO_2 lasers by two orders of magnitude. Three lines have been measured with a few 10^{-12} uncertainty [4] and ≈70 lines, identified as ν_3 fundamentals, have been measured with a few 10^{-11} uncertainty [5,6]. The entire ν_3 spectrum is now known from Fourier transform spectra and from theoretical predictions with an accuracy of a few MHz [6].

II- MOTIVATION OF THE EXPERIMENT

The present work has three objectives:

- The first one is the comparison of the metrological qualities of two completely different systems. One consists of a Fabry-Perot cavity locked to a laser, itself locked to the third derivative of a molecular saturation peak. It is the "F.P./OsO_4" system (LPTF). The second system "Ext Cell/OsO_4" system (LPL) consists of a laser which creates a standing wave in a large absorption cell and whose frequency is scanned in order to record the first derivative of the saturation peak of a molecular line. In this system, the line center is obtained by means of a best fit of the lineshape to the derivative of a Lorentzian shape. The line used for this comparison is that selected for the OsO_4 frequency grid in quasi-coincidence with the $^{II}R(10)$ CO_2 line. This line has been chosen because its absolute frequency is presently the best known in the 9-12 μm region (fractional accuracy : 1.7 10^{-12}).

-The second objective is to improve the determination of the absolute frequency of the $^{II}R(10)$ CO_2 line whose saturation peak has been recently recorded in the large cell [7]. This has been done by measuring the difference between the CO_2 line center and the frequency given by the "F.P./OsO_4" system.

-Finally, the third objective is the comparison of the respective merits of the OsO_4 and CO_2 molecules for the realization of a frequency standard.

III- RELEVANT MOLECULAR PARAMETERS

The absorption of the ν_3 band of OsO_4 covers the region from 940 to 980 cm^{-1}. In the 10μm region this range can be extended by using XeO_4 and RuO_4[5]. As for CO_2, several hundred lines can be obtained if various isotopic species are used. They can be especially useful as reference lines in the 9μm region.

Frequency Standards and Metrology Editor: A. De Marchi

Table I	OsO_4	CO_2
Hyperfine structures:		
- absent for:	Os = 192,190,188	C = 16,18 ; O = 12,14
- resolved for:	Os = 189,187	?
Mass:	≈ 256	≈ 44
Second-order Doppler shift:	≈ - 1.1 10^{-13}	≈ -6.2 10^{-13}
Recoil splitting:	≈ 4.9 10^{-13}	≈ 2.8 10^{-12}
$P_{\pi/2} = (\pi/2^6)\, hu^2/2\pi\alpha(\mu/e)^2$ (*):	≈ 3.4 10^{-6}W [P(39)line]	≈ 7 10^{-4}W [P(20)line]
Absorption coefficient($m^{-1}Pa^{-1}$):	≈ 4.5 10^{-1} [P(39)line]	≈ 2.25 10^{-4} [P(20)line]
Zeeman effect:	≤ 10^{-12} / Gauss	negligible (quadratic effect)
Pressure shift:	< 225 Hz Pa^{-1}	See Fig.1
Pressure broadening (HWHM):	See Fig. 2	7.3 10^4 Hz Pa^{-1} (for p≤2 Pa)

(*) Saturation power in the transit limit for a Gaussian beam; $\alpha = e^2 / 2\varepsilon_0 hc$

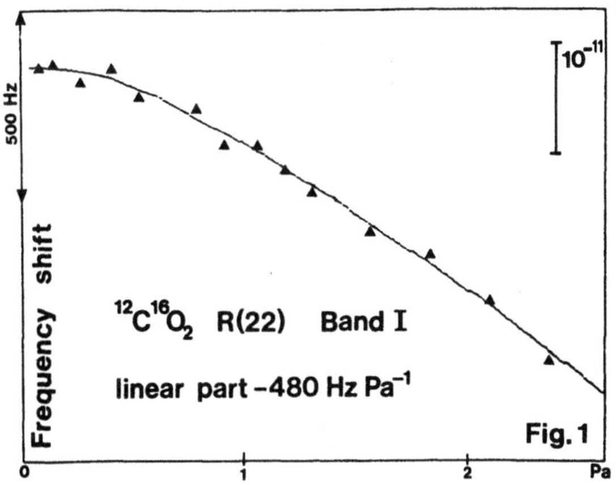

Fig. 1

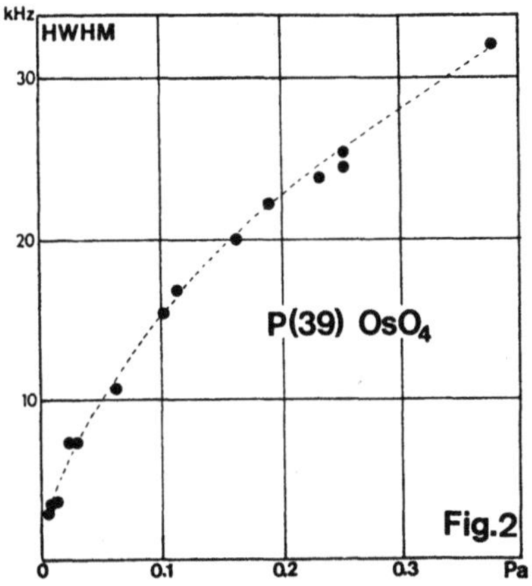

Fig.2

IV - EXPERIMENTAL SET-UP

a)- "F.P./OsO_4" system (Fig. 3)

The saturated absorption is realized in an external symmetrical FP cavity. The cavity is fixed in the absorption cell by means of thin metallic strips to reduce mechanical stresses and vibrations. The length of the FP is 1.5 m and the beam confocal parameter is b ≈ 12.5 m. The Fresnel number N of the cavity is 18. The cavity finesse F_0 (without absorber) is 100-150. The real cavity finesse is given by:

$F = F_0 / [1 + 4\,\beta_0 L\, F_0/\pi\,]$, where $\beta_0 L \ll 1$ is the linear absorption in the cavity. and $\beta_0 L \ll \pi/F_0$ (1)

For frequency stabilization of the overall system, the FP length is weakly modulated at 15 kHz ($\Delta L / L \approx 2.3\ 10^{-10}$) and the first harmonic component of the F.P. transmission fringe is used to lock the cavity F.P. to the laser frequency. The laser frequency is modulated at 2.2 kHz, and the third harmonic component of the saturated absorption signal is used for stabilization of the laser cavity. When locked, the CO_2 laser linewidth is 100 Hz. The short term stability is 1.2 $10^{-13}\,\tau^{-1/2}$ for τ ≈ 0.1 s to 100 s. Two similar systems have been built to evaluate their accuracy. The overall set-up can be used as a "phase-locked spectrometer" for line shape studies, pressure broadening and frequency shifts measurements.

b) "Ext cell " system (Fig. 4)

The characteristics of the laser beam in the cell are as follows : waist radius w_0 = 3.6 cm, confocal parameter b ≈ 800 m, absorption path length, 2L = 108 m. In order to record spectra, another laser called the "reference laser", is needed. This laser is locked to the third derivative of a saturation peak of a so-called reference line; in the present case, it is an $^{192}OsO_4$ line. The beat note of the two CO_2 lasers is

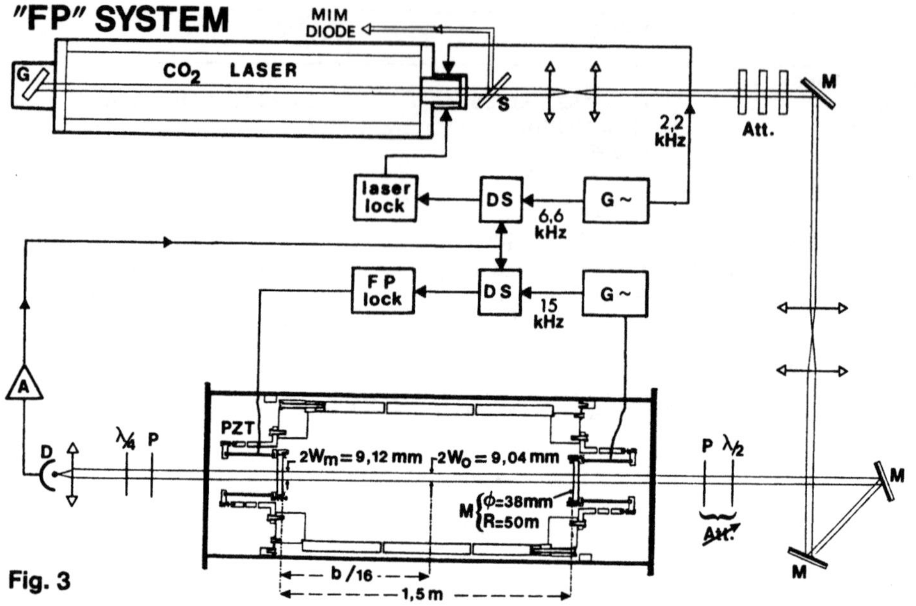

Fig. 3

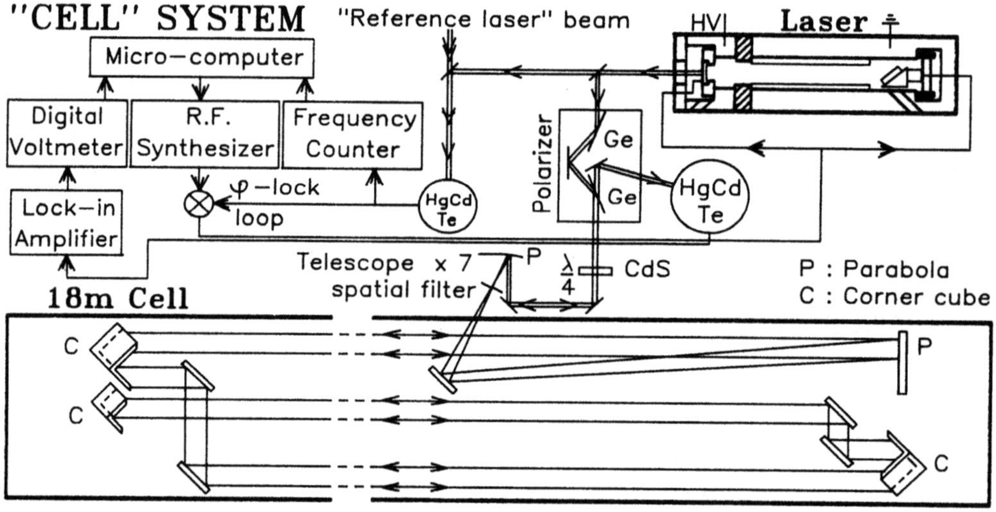

Fig. 4

phase-locked to a radio-frequency synthesizer whose frequency is driven by a microcomputer. Actually, the reference laser plays two roles : it gives a reference frequency and, via the phase-lock loop, it improves the second laser spectral purity to about 10 Hz.

For the recording of OsO_4 and CO_2 lines studied here, we have defined standard conditions which are specified in Figs 7 and 9. These conditions have been chosen because a good signal-to-noise and small systematic effects are observed. The spectrum of an OsO_4 line is obtained after 4 to 6 accumulations of records of duration 1mn while a spectrum of CO_2 requires 40 accumulations of duration 30s. The records are obtained by alternatively increasing and decreasing the frequency to eliminate all delay effects for the line center position.

V - PRINCIPLE OF THE COMPARISON AND RESULTS

The frequencies of the OsO_4 and CO_2 lines recorded in the cell were compared to the frequency of the "F.P./OsO_4" system, whose frequency was measured against a Cesium clock with an accuracy of 50 Hz. For this, two lasers were used : "the reference laser" and a second laser whose beam was alternatively sent into the Fabry-Perot and into the cell. A typical measurement, summarized in Fig. 5, is then done in three steps :

-First step : A counter measures over a period of one second the beat note between the reference laser and the laser locked to the Fabry-Perot. We repeat this operation ten times in order to obtain an estimate of the reproducibility.

-Second step : The second laser is phase-locked to the reference laser and the OsO_4 or CO_2 line in the large cell is recorded.

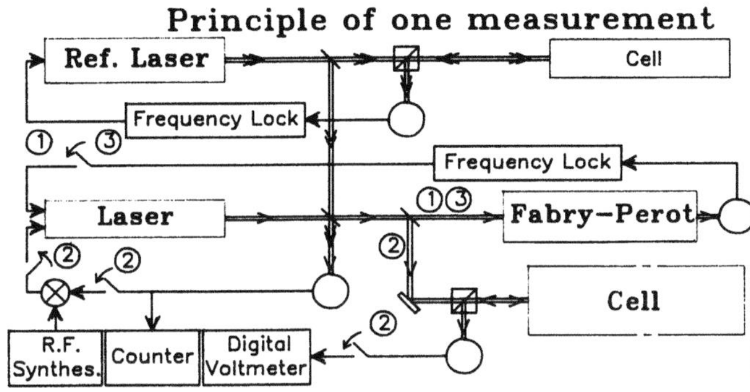

Step ① – ③ : The Laser, to which The Fabry–Perot is locked, is itself locked to the 3rd derivative of the OsO_4 line; the beat note with the "reference Laser" is then counted 10 times (duration : 10s).
Step ② : The beat note of the 2 Lasers is phase–locked to the tunable synthesizer; the first derivative of the line is then recorded and fitted (duration : a few min).

Fig. 5

The line fit gives directly the difference between the line center frequency and the reference laser frequency, as well as the statistical uncertainty of the result.

-Third step : In practice, the reference laser frequency drifts slightly, essentially owing to the variation of the pressure in the cell providing the reference line. Fig. 11 illustrates the reference laser frequency drift during one day. It is well fitted by a linear and a quadratic term. The drift during a record is usually of 10 to 100 Hz but, in the comparison, it can be taken into account with great precision. For this, we repeat the first step of the measurement. Then, a precise knowledge of the times at which the two series of beat note measurements have taken place as well as the time and the duration of the record, allow an evaluation of the frequency drift of the reference, and its compensation in the comparison of the two frequencies. An elementary statistical analysis of the results gives the standard deviation of the measurement.

a) Results for OsO_4

We have tested the sensitivity of the two systems to several parameters : for the "F.P./OsO_4" system, we have noted a systematic frequency shift of 50 Hz when the modulation index was increased to twice its value used in the standard conditions. For the "Ext Cell /OsO_4" system, in the range of pressure $p<1.3 10^{-2}$Pa. and power $P \approx 50 \mu W$, no systematic effect was visible. But at a pressure $p \approx 7\ 10^{-2}$ Pa a small red frequency shift of 200 Hz was noticeable at low laser field while a large blue frequency shift of up to 1 kHz appears at a laser field corresponding to a power of 1 mW. This behavior is well interpreted by the curvature frequency shift : at high pressure and low intensity, the main large and slowly converging laser beam cannot saturate the medium; this is not the case for the diverging part of the beam at the entry of the cell because it is strongly focused (Fig. 4). Consequently, an important contribution to the signal comes from the part of the laser beam which is affected by a red curvature shift. On the contrary, when the laser intensity increases, the lineshape becomes the sum of a large and a narrow slightly blue-shifted line and of a weak red-shifted pedestal whose contribution is negligible.

In fact, when the laser intensity is increased to more than ten times the value for a $\pi/2$ pulse, the curvature affects dramatically the lineshape which presents a sharp dispersive pattern perfectly centered at the line position for a weak field. This behavior, now well understood, is illustrated in Fig. 6. On the right hand side of the distorted line, a line can be seen which is unaffected because of its small transition moment; in standard conditions for the "F.P./OsO_4" system and, moreover, for the "Ext Cell/OsO_4" system, this line is 10^5 times smaller than the OsO_4 line that we have studied, and the variation of the parameters has a negligible effect on its measured frequency in weak field conditions.

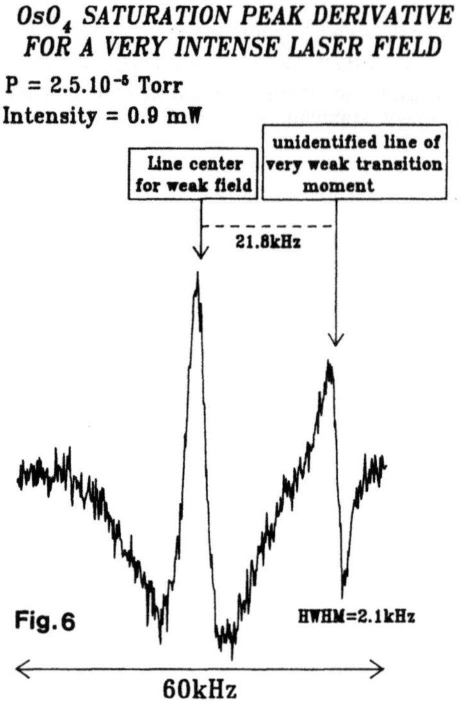

Fig.6

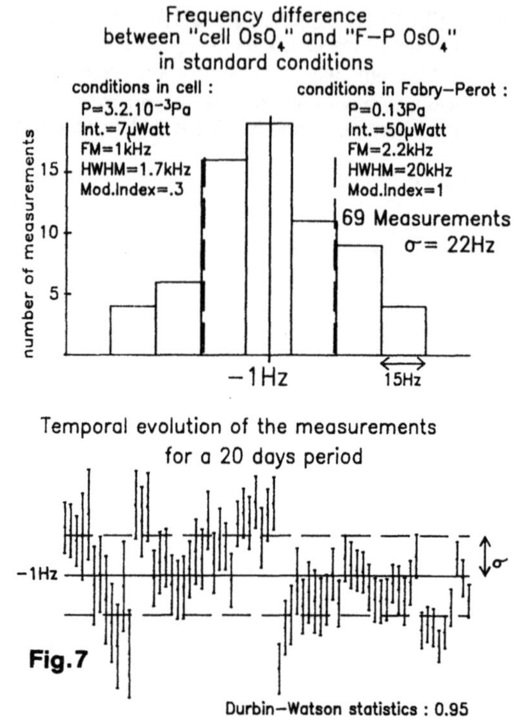

Fig.7

The results are summarized in Figs 7 and 8. A comparison of the two systems for the standard conditions given above has given a difference of - 1Hz with a standard deviation of 22 Hz ($\Delta\nu/\nu = 6\ 10^{-13}$) for 69 measurements. When this sample is increased by including measurements in less restrictive conditions (non-standard), e.g. : $p_{F.P.} \neq$ 0.13 Pa, variation of the modulation index for the "FP/OsO_4" system , $6\ 10^{-4}$ Pa < p_{cell} < 10^{-2} Pa and 1 μW < P_{cell} < 50 μW , a difference of + 1 Hz was found with the same standard deviation for 137 measurements. This remarkable result calls for several comments :

-the first one is that the two radically different systems give the same frequency for the OsO_4 line studied; this increases the degree of confidence we may have in its absolute frequency because unexpected important systematic errors are thus highly improbable,

-the second point is that the line position is not sensitive to the precise value of the experimental parameters around the standard conditions : the systematic effects are lower than or are of the same order of magnitude as the statistical errors.

In practice, it would not be reasonable to neglect systematic effects as can be seen on the time evolution of the measurements since correlations can be seen between successive measurements. These systematic effects are analyzed in the next section.

b) Results for CO_2

The measurements for the determination of the $^{II}R(10)$ CO_2 absolute frequency are summarized in Fig. 9. The difference between the CO_2 line and the OsO_4 line frequency was found to be : +15 254 295±36 Hz ($\Delta\nu/\nu = 1.2\ 10^{-12}$), as obtained from 41 measurements. It is believed that the small number of measurements explains the non-Gaussian character of the results (bar-chart). For the determination of the absolute frequency, this difference must be corrected by : +10 ± 10 Hz due to the second-order Doppler effect. Consequently, from the OsO_4 line absolute frequency, we find for the $^{II}R(10)$ CO_2 line the value :

$\nu[^{II}R(10)] = 29\ 054\ 072\ 700\ 965 \pm 50$ Hz ($\Delta\nu/\nu = 1.7\ 10^{-12}$).

This value is more accurate than the first determination recently obtained in L.P.L by a similar technique [7], but is compatible with it. It constitutes an accuracy improvement of two orders of magnitude with regard to the results of the fluorescence technique.

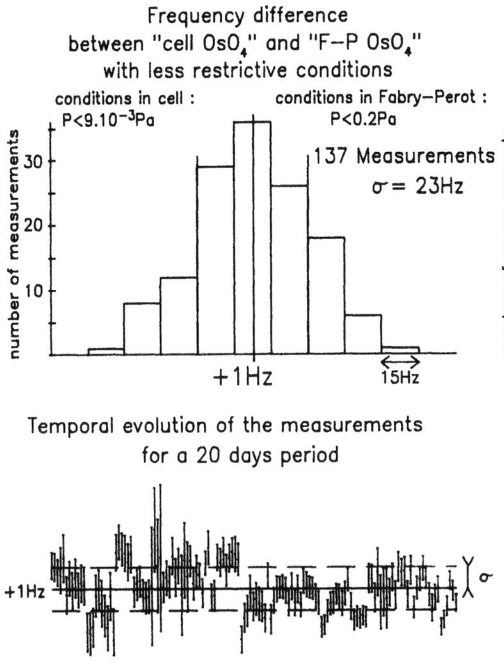

Fig. 8 Durbin–Watson statistics : 0.93

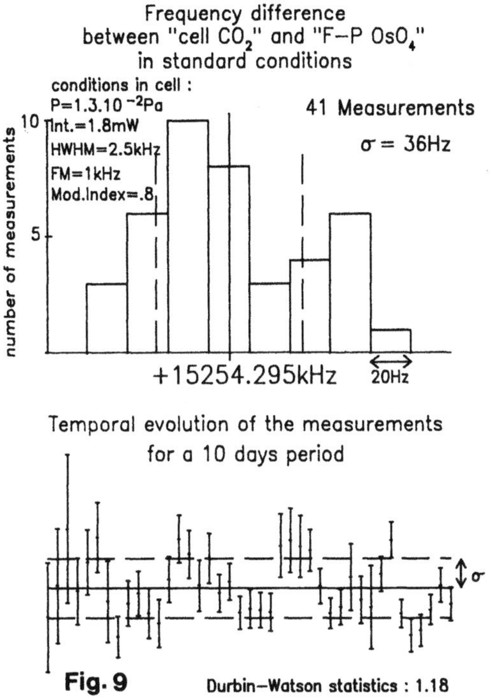

Fig. 9 Durbin–Watson statistics : 1.18

VI- ANALYSIS OF PRESENT PERFORMANCES OF BOTH SYSTEMS

a) Signal-to-noise ratio:

For identical absorber pressures, lengths, and power in a cell and in a cavity, the absorbed power are obviously identical; the main effect of the cavity is to increase the saturation peak contrast. The contrast of the peak can be written as :

$$C_{F\text{-}P} = C_{cell} \times 2F/\pi = (2F/\pi)\,\beta_0 LS \text{ if } \beta_0 L \ll 1 \text{ and the saturation parameter } S = P/P_s \ll 1. \qquad (2)$$

The signal-to-noise ratio will be different for both systems, if the noise is power dependent. If the preamplifier noise is neglected, the signal-to-noise ratio for a photovoltaic detector is :

$$(I_s/I_n)_{1Hz} = [eQSP_s\,\beta_0 LS/h\nu] \,/\, [2e^2Q(P_b+SP_s\pi/2F)/h\nu + 2k_BT/R + (\alpha eQSP_s\pi/2Fh\nu)^2\,]^{1/2}. \qquad (3)$$

For a simple external cell we have $\pi/2F = 1$, Q is the quantum efficiency, R the dark resistance of the photodiode. The first term in the denominator represents the shot noise of the background radiation. α is an experimental factor whose value depends on the laser AM noise and the detection technique. Its contribution is negligible if FM spectroscopy is used. The signal-to-noise ratio depends on the molecule and on the system used. The results are summarized in table II . For both molecules, the external cell is limited by AM laser noise. The signal-to-noise ratio is limited by detector noise for the OsO_4 molecule and by nearly equal contribution of the various noise sources for CO_2 if a F.P. cavity is used.

To reach shot noise limited operation, it is necessary to use simultaneously heterodyne detection and FM spectroscopy. In this case the signal-to-noise ratio with the F.P. cavity is given by :

$$I_s/I_n = \beta_0 LS\,(QSP_sF/\pi h\nu)^{1/2}. \qquad (4)$$

Consequently, it is comparable to that of a cell of length $L_{cell} = L\,(2F/\pi)^{1/2}$.

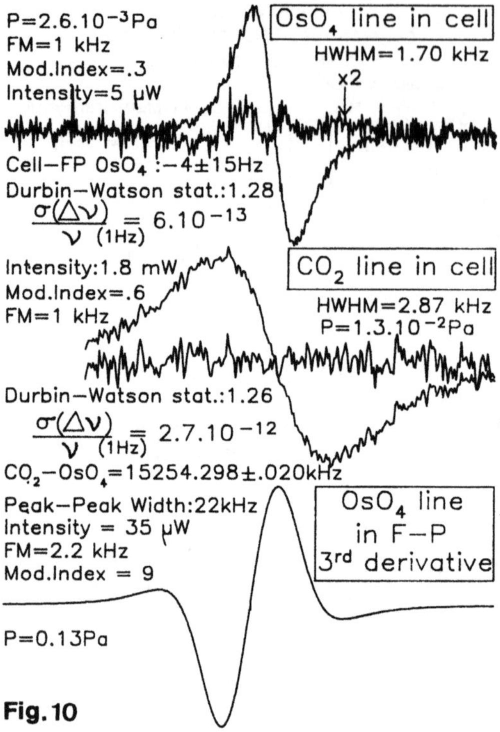

Fig. 10

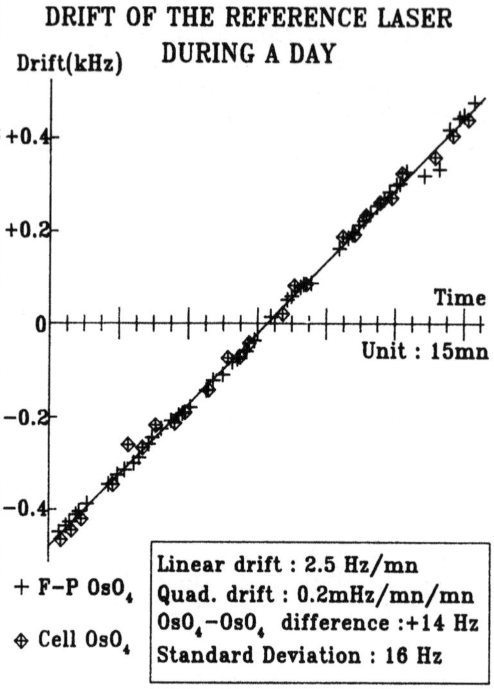

Fig. 11

Table IIa : OsO_4	F.P.	Ext cell
Physical parameters	p = 1.3 10^{-2} Pa [P(39) line] ; L =1.5 m; b =12.5 m; F_0 =100 ;	p = 2.7 10^{-3} Pa [P(39) line] ; L =54 m; b =800 m
	S = 0.3 ; P = 1 μW ; η = 0.5	S = 0.3 ; P = 1 μW ; η = 0.5
Peak contrast	≈ 7 10^{-2}	≈ 4 10^{-3}
HWHM	≈ 7 kHz	≈ 1.6 kHz
Theoretical S/N (1 Hz)	≈ 1.5 10^3	≈ 5 10^3
Limitation	Detector noise	AM laser noise
Experimental S/N (1 Hz)	≈ 10^3	≈ 10^3
Theoretical $\sigma_y(\tau)$	≈ 1.4 $10^{-13}\,\tau^{-1/2}$	≈ $10^{-14}\,\tau^{-1/2}$

Table IIb : CO_2	F.P	Ext cell
Physical parameters	p = 0.13 Pa [P(20) line] ; L = 1.5 m ; F_0 = 100 ;	p = 0.013 Pa [P(20) line] ; L = 54 m ;
	S = 0.3 ; P = 210 μW ; η = 1	S = 0.3 ; P = 210 μW ; η = 1
Peak contrast	≈ 10^{-3}	≈ 5 10^{-5}
HWHM	≈ 20 kHz	≈ 3 kHz
Theoretical S/N (1 Hz)	≈ 4 10^3	≈ 3 10^3
Limitation	Detector noise ≈ shot noise ≈ AM laser noise	AM laser noise
Experimental S/N (1 Hz)	≈ 10^3	≈ 10^2
Theoretical $\sigma_y(\tau)$	≈ 1.7 $10^{-13}\,\tau^{-1/2}$	≈ 4 $10^{-14}\,\tau^{-1/2}$

b) Accuracy limitations:

Gas-lens effect

As explained by BORDE [8], the transverse Gaussian beam field and the non-linear dispersion of the saturated absorption line induce modifications of the beam geometry. To simplify we can consider the absorber as a "lens like medium" , whose non-linear index has a quadratic term given by :

$$n_2 \;\alpha\; \beta_0 S \Delta\gamma / (\Delta^2+\gamma^2) = \beta_0 S\, D(\Delta), \quad \text{with } \Delta = \omega - \omega_0. \tag{5}$$

The beam radius variation at the end of the cell is : $\Delta w / w = - (\beta_0 LS) \times (2L/b) \times D(\Delta)$, where b is the confocal parameter. In an external cavity filled with an absorber, a different expression is obtained :

$\Delta w / w = - (\beta_0 LS) \times (b/8L) \times D(\Delta)$, valid for $b >> L$.

In a F.P. cavity the diffraction losses are frequency-dependent and can be written as:

$L_d \approx \exp[-2\pi N (1-g^2)^{1/2} / (\sin\alpha / \alpha)]$, where $\alpha = \sqrt{n_2}$; N is the Fresnel number and the stability factor $g = \cos\alpha - (\sin\alpha / \alpha) L/R$.

The F.P. transmitted power is proportional to $(1- FL_d)$, and the lineshape of the saturation line can be distorted. In our case we use N = 18 and g = 0.97, and even at 0.13 Pa OsO_4 pressure, the diffraction losses are less than $2\ 10^{-10}$. This effect can be important for a cell situated inside the laser cavity since, in this case, diffraction losses must be high enough to ensure single-mode operation.

In any case, if the linear absorption is neglected, the total power detected is proportional to :

$\int ds_{detector}\ [(\beta_0 LS/2) L(\Delta) + 2 (\pi/2F) (w_0 / w(\Delta))^2] \exp[-2 [r/w(\Delta)]^2]$, $\pi/2F = 1$ for an external cell.

This expression shows that, for small $\Delta w/w$, the lineshape is the sum of dispersion (D) and of absorption (L) terms if the power is only partially detected. As an example, if a centered detector of radius R_d is used, the frequency shift of the stabilized laser is :

$$\delta\omega / \gamma \approx (8L/b)(R_d/w)^2 \exp[-2(R_d/w)^2] / [1-\exp[-2(R_d/w)^2]] \quad \text{if a cell is used, and}$$

$$\delta\omega / \gamma \approx (\pi b/8LF)(R_d/w)^2 \exp[-2(R_d/w)^2] / [1-\exp[-2(R_d/w)^2]] \quad \text{if an external cavity is used.} \tag{6}$$

In table III, the relative frequency shifts are given for L.P.L and L.P.T.F set-ups.

Table III	L.P.L	L.P.T.F
OsO_4: line (HWHM)	1.6 kHz	20 kHz
$\Delta\nu/\nu$ for R_d/w=1.5	$7\ 10^{-13}$	$5\ 10^{-13}$
$\Delta\nu/\nu$ for R_d/w=2	$4\ 10^{-14}$	$3\ 10^{-14}$
CO_2: line (HWHM)	2.9 kHz	20 kHz
$\Delta\nu/\nu$ for R_d/w= 1.5	$1.25\ 10^{-12}$	$5\ 10^{-13}$
$\Delta\nu/\nu$ for R_d/w=2	$7\ 10^{-14}$	$3\ 10^{-14}$

It is interesting to note that with this simple model, the gas-lens frequency shift depends only on the geometry of the experiment, as long as the finesse of the cavity is not limited by absorption. The sign and the exact value of the shift are strongly dependent on the optical alignments; this fact can probably explain the day-to-day frequency fluctuations observed during the L.P.L - L.P.T.F comparison.

Second-order Doppler effect:

The second-order Doppler shifts for CO_2 and OsO_4 are much less than for CH_4. As it is well known, the exact value of this shift is dependent on the parameter $\eta = \gamma w_0/u$, the laser power and the laser frequency modulation. During the L.P.L - L.P.T.F comparison, the η parameter in the cell was ≈ 1 for CO_2 and OsO_4, and 3.5 in the "F.P./OsO_4". With these experimental conditions the second-order Doppler effect is not easily calculated and the uncertainty on its evaluation is nearly equal to its value i.e. :$1.1\ 10^{-13}$ for OsO_4 and $6.2\ 10^{-13}$ for CO_2.

Residual first-order Doppler effect:

Owing to the symetry of the F.P. cavity, the quality of the TEM_{00} mode and the high value of the standing wave ratio, the frequency shift in this apparatus is less than 10^{-14}, at low laser power . In the L.P.L cell the situation is more complex : counter-propagating laser beams are not usually matched, the beam geometry changes along the absorption path and, consequently, we can only estimate that the shift, at low laser power and low pressure, is of the order of 10 Hz.

The accuracy of the "F.P./OsO_4" system is estimated to be 50 Hz, taking into account pressure-shift, electronic offsets, modulation distortion, and various effects previously discussed.

As a general conclusion about systematic effects, the previous study of both systems gives 50 Hz for the "F.P./OsO_4" and 25 Hz for "Ext Cell/OsO_4" as upper bounds. Consequently it is correct to take the smaller value and to add it quadratically to the standard deviation (23 Hz) of the comparison to obtain the final accuracy for the "Ext Cell/OsO_4" : $\Delta\nu = 34$ Hz ($\Delta\nu/\nu = 1.1\ 10^{-12}$). This value is probably conservative because the standard deviation already includes part of the systematic effects.

VII- FUTURE IMPROVEMENTS:

The main source of uncertainty is presently the gas-lens effect. If a F.P. cavity is used, it is possible, in order to reduce this effect, to increase the cavity finesse to 500 - 1 000 if the pressure is simultaneously reduced to (1-2) 10^{-3} Pa. The η parameter is $\approx$ 0.1 in this case. It is tempting to try to reduce, in the same way, the second-order Doppler effect by using very low saturating power ($P \approx \eta^2 P_{\pi/2}$), thus increasing the contribution of slow molecules to the saturation line [8]. The second-order Doppler shift is in this situation $\approx \eta^2 (u/c)^2 \approx 4\ 10^{-15}$ for the OsO_4 molecule, if FM spectroscopy is used. In this case, the linewidth is given by $\Delta\nu = (1.51\ u/2\pi w_0)\ \eta^{1/2} \approx 2$ kHz. The gas-lens frequency shift will be reduced to $7\ 10^{-15}$ and the pressure shift to less than 10^{-14}. The total transmitted power will be $3\ 10^{-11}$W for the P(39) OsO_4 line and the saturation peak contrast $\approx 10^{-2}$. In order to get a reasonable signal-to-noise ratio,heterodyne detection must be used. In the low pressure regime, the signal-to-noise ratio will be given by :

$I_s / I_n = \eta^3\ \beta_0 L(\ QP_{\pi/2}\ F / \pi h\nu\ )^{1/2} \approx 200$ for the P(39) OsO_4 line (a first attempt to achieve this selection of very slow molecules has resulted in the peaked and narrow lineshape of Fig. 12).

The resulting short term stability $\sigma_y(\tau) \approx 3\ 10^{-13}\ \tau^{-1/2}$ will be high enough to reach effectively the accuracy level in less than one hour. With the same parameters, the signal-to-noise ratio obtainable with CO_2 will be only 20 and the linewidth $\approx$ 5 kHz. Consequently, it seems difficult to use this method to increase the accuracy of the CO_2 reference. The same reasoning can be applied to an external cell. In this case, if we use the L.P.L cell parameters, the signal-to-noise ratio will be $\approx$50 for the OsO_4 line with a HWHM $\approx$ 250 Hz and the short term stability $\approx 2\ 10^{-13}\ \tau^{-1/2}$. The optical set-up has to be modified to reduce the gas-lens frequency shift below 10^{-13}. Curvature effects, on the other hand, can be almost completely suppressed in low field by cancelling the dispersion resonance obtained in a strong field through a fine adjustment of the telescope as illustrated in Fig. 13 and by using an external telescope in the future.

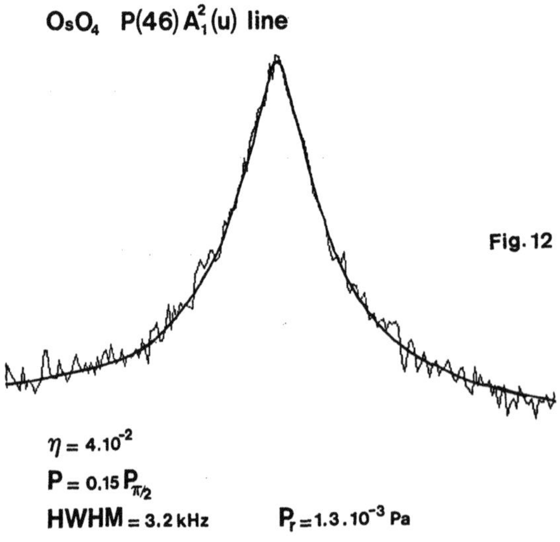

Fig. 12

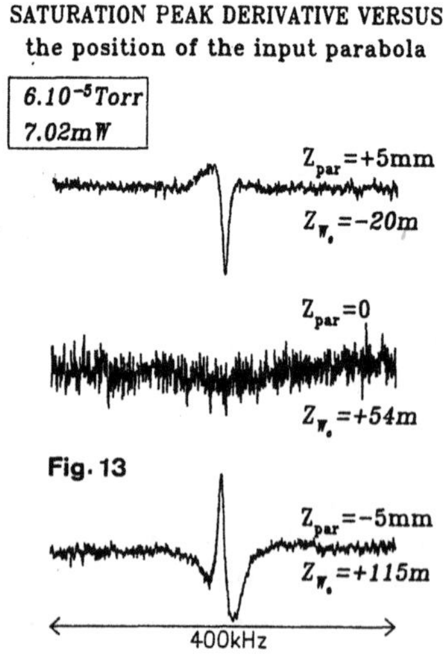

Fig. 13

VIII- CONCLUSION

The present accuracy of CO_2 lasers, stabilized by either OsO_4 or CO_2 molecules, is $\approx 10^{-12}$. This is clearly demonstrated by the L.P.L - L.P.T.F comparison. The accuracy capabilities of the CO_2 reference seems limited to a few 10^{-13}. On the contrary, improvements of the "Ext Cell/OsO_4" or the "F.P./OsO_4" standards should allow us to reach an accuracy of a few times 10^{-14} in the future.

Work supported by DRET, BNM and CNRS.

References

1. Ch.J. Bordé and L. Henry, IEEE J. of Qu.Electron. QE-4, 874-880 (1968)
2. C. Freed and A. Javan, Appl. Phys. Lett. 17, 4854 (1968)
3. O.N. Kompanets, et al., Proceedings of the 2nd Symposium on Frequency Standards and Metrology, Copper Mountain (1976) p 167 and references therein
4. A. Clairon, B. Dahmani, A. Filimon and J. Rutman, IEEE trans. on Instrum. and Measur. IM-34, 265 (1985)
5. A. Clairon et.al., J. de Physique Coll. 42, C8-127-135 (1981) see also Opt. Comm. 35, 368 (1980)
6. -B. Bobin et al., Proceedings of the Tenth Colloquium on High Resolution Molecular Spectroscopy, Dijon 1987 p. 115
   -Ch.J. Bordé et al. in Laser Spectroscopy VII, Springer 1985 p. 108
7. Ch. Chardonnet, A. Van Lerberghe and Ch.J. Bordé, Opt. Comm. 58, 333-337 (1986)
8. Ch.J. Bordé, this Issue and references therein

Part X

Solid State Lasers

Solid State Lasers – Introduction

H.R. Telle

Physikalisch-Technische Bundesanstalt, Bundesallee 100, D-3300 Braunschweig, Fed. Rep. of Germany

The scientist or engineer, who needs a laser for his application, usually wants a high performance system which combines high power, good frequency stability, large tuning range, small size, low costs and so on. Unfortunately, only very few of these criteria can be fulfilled simultaneously by the systems existing today.

Solid state lasers are not the worst choice in many cases. For example, emission wavelengths ranging from the visible (Cr or Ti doped vibronic lasers, laser diodes) to the IR region (different types of color center and vibronic lasers, laser diodes) are available as well as terawatts of output peak power (Nd:glass), hundreds of watts of CW output power (Nd:YAG) or pulse lengths of the femtosecond range (IR color center lasers). Furthermore, systems have been reported, which need extremely low electrical supply powers (MQW laser diodes) or can realize ultra-wide tuning ranges (Ti:Sapphire) or almost eye-safe high power output pulses (rare-earth lasers). Most of these systems provide better mechanical stability than for example dye or gas lasers, which is of great importance in many practical cases. From this point of view, a subgroup of the solid state lasers, laser diodes (LDs) and LD-pumped Neodymium lasers, have even superior properties: they are smaller, more efficient, more reliable and mostly cheaper than many of their competitors. Promising work is in progress at many institutes around the world to overcome the main disadvantages of these systems: limited output power and emission frequency range.

From a metrological point of view, it is more important to improve the coherence properties of these lasers, i. e. the long and short term frequency stability. Although impressing results have been reported in the past about this topic, improvements by orders of magnitude can be expected within the next few years. Solid state lasers having emission linewidths well below 1 kHz and long term stabilities much better than 10^{-9} seems to be not too unrealistic.

Stimulation of various fields of application, including ultra-high resolution spectroscopy, optical sensing and communication technology, can be expected from these solid state lasers of the near future.

Frequency Standards and Metrology Editor: A. De Marchi

Narrow-Linewidth Operation of Diode-Laser-Pumped Nonplanar Ring Oscillators

A.C. Nilsson, T. Day, A.D. Farinas, E.K. Gustafson, and R.L. Byer

Edward L. Ginzton Laboratory, Stanford University,
Stanford, CA 94305, USA

1. Introduction

The search for gravity waves, improved tests of relativity theory, advances in high resolution spectroscopy, and optical frequency standard development impose stringent requirements on the linewidth and frequency stability of lasers. Most efforts to produce narrow linewidth lasers have focused on He-Ne, dye, argon ion, or semiconductor lasers. These lasers exhibit free-running linewidths ranging from tens of kilohertz to several gigahertz and thus require wideband servo techniques for narrow linewidth operation [1]. Diode-laser-pumped monolithic solid-state lasers, on the other hand, can have free-running linewidths of a few kilohertz [2, 3], which makes them attractive candidates for narrow linewidth operation using low bandwith servo techniques. The short-term free-running stability is attributed to the small size and rigidity of the monolithic laser, which makes the optical cavity resistant to acoustical excitation, and to the low noise and efficiency of the diode laser pumping. Of particular interest are the diode-laser-pumped monolithic NonPlanar Ring Oscillators (NPROs) that overcome the problems of spatial hole burning and sensitivity to optical feedback inherent in linear cavity lasers [3, 4, 5]. Here we present our current NPRO design and explain its properties, discuss our recent narrow linewidth results obtained by locking a pair of diode-laser-pumped Nd:GGG NPROs to an optical cavity, and speculate about future developments.

2. Current NPRO Design

A schematic of our current NPRO design is shown in Fig. 1. The NPRO is a monolithic crystal (Nd:YAG or Nd:GGG) in which a nonplanar ring light path is formed by three flat total internal reflection facets and one multilayer-dielectric-coated spherical surface that serves as the output coupler. Figure 1 shows the nonplanar ring light path as a dashed line within the crystal. The light travels in two planes that subtend an angle of 3 degrees in recent Nd:GGG designs. This small angle contrasts with the 90 degree dihedral angle of the original invention [4]. Pump radiation from a single-stripe GaAlAs diode laser operating at a wavelength near 809 nm is focused into the crystal through the center of the output coupler as shown. Proper focusing of the diode laser radiation excites only the TEM_{00} mode of the NPRO. The crystal size is

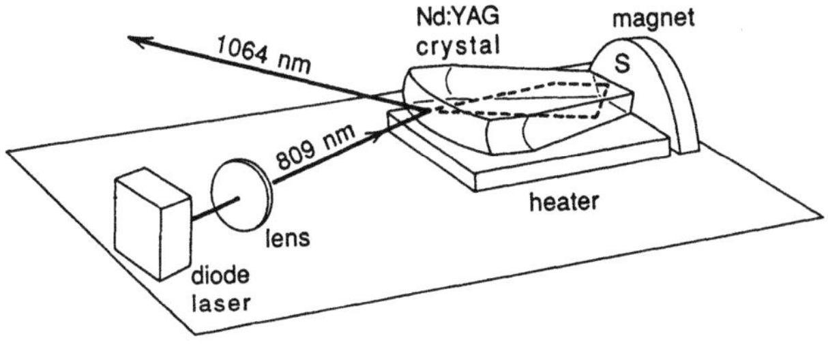

Fig. 1 Diode-laser-pumped monolithic nonplanar ring oscillator (NPRO)

Frequency Standards and Metrology Editor: A. De Marchi

typically 5x4x2 mm^3. The crystal is temperature-controlled through the bottom face. A magnetic induction of 0.4 T, provided by a small permanent magnet, causes Faraday rotation in the crystal.

3. Theory of NPRO Operation

The NPRO, owing to its integral optical diode [6], offers two major advantages over monolithic linear cavity lasers: unidirectionality, and improved resistance to the destabilizing effects of optical feedback. As with a discrete-element unidirectional ring the NPRO's optical diode contains the analogs of optical activity, Faraday rotation, and a partial polarizer. The difference is that in the NPRO all of these effects are elegantly embodied in a monolithic crystal. Reciprocal polarization rotation is provided by the nonplanar ring light path, Faraday rotation occurs in the crystal because of the applied magnetic field, and the partial polarizer results from the oblique reflection from the output coupler. In general there are four possible eigenpolarizations (two for each direction of propagation around the ring) having four different round trip losses and four different eigenfrequencies.

3.1 Unidirectionality

Since the laser transitions in Nd:YAG are homogeneously broadened, the lowest loss eigenpolarization oscillates as a traveling wave and uniformly saturates the gain, preventing competing modes from oscillating. In terms of round trip loss the closest competitor to the oscillating mode is the low loss mode for the nonoscillating direction. The loss difference of the resonator is defined as the difference between the losses of the oscillating mode and its closest competitor. To enforce and maintain unidirectional operation, the loss difference must be sufficiently large to overcome coupling between the two directions of propagation caused by scatter and extracavity optical feedback. Loss differences of 10^{-4} are more than adequate to produce unidirectional operation, even in discrete-element Nd:YAG ring lasers [7].

3.2 Resistance to Optical Feedback

Imagine that the NPRO is operating unidirectionally in the CCW direction and that some of the output radiation is reflected back into the NPRO cavity. We want to know how strongly the closest CW competitor mode is excited by this optical feedback. Apart from the question of spatial modematching, there are two additional factors to consider: polarization state, and frequency. The optical feedback generally arrives in a high loss polarization state that differs in frequency from the resonant frequency for its direction of travel, hence the optical feedback couples weakly to the CW competitor. This heuristic argument suggests that resistance to optical feedback improves with increasing loss difference and frequency splitting [8].

For diode-laser-pumped operation we have designed NPROs that simultaneously yield low output coupling (so that thresholds are low) and large loss differences (for improved resistance to optical feedback). A complete discussion of this problem can be found in [9]. The main result of the theoretical treatment is that in a given applied magnetic field there is a best choice for the geometry of the nonplanar ring light path and a best choice for the output coupler in order to obtain the lowest output coupling and a large loss difference.

4. Locking of NPROS to a Fabry-Perot Cavity

Previous experiments by KANE et al. have demonstrated 3 kHz free-running linewidths of diode-laser-pumped NPROs [3]. The NPROs used in these early experiments were large (12x9x3 mm^3) and thus had high thresholds that necessitated the use of diode-laser-arrays for pumping. Two improvements suggested themselves: reduction of the crystal size to permit pumping by stable, single-stripe diode lasers, and discovery of some nonthermal means of controlling the frequency of the NPROs.

Both of these changes have been implemented in our current experiments, which are based on modified Model 120 monolithic NPROs from Lightwave Electronics [10]. The Nd:YAG crystals have been replaced with custom Nd:GGG crystals. Typical pump power thresholds are below 5 mW for these devices, easily within the range of the single-stripe GaAlAs diode lasers used here. Rapid frequency tuning is implemented by bonding a piezoelectric transducer to the nonoptical top face of the NPRO crystal. Voltage applied to the transducer strains the NPRO crystal and changes the frequency of the laser. The tuning coefficient for our Nd:GGG crystals is 450 kHz/V, and the bandwidth for effective modulation is 400 kHz. KANE and CHENG have recently phase-locked two such NPROs together, demonstrating that low bandwidth servo techniques should be capable of producing narrow linewidth operation of such lasers [12].

4.1 System

To investigate this possibility, we have locked two Nd:GGG NPRO lasers to a single confocal Fabry Perot cavity used in transmission [12], as shown schematically in Fig. 2. Output from the two NPROs is combined in a polarizing beam splitter cube. One output port of the cube is used to mix the two beams on a photodiode to generate a heterodyne signal for spectral analysis. The other output port of the cube sends the two collinear but orthogonally polarized beams into a commercial Fabry-Perot optical cavity (finesse 300). The lasers are separately locked to the sides of two adjacent transmission fringes of the Fabry-Perot, separated in frequency by the 300 MHz free spectral range of the cavity. The slope of the fringe at the lock point is 120kHz/V. In response to a change in the amplitude of the transmitted light a voltage is applied to the piezoelectric transducer of the NPRO to restore the transmitted amplitude to its initial point. The servo loop consists of proportional and integral control, has unity gain at 6 kHz, and has a closed loop gain approaching 10^5 for low frequencies. In this simple servo system amplitude instability of the NPROs is

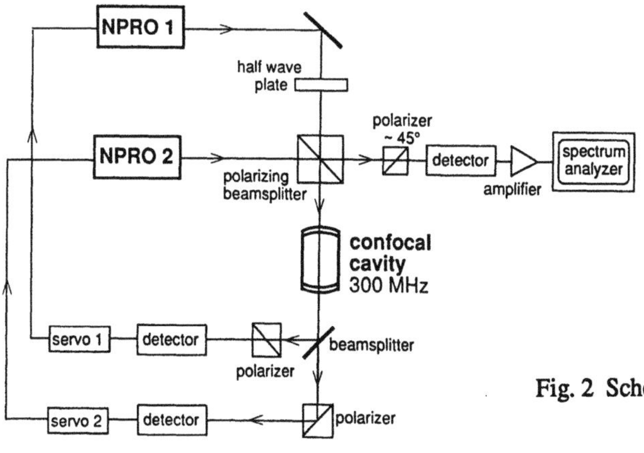

Fig. 2 Schematic of NPRO locking experiment

interpreted as frequency instability, so our linewidth results thus far only set preliminary upper bounds on the linewidth of the heterodyne signal.

Two important diagnostic tools have been explored thus far. First, spectral analysis of the frequency discriminator signals under locked condition indicates how tightly the servo loops are able to lock the lasers to the set points on the transmission fringes. Second, spectral analysis of the heterodyne signal reveals information about the relative short-term linewidths of the two lasers under lock. Since the free-running NPROs already have good short-term linewidths, the process of locking the lasers to an undoubtedly less stable external cavity degrades the frequency stability of the output radiation of each individual laser. Nonetheless, the information gained about our ability to lock the lasers to an external standard is crucial, and improvements of the overall stability of the system will require careful attention to the stability of the external standard [1].

4.2 Results

The importance of the power spectrum of the discriminator signal for diagnosing locking efficacy has been emphasized by SALOMON et al. [1]. Figure 3 is a spectrum analyzer trace of the discriminator signal for one of the two NPROs under lock. The lower frequency limit is imposed by the rf spectrum analyzer. In order to have high enough signal gain from the detector and to limit the bandwidth of the control loop, the cutoff frequency of the detector is 6.25 kHz. Interpretation of the noise spectrum above this cutoff frequency requires caution. If the RMS power spectral density of the discriminator signal were well represented across the entire spectrum by the behavior observed in Fig. 3 the analysis of ELLIOTT et al. [13] would predict a heterodyne linewidth of order 1 Hz. In fact, however, there is structure in the spectrum caused by residual 60-cycle signals and by inadequacies in the control loops.

The power spectrum of the heterodyne signal is shown in Fig. 4. This signal has a resolution-bandwidth-limited full width at half maximum of 511 Hz, which places an upper limit on the short-term

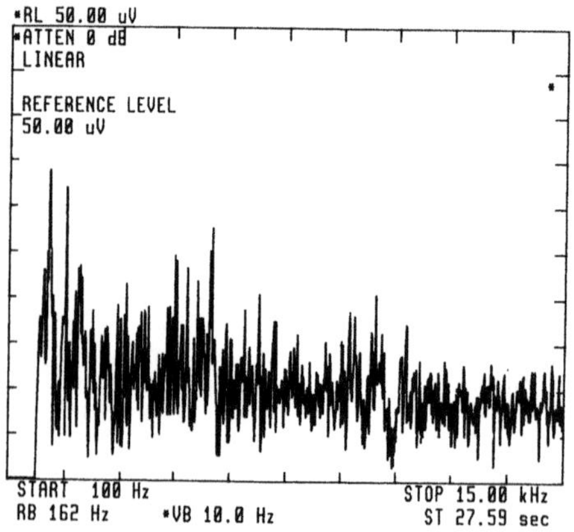

Fig. 3 Spectrum analyzer trace of the frequency discriminator signal for a locked NPRO
Vertical scale: 50 μV = 600 mHz/√Hz

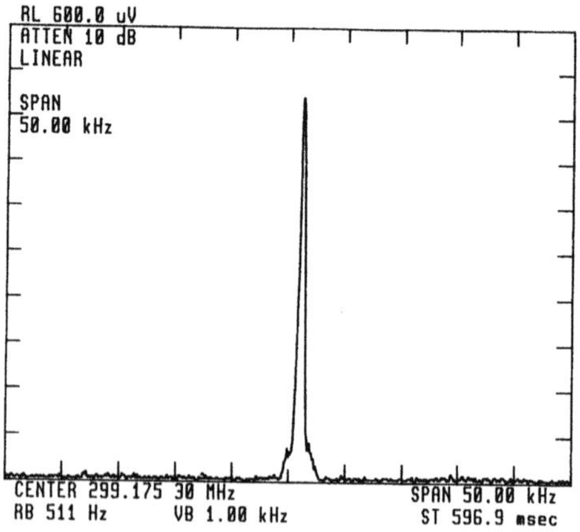

Fig. 4 Spectrum analyzer trace of heterodyne signal of NPROs locked to Fabry-Perot cavity

relative linewidth of the NPROs under lock. In the near future we expect to have improved means of characterizing these linewidths, including the use of the Allan variance.

With improved detection and careful attention to the noise properties of the servo loops we anticipate continued progress in narrowing the linewidths of NPROs. The preliminary results reported here already show six-fold improvement over the free-running linewidths previously attained by KANE et al. [3]. Moreover, the relative frequency drift between the two NPROs is eliminated in our present locking scheme.

5. Speculation About the Future

Immediate plans for improving on these results involve isolation of the external optical cavity from acoustical disturbance. Once the cavity is well isolated, we will benefit from using the rf Pound/Drever technique of locking to the cavity in reflection. To pin down the center frequency of the NPRO we plan to use external second harmonic generation in a monolithic MgO:$LiNbO_3$ ring doubler to produce narrow linewidth green light that can be locked to hyperfine transitions in I_2. Ultimately, NPRO-pumped optical parametric oscillators or new laser transitions in other solid-state gain media will make it possible to interrogate long-lived transitions in trapped ions [14, 15].

6. Acknowledgments

We are grateful to Thomas J. Kane and Lightwave Electronics for their collaboration in providing the customized Model 120 NPROs. We thank Martin M. Fejer and C. David Nabors for many useful discussions. This work was supported by NASA Langley under contract 1NAG839.

7. References

1. Ch. Salomon, D. Hils, and J. L. Hall, "Laser stabilization at the millihertz level," J. Opt. Soc. Am. B5, 1576-1587 (1988)
2. B. Zhou, T. J. Kane, G. J. Dixon, and R. L. Byer, "Efficient, frequency-stable diode laser-pumped Nd:YAG laser," Opt. Lett. 10, 62-64 (1985)
3. T. J. Kane, A. C. Nilsson, R. L. Byer, "Frequency stability and offset locking of a laser-diode-pumped Nd:YAG monolithic nonplanar ring oscillator," Opt. Lett. 12, 175-177 (1987)
4. T. J. Kane and R. L. Byer, "Monolithic, unidirectional single-mode Nd:YAG ring laser," Opt. Lett. 10, 65-67 (1985)
5. W. R. Trutna, Jr., D. K. Donald, and M. Nazarathy, "Unidirectional diode-laser-pumped Nd:YAG ring laser with a small magnetic field," Opt. Lett. 12, 248-250 (1987)
6. T. F. Johnston, Jr. and W. Proffitt, "Design and performance of a broad-band optical diode to enforce one-direction traveling-wave operation of a ring laser," IEEE J. Quantum Electron. 16, 483-488 (1980)
7. A. R. Clobes and M. J. Brienza, "Single-frequency traveling-wave Nd:YAG laser," Appl. Phys. Lett 21, 265-267 (1972)
8. A. C. Nilsson, T. J. Kane, and R. L. Byer, "Monolithic nonplanar ring lasers: resistance to optical feedback," Pulsed Single-Frequency Lasers: Technology and Applications, William K. Bischel, Larray A. Rahn, Editors, Proc. SPIE 912, pp. 13-18 (1988)

9. A. C. Nilsson, E. K. Gustafson, and R. L. Byer, "Eigenpolarization theory of monolithic nonplanar ring oscillators," submitted to IEEE J. Quantum Electron.
10. G. T. Forrest, "Diode-pumped solid-state lasers have become a mainstream technology," Laser Focus/Electro-Optics 23, 62-74 (1987)
11. T. J. Kane and E. A. P. Cheng, "Fast frequency-tuning and phase-locking of diode-pumped Nd:YAG ring lasers," submitted to Opt. Lett.
12. R. L. Barger, M. S. Sorem, and J. L. Hall, "Frequency stabilization of a cw dye laser," Appl. Phys. Lett. 22, 573-575 (1973)
13. D. S. Elliott, R. Roy, and S. J. Smith, "Extracavity laser band-shape and bandwidth modification," Phys. Rev. A26, 12-18 (1982)
14. D. J. Wineland, W. M. Itano, J. C. Bergquist, J. J. Bollinger, and J. D. Prestage, "Spectroscopy of stored atomic ions," Atomic Physics 9, R. S. Vandyck, Jr. and E. N. Fortson, eds., World Scientific Singapore, pp. 3-27 (1985)
15. H. G.Dehmelt, "Mono-ion oscillator as potential ultimate laser frequency standard," IEEE Trans. Instrum. Meas. 31, 83-87 (1982)

Optical Stabilization of Semiconductor Lasers *

L. Hollberg

Time and Frequency Div., National Institute of Standards and Technology, Boulder, CO 80303, USA

The method of using resonant optical feedback to stabilize the frequency of semiconductor lasers has proved valuable for a number of applications [1-3]. It is one of a number of possible optical stabilization methods, as opposed to electronic stabilization methods, and contrasts from other optical methods in that the optical feedback comes from a high Q (quality factor) optical resonator. This technique for frequency stabilization uses commercially available lasers and Fabry-Perot cavities and can reduce diode laser linewidths to as low as 3 kHz. The resonance of the Fabry-Perot cavity serves as the frequency reference which stabilizes the lasers oscillation frequency and narrows its linewidth.

A specific optical feedback locking system is diagramed in Fig 1. This system uses a separated high Q (quality factor) Fabry-Perot resonator as the feedback element. Typically 10^{-2} of the laser output power is split off from the laser beam and is coupled off-axis into the Fabry-Perot resonator in such a way that the incident beam reflects off the cavity input mirror at an angle and does not return to the laser. On the other hand when the laser frequency is tuned to a cavity resonance, optical power is built up in the cavity, some of which leaks back through the input coupling mirror and returns to the laser. This type of resonant optical feedback is labeled type II and it has the characteristic that the feedback power is maximum when the laser frequency matches a cavity resonance. This type II feedback alters the characteristics of the diode laser resonator by coupling it optically to the higher-Q Fabry-Perot cavity. At a resonance of the Fabry-Perot the combined diode laser plus cavity system has lower loss. The laser then naturally oscillates at the point of minimum loss and hence "locks" to the Fabry-Perot cavity resonance. We note in Fig. 1 that the directly reflected type I beam, which has a power minimum at a cavity resonance, is deflected so that it does not return to the laser.

A characteristic of the optical lock is that the frequency of the laser is controlled as usual by the lasers injection current and temperature unless the oscillation frequency comes close to a resonance of the Fabry-Perot. In the latter case the laser frequency locks to the cavity resonance. The tuning range between the laser oscillation frequency and the cavity resonance frequency over which the

*Contribution of the U.S. Government, not subject to copyright.

Frequency Standards and Metrology Editor: A. De Marchi

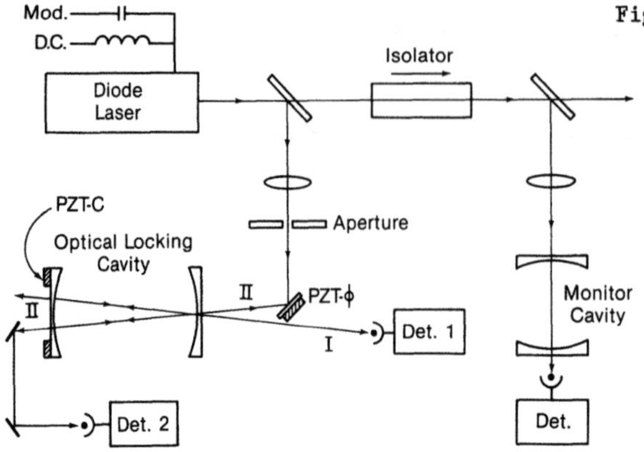

Fig 1. Optical feedback locking system diagram.

lock occurs is typically a few hundred megahertz. When the laser optically locks, its tuning sensitivity with respect to injection current is reduced by a factor of 100 or so. Both the locking range and the tuning sensitivity for injection current depends on the exact parameters of the lock including the cavity Q and feedback power level. P. Laurent and collaborators [2] have done an analysis of this optical locking method and they find good agreement between experiment and theory for most aspects of the lock.

Excellent performance can be achieved with this optical locking method in terms the spectral narrowing and center frequency stabilization of the diode laser. The commercial diode lasers we are using have free running 3 dB linewidths that are about 20 MHz, which can be narrowed with the optical lock to less than 5 kHz. This linewidth reduction represents an increase in the spectral density of the optical power and the laser's coherence length of more than a factor of 1000. Some care must be taken in evaluating the linewidth of semiconductor lasers because some of the common and popular methods are misleading and because the laser's lineshapes can be non-Lorentzian [2]. In fact, the optical lock narrows the linewidth by removing the frequency noise for low Fourier frequencies and out to a couple hundred megahertz. For semiconductor lasers this frequency range contains the most important contribution to the laser,s linewidth even though the frequency fluctuation spectrum extends out as far as the relaxation oscillation frequency at about 3 GHz.

Figure 2 shows the tremendous reduction in the laser linewidths that can be achieved using this optical feedback locking method. The figure shows the beat note generated on a fast photodiode between two independent diode lasers. The broad beat note displays the spectral width when one of the lasers is not optically stabilized whereas the narrow beat note is measured when both lasers are optically stabilized to independent cavities. The remaining width in the sharp beat note trace is due to the 300 kHz resolution bandwidth of the RF spectrum analyzer.

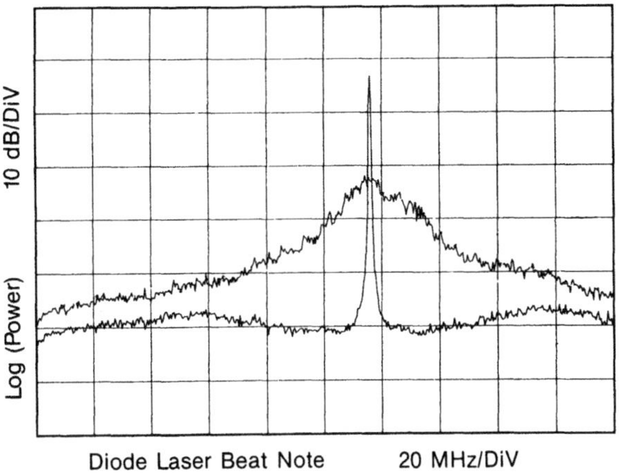

Fig 2. Diode laser beat not spectrum.

Higher resolution measurements have demonstrated beat notes with 3 dB bandwidths of as little as 3 kHz.

One of the many advantages of this frequency stabilization method is that the system is relatively simple to build and it does not require that the diode laser be modified with special coatings and one only needs access to one facet of the laser. In addition it is important to note that the system operates in the weak feedback limit, such that the total power feedback to the laser resonator is a small fraction of the laser's output power (usually 10^{-2} to 10^{-4}). This means that the usable laser output power is not significantly depleted and that the lock operates in a regime of optical feedback that is stable relative to small optical perturbations. The diode laser's linewidth decreases smoothly from 20 MHz to a few kHz as the power of the resonant feedback is increased. If the feedback power is further increased the locking system becomes unstable.

It is interesting to note that even though the frequency modulation sensitivity of the laser to injection current is reduced for low frequencies (say less than 100 Mhz) it is still possible to modulate the laser at high frequencies using the injection current [4]. Modulation of the laser at high frequencies is compatible with the optical lock if either the modulation amplitude is small or if the modulation frequency is harmonically or rationally related to the free spectral range of the Fabry-Perot that is used for the lock. Experimental results are shown in Fig. 3 for strong FM modulation of a semiconductor laser at a frequency of 245 MHz which corresponds to the free-spectral-range of the locking Fabry-Perot cavity.

In this case the laser was biased at 97 ma injection current with 7 ma of additional RF modulation applied through the port labeled mod. in Fig 1. The optical lock with modulation thus provides an array of about 20 sidebands spaced at 245 MHz and each of which has a linewidth of less than 10 kHz. This sideband spectrum is detected using an optical spectrum analyzer as shown in Fig. 1. The asymmetry in the spectrum is due to a combination of AM plus FM modulation as

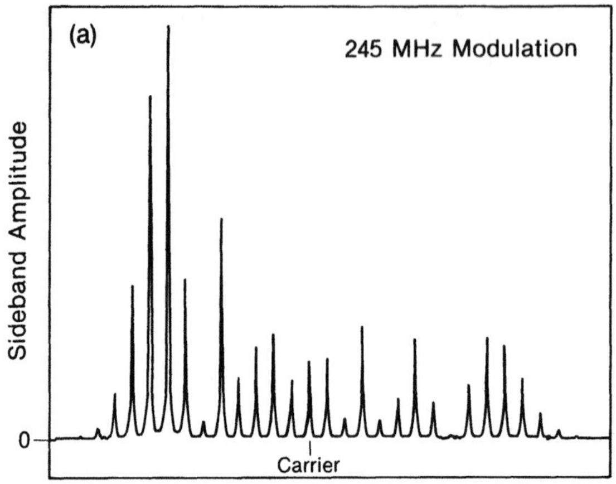

Fig. 3 RF modulation at 245 MHz optically locked diode laser.

well as harmonic distortion in the FM. The vertical scale is linear and the horizontal spacing between peaks is 245 MHz.

In addition to the advantages of this locking technique, it is also important note some of the challenges that must be faced in implementing the method. First of all, to optimize the lock it is necessary to have the proper phase of the light that is fedback to the laser. This is accomplished by adjusting the optical path length between the laser and the cavity using PZT-φ in Fig 1. In addition the laser frequency needs to be within an optical locking range of a cavity resonance. In order to tune and scan the frequency of the optically locked laser over frequency ranges larger than the optical locking range we need to simultaneously scan the Fabry-Perot cavity, the laser injection current and the feedback phase. All of these adjustments can be made automatically using simple electronic servo controls. We also note that there is some frequency pulling of the optically locked laser with variations in the feedback phase, but the pulling is small if the cavity has a high Q.

Lasers with high spectral purity are required in the fields of spectroscopy, precision measurements and for some communication systems applications. Additionally, in some cases the signal to noise ratio for spectroscopic detection is greatly enhanced when the lasers spectral width is reduced. For example laser induced fluorescence, or direct absorbtion spectroscopy of the cesium atom resonance line at 852 nm with diode lasers shows a significant amount of excess noise. This noise, though not well understood in the case of Cs, is typically reduced by a factor of about 100 when the diode lasers spectral width is narrowed by the optical lock. By reducing the laser's linewidth by a factor of 1000 and stabilizing the laser's center frequency this optical locking method can improve the performance characteristics of diode lasers such that new applications become possible.

References

1) B. Dahmani, L. Hollberg and R. Drullinger, Opt. Lett. 12, 876 (1987).
2) P. Laurent, A. Clairon, and Ch. Breant, IEEE J. Quantum Electronics. (To be published). And also contributions in this volume.
3) H.R. Telle and H. Li, IEEE J. Quantum Electron. (To be published).
4) L. Hollberg and M. Ohtsu, Appl. Phys. Lett. 53, 944 (1988).

Frequency Stabilization of Highly Coherent AlGaAs Diode Lasers

A.M. Akulshin, V.V. Nikitin, V.A. Sautenkov, V.V. Vasiliev, V.L. Velichansky, E.K. Yurkin, and A.S. Zibrov

P.N. Lebedev Physical Institute, USSR Academy of Science, Leninsky pr. 53, SU-117924 Moscow, USSR

A solitary diode laser (DL) as compared with a typical gas laser has a large spectral linewidth which limits the attainable frequency stability /1/. One of the most efficient ways to narrow spectral line of a DL is an external optical feedback /2,3/, which makes it possible to decrease the linewidth to 1 kHz /4,5/. An external cavaty diode laser (ECDL) has many advantages but is more complicated than a solitary DL; the first work on DL frequency stabilization /6/ was followed by experiments on ECDL stabilization 12 years later /7/. Here we report on two experiments in which: a fast technique of side-band spectroscopy has been used for an ECDL frequency stabilization to reflection resonance of a confocal interferometer /8/; two ECDLs have been locked to adjacent sub-Doppler resonances in saturated absorption of ^{133}Cs by using conventional lock in servo systems, and DL frequency stability was estimated on the direct measurement of beat frequency fluctuations /9/.

1. Frequency Stabilization by a FM Spectroscopy Technique

In the optical region the technique considered is widely used for a dye laser frequency stabilization and high resolution spectroscopy performed with these lasers /10, 11/. It has also been used with a solitary DL for the detection of Doppler-broadened spectral lines /12/. The most characteristic features of the technique modified for frequency control are a small modulation index (only the carrier and the first side bands should be taken into account) and large modulation frequency f which is greater than the reference linewidth. Passing through an interferometer the carrier and the sidebands get different attenuations and phase shifts, so that their beat notes in square-law detector cease to compensate each other. There appears a signal at f whose phase and amplitude depend on the carrier position relative to the center of reference line. After a proper phase sensitive demodulation the resulting signal vs frequency dependence acquires a sharp despersion-like pattern convenient for frequency stabilization. The most attractive features of the technique are: a possibility to have a servosystem bandpass comparable or even greater

Frequency Standards and Metrology Editor: A. De Marchi

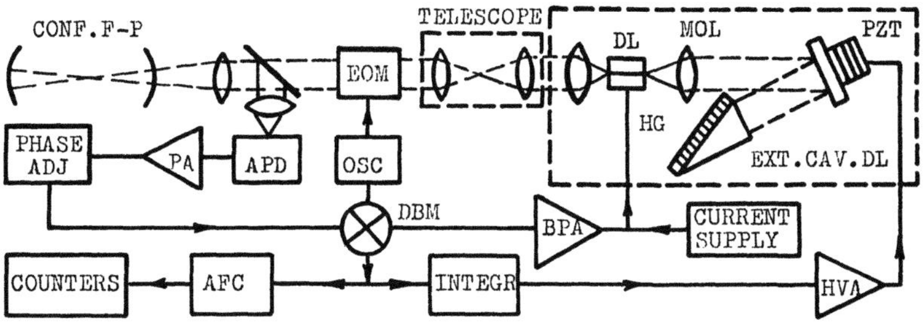

Fig. 1. Block diagram of the ECDL and the experimental setup. MOL, microscope objective lens; HG, holographic grating; EOM, electro-optical modulator; PA, preamplifier; DBM, double balanced mixer; BPA, bandpass amplifier; HVA, high voltage amplifier; AFC, analog frequency converter.

than the linewidth of the reference line; the low frequency noise of the laser can be suppressed by using rather high frequency of modulation and an appropriate bandpass filter. Both factors are important for a DL frequency stabilization. We have used f=9.6 MHz which exceeds the cutoff frequency of the current fluctuation noise (1 - 3 MHz).

The block diagram of the laser and experimental setup is shown in Fig.1. A GaAlAs DHS stripe laser was mounted on a Peltie cooler with a passive temperature stabilization. A microscope objective lens (NA = 0.65, F = 6.3) couples the active region of the DL with a holographic grating (3200 mm^{-1}) via a mirror attached to a PZT driver. The length of the cavity was 27 cm. Coarse tuning over about 150 Å was accomplished with the grating. For fine tuning the length of the cavity was changed with continuous tuning range greater (up to 8 GHz) than the distance between the two adjacent modes of ECDL (550 MHz) due to autostabilization /13,14/. High selectivity of the grating secured stable single-mode operation with a suppression of the other modes by a factor of no less than 100. The linewidth of the laser was 0.1-0.2 MHz and was narrowed to about 1 kHz as the frequency jitter was suppressed by the feedback /4/. The time of the laser operation without any mode hopping varied from thirty minutes to two hours depending on the environmental temperature drift.

A collimated laser beam of about 3 mW power passed the telescope and phase modulator, reflected from a confocal interferometer with FSR and finness 240 and 4 MHz, respectively, and was incident on avalanche PD. The signal from the APD was fed to the preamplifier and after a phase adjuster detected with a double balanced mixer.The reference signal was taken from an oscillator which serves also as a source for EOM. The output signal of DBM is shown in Fig.2 together with the FM modulated laser spectrum displayed with the interferometer. The DBM output is

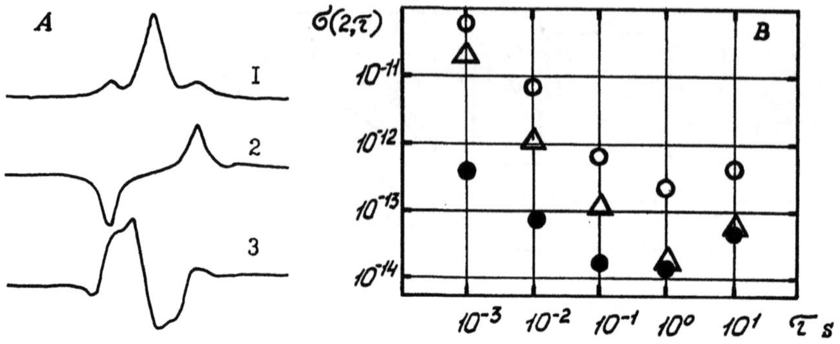

Fig.2. A: Spectrum of the phase modulated laser radiation as displayed by the interferometer /1/. The carrier - sideband offset is 9.6 MHz. DBM output signals for in-phase (2) and quadrature (3) detection. B: Frequency stability vs integration time: O only the slow feedback loop is operating; △ both feedback channels are put into operation; ● the noise level of the measuring system.

divided into three channels, two of which serves as feedback loops. One includes an integrator and high-voltage amplifier and controls the length of the cavity with a PZT driver. It is a slow channel with a large dynamic range (linear response is about 1 GHz) and a 1 kHz bandwidth. The other channel is a fast one with a bandpass amplifier (50 Hz - 130 kHz) whose current output is added to the pumping current of the DL.

The evaluation of the Allan variance was based on the error signal. For this purpose the output voltage in the third channel was fed to a voltage-frequency converter. Two frequency meters measured the number of periods in the signal of the converter falling within the integration time τ. The delay time operation of the frequency meters was also τ. The number of measurements in the sample was thirty. The dependence of the square root of the Allan variance on the integration time is shown in Fig. 2B. To our knowledge the stability of $1.2 \cdot 10^{-11} \geq \sigma \geq 3 \cdot 10^{-14}$ at integration times $10^{-3} \leq \tau \leq 10$ s is the highest achieved with semiconductor lasers. This result is based on the narrowed linewidth of the ECDL (and, hence, suppressed frequency fluctuation noise) and the increased Q merit of the reference cavity. Both factors according to /1/ lead to a higher level of attainable stability. Still, as the estimations show, the result is not quantum limited, and further improvements in electronics and optics are possible. We should also emphasize that we have measured relative stability - against the interferometer resonance stabilized only passively.

2. Frequency Stabilization to Doppler-Free Saturated Absorption of D_2 Line in Cesium

The resolution of a solitary DL in Doppler-free saturation spectroscopy of alkalines is still not enough to come close to the natural linewidth limit /15/. The ECDL makes it possible /16/. Doppler-free spectrum of F=3 to F=2,3,4 component of D_2 Cs line is demonstrated in Fig. 3A. Three dips at atomic eigenfrequencies and three crossco-vers are identified there. In the experiment described here the two lasers have been locked to the most closely separated pair of dips with nominal frequency difference equal $(v_{32}+v_{34})/2 - v_{33}=25.5$ MHz. The ECDLs were of the same type as in Part 1. Two identical schemes of saturated absorption spectrosocopy were used for each laser stabilization. They included two independent Cs filled 3 cm long cells and servosystems. The crossection of the saturated beam was 6 10mm^2 while the probe beam was somewhat smaller to be totally covered by the strong beam within the cell volume. Both beams have linear polarization parallel to magnetic field of about 1 Gauss.

The resonant lines of the alkalines are easily saturated due to optical pumping and the intensities of the beams were attenuated to decrease the saturation broadening. The total width of the resonances was 10 MHz with the Doppler contribution of 1.5 MHz due to the angle between the beams (3 10^{-3} rad), and their divergency and natural width being 5.3 MHz. The frequency modulation was performed at 6.6 and 7 kHz with the deviation amplitude about 1 MHz for both lasers. The servoloop comprised a phase detector, an integrator and a high-voltage amplifier whose output was fed to PZT. The band width of servosystems was 150 Hz. The split parts of the lasers' beams were superimposed on APD to produce a heterodyne beat signal. The latter passed to the spectrum analyser and the counter. The spectrum of the beating is shown in Fig. 2B. It was possible to narrow its width by decreasing the frequency deviation of both lasers, but it degraded the servocontrol.

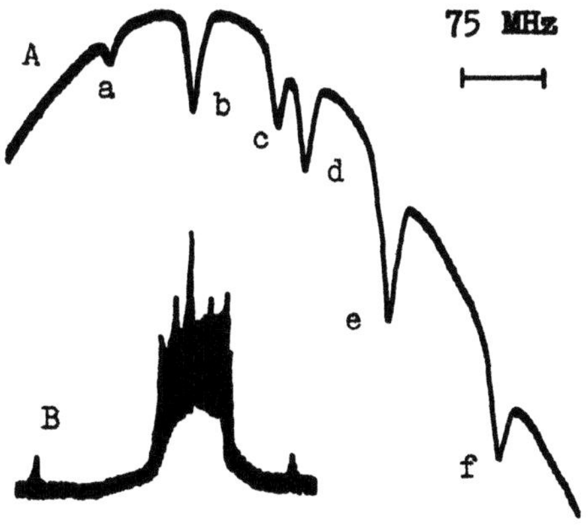

Fig. 3. A: Saturated absorption spectrum of the high frequency component of Cs D_2 line. The dips from "a" to "b" occur at frequencies : v_{32}; $(v_{32}+v_{33})/2$; $(v_{33}+v_{34})/2$; v_{33}; $(v_{32}+v_{34})/2$; v_{34}, respectively. The indices are the F numbers of the ground- and excited states. B: The beat spectrum of the lasers stabilized to "c" and "d" dips. The distance between the marks is 10 MHz.

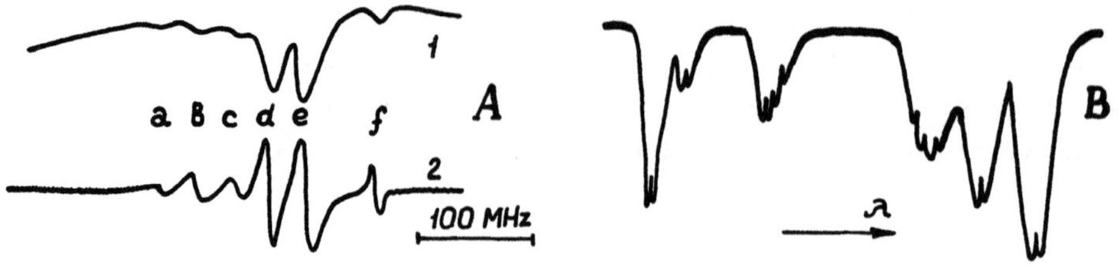

Fig. 4 A: Saturated absorption spectrum of the low frequency component of ^{85}Rb D_2 line. Both the APD signal (1) and the quadrature component of DBM output signal with suppressed Doppler background (2) are shown. Only the probe beam was phase-modulated here. The resonances from "a" to "f" correspond to frequences : $\nu_{32}, (\nu_{32}+\nu_{33})/2, \nu_{33}, (\nu_{32}+\nu_{34})/2, (\nu_{33}+\nu_{34})/2, \nu_{34}$.

B: Saturated absorption spectrum of ^{85,87}Rb at D_1 line. The four lines of ^{87}Rb, each with a single dip, enclose the two lines of the ^{85}Rb atoms which reveal two additional crossovers since the sublevels of the upper state are more closely spaced.

The square root of the Allan variance was determined by measuring beat frequency fluctuations. The results are $4\ 10^{-11}$, $1.4\ 10^{-11}$, $1.9\ 10^{-11}$ for N=15 and integration times of 0.1, 1 and 10 s, respectively. The main source of fluctuations leading to poor stability results from power-dependent shifts as the Doppler background was not eliminated. For intensity changes around 0.5 mW/cm^2 the frequency shifts of "c" - and "d" resonances were found to be +0.25 and -0.2 MHz cm^2/mW, respectively. The different signs of the shifts reveal the mutual attraction of the peaks. The stability can be further increased when passing to D_1 line of Cs of ^{87}Rb since there the sub-Doppler peaks are at the center of the Doppler line.

The next stage of the work is to combine the fast method of the sideband heterodying frequency stabilization and the atomic reference lines. With that end in view we have performed the FM spectroscopy experiments with all $D_{1,2}$ lines of ^{85,87}Rb. An example (low frequency component of D_2 line of ^{85}Rb) is given in Fig. 4A. As has been mentioned above the sub-Doppler dips for ^{87}Rb D_1 line are close to the peak of the Doppler line and should exibit smaller power-dependent shifts. This is demonstrated in Fig. 4 B where the saturated absorption at D_1 lines of ^{85,87}Rb are displayed with the ECDL operating in autostabilization regime.

3. Summary

An external cavity diode laser was stabilized to the reflection resonance of an interferometer by the sideband heterodyne technique. The result is $1.2\cdot10^{-11} \geq \sigma \geq 3\cdot10^{-14}$ at $10^{-3} \leq \tau \leq 10$ s. The same technique was applied for Doppler-free spectroscopy of Rb resonance lines.

By using conventional locking systems two ECDL were stabilized to sub-Doppler resonances of cesium. The frequency stability estimated on the direct measurement of beat frequency fluctuations was found to be $4 \cdot 10^{-11}$, $1.4 \cdot 10^{-11}$, $1.9 \cdot 10^{-11}$ at integration times 0.1, 1 and 10 s, respectively.

References

1. H. Tsuchida and T. Tako: Jap. J. of Appl. Phys., 22, 1970 (1983)
2. V.L.Velichansky, A.S. Zibrov et al : Zh. Tekh. Letts., 4 1087 (1978)
3. V.Yu. Bazhenov, A.P. Bogatov et al.: Sov. J. Quant. Electr. 10, 1547 (1980)
4. A.M. Akulshin, N.G. Basov et al : Sov. J. Quant. Electr. 13, 1003 (1983)
5. R. Wyatt : Electr. Lett. 21, 658 (1985)
6. Yu. A. Bykovsky, V.L. Velichansky et al: Sov. Phys.-JETP 30, 605 (1970)
7. A.S. Zibrov, A.M. Akulshin et al: Sov. J.Quant. Electr. 12, 502 (1982)
8. E.K. Yurkin, A.S. Zibrov et al: P.N.Lebedev Physical Institute, Preprint № 34, Moscow (1988)
9. A.M. Akulshin, V.L. Velichansky et al: to be published in Sov. J., Quant. Electr., 15, № 10 (1988)
10. R.W. Drever, T.L. Hall et al: Appl. Phys. B 31, 97 (1983)
11. M. Gehrtz, G.C. Bjorklund et al: Appl. Phys. B 32, 145 (1983)
12. W. Lenth: Opt. Lett., 8, 575 (1983)
13. A.P. Bogatov, P.G. Eliseev et al: Sov. J. Quant. Electr. 13, 1221 (1983)
14. A.M. Akulshin, V.Yu., Bazhenov et al: Sov. J. Quant. Electr. 16, 912 (1986)
15. S. Ohshima, Y. Nakadan, and Y. Koga: IEEE J. of Quant. Electr. 23, 473 (1987); G.P. Barwood, P.Gill, and W.R.C. Rowley: El. Letts. 24, 769 (1988)
16. A.M.Akulshin, V,L.Velichansky et al: P.N.Lebedev Physical Institute, Preprint № 99, Moscow (1985); A.N. Akulshin, V.L.Velichansky et al: "Proc. of the Seminar on the Optical Orientation of Atoms and Molecules" p. 152, Leningrad

Frequency Control of Semiconductor Lasers

M. Ohtsu, K. Kuboki, C.H. Shin, and M. Murata

Graduate School at Nagatsuta, Tokyo Institute of Technology,
4259 Nagatsuta, Midori-ku, Yokohama, Kanagawa 227, Japan

1. Introduction

Frequency control of semiconductor lasers has become very important in recent years for several applications such as frequency standards, optical metrology, optical communication, and so on[1]. In the present work, a technique of electrical feedback was employed for the frequency control, because (a) stability and reliability of the feedback is high, (b) computer-aided-design (CAD) can be employed to realize the optimum feedback loop, and (c) magnitude of residual frequency fluctuations under the feedback can be reduced to a value which is far lower than that of the quantum noise limit of the free-running laser[2].

For practical applications, the following four problems should be solved simultaneously to improve the coherence of semiconductor lasers : (1) Stabilization of the center frequency of field spectrum of the master laser. (2) Linewidth reduction of field spectrum of the master laser. (3) Accurate frequency tracking of the slave laser to the master laser. (4) Stable and wideband frequency sweep of the slave laser. It should be important to develop a synthesized feedback system for solving these four problems simultaneously, which is possible by employing a technique of electrical feedback. In the following sections, experimental results for 0.8 μm wavelength AlGaAs laser (CSP-type, Hitachi HL8312E) on these four problems are presented.

2. Frequency Stabilization of the Master Laser

A slow electrical feedback loop with the bandwidth of 1 MHz was provided to the master laser to stabilize the center frequency of its field spectrum. The stability of 2×10^{-12} was obtained at the integration time τ of 100 s by using the absorption spectral line in Rb vapor as a frequency reference. The frequency stability limited by the shot noise of the photodetector ($= 1.7 \times 10^{-14} \tau^{-1/2}$) can be expected to accomplish by improving the performances of the feedback loop[2].

Frequency Standards and Metrology Editor: A. De Marchi

3. Linewidth Reduction of the Master Laser

The linewidth of the field spectrum of the master laser can be reduced by a fast electrical feedback loop if its bandwidth is wider than the linewidth of the free-running master laser. A slope of the resonance curve of the reflected light from a high finesse Fabry-Perot cavity was used as a sensitive frequency discriminator to realize a wideband feedback loop. Comparing the transfer characteristics of frequency discrimination of the reflected light with that of the transmitted light, the passband of frequency response is wider, and the slope of the transition band is improved by 20 dB/decade, and also phase-lead compensation is shown in the transition band, which allows to increase the feedback gain substantially. Thereby, it can be considered that the reflected light acts as the first order differentiator. This is why the reflected mode was chosen as the frequency discriminator.

The curve A in Fig. 1 represents the experimental result of the power spectral density S_{FB} of frequency fluctuations obtained by using the feedback loop which was optimized by CAD. In this figure S_{FB} was normalized to the power spectral density S_{FR} of the free-running laser. The linewidth $\Delta\nu_{FP}$ of the resonant curve of the Fabry-Perot cavity used was 0.7 MHz (finesse = 8,600). The curves B and C represent the results of computer simulations, where the values of $\Delta\nu_{FP}$ were assumed to be 10 MHz, and 0.1 MHz, respectively. It is seen from this figure that a wideband and large reduction of frequency fluctuations was realized, and that the magnitude of the reduction is larger by using a narrower $\Delta\nu_{FP}$. The minimum on the curve A was - 60 dB, and from such a large reduction of frequency fluctuations, a very narrow laser linewidth can be estimated. FFT program

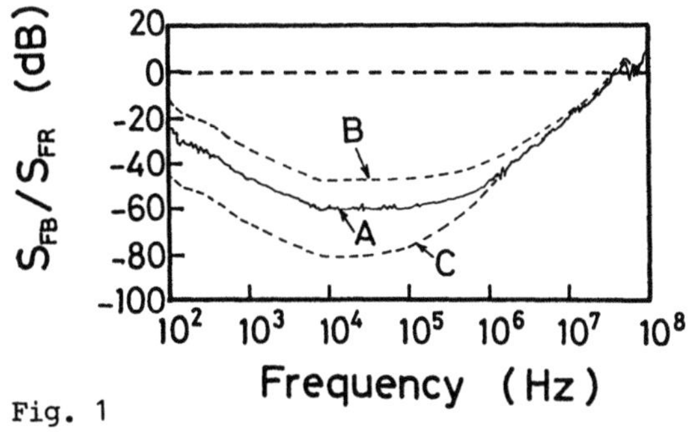

Fig. 1

Power spectral density S_{FB} of frequency fluctuations under feedback, which is normalized to that of the free-running laser S_{FR}. A ; experimental result by using a Fabry-Perot cavity of 0.7 MHz linewidth. B, C ; the results of computer simulation by assuming the linewidths $\Delta\nu_{FP}$ of the Fabry-Perot cavity to be 10 MHz and 0.1 MHz, respectively.

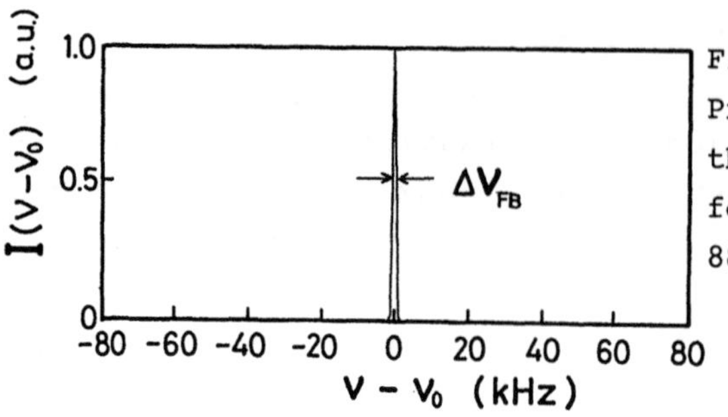

Fig. 2
Profile of the field spectrum of the laser obtained by electrical feedback. The linewidth $\Delta\nu_{FB}$ is 880 Hz.

was employed to estimate such a narrow linewidth by using the value of the curve A. The result is shown by Fig. 2. This figure shows that the linewidth of the field spectrum of the laser was reduced to 880 Hz, which is, to our knowledges, the world's record of the narrowest linewidth of semiconductor lasers. It can be estimated that the linewidth will be reduced to as narrow as 1 Hz by improving the performance of the electrical feedback loop and by utilizing the advanced technology of Opto-Electronic Integrated Circuit (OEIC) in the future.

4. Frequency Tracking of the Slave Laser

Frequency of the slave laser can be tracked to that of the master laser by employing the technique of frequency offset locking[3]. In the present experiment, the slave laser frequency was controlled by electrical feedback in order that the phase of the beat signal between the two lasers was locked to that of the signal from the local microwave oscillator. A slow and fast feedback loops were used simultaneously. The slow feedback loop was used to suppress the drift in the beat frequency. This loop had a digital phase comparator with a dynamic range as wide as $2\pi \times 2^{11}$ (rad) to realize a wide capture range of the tracking loop. The fast feedback loop had an analog phase comparator of the 5 MHz bandwidth. Figure 3 shows the experimental result of the square root of the Allan variance $\sigma_{yB}{}^2(\tau)$ of the beat frequency fluctuations. The minimum of $\sigma_{yB}(\tau)$ was 1.7×10^{-15} at τ = 1000 (s), which corresponded to 0.6 Hz fluctuation. Since this value was far lower than the value of frequency fluctuations of the master laser (see section 1), it was confirmed that the slave laser frequency tracked the master laser frequency accurately, or, in other words, high coherence of the master laser was transferred to the slave laser.

Capture range of this feedback loop depended on the division rate M of the beat frequency provided by the prescaler, and the maximum capture range was 1.8 GHz which was obtained at M = 2,000.

Furthermore, as will be shown in the next section, the locking range was 2GHz. The broken curve in this figure represents the beat frequency

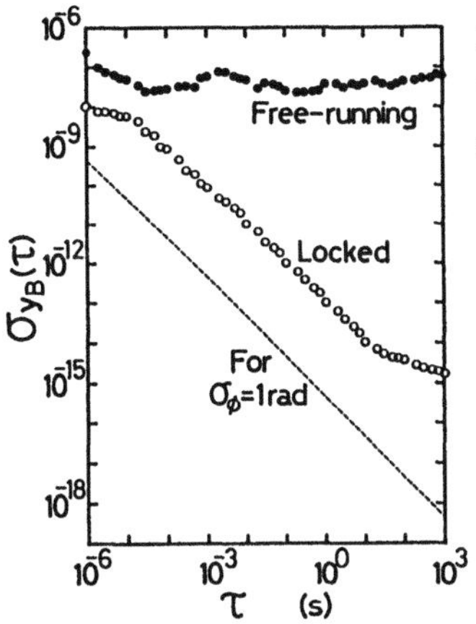

Fig. 3
The square root of the Allan variance $\sigma_{yB}{}^2(\tau)$ of the frequency fluctuations of the beat signal obtained by frequency tracking.

fluctuations which is required to realize 1 radian phase fluctuations of the beat signal. That is, the heterodyne-type optical phase locked loop can be realized if the beat frequency fluctuations will be reduced to this value. This experiment is now in progress[4].

5. Frequency Sweep of the Slave Laser

The beat frequency and the slave laser frequency can be swept accurately and in a very stable manner if the frequency of the local microwave oscillator is swept under the condition of the frequency tracking of the section 4. The beat frequency was swept continuously for the range of 2GHz in the present experiment. It corresponds to the locking range of the frequency tracking loop, which was limited by the bandwidth of the photodetector for the beat signal detection. The stability of the beat frequency was as high as that of Fig. 3, which was independent of the value of the beat frequency within this locking range.

The range of stable frequency sweep of the slave laser can be increased by achieving discrete tuning of the master laser frequency by locking it to successive frequencies of a frequency reference grid which are nearly equally spaced. For each locked frequency of the master laser, frequency tracking loop is applied to the slave laser. By this control scheme, a stable and accurate sweep of the slave laser frequency can be realized. In the present experiment, equally spaced resonance frequencies of a Fabry-Perot interferometer were used as a simple frequency reference grid. The master laser frequency was locked to 16 successive resonance frequencies, and as a result of it, the total range of the continuous sweep of the slave laser frequency was increased to 60 GHz. This range was limited by the discontinuous mode-hopping phenomenon of the AlGaAs laser[5]. If the improved single-longitudinal-mode laser, e.g., the DFB laser will be used, it can be estimated that this range will be increased to as wide as 1 THz.

6. Summary

Table 1 summarizes the present status of the experimental results for the four problems and the estimated values which may be realized in the future. It is seen from this table that the performances of the semiconductor lasers have been improved to as high as, or even higher than, those of carefully designed gas or dye lasers. Further improvements can be expected by improving the electrical feedback loop and laser devices.

Table 1.
Present status of the experimental results for the four problems, and their estimated values which may be realized in the future.

Problem	Present	Future
(1) Center frequency stabilization	2×10^{-12} (at τ = 100 s)	1×10^{-15}
(2) Linewidth reduction	880 Hz	1 Hz
(3) Frequency tracking	1.7×10^{-15} (at τ = 1000 s)	1×10^{-17}
(4) Frequency sweep	60 GHz	1 THz

Acknowledgments

The authors wish to thank Dr. L. Hollberg of NBS, Boulder, for many useful discussions regarding a high finesse Fabry-Perot cavity.

References

1. M. Ohtsu and T. Tako, "Coherence in semiconductor lasers", in Progress in Optics, ed. by E. Wolf, Vol. XXV, (Elsevier, Amsterdam, 1988)
2. M. Ohtsu, J. Lightwave Technol., 6, 245 (1988)
3. K. Kuboki and M. Ohtsu, IEEE J. Quantum Electron., QE-23, 388 (1987)
4. C.H. Shin, K. Kuboki, and M. Ohtsu, Trans. IEE of Japan, 108-C, (1988), August issue, in press (in Japanese)
5. M. Ohtsu, Y. Teramachi, Y. Otsuka, and A. Osaki, IEEE J. Quantum Electron., QE-22, 535 (1986)

Part XI

Containment of Neutrals

Recent Progress in the Control of the Motion of Atomic Beams by Laser Fields

V.I. Balykin, V.S. Letokhov, Y.B. Ovchinnikov, A.I. Sidorov, and S.V. Shul'ga

Institute of Spectroscopy, USSR Academy of Sciences,
SU-142092 Troitsk, Moscow Region, USSR

1. Introduction

Considerable attention is being given currently to the problem of the interaction of resonance laser radiation with translational state of atoms. One of the main reason for this interest is the attempt to develop methods for controlling spatial motion of atoms with the aid of resonance light pressure. The experiments have been reported on the deflection, the deceleration of the longitudinal motion of an atomic beam, the cooling of atomic beam, the coolimation and focusing of atomic beam, and finally, the trapping of atoms (see a review in Ref.1).

In present paper we consider two methods of controlling the motion of atoms in a beam by means of resonance laser radiation (mirror reflection of atoms and channeling of atoms) which are required in order to realize various sorts of traps for slow neutral atoms.

II. Mirror reflection of atoms

The idea of the "atomic mirror" is as follows (Fig. 1a /2/. When a plane traveling light wave is totally reflected internally at the surface of a dielectric in a vacuum, a thin wave is generated on the surface. In the direction OY normal to the surface, this wave decays at a distance of $d = \lambda(\sin^2\vartheta - n^{-2})^{-1/2}/2\pi \simeq \lambda/2$ /3/, where λ is the length of the light wave, ϑ the angle of incidence of the wave, and n the refractive index of the dielectric. It is this surface wave that serves as an

Frequency Standards and Metrology Editor: A. De Marchi

"atomic mirror" for an atom running into it. Indeed, a twolevel atom near the surface experiences the gradient force /2/

$$F_{grad}(y)=\alpha\hbar(\Delta-k_1v_x)G(y)/\left[1+G(y)+(\Delta-k_1v_x)^2/\gamma^2\right], \quad (1)$$

where $\alpha = k(\sin^2\vartheta-n^{-2})^{1/2}$ and $k_1=k\sin\vartheta$ are the imaginary and real components of the wave vector k, $\Delta - k_1v_x = \omega-\omega_0 -k_1v_x$ is the detuning of the laser frequency ω relative to the transition frequency ω_0 with account being taken of the Doppler shift k_1v_x, v_x is the x- component of the atomic velocity, 2γ is the natural linewidth, and $G=I(y)/I_{sat}$ is the saturation parameter, I(y) being the laser intensity and I_{sat} the transition saturation intensity.

With a positive detuning ($\Delta -k_1v_x>0$), the gradient force expels the atom out of the field. The reflection of the atom at the "mirror" may be considered a result of its being expelled out of a potential field whose energy is

$$U(y)=(1/2)\hbar(\Delta-k_1v_x)\ln\left\{1+G(y)/\left[1+(\Delta-k_1v_x)^2/\gamma^2\right]\right\} \quad (2)$$

It can be easily demonstrated from consideration of the atomic motion in field (2) that the angle of reflection of the atom is equal to the angle of incidence, which means that what we have here is the mirror reflection of the atom. With the transverse atomic velocity component v_y being great enough ($v_y > v_y^{max}= [2U(0)/M]^{1/2}$, M being the atomic mass), the atom reaches the surface. In that case, it either adheres to the surface or is reflected from the surface in a diffuse manner /4/. For typical atomic parameters (e.g., for the sodium atom) $M=4\times10^{-23}$g, $v_x=6\times10^4$ cm/s, $\gamma/2\pi$ =5 MHz, and tunable laser radiation parameters $k_1=10^5$ cm^{-1}, $G=10^5$, $\Delta/2\pi$ =+2.6 GHz,we have v_y^{max} =430 cm/s. Accordingly, for thermal atomic velocities, the maximum grazing angle $\psi_{max}=v_y^{max}/v_x=7\times10^{-3}$

rad. The atomic mirror was realized for sodium atoms in /5/. At small grazing angles the reflection coefficient approaches 100%.

That the atomic reflection is quantum-state-selective follows from the character of the relationship between gradient force (1) and the detuning $\Delta - k_1 v_x$. When the detuning is positive, the gradient force repels the atom from the surface and thus mirror reflection of the atom is effected. With a negative detuning, the force attracts the atom to the surface, and so diffuse reflection is observed.

Let an atom (or molecule) has several sublevels in the ground state. For atoms, these may be, for example, fine and hyperfine groundstate splitting components (and for molecules, the vibrational-rotational sublevels of the groung state). Atoms (molecules) on a sublevel for which the frequency of transition to an excited state is lower than the laser frequency are reflected from the "atomic mirror". Thus, if a beam of atoms (molecules) distributed among several groung state sublevels is incident upon the "mirror", the reflected beam will contain only atoms in one and the same quantum state. Since the expression for the gradient force (1) includes the Doppler shift, the minimum distance between the lower sublevels at which the quantum-state-selective reflection will still obtain is determined by the magnitude of the Doppler broadening.

We have studied the quantum-state-selective reflection of the sodium atoms. The ground state of Na, $3S_{1/2}$, is split, because of hyperfine interaction, into two sublevels, one with the quantum number F=2 and the other with F=1, the distance between which is equal to 1772 MHz. According to the statistical weights of these sublevels, 62.5% of sodium atoms in a thermal beam are on the sublevel with F=2 and 37.5% on that with F=1.

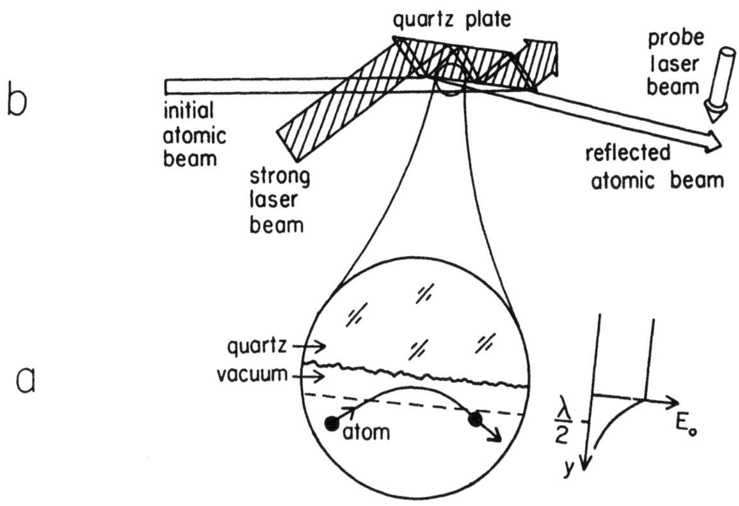

Fig.1. (a) Principle of mirror reflection of atoms by nonuniform laser field of surface wave; (b) schematic diagram of experimental setup.

The beam of sodium atoms (Fig.1b) was formed by a 400-μm-dia, hole in the source and a square slit measuring 85 μm horizontally by 400 μm vertically and placed at a distance of 220 mm from the oven. The "atomic mirror" reflecting the atoms in a horizontal plane was at a distance of 15 mm from the slit. It was a plane-parallel plate of fused quartz 0.4 mm thick and 25 mm long into which a single-frequency laser beam was directed through a bevelled side. It was made a multiple total internal reflection in the plate. The laser beam diameter was 0.5 mm at the $1/e^2$ level. The laser power was 300 mW. The plate surface was polished to an accuracy of $\lambda/10$. To prevent the atoms from being deposited onto the surface, the plate was heated to around 150°C. The residual gas pressure in the vacuum chamber was no worse than 5×10^{-6} Torr.

The detection region was 330 mm distant from the slit. The atomic intensity in the beam was registered by means of a single-frequency probe laser beam which crossed the reflected atomic beam at an angle of 76° in a vertical plane. The probe laser frequency was scanned over the $3S_{1/2} \rightarrow 3P_{3/2}$ transition, which made it possible to monitor the presence in the beam of atoms on the sublevels with F=2 and F=1.

Figure 2 demonstrates the results of the quantum-state-selective reflection of sodium atoms at the "atomic mirror". The frequency ω of the intense laser radiation is measured relative to the frequency $\omega_{20} + k_1 v_p$ of the $3S_{1/2}(F=2) \rightarrow 3P_{3/2}$ transition for the atom with the most probable velocity v_p equal to 8.2×10^4 cm/s at a source temperature of T=630K. Figure 2a corresponds to the case in which the laser frequency detuning is selected so as to be positive only with respect to the frequency of the transition $3S_{1/2}(F=2) \rightarrow 3P_{3/2}$ $\Delta\omega_2/2\pi = [\omega - (\omega_{20} + k_1 v_p)]/2\pi = +0.5$ GHz. In that case, it is only the atoms on the sublevel with F=2 that are being reflected. The presence of a small number of the atoms on the sublevel with F=1 is explained by the fact that the sodium atoms which are being scattered on the residual gas molecules reach the detection region without being reflected at the mirror (Fig.2c, no intense laser field). Subtracting the signal due to the scattered atoms (dashed line in Fig.2c) shows that there are no (accurate to within measurement error) reflected atoms on the sublevel with F=1. If the detuning of the laser frequency is positive with respect to the frequencies of the both transitions ($\Delta\omega_1/2\pi = +2.9$ GHz, $\Delta\omega_2/2\pi = +4.6$ GHz), the reflected atomic beam contains both the atoms on the sublevel with F = 2 and those on the sublevel with F=1 (Fig.2b).

Figure 3 presents the ratio between the reflection coefficients of the atoms on the sublevels with F=2 and F=1 (reflection selectivity) as a function of the detuning of the intense laser radiation frequency relative to

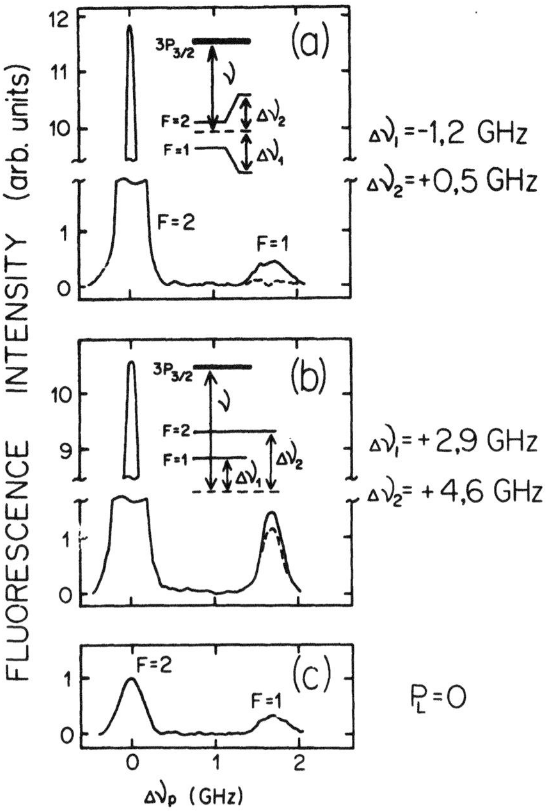

Fig.2. Quantum-state-selective reflection of atomic sodium beam at "atomic mirror": (a) absorption spectrum of reflected atomic beam at detunings of intense laser field forming "atomic mirror" $[\omega - (\omega_{10} + k_1 v_p)]/2\pi = -1.2$ GHz, $[\omega - (\omega_{20} + k_1 v_p)]/2\pi = +0.5$ GHz ($v_p = 8.2\times10^4$ cm/s); (b) same for the case $\Delta\omega_1/2\pi = +2.9$ GHz, $\Delta\omega_2/2\pi = +4.6$ GHz; (c) same at $P_L = 0$ (scattered atoms); dashed line - result of subtraction of background signal due to scattered atoms.

the frequency of the $3S_{1/2}(F=2) \rightarrow 3P_{3/2}$ transition of the atom with the most probable velocity in the beam. The reflection coefficient of the atoms in the state with F=2 (F=1) was found as the ratio of the number of reflec-

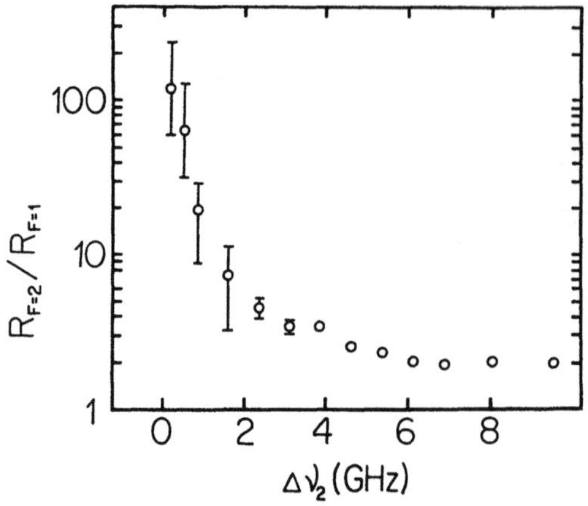

Fig.3. Selectivity of reflection of sodium atoms on sublevels with F=2 and F=1 (ratio between reflection coefficients $R_{F=2}/R_{F=1}$) as a function of laser frequency detuning relative to $3S_{1/2}$ (F=2) $\rightarrow 3P_{3/2}$ transition frequency of atoms with the most probable velocity v_p = 8.2x10^4 cm/s in the beam.

ted atoms to the number of the atoms in the state with F=2 (F=1) incident upon the "mirror". The reflection coefficients were calculated including the effect of redistribution of the atoms between the sublevels due to optical pimping.

It is evident from Fig.3 that there is an interval of detunings of the laser frequency relative to the atomic transition frequency wherein the reflection selectivity reaches a sufficiently high value. The maximum experimentally measured reflection selectivity is 10^2. In our case, it was determined by the noise of the registration system. It can also be seen fron Fig.3 that the characteristic frequency interval within which selective reflection is observed amounts to a few gigahertz and coincides with the frequency interval wherein the gradient force of the surface wave of the "atomic mirror" acts effectively on the atoms.

It should be noted that such a quantum-state-selective reflection can be expected in the case of molecule as well. This will offer unique possibilities of preparing and spectroscopically analyzing beams of molecules in one and the same vibrational state which can be changed by varying the frequency of the reflecting laser radiation.

The successfull experimental observation of the mirror reflection of neutral atoms allows for new approach to the problem of localization of atoms and makes it possible to propose new schemes of traps in which atoms would reside for a long time a field-free region.

Figure 4 shows two schemes of atomic traps based on the mirror reflection of atoms. The scheme of Fig.4a uses atomic reflection at an angle of 45° to the mirror surface, while that of Fig.4b, normal reflection. The first scheme is similar to a ring laser cavity with the laser mirrors replaced by atomic mirrors and using atoms instead of photons. The second is a confocal Fabri-Perot atomic interferometer. To create an atomic mirror, use is made of the total internal reflection of atoms at a dielectric-vacuum interface. One can suggest two ways to introduce atoms into the trap. The first, pulse method is carried into effect by turning one of the atomic mirrors as shown in Fig.4. The second, continuous method can be realized by providing one of the trap mirrors with a small hole.

Let us estimate the parameters of the trap. The maximum velocity component of an atom, normal to the atomic mirror surface, at which the mirror reflection of the atom is observed, is given by

$$v_{\perp} = (2U_o/M)^{1/2} = [(\hbar\Delta/M)\ \ln(1+p)]^{1/2} \qquad (3)$$

where $p = \Omega_R^2/2(\Delta^2 + \gamma^2)$ is the atomic transition saturation parameter, Ω_R the Rabi frequency, Δ the detu-

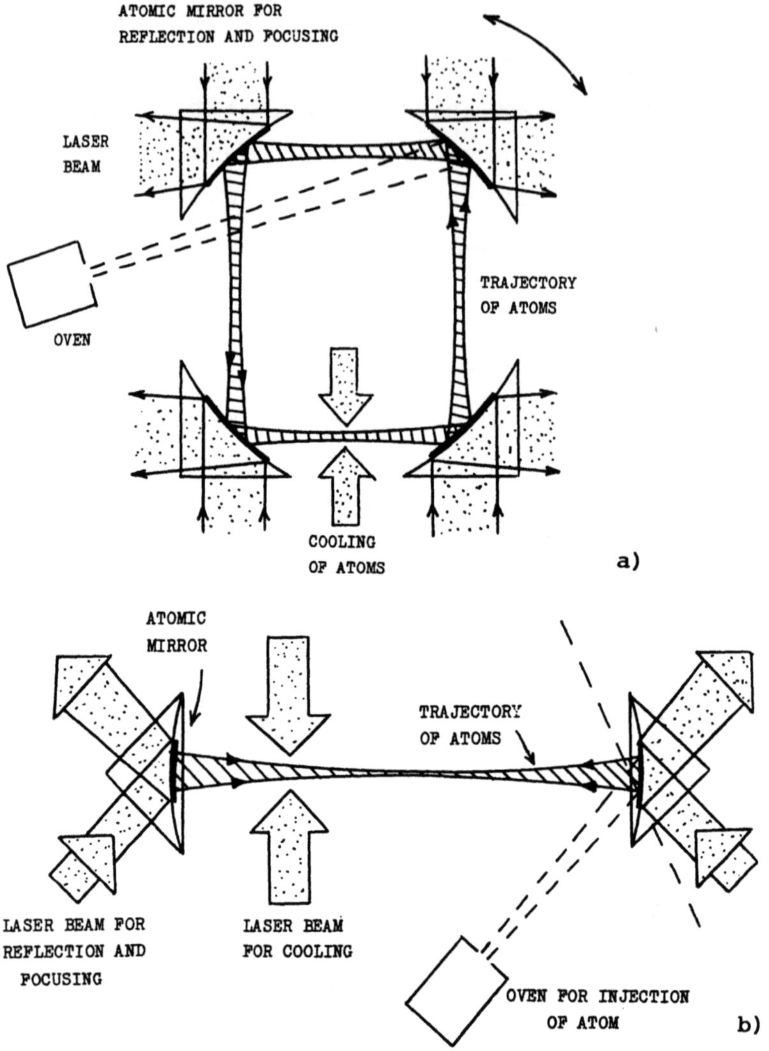

Fig.4. Optical trap versions using the phenomenon of atomic reflection from a laser mirror: (a) ring atomic trap configuration. The trap is formed by four atomic mirrors. To compensate for the diffusive heating of the atoms upon their reflection, use is made of an additional laser cooling; (b) "atomic confocal interferometer". The main elements are the same as in Fig.4a.

ning of the laser frequency relative to the atomic transition frequency, 2γ the radiative atomic absorption line width, and M the atomic mass. At $\Omega_R \gg \gamma$ the optimum detuning at which the atomic mirror reflects the atoms with the maximum normal velocity component $v_\perp^{max}$ is equal to $\Delta = \Omega_R/8^{1/2}$ ($p=4$). In this case

$$v_\perp^{max} = \left[(0.6)\,(\hbar\Omega_R/M)\right]^{1/2} = v_{rec}\left[(0.6)(\Omega_R/\omega_{rec})\right]^{1/2} \quad (4)$$

where $v_{rec} = \hbar k/M$ is the atomic recoil velocity and $\omega_{rec} = kv_{rec}$. The maximum reflected velocity expressed in terms of the laser power P and the atomic mirror size is

$$v_\perp^{max} = v_{rec}\left[0.8\,(\gamma/\omega_{rec})\right]^{1/2}(P/\pi w_0^2 I_s)^{1/4} \quad (5)$$

where I_s is the atomic transition intensity. For the sodium atom at a laser power of $P_0=1$W and a laser beam radius of $w_0=0.1$ mm, $v^{max}=650$ cm/s. Increasing the beam radius to $w_0=1$mm brings the maximum velocity to $v^{max} = 210$ cm/s.

As the atoms reflect from the atomic mirror, they heat as a result of momentum diffusion, and their transverse velocity distribution increases. This heating is easy to compensate for by means of an additional cooling laser radiation, and the diffusive spatial broadening can be compensated by using focusing atomic mirrors.

III. Channeling of Atoms in a Standing Spherical Light Wave

The localization of atoms in a nonresonant light wave has been suggested in /6/. In recent works, it has been inferred from observations of atomic fluorescence /7/ and absorption /8/ in a quasiresonant standing light wave that atoms placed in such a wave undergo density redistribution /7/ and channeling /8/ under the action of the gradient force. In the present paper, we report the observation of the localization of atoms in the vicinity of the nodes or loops of a standing spherical light wave in the course of their being channeled along the curvilinear wave front of the laser field.

The basic idea and schematic of the experiment are illustrated in Fig.5. When atoms are being channeled along the nodes (or loops) of a light wave, the atomic trajectory must follow the wave front accurate to within $\lambda/2$. Therefore, if the velocity of an atom entering a standing spherical light wave is tangent to the wave front, the localization of the atom must cause it to change the direction of its motion through an angle determined by the diameter of the laser beam forming the light wave and the wave front curvature. Assuming that the atomic motion is determined mainly by the gradient force produced by the standing light wave, the angle α through

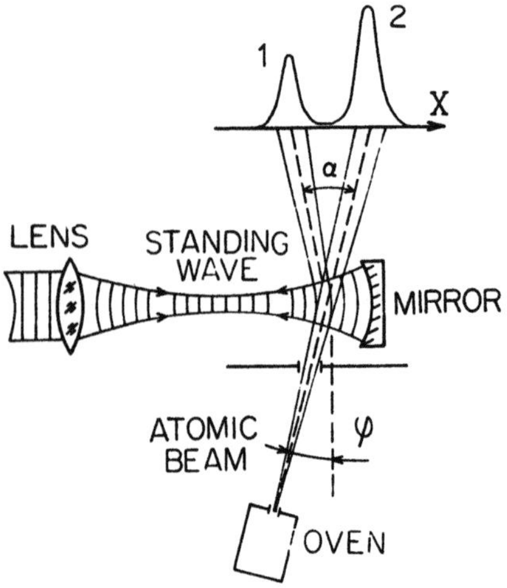

Fig.5. Illustrating the localization of atoms along the nodes (loops) of a standing light wave: 1-atomic beam intensity peak corresponding to localized atoms; 2 - peak corresponding to nonlocalized atoms.

which the atoms will be deflected from their initial direction of motion may be defines as

$$\alpha = \omega_o z / \left[z_R \, (z^2 + z_R^2)^{1/2} \right], \qquad (6)$$

where ω_o is the waist radius of the Gaussian laser beam forming the standing light wave, $z_R = \pi \omega_o^2 / \lambda$, and z is the distance from the beam waist to the point of intersection between the atom and the light wave.

The maximum angular divergence of the atomic beam, within which the atoms still can be localized in the standing light wave is given by

$$\Delta\varphi = 2(U_{max}/E)^{1/2} = 2(2U_{max}/M)^{1/2}/v, \qquad (7)$$

where v is the atomic velocity, M the atomic mass, E the kinetic energy of the atom, and U_{max} the maximum atomic potential energy in the, standing light wave, defined as

$$U_{max} = \tfrac{1}{2}\hbar\Delta \,\ell n\,(1 + p_o), \qquad (8)$$

where $p_o = 8 g_o/(\Delta^2 + \gamma^2)$ is the atomic transition saturation parameter in the standing wave loop and g_o the Rabi frequency of the traveling light wave forming the standing light wave /9/. Expression (7) also defines the angular divergence the localized atoms will have after leaving the standing light wave.

To produce a standing spherical light wave, use was made of a single-frequency radiation from a cw dye laser. The laser beam was focused with a lens L (F=80 mm) into the center of curvature of a spherical mirror M (R=50mm). The beam waist radius ω_o in that case amounted to 0.015mm. The size of the standing light wave at the point of intersection with the atomic beam was 1 mm, and the radius of curvature of the wave front was 40 mm. The beam of sodium atoms was shaped by two diaphragms. The diaphragm used in the atomic source was a round hole with a diameter of d=0.4 mm, and the other one was a slit with dimensions l_1 = 0.17 mm and l_2 = 0.5 mm. The distance between the diaphragms was L_1 = 230 mm. In that case, the angular divergence of the atomic beam was 4×10^{-3} rad.

The distance from the slit diaphragm to the laser beam was L_2=10 mm, the long side of the slit being parallel to the beam. The radiation power in a single traveling wave was P_1 = 0.11 W. The atoms angle of entry into the laser beam (Fig.5) was $\varphi = 1.25\times10^{-2}$ rad. The atomic beam profile was scanned by means of a probe single-frequency radiation tuned to resonance with the $3S_{1/2}$, F = 2 $\rightarrow$ $3P_{3/2}$, F = 3 transition. The probe beam traversed the atomic beam at an angle of 76°. The detection region was at a distance of L_3=290 mm from the standing light wave. The probe radiation frequency was set to fall within the Doppler absorption line contour of the atomic beam, and the fluorescence of the atoms moving with a certain longitudinal velocity was detected. The velocity of the atoms subject to probing was 450 m/s.

Figure 6a shows the experimental atomic beam profile resulting from the interaction between the beam and the standing spherical light wave. The laser frequency ω was detuned from the frequency of the $3S_{1/2}$, F = = 2 $\rightarrow 3P_{3/2}$, F = 3 transition, ω_0, by an amount of $\Delta/2\pi$ = $(\omega-\omega_0)/2\pi$= 300 MHz. The interaction gave rise to the deflection of some atoms through an angle of a relative to their initial direction of motion and the appearance in the detection region of an additional atomic intensity peak (Fig.6a). The distance from the maximum of this peak to the center of the beam profile in the absence of the standing light wave (dashed line in Fig.6a) was 7.1 mm, which corresponded to a deflection angle of α =2.5×10^{-2}

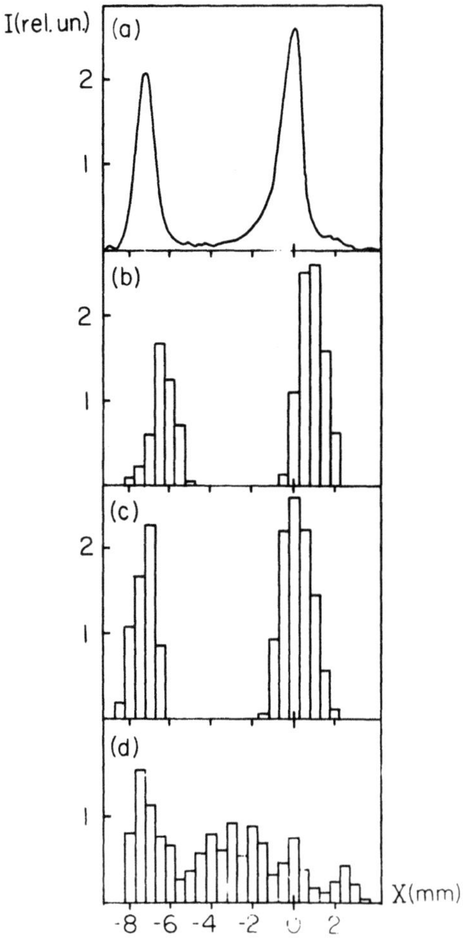

Fig.6. Atomic beam profile in the registration region after interaction with the standing spherical light wave: (a) experiment; (b), (c), and (d) calculation; 1 - atomic beam intensity peak corresponding to localized atoms; 2 - peak corresponding to nonlocalized atoms.

rad. Estimation of the atomic deflection angle by formula (1) yields $\alpha = 2.4 \times 10^{-2}$ rad., which agrees well with the measured value.

The measured angular divergence $\Delta\varphi$ of the atoms subjected to localization also agrees with the estimate by expression (7) and is equal 5×10^{-3} rad.

The atoms entering the standing light wave at a certain angle of $\delta\varphi$ comparable with the angle $\Delta\varphi$ and not along the tangent to the wave front will fail to become localized. In our experiment, this was the case with $\delta\varphi = \pm 3.5 \times 10^{-3}$ rad.

Figures 6b, c, and d show beam profiles in the detection region after interaction with the standing light wave, calculated with allowance being made for the gradient force (Fig.6b), gradient force and friction force /10/ (Fig.6c), and gradient force, friction force and momentum diffusion (Fig.6d) /11/. The displacement of the beam profile produced by the localized atoms relative to that due to nonlocalized atoms (Figs.6c and d) was used to determine the deflection angle α which proved to be close to both the measured value and also the value found by expression (6). Note that failure to include the friction force in calculations causes the beam profile produced by nonlocalized atoms to displace from its initial position (Fig.6b). The better agreement between theory and experiment occurs when both the gradient and friction forces are taken into account (Fig.6c). Inclusion of momentum diffusion /10-12/ makes the calculated beam profiles (Fig.6d) wider than their experimental counterparts. This effect cannot as yet be explained comprehensively. In our experiment, we also observed the localization of atoms along the loops of the standing waves at $\Delta < 0$.

References

1. A.Ashkin, Science, 210, 1081 (1980). A.P.Kazantsev, G.A.Ryabenko, G.I.Surdutovich, V.P.Yakovlev, Phys. Rep., 129, 75 (1985); Mechanical Effects of Light, Special Issue of J.Opt.Soc.Am., B2 (1985); S.Stenholm, Rev. Mod.Phys., 58, 699, (1986); V.G.Minogin , V.S.Letokhov,

Laser Light Pressure on Atoms, (Gordon and Breach, New York, 1987).

2. R.J.Cook, R.K.Hill, Opt.Comm.,43,258 (1982).

3. M.Born, E.Wolf, Principles of Optics (Pergamon Press, London, 1970) p.47.

4. N.F.Ramsey, Molecular Beams (Cla rendon Press, Oxford, 1956).

5. V.I.Balykin, V.S.Letokhov, Yu.B.Ovchinnikov, A.I.Sidorov, Pis'ma ZhETF, 45,282 (1987) (in Russian).

6. V.S.Letokhov, Pis'ma Zh.Eksp.Teor.Fiz.7,348 (1968) (in Russian).

7. M.G.Prentiss, S.Ezekiel, Phys.Rev.Lett.56,46 (1986).

8. C.Salomon, J.Dalibard, A.Aspect, H.Metcalf, C.Tannoudji, Phys.Rev.Lett.59, 1969 (1987).

9. V.G.Minogin, V.S.Letokhov, Laser Light Pressure on Atoms (Gordon and Breach Science Publishers, New York, 1987).

10. A.P.Kazantsev, Zh.Eksp.Teor.Fiz.66,1599 (1974) (in Russian).

11. J.P.Gordon, A.S.Ashkin, Phys.Rev.Lett.21, 1606 (1980).

12. V.G.Minogin, Yu.V.Rozhdestvenskii, Zh.Eksp.Teor. Fiz.93,1173 (1987).

Atoms Laser-Cooled Below the Doppler-Cooling Limit

P.D. Lett, C.I. Westbrook[a]*, R.N. Watts*[a]*, S.L. Rolston, P.L. Gould*[b]*, H.J. Metcalf*[c]*, and W.D. Phillips*

National Institute of Standards and Technology,
(formerly National Bureau of Standards), Gaithersburg, MD 20899, USA

We have previously reported [1] laser cooling of atoms in optical molasses to temperatures as low as 40 μK; more recently, we have measured the temperature of this three-dimensional gas of Na atoms to be as low as $10\,^{+20}_{-10}$ μK. These low temperatures are in strong disagreement with the traditional theory of laser cooling, which predicted a lower limit of 240 μK. Temperatures below the "cooling limit" were surprising and unexpected. They pointed out the inadequacy of the usual theory for explaining a rather simple experiment. The results have great practical interest because it now appears to be relatively easy to obtain temperatures much lower than previously thought possible.

Here we describe the measurement of these ultra-low temperatures and report the dependence of the temperature on magnetic field, laser intensity, laser polarization, and detuning. These are in qualitative agreement with theories recently advanced to explain the anomalously low temperatures.

The traditional and generally accepted theory of laser cooling of free, 2-state atoms predicts that the lowest achievable temperature is given by $k_B T = \hbar\gamma/2$, where k_B is Boltzmann's constant and γ is the natural linewidth of the transition for laser cooling. This "Doppler-cooling limit" results from the minimization of the detuning-dependent temperature at low laser power [2]:

$$k_B T = \frac{\hbar\gamma}{4}\frac{1 + (2\Delta/\gamma)^2}{2|\Delta|/\gamma}, \qquad (1)$$

where $\Delta < 0$ is the detuning of the laser from resonance. Eq. 1 is minimized at $\Delta = -\gamma/2$. For Na cooled on the D2 line at λ = 589 nm, where $\gamma/2\pi$ = 10 MHz, the Doppler-cooling limit is 240 μK. Although the theory was derived assuming 2-state atoms, it was believed that this simplification was applicable to cases such as laser cooling of alkalis.

Our experiment uses Na atoms in "optical molasses" [2,3] formed at the intersection of three orthogonal pairs of counterpropagating laser beams apertured to 9 mm diameter. The strong laser cooling in the intersection region results in a "viscous confinement" of atoms. We

---

This work is supported by the Office of Naval Research.

[a] NBS-NRC Post-doctoral Fellow.

[b] Permanent address: Department of Physics, University of Connecticut 06268.

[c] Permanent address: Dept. of Physics, S.U.N.Y., Stony Brook, NY.11794.

Frequency Standards and Metrology Editor: A. De Marchi

continuously load the molasses with slow atoms from a laser cooled atomic beam operating on the Zeeman tuning principle [4]. The atomic density in the molasses builds up to an equilibrium value determined by the balance between the diffusion of atoms out of the molasses and the flux of atoms into it. We then turn off the molasses laser beams, releasing the atoms to travel ballistically. A 4 mm diameter retroreflected probe laser beam is placed 1-2 cm beneath the molasses. The fluorescence from atoms falling through the probe is imaged onto a photomultiplier which is used to record the atoms' arrival times at the probe. The temperature of the atoms released from the molasses is determined by comparing the arrival time distributions with calculations of the arrival time distribution obtained by numerically integrating over an assumed 3-D Maxwell-Boltzmann velocity distribution, taking into account the initial spatial distribution, the geometry of the probe, and the effects of gravity.

Figure 1 shows the results of these time-of-flight (TOF) temperature measurements as a function of Δ for various probe-molasses separations *d*. The molasses laser power was 5 mW in each of the six beams. The data are radically different from Eq.(1) (shown as a solid line in the figure); the temperature is minimized for $|\Delta|$ much larger than the predicted $\gamma/2$ and is always below the cooling limit except at small detuning. The uncertainty in the lowest temperatures, including systematic uncertainties, is 20 μK. These data were previously presented in ref. 1.

The measured atomic temperature depends not only on detuning, but also on the laser polarization, the applied magnetic field and the laser intensity. The lowest temperature is achieved when the linear polarizations of each of the pairs of laser beams are mutually orthogonal and the magnetic field is zero. Even a field as small as 10^{-4} T is sufficient to significantly raise the temperature. This is particularly surprising in view of the fact that the Zeeman shifts associated with such a field are small compared to the linewidth of the transition or the detuning of the laser from resonance.

The dependence of the observed temperature on magnetic field and laser polarization, suggests that the magnetic sublevel structure is responsible for the anomalously low temperature, and that the 2-level simplification of Eq. (1) is inadequate. A theory developed by Jean Dalibard and Claude Cohen-Tannoudji of Ecole Normale Superieure [5], and presented

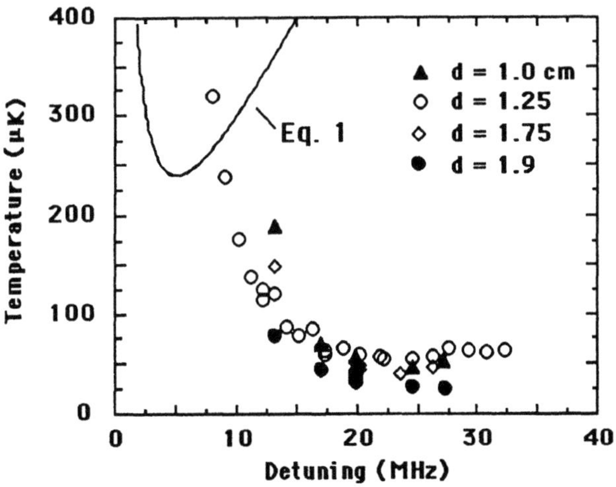

Figure 1. T vs. Δ by TOF for various separations d. The solid line is the temperature given by the 2-level theory of Equation (1).

at this conference, has offered an explanation of this phenomenon. S. Chu of Stanford University has also given a similar explanation [6]. (These laboratories have also observed below-Doppler-limit cooling in cesium and sodium respectively.) The essence of the idea is that atoms are optically pumped into those sublevels that are most strongly light-shifted to lower energy by the optical field. The exact distribution of sublevels into which the atom is pumped depends on the local polarization of the light. As the atom moves through the optical molasses the local polarization of the 3-D standing wave changes rapidly while the atom tries to adjust by optical pumping. The existence of a lag between the actual state of the atom and the state into which it would be pumped if it were at rest (the lowest energy state) implies that the light-shifted potential energy of the atom is always increasing as it moves, so the kinetic energy must decrease. The increased potential energy is continually dissipated radiatively. The atom loses kinetic energy in proportion to the difference in light shifts between the different sublevels and to the time lag of the optical pumping. The shift is proportional to the laser intensity, and the optical pumping time is inversely proportional to it. At low laser power the damping of atomic velocity is therefore independent of intensity when the velocity is low enough.

In ref. [1] we reported the sensitivity of temperature to magnetic field. We now understand this sensitivity as being due to mixing of magnetic sublevels, which interferes with the optical pumping process described above. We have measured the dependence of temperature on the current in each of the three sets of orthogonal field-compensating coils used to adjust the field at the molasses. In this way we determine the optimum field in each direction. The field used in ref. [1] was found to be 50 μT different from this optimum. Measurement with a flux gate magnetometer confirmed that this optimum field was in fact zero to within an uncertainly of 3 μT. We also note that the field gradient in any direction is less than 30 μT/cm and that increasing this gradient by at least a factor of two had no measurable effect on the temperature. We had previously estimated, based on Hall probe measurements, that the field in [1] was zero to within 20 μT, so we were apparently in error by more than twice our estimated uncertainty.

Once the magnetic field was adjusted to be optimum (and presumably near zero) in each direction, we found that the temperature was significantly lower than reported in ref. [1]. Figure 2 shows the measured dependence of temperature on the field along one axis, with the zero of the field axis set to correspond to the minimum temperature. Between 10 μT and 100 μT the temperature varies linearly with the field. We find that the slope of k_BT vs. μ_BB is 1.2, where μ_B is the Bohr magneton.

We emphasize that the very low temperatures, of 10 μK or less, shown in Fig. 2, have a rather large systematic uncertainty. The temperatures are determined by matching the peak of the observed time-of-flight distribution to that of TOF distributions calculated from measured molasses and probe geometry and an assumed temperature. Uncertainties in the initial distribution of atoms in the molasses before release, the distance between the molasses and probe, and the time the atoms spend in the probe contribute to the uncertainty. The fact that the width in time of the observed time-of-flight spectrum is generally larger than the calculated width also adds to the uncertainty. Our best estimate of the lowest temperature achieved is

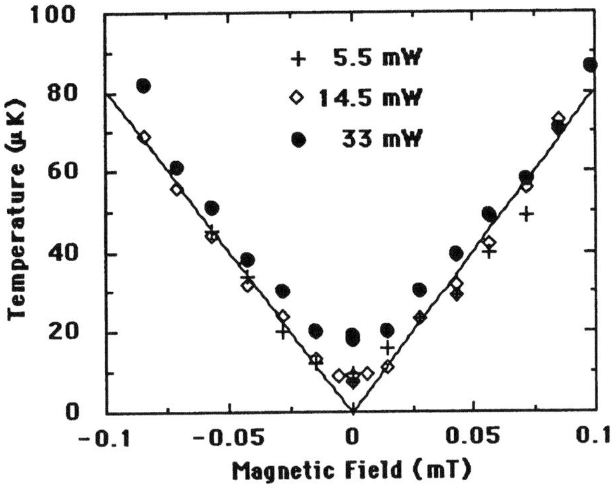

Figure 2. Temperature versus magnetic field at three different laser powers, for Δ = -25 MHz. The solid lines have slopes of ±0.8 K/T. Powers indicated must be divided by three to obtain power per beam.

10 µK, and we include the region from zero to 30 µK in our one-standard-deviation estimate of the statistical and systematic uncertainty interval. Note that in spite of this large systematic uncertainty, the statistical scatter of points is only several microkelvin.

Although we reported earlier that the temperature was independent of intensity, within experimental error, over a limited range of powers, it is apparent from Figure 2 that the temperature is lowest at lowest intensity. Fig. 3 shows the dependence of temperature on laser power both for zero magnetic field (circles) and, over a wider range of powers, some earlier data when the magnetic field was presumably about 50 µT (squares). The power indicated is the total power delivered to the molasses, and should be divided by three to obtain the power per beam. The detuning for these data was $\Delta = -25$ MHz. The temperature is seen to be linearly dependent on the intensity, within experimental uncertainty. This linear dependence is consistent with the new theories, since in these the damping does not change with intensity, but the heating rate is proportional to intensity.

The theory also predicts that the temperature varies as $1/\Delta$. This comes from the fact that the damping force varies as $1/\Delta$, while the heating rate goes as $1/\Delta^2$. Figure 4 shows the

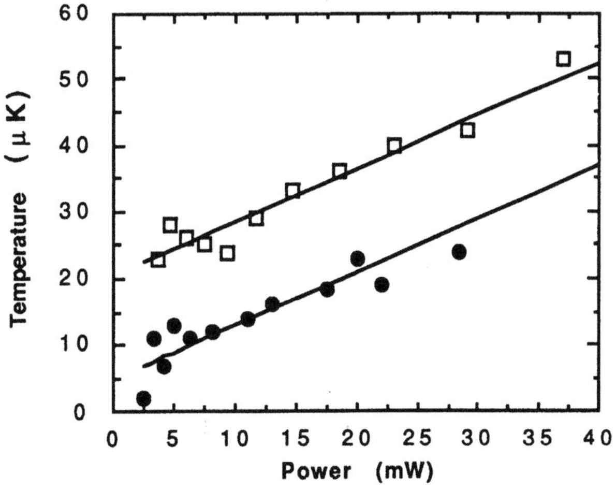

Figure 3. Molasses temperature versus laser power for two different values of the bias magnetic field (see text). Straight lines with the same slope have been drawn through the two data sets.

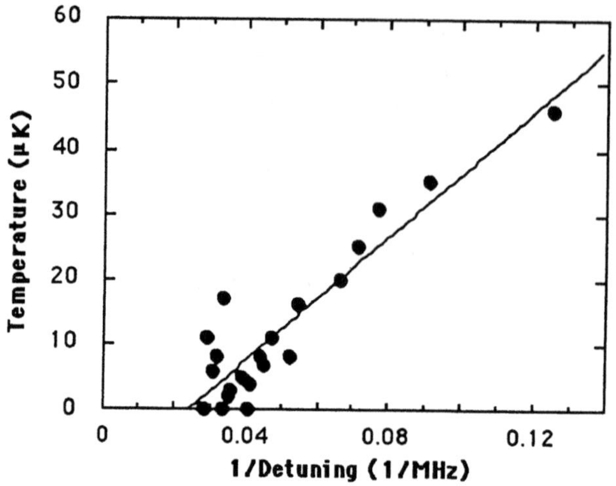

Figure 4. Temperature versus inverse detuning.

temperature plotted against $1/\Delta$, and fitted to a straight line. Within experimental uncertainty, it confirms the expected dependence of temperature on detuning from 8 to 40 MHz.

The results for temperature as a function of detuning and of laser power confirm the functional dependence predicted by the theory of Dalibard and Cohen-Tannoudji. The dependence of temperature on magnetic field also is consistent with that theory, although at this writing there is not a specific and firm prediction of the functional form of that dependence. We have observed that rotating the linear polarization of one of the three pairs of molasses beams (normally mutually orthogonal) by $\pi/2$ raises the temperature by about 20 μK. This is also consistent with the new theories, because when two of the beam pairs have parallel polarization, the spatial variation of polarization is not as rapid. We feel that these observations provide strong experimental support for the hypothesis that the damping mechanism based on polarization gradient and optical pumping is responsible for our very low temperatures.

A very important remaining question is: What is the new cooling limit? Experimentally, our results are consistent with zero temperature. We feel, however, that it is quite unlikely that we have temperatures below the "recoil limit" of $k_BT = \hbar^2k^2/M$, and it may be that the temperature is much higher than that. Furthermore, we have no direct experimental evidence that the molasses velocity distribution is even Maxwell-Boltzmann.

It will be essential in the future to reduce the systematic error associated with temperature measurement by our time-of-flight technique. One way to do this may be to study the present sources of error and reduce them as much as possible. This has some intrinsic limitations. Consider the fact that while we are trying to measure a temperature near 10 μK, the gravitational energy difference between the top and bottom of the molasses volume (about 1 cm) is equivalent to a temperature difference (defined by $k_BT = mgh$) of 250 μK. This gravitational energy difference is of course taken into account in our analysis, but it makes the temperature determination rather difficult. Another approach to temperature measurement would be the "shower" method described in [1]. There we measured the transverse spreading of atoms released from the molasses, a technique which is sensitive to the transverse velocity rather than the vertical velocity, and becomes more sensitive at lower temperature. Still another possibility would be to measure the Doppler width spectroscopically. This could be done by measuring the frequency spectrum of the fluorescence from the optical molasses or the width of a narrow Raman resonance between ground hyperfine states.

References

[1] P. D. Lett *et al.*, Phys. Rev. Lett., **61**, 169 (1988).

[2] P Gould *et al.*, in *Laser Spectroscopy VIII*, S. Svanberg and W. Persson, Eds. (Springer Verlag, Berlin, 1987) p. 64. and references therein.

[3] S. Chu *et al.*, Phys. Rev. Lett. **55**, 48 (1985).

[4] W. Phillips and H. Metcalf, Phys. Rev. Lett., **48**, 596 (1982).

[5] J. Dalibard and C. Cohen-Tannoudji, presented at the Eleventh International Conference on Atomic Physics, Paris, July 1988; presented at this conference; and private communications.

[6]. S. Chu, presented at the Eleventh International Conference on Atomic Physics, Paris, July 1988.

High-Contrast, High-Resolution Single Recoil Component Ramsey Fringes in Ca

J. Helmcke[1], *J. Ishikawa*[2], *and F. Riehle*[1]

[1]Physikalisch-Technische Bundesanstalt, Bundesallee 100, D-3300 Braunschweig, Fed. Rep. of Germany

[2]National Research Laboratory of Metrology, 1-4 Umezono, 1-Chome, Tsukuba-Shi 305, Japan

1. Introduction

Frequency standards operating in the optical range promise very low frequency uncertainties due to the high spectral resolution and the small sensitivity to external fields. The intercombination line $^3P_1 - {}^1S_0$ of ^{40}Ca at $\lambda = 657.46$ nm provides an attractive reference frequency to develop an optical frequency standard /1/. It has a narrow linewidth of $\delta\nu \cong 400$ Hz $(\delta\nu/\nu \cong 9 \cdot 10^{-13})$ and its frequency depends only very little upon electric and magnetic fields. The transition can be excited by dye lasers utilizing longlived dyes and probably by semiconductor lasers in the near future. To achieve high spectral resolution, the transit–time broadening may be reduced by means of separated field (Ramsey) excitation /2,3/. The highest fringe contrast in nonlinear optical Ramsey spectroscopy is observed by applying an excitation geometry consisting of two pairs of counter – propagating travelling waves /4 – 6/. However, the position of the center of the Ramsey fringe depends critically upon the alignment of the optical phases between the interaction zones. Furthermore, if unresolved, the doublet structure due to the recoil effect may lead to line asymmetries caused by different size and broadening of each component. In this paper we therefore present methods to reduce largely the uncertainties associated with these effects in order to find exactly the center of the reference line.

2. Experimental Setup

The experiments were performed using a thermal (T $\cong$ 1000 K) beam of calcium atoms (Fig. 1) interacting with the radiation of a high–resolution dye laser spectrometer /7/ tuned to the wavelength of the Ca intercombination line. Two cat's eye retroreflectors were employed to have two equally spaced parallel pairs of counterpropagating travelling waves with the laser beam waists projected onto the axis of the atomic beam. The fluorescence emitted by the decay of the $^3P_1$ level to the $^1S_0$ ground state is monitored approximately one decay length (20 cm) downstream by means of a photomultiplier tube (pmt).

3. Phase Alignments in Travelling Wave Ramsey Excitation

Using separated field excitation with travelling waves, the Ramsey signal $I_R(\nu)$ can be described (simplifying the expression given by Bordé et al. /6/) by

Frequency Standards and Metrology Editor: A. De Marchi

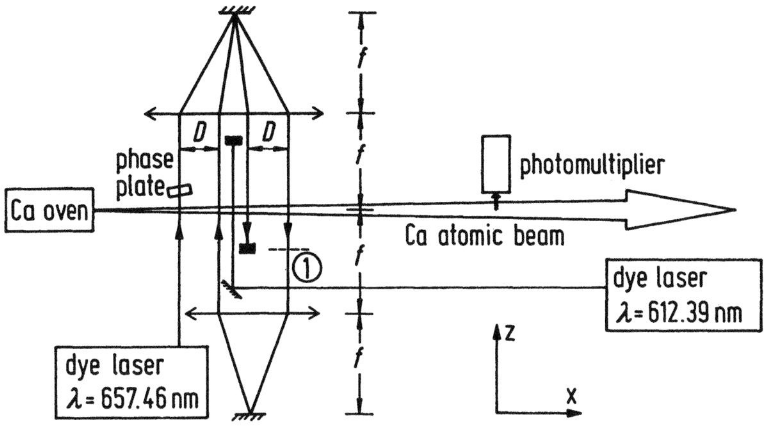

Fig.1: Experimental setup to detect optical Ramsey fringes in a Ca atomic beam by means of travelling wave excitation. The second dye laser ($\lambda \cong 612$ nm) is used for suppression of a recoil component (sect.5).

$$I_R(\nu) \simeq \int_0^\infty A(P,t,\nu)\cdot f(v_x)\cdot\cos(4\pi\Delta\nu D/v_x + \Delta\Phi)\cdot dv_x + B(P,t,\nu) \tag{1a}$$

with

$$\Delta\nu = \nu - \nu_0\left[1 - v^2/(2c^2) \pm h\nu_0/(2mc^2)\right]. \tag{1b}$$

In these equations, $A(P,t,\nu)$ represents the signal amplitude and $B(P,t,\nu)$ the background of the Doppler broadened line including the Lamb dip, both depending on laser power P, interaction time t and weakly on the laser frequency ν. $f(v_x)$ is a modified velocity distribution, D the separation between copropagating beams (Fig. 1) and $\Delta\phi$ the phase difference between the excitation fields. The detuning term $\Delta\nu$ contains shifts caused by the 2nd order Doppler effect $\delta\nu/\nu_0 = v^2/(2c^2)$ ($\cong 3\cdot10^{-12}$ in the present experiment) and the photon recoil $\delta\nu/\nu_0 = \pm h\nu_0/(2mc^2)$ $\cong \pm$ 11.6 kHz in the case of Ca. The Doppler effect causes velocity and resolution dependent shifts resulting in distortions of the fringe structure at resolutions comparable to the 2nd order Doppler broadening /8/ whereas the photon recoil generates a symmetric line splitting.

If the phases of the laser beams in the interaction zones are Φ_1, Φ_2, Φ_3, and Φ_4, respectively, $\Delta\Phi$ can be expressed as $\Delta\Phi = \Phi_2 - \Phi_1 + \Phi_4 - \Phi_3$ /6/. According to (1a) phase differences $\Delta\Phi \neq 0$ linearly shift the Ramsey fringes and lead to frequency offsets from the atomic line center. In our experiment the phase of the laser field in the first interaction zone (and therefore $\Delta\phi$) can be aligned in reference to the other interactions by rotating a 2 mm thick plane parallel glass plate (phase plate in fig. 1).

The measurements were performed at a fringe width of 46 kHz (FWHM) where the recoil splitting was not resolved. The corresponding field separations were D = 0.75 cm (see Fig. 1). With increasing $\Delta\phi$, the fringes shift towards lower frequencies (Fig. 2). However, if the propagation directions of the interacting laser beams are reversed by blocking either one of the incident beams (direction "1" or "2") obtained by means of the beamsplitter (BS), the phase

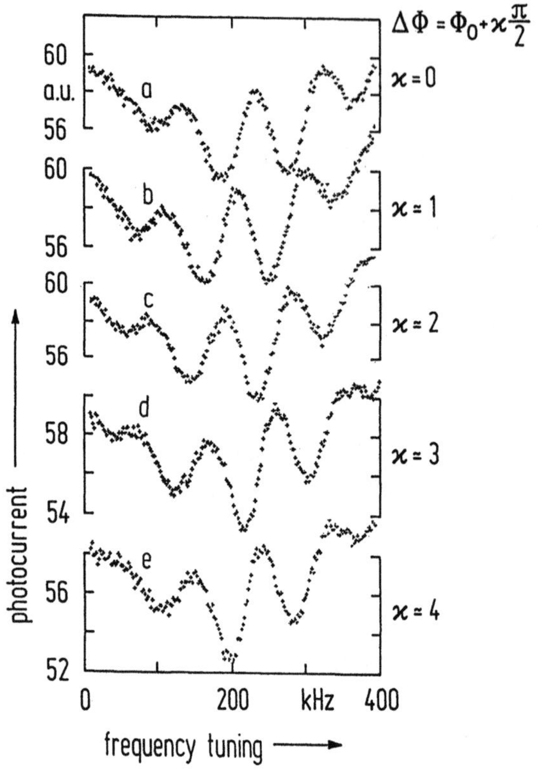

Fig. 2: Ramsey fringes observed with travelling wave excitation for different values of $\Delta\phi$ (direction of the laser beams: "1", see fig. 1).

difference and the shift are also reversed. Consequently, the influence of phase errors should cancel if the signals observed for both directions at the same phase difference are summed. In this case we obtained symmetrical, unshifted Ramsey fringes at any arbitrary phase setting $\Delta\Phi$. Within our experimental resolution of approximately a tenth of a fringe width –limited by the frequency stability of the laser and the signal–to–noise ratio of the Ramsey structure– the center frequencies of the fringe signals coincide for all $\Delta\Phi$. The fringe contrast, however, depends upon $\Delta\Phi$ and the signal changes sign at $\Delta\Phi = \pi/2$. Background free symmetrical Ramsey structures can be obtained from the difference between the summed signals of a phase difference of $\Delta\phi = 0$ and $\Delta\phi = \pi$, respectively.

4. Frequency Stabilization by Means of Ramsey Resonances

For frequency stabilization, however, an antisymmetric Ramsey structure free of background is more suitable than the symmetrical structure described above. Such a discriminant curve can be obtained at fixed values of $\Delta\phi$ if the Ramsey signals recorded for the two laser beam directions are subtracted and if the two signals have equal sizes. The maximum slope at line center is obtained for $\Delta\Phi = \pi/2$ (Fig. 3). In this example, the directions of propagation of the interacting laser beams were reversed repeatedly by chopping the two directions (see Fig. 1) alternatingly at a frequency of about 50 Hz and phase–sensitively detecting the signal of the photomultiplier. The powers of the two partial beams "1" and "2" (see Fig.1) had to be carefully aligned to be equal.

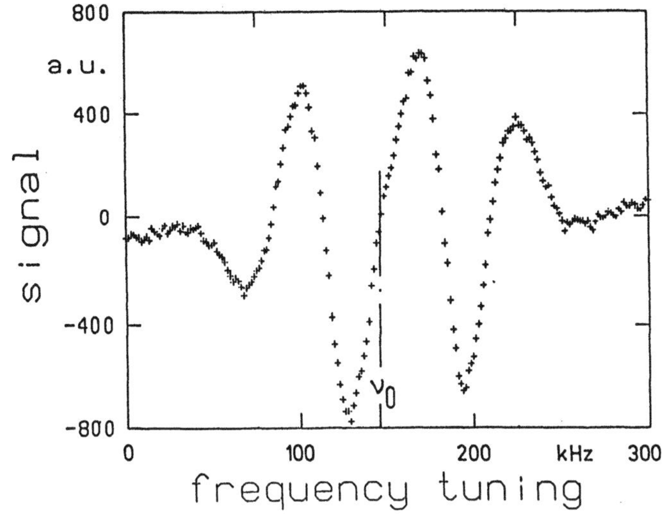

Fig. 3: Error signal generated by subtracting the signals observed for $\Delta\phi \cong +\pi/2$ and $\Delta\phi \cong -\pi/2$ (directions "1" and "2", respectively, of Fig. 1).

The error signal was digitally integrated and used to correct the frequency of the dye laser spectrometer /7/.

As a first application of the "Ramsey fringe stabilized" dye laser, the small frequency drift of the optical reference resonator used in the spectrometer for fast frequency stabilization was monitored. This drift was determined also by measuring the beat frequency between a HeNe offset laser stabilized to a suitable resonance of the resonance resonator and an I_2–stabilized HeNe laser. The standard deviations of the measurements obtained with the Ca stabilized dye laser were smaller by an order of magnitude compared to the I_2–stabilized HeNe laser.

The present stabilization scheme utilizing an unmodulated laser beam depends critically upon the alignment of the two incident beam directions to equal powers. Presently, a new optical setup is being developed utilizing acoustooptic frequency modulation combined with an active intensity stabilization.

5. Suppression of a Recoil Component

Frequently the Ramsey fringes of the two recoil components overlap. This overlap is further complicated by the distortion due to the second order Doppler broadening of the thermal absorbers /8/. Attempts to reduce this distortion either by velocity selection /9,10/ or by cooling the absorbers /11/ leads to an increased number of fringes further complicating the fringe pattern. Because of these difficulties efforts have been made to suppress one recoil component /12–14/.

In saturated absorption one of the recoil components can be suppressed either by removing the population holes in the lower state or the population enhancement in the higher state before the probe beam is applied. In our method we depopulate the $^3P_1$ state of calcium by optically pumping the atoms via the shortlived $^3S_1$ state into the metastable $^3P_0$ and $^3P_2$ states, respectively.

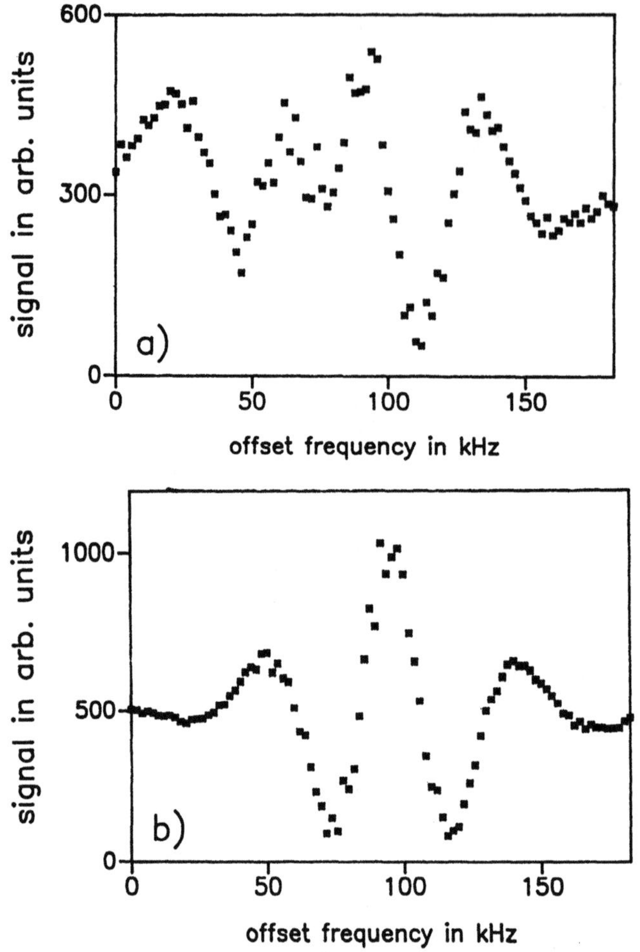

Fig. 4 Ramsey fringes observed (integration time / point: 1s) in a Ca atomic beam without a) and with b) the pump laser beam to suppress the low frequency recoil component.

We recorded Ramsey fringes at field separations of D = 9 mm which are large enough to resolve partially the recoil splitting. The Ramsey fringe pattern shown in Fig. 4a was obtained without the pumping beam. When the pump laser ($\lambda \cong 612$ nm) depopulated the $^3P_1$ level of the atoms in a region between the innermost interaction zones of Fig. 1, only the Ramsey structure of the high frequency recoil component remained (Fig. 4b). If the single recoil component Ramsey fringes of Fig. 4b are shifted by 23.1 kHz towards lower frequencies and then added to the unshifted fringe pattern of Fig. 4b, a Ramsey fringe pattern similar to that in Fig. 4a is simulated. The suppression of a recoil component makes the use of a laser cooled atomic beam /11/ extremely attractive in order to reduce the second order shift and broadening, the utilization of which is normally hampered by the overlap of the large numbers of Ramsey fringes of both recoil components. In the following we therefore estimate the uncertainty to realize the true line center of a calcium atom in a *thermal* and in a *laser cooled* atomic beam.

6. Uncertainty to Realize the Line Center

The most significant contributions to the uncertainty to realize the line center are listed in table 1. Similar as in microwave standards, the main contribution is caused by the uncertainty of the phase difference $\Delta\phi$. At the present state of our experiments, we estimate this effect to be approximately one tenth of a fringe width corresponding to 1 kHz at the highest resolution obtained in our experiment. Compensation of phase errors is mostly limited by the stability of the I_2–stabilized HeNe laser used as frequency reference in our dye laser spectrometer. Replacing this reference by the transition frequency of the Ca intercombination line itself will probably improve this situation by at least one order of magnitude. Second order Doppler broadening being about 3 kHz for our effusive Ca beam can be corrected to better than 300 Hz. The effects of magnetic and electric fields as well as of wavefront curvature are negligible at the present level of uncertainty. We estimate that the present uncertainty of 1.05 kHz ($\delta\nu/\nu$ = 2.3 10^{-12}) can be reduced considerably utilizing improved phase alignment and slow, laser cooled atomic beams. A value of about 6 Hz ($\delta\nu/\nu \cong 10^{-14}$), for an optical frequency standard based on a laser cooled Ca beam (v = 40 m/s) can be estimated by scaling the values for a thermal beam. This estimate, however, results from experimental results and does not represent the theoretical limits.

The absolute frequency uncertainty of our Ca standard, however, depends on the uncertainty to which optical frequencies can be measured. Wavelength comparisons with I_2–stabilized lasers /15/ are in preparation and optical frequency measurements are planned for the future /16/.

Table 1 Estimated uncertainty to realize the true line center of the Ca intercombination line

effects	uncertainty effusive beam (v $\cong$ 780 m/s)	uncertainty cooled beam (v = 40 m/s)
optical phase	0.1 fringe width 1 kHz	$\lesssim$ 5 Hz
2nd order Doppler effect	10% of broadening 300 Hz	$\lesssim$ 1.5 Hz
magnetic field ($3\cdot10^{-4}$ T)	$\lesssim$ 1 Hz	$\lesssim$ 1 Hz
electric field ($\lesssim$ 1 V/cm)	$\lesssim$ 0.1 Hz	$\lesssim$ 0.1 Hz
Total sum in quadrature	<1.05 kHz	< 6 Hz

We thank Mr. Baldau of Spectra Physics for generously loaning us the stabilized ring dye laser used for optical pumping. Stimulating discussions with Ch. Bordé on the effect of laser beam reversal and the Ramsey phase shift problem are gratefully acknowledged. J.Ishikawa was supported by the Deutscher Akademischer Austauschdienst (DAAD).

References

1. J.C. Bergquist, R.L. Barger, D.J. Glaze: in Laser Spectroscopy IV, Springer Ser. Opt. Sci. 21, (Springer, Berlin, Heidelberg, N. York) H. Walther, K.W. Rothe eds. p. 120 (1979)
2. N.F. Ramsey: Molecular Beams (Oxford University Press, London 1956) Chap. 2
3. Ye.V.Baklanov,B.Ya.Dubetsky, and V.P.Chebotayev: Appl. Phys. 9, 171 (1976)
4. Ch. J. Bordé, S. Avrillier, A. van Lerberghe, Ch. Salomon, Ch. Bréant, D. Bassi, G. Scoles : Appl. Phys. B 28, 82 (1982)
5. J. Helmcke, D. Zevgolis, B.Ü. Yen: Appl. Phys. B 28, 83 (1982)
6. Ch. J. Bordé, Ch. Salomon, S. Avrillier, A. van Lerberghe, Ch. Bréant, D. Bassi, G. Scoles: Phys. Rev. A 30, 1836 (1984)
7. J. Helmcke, J.J. Snyder, A. Morinaga, F. Mensing, M. Gläser: Appl. Phys. B 43, 85, (1987)
8. R.L. Barger: Opt. Lett. 6, 145 (1981)
9. H.Hellwig, S.Jarvis, D.J.Glaze, D.Halford, and H.E.Bell, in Proceedings of the 27th Annual Symposium on Frequency Control, 12–14 June 1973, Fort Monmouth, N.J. (Electronic Industries Assoc., Washington D.C.,1973) p.357 H.F.Foster and H.K.Ziegler (eds.)
10. A. Morinaga, J. Helmcke: Appl. Phys. B 45, 273 (1988)
11. See e.g. W.D.Phillips, J.V.Prodan, and H.J.Metcalf, J. Opt. Soc. Am. B 2,1751 (1985)
12. Ch.J.Bordé in *Laser Spectroscopy III* (Springer–Verlag,Berlin, Heidelberg, New York, 1977), p.121, J.L.Hall and J.L.Carlsten (eds.)
13. A.N. Goncharov, M.N.Skvortsov, and V.P.Chebotayev, Sov. J. Quantum Electron. 13, 1429 (1983)
14. G.Camy, N.Courtier, and J.Helmcke, in *Laser Spectroscopy VIII* (Springer–Verlag, Berlin, Heidelberg, New York,1987), p.386, W.Persson, S.Svanberg (eds.)
15. See e.g.: G. Bönsch: IEEE Trans. Instr. Meas. IM–34, 248 (1985)
16. H.H. Klingenberg, H.R. Telle, J. Helmcke: IEEE Trans. Instr. Meas. IM–34, 268 (1985)

Obtaining of Supernarrow Saturated Absorption Resonances on Cold Particles and Their Use for Optical Frequency Standards

S.N. Bagayev, V.P. Chebotayev, A.K. Dmitriyev, A.E. Om, Y.V. Nekrasov, and B.N. Skvortsov

Institute of Thermophysics, Siberian Branch of the USSR Academy of Sciences, SU-630090 Novosibirsk, USSR

Several years ago we started studies of the optical selection of cold particles in a gas. Our program involved theoretical and experimental investigations of the conditions of effective selection of slow particles in a low-pressure gas, when a free path length of particles largely exceeded the field dimensions (this pressure range will further be called a transit-time region), of the behaviour of intensity and width of saturation resonances in the transit-time region /1, 2/. As a result, in the transit-time conditions, when the transit-time parameter is $\beta = \Gamma\tau_0 \ll 1$ (Γ is a homogeneous linewidth, $\tau_0 = a/v_0$ is the time of flight of a particle with an average thermal velocity v_0 through a light beam with radius a), we obtained a saturation resonance at the $F_2^{(2)}$ absorption line of methane (λ = 3.39 μm) with a homogeneous width of ~ 1 kHz in the light beam of ~ 2 mm in size /2/. The main contribution to the saturation resonance in this case is provided by cold CH_4 molecules whose velocity ($\lesssim$ 10 m/s) is much lower than an average thermal one ($v_0 \approx$ 500 m/s). The effective temperature of slow particles $T_{eff} = \beta^2 T_0$ (T_0 is the gas temperature) is about 0.1 K.

Main difficulties of application of cold particles in spectroscopy and for laser frequency stabilization are due to an extremely low intensity of saturation resonances in the transit-time region /1, 2/. For the method of saturated absorption the intensity of saturation resonances of cold particles is increased by using intracavity telescopic beam expanders.

Here we report on start of the installation for observation of optical saturation resonances of cold particles of about 10 Hz width and for production of a new optical frequency standard on the basis of these resonances. In the first experiments we obtained saturation resonances at the $F_2^{(2)}$ line of methane (λ = 3.39 μm) with width of about 100 Hz and observed the influence of a second-order Doppler effect on their shape.

Our installation comprises three lasers: a stable He-Ne/CH_4 laser with narrow radiation line, a heterodyne laser and the investigated

Frequency Standards and Metrology Editor: A. De Marchi

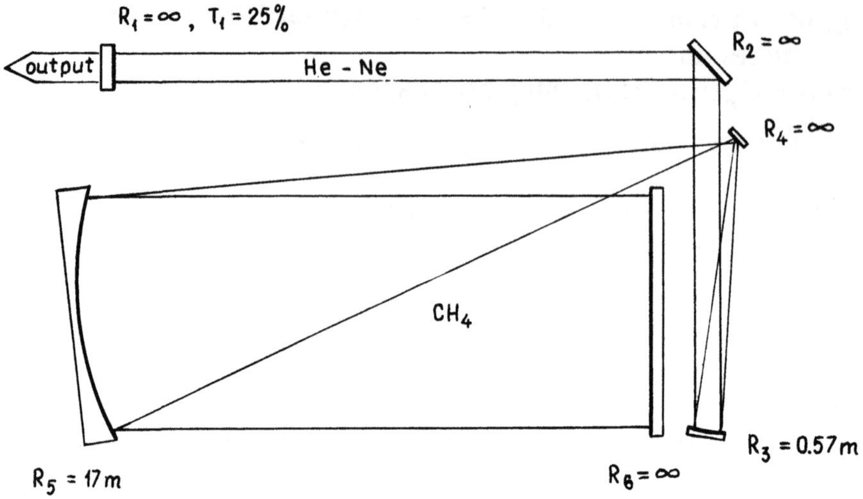

Fig. 1. Scheme of the telescopic laser with an intracavity beam expander

He-Ne/CH_4 laser with a telescopic beam expander. The cavity of the stable laser was formed by flat and spherical (R = 50 m) mirrors. Its length was about 8 m, the absorption cell length was 4.5 m. As a heterodyne laser we used the He-Ne laser with power of about 10 mW.

The six-mirror cavity of the telescopic laser is schematically presented in Fig. 1. Such a cavity geometry allowed practically complete elimination of the influence of astigmatism on the shape of a light beam in the cell. The light beam diameter in the absorption cell was 30 cm, the cell length 8.5 m. The operating methane pressure was 10^{-5} - 10^{-6} Torr. The absorption cell was protected from the magnetic Earth field and other external magnetic field by a multilayer permalloy screen. A special system of gas cooling by liquid nitrogen down to 77 K was put inside the cell. All the lasers were mounted in the same frame. The installation was placed on a 20-ton concrete plate lying on air dampers. A narrow radiation line of the telescopic laser was provided by its phase locking to the stable laser.

The resonance was recorded by synchronous detection of the signal of the first and second harmonics in the laser power.

Figure 2 shows the record of the recoil doublet of the 7 - 6 component of the magnetic hyperfine structure of the $F_2^{(2)}$ methane line (transition P(7), ν_3). The experimental data are marked with dots. The continuous curve is the theoretical form of the second resonance derivative with allowance for the second-order Doppler effect. The methane pressure in the cell is about 10 μTorr, the modulation

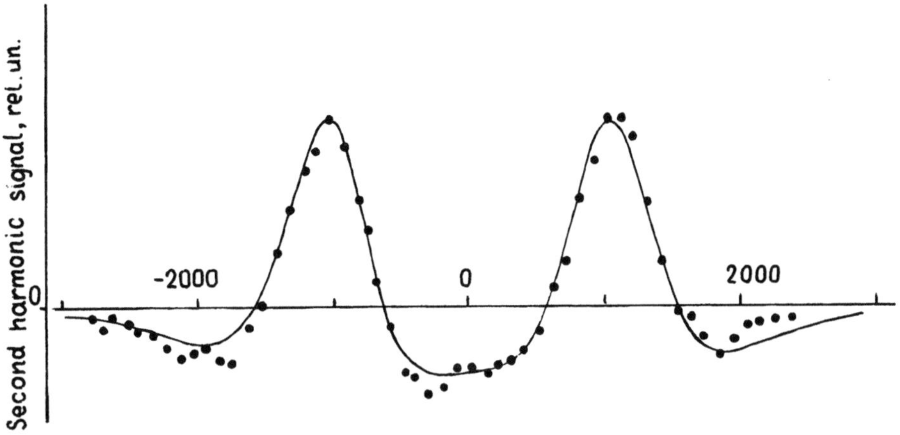

Frequency detuning from the center of the 7-6 component, Hz

Fig. 2. Record of the recoil doublet of the 7-6 component of the MHS of the $F_2^{(2)}$ transition in methane. • - experiment, —— - theory

frequency and the deviation amplitude 230 and 250 Hz, respectively. The resonance halfwidth is about 300 Hz. There is a good agreement of theory and experiment. The asymmetry of the recoil doublet wings near the line centre is due to the influence of the second-order Doppler effect. Fig. 3 presents the record of the low-frequency component of the recoil doublet of the 7 - 6 transition at a pressure about 5×10^{-6} Torr. The resonance halfwidth is about 200 Hz, the resonance intensity 10^{-7}W. The signal-to-noise ratio is here determined by

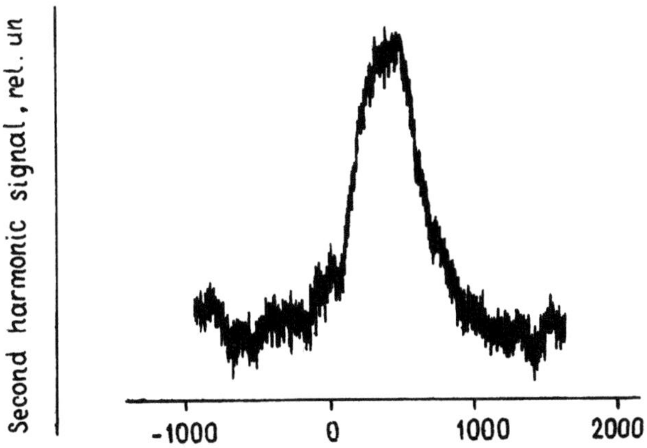

Frequency detuning with respect to stable laser, Hz

Fig. 3. Record of the low-frequency recoil component. The integration constant is 4 s, the recording time 10 min, the resonance halfwidth 200 Hz low-frequency fluctuations of the laser intensity due to angular detuning of the cavity. The magnitude of these fluctuations was several orders higher than the noises from an InSb photodetector cooled by liquid nitrogen. Elimination of this amplitude noise by active power stabilization will enable us even in the nearest future to operate with the methane pressure of 10^{-6} Torr and to obtain resonances with an absolute width down to 10 Hz.

Fig. 4 shows the possible spectral resolution of the made He-Ne/CH_4 laser with a telescopic beam expander at a pressure of 10^{-6} Torr and a gas temperature in the cell 77 K. The calculated form of the second saturation resonance derivative in methane for one recoil component at $\beta \simeq 3 \times 10^{-2}$ is given here with no allowance for radiative width. The resonance halfwidth is about 10 Hz. The effective temperature of cold methane molecules determining the resonance shape is $T_{eff} \simeq 0.07$ K. The according second-order Doppler resonance shift $\Delta \simeq (\Gamma \tau_0)^2 \Delta_0$ ($\Delta_0 = 40$ Hz is the shift corresponding to the gas temperature T =77 K) is $\Delta \simeq 4 \times 10^{-2}$ Hz.

These resonances with a relative width of 10^{-13} and lower open up new possibilities for making precision physical experiments and producing a new generation of laser frequency standards on cold particles with reproducibility of 10^{-15} - 10^{-16}.

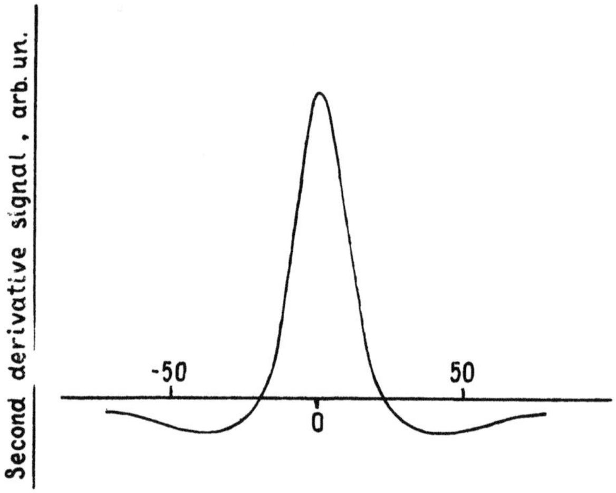

Frequency detuning with respect to resonance center, Hz

Fig. 4. Theoretical shape of the saturation resonance of cold CH_4 molecules: pressure 10^{-6} Torr, gas temperature T_0=77 K, beam diameter 30 cm, transit parameter $\beta = 3 \times 10^{-2}$, effective temperature T_{eff} = 7×10^{-2} K.

References

1. S.N.Bagayev, A.E.Baklanov, A.S.Dychkov, P.V.Pokasov, V.M.Semibalamut, E.A.Titov, V.P.Chebotayev: Preprint No. 125 (Institute of Thermophysics, Novosibirsk 1985); Optika i Spektr. 59, 481 (1985)
2. S.N.Bagayev, A.E.Baklanov, V.P.Chebotayev, A.S.Dychkov. ZhETF Pisma Red. 45, 371 (1987); Rev.Roum.Phys. 33, 361 (1988)

Laser Cooling in Calcium and Magnesium Atomic Beams

N. Beverini, E. Maccioni, D. Pereira, F. Strumia, and G. Vissani
Dipartimento di Fisica dell'Università di Pisa (CISM),
I-56126 Pisa, Italy, and
INFN, Sezione di Pisa, P. Torricelli 2, I-56126 Pisa, Italy

The extension of the laser cooling technique from Na and Cs to Ca and Mg is of great interest for time and frequenciy metrology [1]. The laser deceleration (cooling) can be achieved by irradiating a thermal Ca or Mg atomic beam with a counterpropagating laser beam in resonance with the $^1S_0$–$^1P_1$ transition at 422 nm and 285 nm, for Ca and Mg respectively. In particular, atoms with an initial velocity v_0 are stopped, in a laser field of constant intensity I, after a distance

$$L(S) = \frac{v_0^2 M\lambda\tau}{h}\left(1 + \frac{1}{2S}\right) \tag{1}$$

where τ is the radiative lifetime of the upper $^1P_1$ level, λ the resonance wavelength, M the atom mass, and S=I/Is the saturation parameter. Only a small fraction of the atoms in the thermal velocity distribution can absorb the laser radiation when a monochromatic single-frequency laser beam is used. It consists of those atoms for which the laser frequency offset is compensated within the natural linewidth by the Doppler effect. As soon the deceleration process starts, the velocity is reduced, the atoms are Doppler shifted out of resonance, and absorption become negligible. The Zeeman tuning of the atomic absorption was used to overcome this difficulty [2]. The Zeeman frequency shift is produced by a magnetic field B(z) of appropriate intensity along the atomic beam axis. This way a continuous beam of cooled atoms can be produced. For a linear Zeeman effect, as in Mg and Ca, and a constant laser intensity, the magnetic field intensity required to keep the atoms in resonance along the beam is

$$B(z) = B(0)\sqrt{1 - \frac{z}{L(S)}} \tag{2}$$

In fact the above equation is just giving the minimum value of the field B(z) for any initial value B(0). Other field profiles are allowed but, since the acceleration has a finite value, there is also an upper limit on the field gradient, that must satisfy the boundary condition

$$\frac{dB}{dz} \le \frac{\dot{v}}{\mu\lambda v} \tag{3}$$

where μ= 1.4 MHz/gauss for Mg and Ca. If the above constraint is not satisfied, the Zeeman frequency tuning is too fast for the cooling rate and the atoms are shifted out of resonance. Of course a CW , frequency stabilized laser beam of enough power must be available. At 285.2 nm such a CW laser beam was obtained by frequency doubling the

Frequency Standards and Metrology Editor: A. De Marchi

radiation at 570.4 nm obtained in a ring Dye laser with Rhodamine 6G dye ; at 422.6 nm the laser beam was obtained by using the Stilben 3 dye pumped with an UV Ar^+ laser. The deceleration of a Mg atomic beam was recently demonstrated [3], and here we will report the results obtained in the case of a Calcium atomic beam.

The experimental apparatus is shown in fig.1. A thermal beam of Ca atoms is produced from an effusive source at a temperature of about 900°K, and interacts with a counterpropagating laser beam, resonant with the $^1S_0$ - $^1P_1$ transition . A single frequency, CW laser beam was produced from a Coherent 699/21 ring dye laser with a maximum power of about 25 mW , single mode TEM_{00} operation, and r.m.s. line width of ≈1 MHz . The modified velocity distribution was analyzed with a second laser beam at 422 nm, scanning the frequency around the Ca resonance line. This laser beam was produced by a frequency stabilized Coherent 599/21 dye laser, and its intensity was chopped at about 1 kHz in order to discriminate, with a lock-in amplifier, against the stronger fluorescence signal from the cooling laser beam.

Typical experimental results are shown in fig.2. Three identical scans of the analyzing laser beam are shown. In the upper part of fig.2 the cooling laser beam was offset thus showing the unperturbed velocity profile, while in the lower part it was set on resonance at a frequency corresponding to a small velocity value. To show the effect of the Zeeman tuning, an axial magnetic field, whose intensity profile adequately fitted the theoretical requirements, was generated in the deceleration region by an ensemble of three solenoids and properly designed iron poles [1]. The magnetic field intensity B(z) decreased from its maximum to zero in about 25 cm, which was the maximum cooling length with the field on. The results given in fig.2 were obtained in the same experimental conditions and with B(0)=730 Gauss. Laser deceleration down to v=75 m/s and v=21 m/s (corresponding to a temperature T<1°K) was obtained with two different frequency settings of the cooling laser beam, in agreement with the predictions. As a

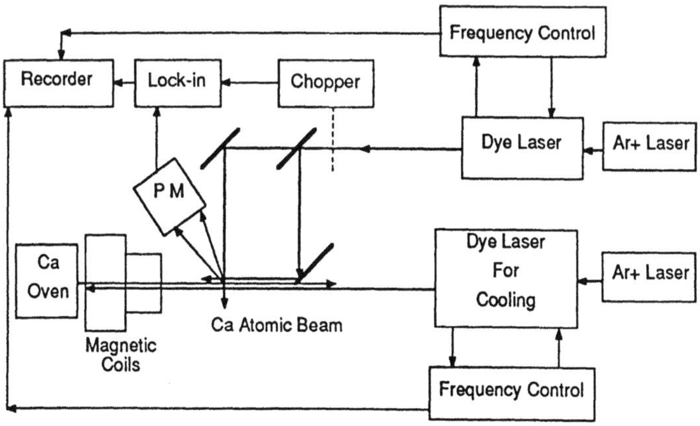

Fig. 1

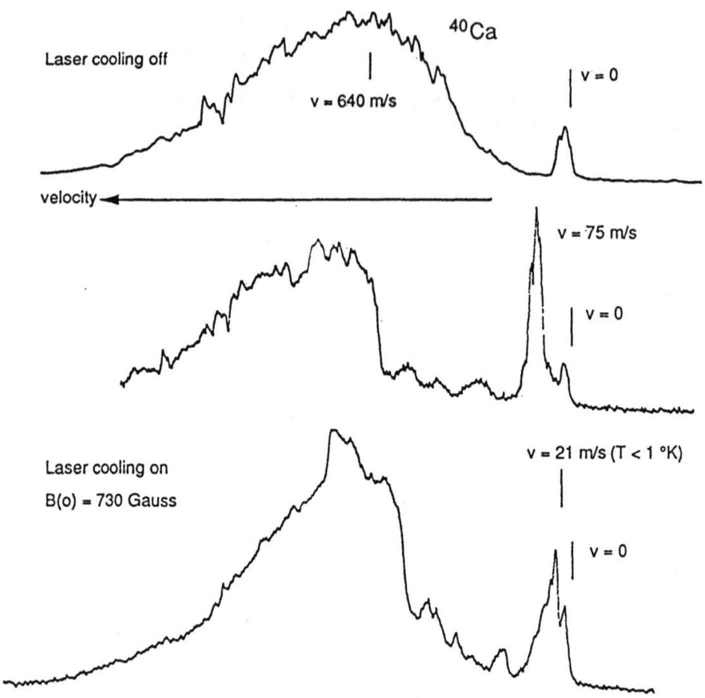

Fig. 2 - Laser cooling observed in presence of magnetic field (Zeeman tuning), and for two different frequency settings of the cooling laser beam. Maximum initial field intensity B(0) = 0.073 T.

final remark it is worth noting that the benefits of laser cooling for atomic frequency standards are not limited to the increase of the Q of the reference line. In particular, as pointed out in [1], very important benefits are: a- the decrease of the influence of many small effects, which limit the final precision and reproducibility of the standards and are velocity dependent; b- the reduction of the microwave power level needed in the interrogation region, that scales as $1/v^2$. This last benefit is crucial for standards, as Mg and Ca, with the reference line in the submillimeter region where the available power from high spectral purity oscillators is not large enough [4].

1- F. Strumia ."Application of laser cooling to the atomic frequency standards".in " Laser science and technology " Eds. A.N.Chester and S.Martellucci ,(Plenum Press, November 1988).
2- W.D.Phillips, J.V.Prodan, H.J.Metcalf, A.Migdall,I.So,J.Dalibard : Phys.Rev.Lett. 54,992 (1985)
3- N. Beverini, S.De Pascalis, E. Maccioni, D. Pereira, F. Strumia,G. Vissani, Y.Z. Wang, C. Novero : Europhys. Lett. , in press
4- E.Bava, A.Godone, G.Giusfredi, C.Novero : IEEE J. Quant. Electr., QE-23, 455-457, (1987)

Part XII

Trapped Ions

Introduction to the Session on Trapped Ions

H. Dehmelt

Department of Physics, University of Washington,
Seattle, WA 98195, USA

'Less is more'

The excellent papers in this session might just convince the expert audience, that only an individual trapped ion offers solid promise for any kind of optical frequency standard with an accuracy and 1000-day reproducibility of 10^{-18}. While such an ion, say Ga^+, is only a passive standard, its frequency would be exactly that of a free ion in the Andromeda nebula and a measurement need take no longer than a few hours. All this is no wonder, because high resolution spectroscopy on anything else but on an individual isolated metastable atomic particle at rest in the laboratory will always be affected by shifts of various kinds. In single ion work highest resolution has been demonstrated in the beautiful laser spectroscopy experiments on Hg^+ in a 1-mm rf trap which JAMES BERGQUIST will report. Similar work on Ba^+, with a potentially even higher resolution, but sofar limited by lack of a good laser flywheel oscillator, will be described by my colleague WARREN NAGOURNEY. My former colleague PETER TOSCHEK has carefully studied the vibrations of a Ba^+ ion in his rf trap. Isolation from the environment is most easily approached in the laboratory when the electronic states between which the (clock) transition of interest occurs have zero electronic angular momentum, as for Thallium$^+$ homologues, but not for Hg^+ or Ba^+. Such states may have life times of hours. GUENTHER WERTH will discuss his extensive measurements of ion life times. If more than one ion are packed into a trap their interactions are important. Uncontrolled plasma oscillations have been reported in rf traps and delight chaos fanciers. Most affected by such cooperative phenomena are microwave frequency standards employing clouds of Hg^+, Yb^+ in rf traps and Be^+ in a Penning trap. In the past the rf trap work was limited by micromotion 2nd Doppler effect. It has used only the standard cap-ring-cap trap geometry and buffer gas cooling, while in the Penning trap work laser cooling has lowered the temperature to $\approx$2K. However, by spreading 10^5 ions out in a 1m-long linear chain at the bottom of the potential valley in a rf "race track" trap it should be possible to laser-cool the ions to $\approx$0.001K, totally suppressing the micromotion. A linear string of charged oil drops in a similar 2-wire ac trap was photographed by H. STRAUBEL as early as 1955, see Fig. 1. Progress with Yb^+, such as a week-long metastable state lifetime, will be discussed by RAINER BLATT, and with Be^+ clouds by JOHN BOLLINGER.

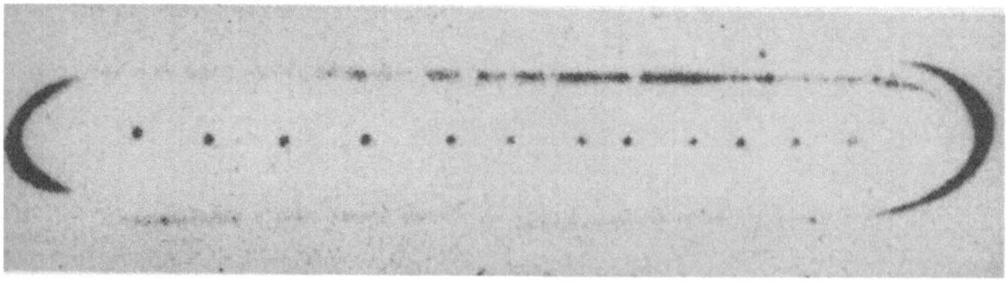

Fig. 1

Frequency Standards and Metrology Editor: A. De Marchi

Hg^+ Single Ion Spectroscopy *

J.C. Bergquist, F. Diedrich†*, W.M. Itano, and D.J. Wineland*

Time and Frequency Div., National Institute of Standards and Technology, Boulder, CO 80303, USA

A single Hg^+ ion that is confined in an rf (Paul) trap can be laser cooled so that the amplitude of its motion is much less than a wavelength (the Dicke limit) for optical transitions. Recently, we have used the technique of optical sideband cooling to reach the zero point of motion. This realizes for the first time the fundamental limit of laser cooling for a bound atom and the ideal of an isolated atomic particle at rest in space to within the quantum mechanical limits imposed by the surrounding apparatus. In both limits, Doppler effects become negligible to all orders, the interrogation time is long and the fundamental shot noise detection limit of a single atom is readily attained.

Introduction

Laser cooling of a single atom has led to a number of important experiments in fundamental physics [1-5] and may allow optical spectroscopy and metrology with a resolution and accuracy that approach 1×10^{-18} [6-9]. One of the candidates for an optical frequency standard that approaches this accuracy is a single, laser-cooled $^{199}Hg^+$ ion confined in a miniature rf trap [8]. At NIST we have begun with an investigation of a single $^{198}Hg^+$ ion, laser cooled into the Dicke limit for optical transitions [8] and, in a second experiment, laser cooled in the sideband limit to the zero-point energy of motion [10,12]. The $^{198}Hg^+$ ion offers the experimental advantage of zero nuclear spin; hence, no hyperfine structure and a simpler optical configuration for cooling. However, all the narrow optical transitions have a magnetic field sensitivity of the order of 10 MHz/mT (1 MHz/G), whereas the $^{199}Hg^+$ ion has several first-order magnetic field-independent transitions.

In the work summarized here we begin to show the spectroscopic resolution and signal-to-noise ratio that are possible with a single atom nearly at rest in space. We have resolved the recoilless optical resonance and the motional sidebands of the $5d^{10}6s\ ^2S_{\frac{1}{2}}(m_J=-\frac{1}{2})$ - $5d^96s^2\ ^2D_{5/2}(m_J=\frac{1}{2})$ electric quadrupole transition ($\lambda\cong282$ nm) on a laser-cooled $^{198}Hg^+$ ion [8]. Since we can detect each transition to the metastable $^2D_{5/2}$ state ($\gamma/2\pi \approx 1.7$ Hz) with nearly 100% efficiency, there is essentially no technical noise. The noise results only from the quantum

*Contribution of the NIST, not subject to copyright in the U.S.

†Permanent address: Max Planck Institut für Quantenoptik, Garching, West Germany.

Frequency Standards and Metrology Editor: A. De Marchi

statistical fluctuations in the transition probability of the single ion [6]. From the sideband structure we are able to make a determination of the amplitude of the motion of the ion and its effective temperature [10,11,13]. The sidebands at $\omega_0 \pm n\omega_v$ (n an integer) are generated by the oscillatory motion at frequency ω_v of the ion in its confining harmonic well. If the ion is driven with narrowband radiation tuned to the first lower sideband, the atom will absorb photons of energy $\hbar(\omega_0-\omega_v)$ and in most cases re-emit photons of energy $\hbar\omega_0$. This reduces the atom's vibrational quantum number by 1 for each scattered photon at frequency ω_0. In this way we can obtain $\langle n_v \rangle \ll 1$, which means that the atom spends most of its time in the ground state level of its confining potential [10,12-14]. To the extent that the atom is in the ground state of its confining potential the fundamental limit of laser cooling for a confined particle has been reached [12].

Experiment

The mercury ion is confined in a miniature rf trap that has internal dimensions of $r_0 \cong 466$ μm and $z_0 \cong 330$ μm [8,15]. The amplitude of the trapping field (frequency $\Omega/2\pi \cong 21\text{-}23$ MHz) could be varied to a peak of 1.2 kV. The ion is laser cooled to a few millikelvins by a few microwatts of cw laser radiation [16] that is frequency tuned below the $^2S_{\frac{1}{2}}$ - $^2P_{\frac{1}{2}}$ first resonance line near 194 nm [2,8]. In order to cool all motional degrees of freedom to near the Doppler cooling limit ($T = \hbar\gamma/2k_B \cong 1.7$ mK) the 194 nm radiation irradiates the ion from 2 orthogonal directions, both of which are at an angle of 55° with respect to the symmetry (z) axis of the trap. The 282 nm radiation that drives the narrow $^2S_{\frac{1}{2}}$ - $^2D_{5/2}$ transition is obtained by frequency-doubling the radiation from a narrowband cw ring dye laser. In the long term, the laser is stabilized by FM optical heterodyne spectroscopy to a saturated absorption hyperfine component in $^{129}I_2$ [17]. The frequency of the laser is scanned by an acousto-optic modulator that is driven by a computer controlled synthesizer. Up to a few microwatts of 282 nm radiation could be focussed onto the ion in a direction counterpropagating with one of the 194 nm light beams.

Optical-optical double resonance (electron shelving) [1-3,8,15,18-20] with a net quantum amplification (including detection efficiency) in excess of 100 at 10 ms was used to detect transitions driven by the 282 nm laser to the metastable $^2D_{5/2}$ state. The fluorescence rate from the laser-cooled ion is high when the ion is cycling between the $^2S_{\frac{1}{2}}$ and $^2P_{\frac{1}{2}}$ states and nearly zero when it is in the metastable $^2D_{5/2}$ state [1-3,8]. Thus the $^2S_{\frac{1}{2}}$ - $^2D_{5/2}$ resonance spectrum was obtained by probing the S-D transition at a particular frequency for the 282 nm radiation for 20 ms, then turning off the 282 nm radiation and turning on the 194 nm radiation to look for the presence or absence of scattered photons at 194 nm (the two radiation fields are alternately applied to avoid light shifts and broadening of the narrow S-D transition). If there was no fluorescence at 194 nm, a transition into the metastable D state had occurred; the presence of 194 nm fluorescence indicated that the ion was in the ground state and no transition was recorded for this frequency of the 282 nm laser. The frequency was then stepped and the measurement cycle repeated. As the frequency was stepped back and forth each new result at a particular frequency of the 282 nm radiation was averaged with the previous measurements at that frequency. Normalization (or quantization) of the signal was obtained by assigning a 1 to each measurement of high fluorescence and a 0 to each measurement of no fluorescence. The high fluorescence level made it possible to determine the state of the atom with almost no ambiguity in a few milliseconds. Thus, it is easy to reach the shot noise limit imposed by the single atomic absorber [8].

The quantized fluorescence signal obtained from an 8-MHz scan of the 282 nm laser through the $^2S_{\frac{1}{2}}(m_J=-\frac{1}{2}) \rightarrow {}^2D_{5/2}(m_J=\frac{1}{2})$ Zeeman component of the electric quadrupole transition is shown in Fig. 1 [8]. The recoilless-absorption resonance (carrier) [21,22] and the motional sidebands due to the secular motion in the harmonic well of the rf trap [10] are completely resolved. The number and amplitudes of the sidebands are a direct measure of the ion's motion and its effective temperature. A careful comparison of the intensities of the sidebands to that of the carrier [9] for several sets of data similar to that shown in Fig. 1, including the effects of saturation, gave temperatures that ranged from 1.6 to 6.7 mK. This is consistent with the theoretical laser-cooling limit for $^{198}Hg^+$ on the first resonance line at 194 nm given by $T_{min} = \hbar\gamma/2k_B \cong 1.7$ mK [10].

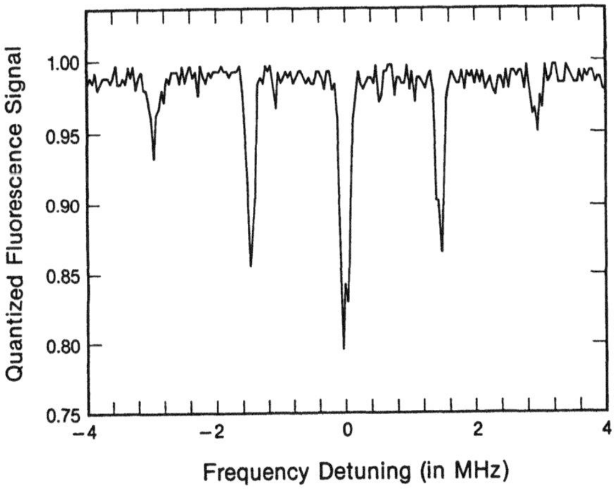

Fig. 1. Quantized signal showing the electric-quadrupole-allowed $5d^{10}6s\ ^2S_{\frac{1}{2}}(m_J=-\frac{1}{2})-5d^96s^2\ ^2D_{5/2}(m_J=\frac{1}{2})$ transition in a single, laser-cooled $^{198}Hg^+$ ion. On the horizontal axis is plotted the relative detuning from line center in frequency units at 282 nm. On the vertical axis is plotted the probability that the fluorescence from the 6s $^2S_{\frac{1}{2}}$-6p $^2P_{\frac{1}{2}}$ first resonance transition, excited by laser radiation at 194 nm, is on. The electric-quadrupole-allowed S-D transition and the first-resonance S-P transition are probed sequentially in order to avoid light shifts and broadening of the narrow S-D transition. Clearly resolved are the recoilless absorption resonance (carrier) and the Doppler sidebands due to the residual secular motion of the laser-cooled ion. The integration time per point is about 16 s (230 measurement cycles).

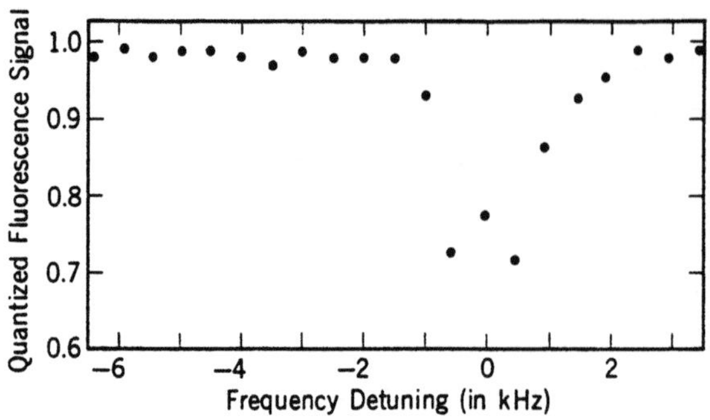

Fig. 2 High resolution scan through the recoilles absorption resonance of the $^2S_{\frac{1}{2}}(m_J=-\frac{1}{2})-^2D_{5/2}(m_J=1)$ transition in a single laser cooled $^{198}Hg^+$ ion. The full width at half maximum is about 1.7 kHz at λ=563 nm (3.4 kHz at λ=282 nm).

In Fig. 2 we show a high resolution scan through the Doppler-free (recoilless) resonance at line center The full width at half maximum (FWHM) is approximately 1.7 kHz at 563 nm (3.4 kHz at 282 nm). This corresponds to a fractional resolution of about 3.4×10^{-12} ($Q\approx3\times10^{11}$). For this trace the laser was spectrally narrowed by locking to a mechanically, acoustically and thermally quiet reference cavity that had a finesse of about 60,000. The frequency of the laser fluctuates less than 1 Hz relative to this reference cavity. This is determined from the measurement of the residual noise in the error signal; the actual frequency fluctuations of the laser are then governed by the stability of the well isolated cavity [23]. The natural linewidth for the $^2S_{\frac{1}{2}}$ - $^2D_{5/2}$ transition is approximately 1.7 Hz [15]. The measured linewidth in Fig. 2 is dominated by fluctuations in the first-order magnetically-sensitive electric quadrupole transition in the $^{198}Hg^+$ ion. The field sensitivity of the $^2S_{\frac{1}{2}}(m_J=-\frac{1}{2})$ to $^2D_{5/2}(m_J=\frac{1}{2})$ transition is on the order of 22.4 MHz/mT (2.24 MHz/G). In our laboratory there are field fluctuations of a few milligausses with a frequency-noise spectrum that is dominated by 60 Hz and its harmonics. Shielding would help but is difficult in the present experiment. The better approach is to switch to the $^{199}Hg^+$ isotope. This isotope has first-order, field-independent transitions near zero magnetic field that should permit us to reach the 1.7 Hz natural linewidth. In this case, the ion will provide a good test of the spectral purity of the 282-nm laser.

The possibility of cooling the Hg^+ ion further, to the zero point energy of motion, is intriguing for several reasons. First, cooling so that the average occupational number $\langle n_v \rangle$ for the motional energy of the bound atom is zero is a fundamental limit to laser-cooling for any bound particle [8]. This limit is imposed by the position-momentum Heisenberg uncertainty principle where the spatial confinement (Δx) is provided by the physical surroundings. Driving a single atom in a macroscopic trap to the zero point energy of motion also exploits the benign environment near the center of an rf trap - certainly a system worthy of consideration for the high resolution and accuracy in optical spectroscopy, metrology and frequency standards. Second, with an ion prepared in the lowest vibrational state (most of the time), experiments such as squeezing the atom's position and momentum can be demonstrated.

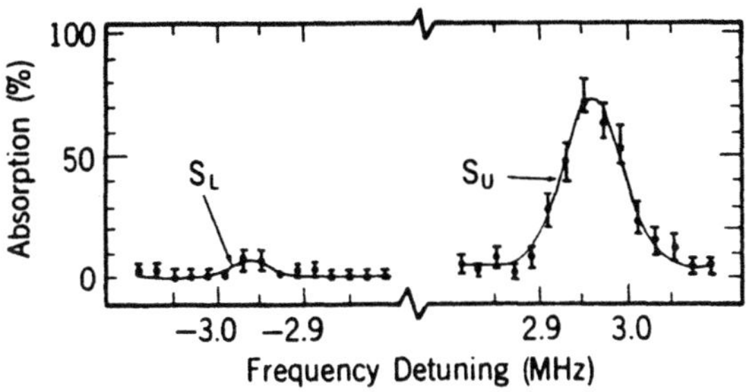

Fig. 3 First lower (S_L) and upper (S_u) motional sidebands of the $^2S_{\frac{1}{2}}$-$^2D_{5/2}$ electric quadrupole transition of a laser-sideband-cooled $^{198}Hg^+$ ion measured 10 ms after cooling. Data points are averages of 41 sweeps.

Cooling to the zero point energy of motion is possible for a bound atom in the resolved sideband limit where the linewidth of the cooling transition ($^2S_{\frac{1}{2}}$ - $^2D_{5/2}$ in our case) is less than motional oscillation frequency of the atom [10,12,13]. When the ion is in the lowest or ground vibrational state (n_v=0) it can no longer resonantly absorb a photon at ω_0-ω_v; at this point the resonant feature at ω_0-ω_v disappears. Experimentally, a measurement of the amplitude of the resonant feature at ω_0-ω_v (for saturating intensity) is a direct measurement of $\langle n_v \rangle$ [13]. The result from ref. 12 is shown in Fig. 3.

From these data we deduce that the y and z degrees of freedom are in the zero point energy state (n_v=0) 95% of the time. Because the 282 nm probe beam is nearly orthogonal to the x-axis, we were not sensitive to the x-motion. The uncertainty in the second-order Doppler shift in this example is dominated by the uncertainty in $\langle n_v \rangle$ and amounts to $\Delta\nu/\nu \approx 10^{-20}$ [13]. It could be made even lower by adiabatically lowering the potential well depth (thereby lowering ω_v) after the ion is cooled into the ground state.

With this system, the absorption of a <u>single</u> quantum of energy at a frequency in the megahertz range would raise n_v by one unit and this could be detected with an efficiency of nearly 100%. Also, it should be possible to produce squeezed states [24] of the atom's motion from the zero point energy state by a sudden, non-adiabatic weakening (and, in general, shifting) of the trap potential or by driving the atomic motion parametrically at 2 ω_v. If after some time the atom could then be returned to the n_v=0 state by reversing the above procedures, the zero point energy state could be unambiguously detected by the absence of the lower sideband. This would convincingly demonstrate the squeezed state of the single atom in the rf trap.

We acknowledge the support of the U.S. Air Force Office of Scientific Research, the U.S. Office of Naval Research and the continued support of the NIST. F. Diedrich thanks the Deutsche Forschungsgemeinschaft for financial support.

References

1. W. Nagourney, J. Sandberg, and H. Dehmelt, Phys. Rev. Lett. 56, 2797 (1986)
2. J.C. Bergquist, R.G. Hulet, W.M. Itano, and D.J. Wineland, Phys. Rev. Lett. 57, 1699 (1986)

3. T. Sauter, W. Neuhauser, R. Blatt, and P.E. Toschek, Phys. Rev. Lett. 57, 1696 (1986)
4. F. Diedrich and H. Walther, Phys. Rev. Lett. 58, 203 (1987); W.M. Itano, J.C. Bergquist, D.J. Wineland, Phys. Rev. A38, 559 (1988)
5. D.J. Wineland, W.M. Itano and J.C. Bergquist, Opt. Lett. 12, 389 (1987)
6. D.J. Wineland, W.M. Itano, J.C. Bergquist, and F.L. Walls, in Proc. of the 35th Ann. Symp. Freq. Control, 1981, p. 602 (copies available from Electronic Industries Association, 2001 Eye Street, N.W., Washington, DC 20006)
7. H. Dehmelt, IEEE Trans. Instrum. Meas. IM-31, 83 (1982); D.J. Wineland, Science 226, 395 (1984).
8. J.C. Bergquist, W.M. Itano, and D.J. Wineland, Phys. Rev. A36, 428 (1988)
9. See articles by D.J. Wineland et. al.; Dehmelt, et. al.; W. Nagourney, et. al.; R. Blatt, et al.; P. Gill et. al. in this volume
10. D.J. Wineland and W.M. Itano, Phys. Rev. A20, 1521 (1979)
11. G. Janik, W. Nagourney, and H. Dehmelt, J. Opt. Soc. Am. B2, 1251 (1985)
12. F. Diedrich, J.C. Bergquist, W.M. Itano, and D.J. Wineland, submitted for publication
13. D.J. Wineland, W.M. Itano, J.C. Bergquist, and R.G. Hulet, Phys. Rev. A36, 2220 (1987)
14. W. Neuhauser, M. Hohenstatt, P. Toschek, and H. Dehmelt, Phys. Rev. Lett. 41, 233 (1978)
15. J.C. Bergquist, D.J. Wineland, W.M. Itano, H. Hemmati, H.-U. Daniel, and G. Leuchs, Phys. Rev. Lett. 55, 1567 (1985)
16. H. Hemmati, J.C. Bergquist, and W.M. Itano, Opt. Lett. 8, 73 (1983)
17. J.L. Hall, L. Hollberg, T. Baer, and H.G. Robinson, Appl. Phys. Lett. 39, 680 (1981)
18. D.J. Wineland, J.C. Bergquist, W.M. Itano, and R.E. Drullinger, Opt. Lett. 5, 245 (1980); D.J. Wineland and W.M. Itano, Phys. Lett. 82A, 75 (1981)
19. H. Dehmelt, J. Phys. (Paris) Colloq. 42, C8-299 (1981)
20. H. Dehmelt, Bull. Am. Phys. Soc. 20, 60 (1975)
21. R.H. Dicke, Phys. Rev. 89, 472 (1953)
22. See, for example, Hans Frauenfelder, The Mössbauer Effect (Benjamin, New York, 1962)
23. J. Hough, D. Hils, M.D. Rayman, Ma L.-S., L. Hollberg, and J.L. Hall, Appl. Phys. B33, 179 (1984)
24. See for example, J. Janszky and Y.Y. Yushin, Opt. Commun. 59, 151 (1986); Fan Hong-Yi and H.R. Zaidi, Phys. Rev. A37, 2985 (1988)

Lifetime Measurements of Metastable States in Ions

G. Werth
Institut für Physik, Universität Mainz, D-6500 Mainz, Fed. Rep. of Germany

As pointed out by Dehmelt in 1973, optical transitions between ground- and metastable states of ions offer intrinsically very narrow natural lines and consequently may be used as frequency references in the optical domain. Several ions are considered as possible candidates and the measurement of their metastable state lifetime is being discussed. In the submillimeter region fine structure transitions between metastable ionic states have natural line-Q's of the order of 10^{12}. We propose to use Ca^+ as a possible candidate to induce and detect such a transition.

1. Introduction

The significance of metastable atomic states for metrology has been pointed out in different ways: In 1974 BARGER et al. /1/ demonstrated that the weak optical transition from the ground state to the metastable $^3P_1$ state of calcium can be measured with high precision and possibly serve as a reference transition for laser stabilization. In 1972 STRUMIA /2/ proposed to use the fine structure splitting of the metastable 3P levels of magnesium or calcium as a frequency reference in the submillimeter region. The instrinsic advantage of metastable states is the fact that due to their long lifetime the natural width may be extremely small and consequently very high Q-values in transitions involving those states may be obtained, which in practice may be limited only by the laser frequency fluctuation.

In neutral atoms the line-Q may in fact also be limited by transient time effects due to the final time of flight of the atoms through the interrogation region of laser or microwave fields. This limitation is completely avoided by the use of ion traps, which extends the interrogation time virtually to infinity, since in numerous examples ions have been confined for many hours to very small regions in space. For ionic ground states frequency standards based on microwave transitions between hyperfine levels are presently

Frequency Standards and Metrology Editor: A. De Marchi

being investigated and long-term frequency instabilities reaching the 10^{-15} level have been reported /3/. Ionic transitions involving metastable states may also serve as frequency references and various schemes have been discussed in the literature.

2. Optical Transitions

The first proposal to use an optical transition between an ionic ground state and a metastable level in an ion trap was made by DEHMELT /4/, who suggested the $6^1S_0-6^3P_0$ transition at 202 nm in Tl^+ as a "clock-transition". The natural line-Q, given by the radiative lifetime of the excited state, would be about 10^{-14} and long-term instabilities down to 10^{-18} are envisaged. The wavelength, however, is rather inconvenient at the present level of technology, as is the 191 nm $6^1S_0-6^3P_1$ transition, which could be used for low cooling of the ion.

Apart from Tl^+ there are a number of different ions which have long-living metastable states. The determination of their lifetime is of importance not only for metrology but also for astrophysics and plasma physics. From the atomic physics point of view ions with simple level structure-like alkali atoms are of particular interest, since they may be used to compare theoretical calculations based on different approximation methods. In this contribution I shall restrict myself to a discussion of those ions. Fig. 1 shows a level diagram which is representative for many of those ions.

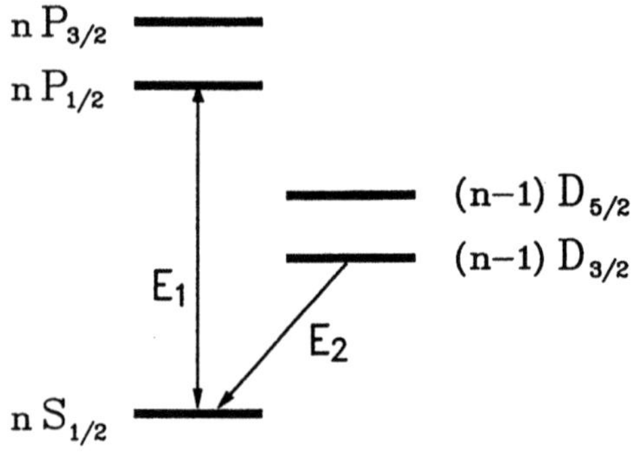

Fig. 1 Level diagram of alkali-like ions Ca^+ (n=4), Sr^+ (n=5), Ba^+ (n=6), Hg^+ (n=6), Yb^+ (n=6)

Table 1 lists a number of metastable ionic states, the experimental determined lifetimes, the wavelength of the transition to the ground state and the corresponding natural line-Q.

Table 1 Metastable states of ions, their lifetimes and transition wavelength to the ground state. Also listed are wavelengths of resonance electric dipole transitions which might serve to populate the metastable state and to cool the ions by photon recoil

Ion	Metastable state	lifetime	transition wavelength to ground state /nm/	cooling wavelength /nm/
Ca^+	$3D_{5/2}$	∿ 1 s /11/	729	393
Ca^+	$3D_{3/2}$	∿ 1 s /11/	733	397
Sr^+	$4D_{5/2}$	345 ms /10/	687	408
Sr^+	$4D_{3/2}$	395 ms /10/	674	422
Ba^+	$5D_{5/2}$	30 s /12/	1762	455
Ba^+	$5D_{3/2}$	17.5s /7/	2051	493
Yb^+	$5D_{3/2}$	52 ms /6/	435	369
Hg^+	$5D_{5/2}$	86 ms /8/	282	165
Hg^+	$5D_{3/2}$	9.2ms /8/	198	194
Pb^+	$6P_{3/2}$	41.2ms /13/	710	168

Of particular interest for frequency standard application is the question whether the ions' temperature in a trap can be reduced by laser cooling to eliminate second order Doppler effects. This requires an optical dipole resonance transition (for example a $S_{1/2}$-$P_{1/2,3/2}$ transition in Fig. 1) in the range of narrow band tunable lasers. When the excited state partially decays into the metastable level, a second laser is needed to repump the ions out of this state because otherwise they are lost from the cooling cycle. In those cases where laser cooling is not feasible other cooling mechanisms like resistive cooling of of the stored ions /5/ may be considered. Buffer gas cooling is obviously prohibited since collisions with background gas molecules rapidly quench the metastable states /6/.

The measurement of the metastable state lifetimes in ion traps is in general straightforward. The ions are excited by a short laser pulse either by an allowed electric dipole transition to an upper state, which partially decays into the metastable state, or directly by the forbidden transition at high laser power. The slow decay of the metastable state can in many cases be observed directly by time resolved photon counting. The experimental difficulty of the low count rate (usually less than 1 detected fluorescence count per laser pulse in a cloud of 10^5 ions) can easily be overcome by averaging over many excitation cycles, since the storage time usually exceeds several hours. Sr^+, Yb^+ and Pb^+ are examples of such measurements. In those cases where the wavelength of the forbidden transition is too long to allow single photon counting, as in the case of Ba^+, the ground state population can be monitored by the electric dipole fluorescence from the laser excited state. One observes a decrease of this population in time due to optical pumping into the metastable state. From the time dependence of the pumping process, which can be expressed analytically in a three level system /7/ one can deduce the metastable state lifetime. A variant of this method has been used in the case of Hg^+, where the observation of "quantum jumps" in a single stored ion allowes to monitore the occupied energy level continuously in time and one determines the metastable level lifetime by the distribution of times, which the ion spends in this state /8/. A first successful step to exploit the advantages of the ion storage technique for high precision optical spectroscopy and possible application for frequency standards has been reported by BERGQUIST et al. /9/, who observed a linewidth of 30 kHz at the 282 nm $6^2S_{1/2}-5^2D_{5/2}$ electric quadrupole transition in a single Hg^+ ion, cooled down by laser irradiation at 194 nm to about 1.7 mk. This corresponds to a line-Q of $3 \cdot 10^{10}$. Most recently the linewidth has been reduced to 1.7 kHz (see contribution of J. BERGQUIST to this conference).

3. Fine Structure Transition

In a similar way, as the early proposal by STRUMIA on Mg /2/, fine structure splittings between metastable energy levels of stored ions may be used for frequency standard application. From the level diagram of Fig. 1 one can see that the laser excitation of the $P_{1/2}$ excited state from the ground state populates the metastable $D_{3/2}$ state, but not the $D_{5/2}$ state. A transition between the two states induced by submillimeter waves could be observed if the fluorescence detection system is sensitive to the forbidden $D_{5/2}-S_{1/2}$ transition. The feasibility of such a detection was demonstraded on Sr^+ where fine structure transitions were induced by collisions with neutral background atoms (Fig. 2) /10/. Table 2 lists a number of ions and the energy differences between metastable fine structure levels as well as the pumping wavelength. The presently available technology makes Ca^+ a promising candidate for such a scheme because of its

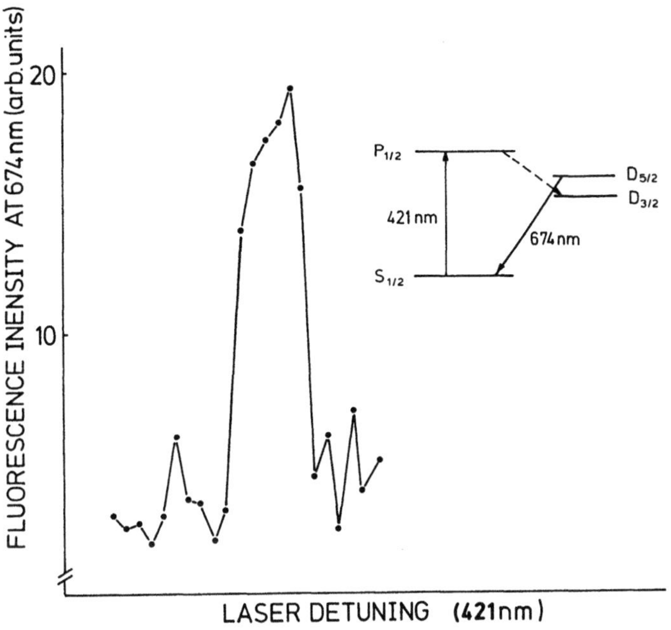

Fig. 2 Collision induced finestructure mixing in Sr^+. The $D_{3/2}$ state was populated by decay of the pulsed laser excited $P_{1/2}$ state and fluorescence from the $D_{5/2}$ into the ground state at background pressures of ca.10^{-5} mbar was observed. Averaging time per data point was 10 seconds. An improvement in signal strength of 1-2 orders of magnitude can be expected for radiation induced $D_{3/2}$-$D_{5/2}$ transition due to more efficient $D_{3/2}$ population by c.w. laser excitation and by the absence of competing processes like collisional quenching of the $D_{3/2}$ state into the ground level

small transition frequency of 1.8 THz. The lifetime of the metastable states is estimated to about 1 s /11/ and consequently an ultimate line- Q of 10^{13} could be expected. The main limitation in precision would come from second order Doppler effect, which amounts to about 1 part in 10^{-11} for large uncooled ion clouds. Ultimately, as in other frequency standard schemes for trapped ions, a single laser cooled ion would give the smallest uncertaincy. Then of course the detection of the metastable $D_{5/2}$ state would be very difficult considering the low counting rate at a 1 s lifetime. A promising scheme provides the amplification by the "quantum jump" process as used in ref. 9: Two lasers L_1 and L_2 at the S-P and P-D transition, resp., would be required to cool the ion, which takes less than 100 ms. Then L_2 would be switched off to populate the metastable $D_{3/2}$ state. A $D_{3/2}$-$D_{5/2}$ infrared transition could be detected by the absence of fluorescence,

Table 2 Fine structure splittings $E(D_{5/2})-E(D_{3/2})$ of metastable ionic states. Also quoted are the electric dipole transition wavelengths to populate the $D_{3/2}$ states, the branching ratio of the excited state decay into the $D_{3/2}$ state with respect to the decay into the ground state, and the transition wavelength from the $D_{5/2}$ level to the ground state

Ion	Metastable state fine structure splitting	excitation wavelength /nm/	branching ratio	detection wavelength /nm/
Ca^+	1.82 THz	396.9	0.10	729.3
Sr^+	8.402 THz	421.6	0.07	673.8
Ba^+	24.00 THz	493.4	0.3	1761.7
Yb^+	41.11 THz	369.4	0.005	410.9
Hg^+	446 THz	193	10^{-7}	281

when L_1 and L_2 are switched on again. A Second problem arises if one uses an even isotope of the element under consideration: For zero nuclear spin the $D_{3/2}$ and $D_{5/2}$ states do not have Zeeman sublevels, which are magnetic field independent to first order. A residual field of 1 mG would give a shift of about 1 kHz corresponding to 10^{-9}. Odd isotopes, which have such "field independent" transitions, require measures to avoid ground state optical hyperfine pumping. This could be avoided by an additional laser, by laser sideband modulation or by simultaneously induced ground state hyperfine transitions in the microwave region.

4. Acknowledgments

The lifetime measurements at Mainz have been performed by Ch. Gerz, Th. Hilberath, A. Roth, J. Roths and D. Wilsdorf. The experiments were supported by the Deutsche Forschungsgemeinschaft.

References

/1/ R.L. Barger: Proc.of 1973 Laser Spectr.Conf., in Laser Spectroscopy (Plenum Press, N.Y., 1974) p.273

/2/ F. Strumia: Metrologia 8, 85 (1972)
/3/ L.S. Cutler, R.P. Giffard, P.J. Wheeler and G. Winkler: Proc.Frequ.Contr.Symp., Philadelphia, (1987)
/4/ H.G. Dehmelt, Bull.Am.Phys.Soc. 18, 1521 (1973)
/5/ D.A. Church and H.G. Dehmelt: J.Appl.Phys. 40, 3421 (1969)
/6/ Ch. Gerz, J. Roths, F. Vedel and G. Werth: Z.Phys. D8, 235 (1988)
/7/ R. Schneider and G. Werth: Z.Phys. A293, 103 (1979)
/8/ W.M. Itano, J.C. Bergquist, R.G. Hulet and D.J. Wineland: Phys. Rev.Lett. 59, 2732 (1987)
/9/ J.C. Bergquist, W.M. Itano and D.J. Wineland: Phys.Rev. A36, 428 (1987)
/10/ Ch. Gerz, Th. Hilberath and G. Werth: Z.Phys. D5, 97 (1987)
/11/ M. Godefroid, Bruxelles, private communication
/12/ W. Nagourney, J. Sandberg and H. Dehmelt: Phys.Rev.Lett. 56, 2797 (1986)
/13/ A. Roth, Ch. Gerz, D. Wilsdorf and G. Werth: to be published

Nonlinear Effects in Trapped Three-Level Ions

H. Gilhaus, T. Sauter, R. Blatt, W. Neuhauser, and P.E. Toschek
Universität Hamburg, D-2000 Hamburg 36, Fed. Rep. of Germany

Single trapped and laser-cooled ions have been considered so far paradigms of the ultimate high-resolution spectroscopy and photon statistics. They also represent individual systems with coupled internal and external degrees of freedom simple enough to allow detailed comparison of their kinetics with current models. - We have observed several metastable vibrational states of a single laser-illuminated Ba^+ ion which become excited upon gently heating the ion. They correspond to linear oscillatory modes in r, z directions, which have been documented by image conversion and video recording. - Suitable normalization of the photon-counted and spatially resolved signals of resonance fluorescence allows us to discriminate the features of the electronic excitation spectra and results in spectra of the ion's vibrational excitation. These spectra reveal two novel types of light force which we ascribe to stimulated two-photon transitions followed by Raman scattering, and to all-stimulated parametric action. - These observations bear upon the interpretation of recent observations on the kinetics of atoms confined in optical molasses.

A single trapped ion cooled to near immobility is usually considered a physical system of *ultimate simplicity:* an individual atomic particle at rest in space, characterised by just one internal degree of freedom. Most types of ions, however, are poor approximations to the archetypical two-level atom. Optical cooling, e.g., was observed with a cloud of Ba^+ ions |1|, which are explicit three-level systems (see Fig. 1), where *two* resonance lines, connecting the $^2S_{1/2}$ and $^2D_{3/2}$ levels with the $^2P_{1/2}$ level, are coupled by recoil to the three external degrees of freedom. Also, the first single ion to be prepared and observed was a Ba^+ ion |2|; it was located in a miniature rf trap as in Fig. 2.

Na atoms have been believed to represent perfect two-level systems when interacting with σ^+ or σ^- light. However, the presence of hfs and spurious excitation of Zeeman coherence leave this claim doubtful.

We have tried to unravel the coupled internal-external dynamics of a single trapped Ba^+ ion and have identified novel phenomena which include multi-stability of the ion oscillation, and two types of nonlinear stimulated light force.

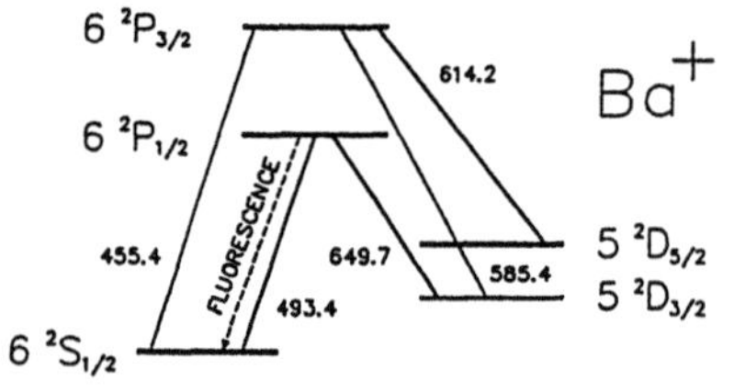

Fig. 1

The basis of our observations |3| is the detection of the ion's laser-excited fluorescence in *two* channels: The first channel images a small volume (35 μm Ø, "local" signal) at trap centre, the second

Frequency Standards and Metrology Editor: A. De Marchi

one a larger volume (200 μm Ø, "total" signal) on the photo-cathode of the respective photo-multiplier used for photon counting (Fig.3). The ratio of the two signals is insensitive to the resonant features of the internal ionic excitation, whereas it indicates any expansion of the ion orbit and records vibrational excitation. Therefore, the signal ratio serves as a monitor of the *kinetic state* of the ion under the mechanical effects of the laser radiation at λ_1 = 493 nm and λ_2 = 650 nm. These effects arise from 1. heating, or cooling, the ion by the illuminating laser light in certain ranges of laser detuning, or 2. heating the ion by rf pulses at the frequency of the secular oscillation, 1.5 MHz. A recording of that signal, i.e. a kinetic history of the ion, is shown in Fig. 4. We recognize three discrete and reproducible states of vibrational excitation. Occasionally, the ion orbit spontaneously shrinks, it seems, when a fluctuation unlocks the oscillation, and optical cooling sets in. Some of these states have been identified as oscillations along the symmetry axes of the trap (z) and the other principal axes (x,y) defined by slight mechanical imperfections of the quadrupole. One of these quasi-one-dimensional oscillatory orbits is shown in Fig. 5. It corresponds to a vibrational quantum number on the order of 10^4.

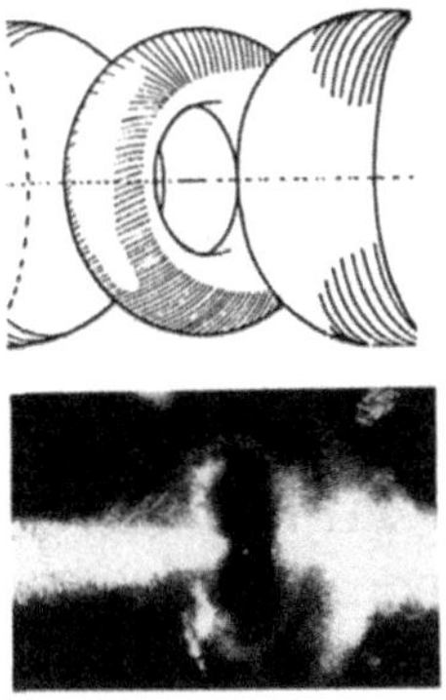

Fig.2: Trap design (top).Three Ba^+ in centre of trap (bottom). From Ref. 2

Oscillators are known to show metastable levels of excitation when they interact with the internal energy of random ensembles of atoms.With modes of the light field, this is the basis of optical solitons |4|. It seems that a model designed for the description of an electromagnetic cavity mode with individual atoms is also applicable to the *vibrational excitation* of a trapped ion tied to the ion's radiative interaction|5|. The multi-stability also affects *excitation spectra* of the ionic resonance fluorescence. Figure 6 shows the total and local signals upon scanning up and down ω_2. Four traces are superimposed. The broad dip on the l.h.s. represents the dark resonance generated by the coherence $^2S_{1/2}$-$^2D_{3/2}$ around the Raman resonance condition $\omega_1 - \omega_{10} = \omega_2 - \omega_{20}$. At the tuning labelled C, the ion orbit quickly expands. Note the reproducible hysteresis upon inversion of the scan, which is a consequence of kinetic multi-stability. Another hysteresis shows up close to

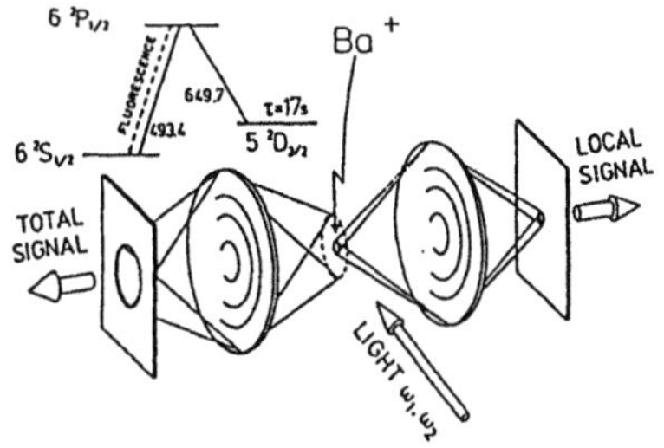

Fig.3: Vibrational spectrometer for single Ba^+ ion. The ratio of total and local signals measures the excitation of the ion oscillation

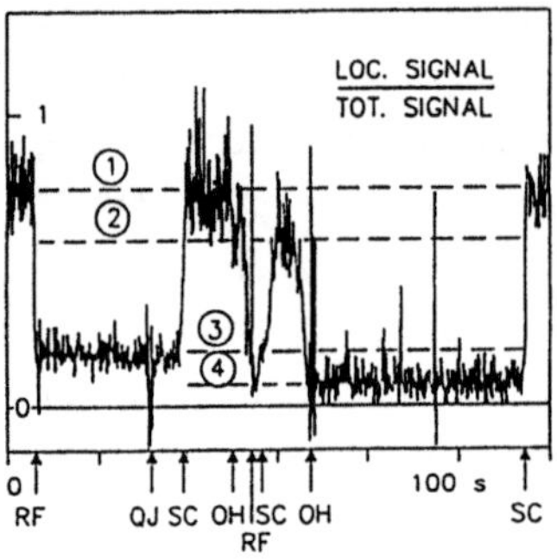

Fig.4: Ratio of local and total signals *vs.* time. Sequence of events (RF:pulsed rf heating,QJ:quantum jump to $^2D_{5/2}$ level, OH:incremental change of ω_2, SC:spontaneous onset of optical cooling) takes ion through three metastable orbits. $I_1 = 12\ W/cm^2$, $I_2 = 20\ W/cm^2$

the Raman resonance|6|, at smaller detuning values $\omega_1 - \omega_{10}$, see Fig. 7a. The ratio of total and local signals indicates resonant expansion of the ion orbit,Fig. 7b,c. These heating resonances which show up when ω_2 is below ω_{20}, close to the Raman resonance, depend on both ω_1 and ω_2 and prove the action of *two-photon* light force. We compare these heating spectra with the heating/cooling function calculated from an analytic model of the light forces upon a three-level atom|7|. This model applies to the ion in the Lamb-Dicke regime, but does not make use of other standard approximation, in particular it does not involve a Fokker-Planck equation. The calculated ratio of heating and cooling rates, q *vs.* ω_2, is shown in Fig. 8. Note the heating peak close to the Raman resonance, and another one about 40 MHz below, and the cooling between these peaks; this structure matches the observed double peaks in Fig.7c. The heating resonance of Fig. 8 with up-tuned ω_2 has a corresponding feature in Fig.7a,b. In 7c it appears farther up-shifted, as it is expected from the distribution of the oscillator in phase space upon exciting the ion into the Doppler regime with partial dephasing of the coherence|6|. Spectra as shown in Fig. 8 are plotted in Fig. 9 *vs.* the Rabi frequency Ω_1. There is an unshifted heating and cooling resonance of dispersive type at the Raman resonance (B), whereas the other resonances are light-shifted by Ω_1. This feature reveals the resonances A and C to involve components of the *dielectric polarisation* close to ω_1 and ω_{10}, respectively. These components are generated when light (ω_1) is *spontaneously* Stokes scattered off the Raman coherence ($^2S_{1/2}$ - $^2D_{3/2}$).(Spontaneous re-emission *incoherent* with the excitation is negligible since the lifetime of the $^2P_{1/2}$ level is short compared with that of the low-lying states.) This stimulated generation and spontaneous read-out of the Raman coherence, visualised in Fig. 10, gives rise to type-I stimulated light force.

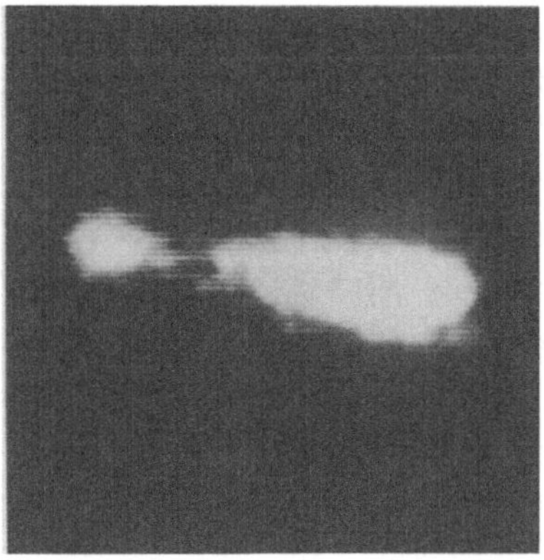

Fig.5: Single Ba^+ oscillating along symmetry axis of trap (z), which extends oblique angle with line of sight

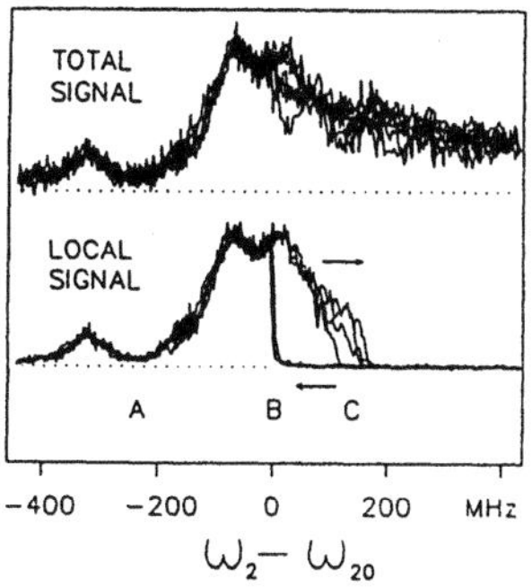

Fig.6: Excitation spectra (4 superimposed) of single-Ba^+ resonance fluorescence *vs.* ω_2. Sudden expansion (C) and compression (B) of the orbit. A: Raman resonance. I_1 = 30 W/cm^2, I_2 = 56 W/cm^2

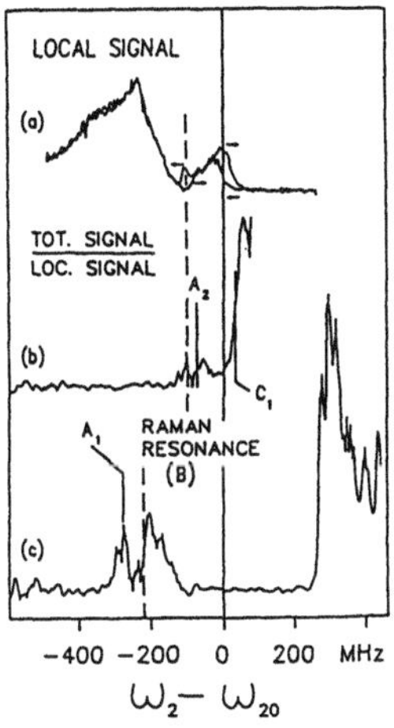

Fig.7: (a) Green fluorescence from trap centre ("local signal", arb. unit) *vs.* tuning of red laser 2. $\omega_1 - \omega_{10}$ = - 103 MHz • 2π, I_1 = 9 W/cm^2, I_2 = 34 W/cm^2. Heating/cooling spectra *vs.* tuning of red laser: (b) As in a. (c) $\omega_1 - \omega_{10}$ = -220 MHz • 2π, I_1 = 18 W/cm^2, I_2 = 50 W/cm^2. After Ref. 6

An individual interaction process including formation and read-out of the Raman coherence is accompanied by a net Doppler shift, which balances the energy transfer ε, where the contribution from spontaneous emission averages to zero over all spatial directions |6|. With the ion velocity scaled by the detuning, $\omega_2 - \omega_{20} = k_2 v$, this balance is

$$(k_1 - k_2)v_B - k_1 v_{C1} = \varepsilon = -2k_1 v_{C1}, \quad (1)$$

$$k_1 v_{A1} - (2k_1 - k_2)v_B = \varepsilon = 0, \quad (2)$$

for resonances C_1 and A_1, respectively. Equation 1 shows net energy transfer on resonance since the two-photon process and the Stokes scattering interact with *opposite* sidebands of the Raman coherence phase-modulated by the vibration. In contrast, (2) describes zero net energy transfer on resonance, i.e. a dispersive resonance, since the interaction with the light in the formation and read-out is via the *same* sideband. The doublet structure of the resonances is due to the Lamb-Dicke limit assumed in the model which only allows for the first pair of modulation sidebands: A_2, e.g., results from the substitution $v_{A1} \rightarrow v_B$, $v_B \rightarrow v_{A2}$ in (2).

The resonances of (1) and (2) are marked as circles in the contour plot of Fig. 9.

The resonance B does not depend upon the light-shifted $^2P_{1/2}$ level; it does not include spontaneous emission, but rather the read-out of the Raman coherence by stimulated two-photon transition at slightly reduced or enhanced ion speed. The heating/cooling ratio q decreases or increases, respectively. The corresponding light force is all-stimulated (type II).

Both specified types of light force include the formation of the long-lived Raman coherence at a particular location in phase space and its subsequent non-adiabatic shift to another location where its read-out takes place by spontaneous or stimulated scattering. This scenario has been proposed as possible cooling and heating mechanism of three-level atomic particles |8|. Sti-

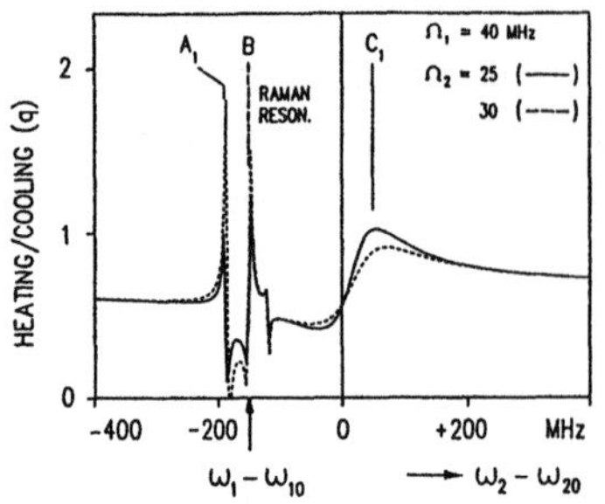

Fig.8: Heating/cooling (q) *vs.* detuning of red laser, calculated after Ref. 7. Ω_1, Ω_2: Rabi flopping frequencies of green, red light at ion site, match values of Fig.7c with reduction for residual 0.1-μm micro-motion by 1/8, 1/17, respectively. From Ref. 6

mulated read-out of the Raman coherence is non-destructive, it involves its shift by the modulation frequency. The resulting all-stimulated process is of parametric nature, and it is tied to the correlation of location and resonance frequency by the Doppler shift, i.e. to the existence of a particle orbit. It is analogous to the interaction of electrons with an electromagnetic wave in the action of a free-electron laser |9|.

The outlined scenario also applies to the dynamics of two-level atoms in a standing light wave, where the optical coherence diffuses in space and is destructively read-out by spontaneous emission preferentially in the anti-nodes of the wave |10|. Also, non-degenerate Zeeman sublevels of atoms in optical molasses give rise to the formation of long-lived Zeeman coherence whose spatial diffusion in a light field of spatially varying amplitude or polarisation is not adiabatic. Its non-adiabatic transport exerts work, either on the waves or on the particle, and makes the particle motion damped or accelerated, respectively. This type of interaction seems to be responsible for final temperature values measured in optical molasses |11,12| which are on order of magnitude lower than predicted from the conventional model of optical cooling of two-level atoms.

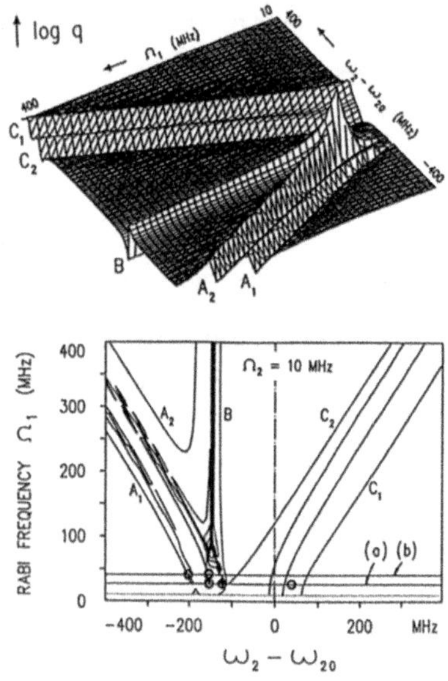

Fig.9: Plot of heating/cooling (q) spectra *vs.* Rabi flopping frequency Ω_1 of green light (top), and corresponding contour map (bottom). Branches A and C show type-I light force, branch B shows type-II light force. (a) $\hat{=}$ (b of Fig.7); (b) $\hat{=}$ (c of Fig.7). Circles mark resonances. From Ref. 6

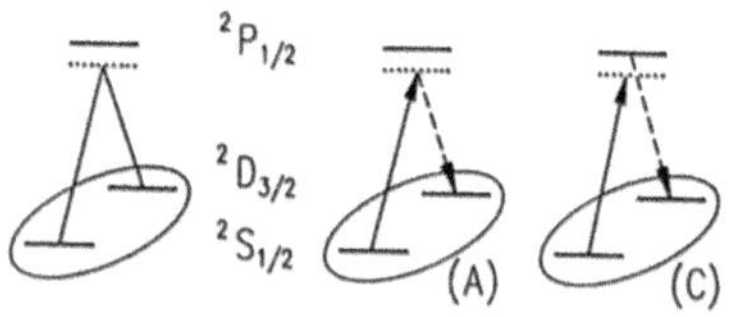

Fig.10: Formation of Raman coherence and destructive read-out by spontaneous Stokes scattering via ω_2 and ω_{20}

1. W. Neuhauser, M. Hohenstatt, P.E. Toschek, H.G. Dehmelt: Appl. Phys. 17, 123 (1978). - Phys.Rev.Lett. 41, 233 (1978)
2. W. Neuhauser, M. Hohenstatt, P.E. Toschek, H.G.Dehmelt: Phys. Rev. A22, 1137 (1980)
3. Th. Sauter, H. Gilhaus, W. Neuhauser, R. Blatt, P.E. Toschek: Europhys.Lett., to be published
4. S.L. McCall, E.L. Hahn: Phys.Rev. 183, 457 (1969). - W.Krieger, G. Gaida, P.E. Toschek: Z.Physik B25, 297 (1976)
5. P. Meystre: Opt.Lett. 12, 669 (1987)
6. H. Gilhaus, Th. Sauter, W. Neuhauser, R. Blatt, P.E. Toschek: Opt. Communic., to be published
7. M. Lindberg, J. Javanainen: J.Opt. Soc. Am. B3, 1008 (1986)
8. P.E. Toschek: Symposium A. Kastler, Jan. 1985. - Ann.Phys.Fr. 10, 761 (1985)
9. G. Dalotti, A. Remeri: In "*Laser Handbook*", Vol. 4. M.L. Stitch, M. Bass, eds., North-Holland, Amsterdam, 1985, p. 1
10. J. Dalibard, C. Cohen-Tannoudji: J. Opt. Soc. Am. B2, 1707(1985)
11. P.D. Lett, R.N. Watts, C.I. Westbrook, W.D. Phillips, Phys.Rev. Lett. 61, 169 (1988)
12. S. Chu: Eleventh Conference on Atomic Physics, ELICAP, Paris, July 1988

New Frequency Standards Based on Yb^+

R. Blatt, R. Casdorff, V. Enders, W. Neuhauser, and P.E. Toschek
Universität Hamburg, D-2000 Hamburg 36, Fed. Rep. of Germany

1. Introduction

Ion storage in both Paul and Penning traps has been the basis for many high-precision experiments during the past years [1,2]. The preparation of atomic samples for spectroscopic experiments which are free of perturbations by collisions with walls or buffer gas and by uncontrolled external fields makes the trapping technique a unique tool for a frequency standard. Moreover, optical cooling [3] eliminates systematic shifts and corrections due to the ion's motion.

A large variety of ions has been proposed and investigated for application as frequency standards using ion traps [4]. Unfortunately, the light at the wavelengths required for excitation and optical pumping of these ions usually has to be supplied by dye lasers which limit the uninterrupted operation of a real clock to a few hours. An alternative method was used when a stable frequency was derived from the measurement of the h.f.s. splitting in the ground state of trapped $^{199}Hg^+$ ions, where a discharge lamp served for optical pumping [5]. However, incoherent light sources are not suitable for optical cooling. Thus, the availability of semiconductor lasers made us reconsider the choice of appropriate ions for possible new frequency standards.

Trapped Yb^+ ions require two wavelengths for continuous resonance excitation, 369.4 nm and 2.44 μm. The UV radiation can be obtained by doubling the output of commercially available diode lasers at 739 nm; 2.44 μm may be achievable with Pb salt laser diodes. Clock transitions of the Yb^+ ion with high frequency resolution exist in the rf region (h.f.s. separation in $^{171}Yb^+$ ions) as well as in the optical domain (weak transitions to metastable D and F levels).

In section 2 the $^{171}Yb^+$ h.f.s. separation is considered as a basis for an rf standard, and in section 3 various concepts for optical frequency standards are discussed.

2. $^{171}Yb^+$ RF Standards

A frequency standard based on a precision measurement of the $^{171}Yb^+$ ground-state h.f.s. separation has been proposed many years ago [6]. In a subsequent experiment carried out on a cloud of trapped Yb^+ ions indeed very narrow resonances with $\Delta\nu/\nu \simeq 10^{-11}$ were achieved at a signal/noise ratio of 10-100 within a few seconds of detection time [7]. Figure 1 shows the relevant energy levels of the the $^{171}Yb^+$ ion. A pulsed laser (linewidth $\simeq$ 1 GHz) was used for optical pumping which is easily achieved by tuning the 369-nm radiation to one of the h.f.s. components. The h.f.s. splitting in the excited P and D states is not indicated in figure 1 since it is smaller than the Doppler width of the observed transition in the ion cloud. However, in the

Frequency Standards and Metrology Editor: A. De Marchi

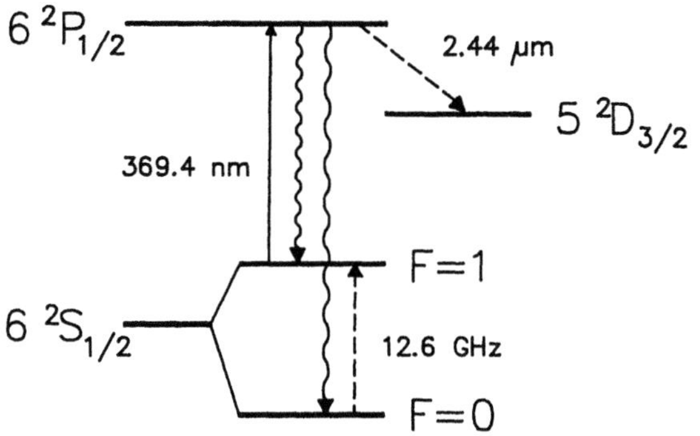

Fig.1 Hyperfine pumping of $^{171}Yb^+$ ions (after [7]).

microwave domain no broadening due to first-order Doppler effect is observed since the particles are confined to spatial dimensions smaller than the wavelength (Dicke criterion). The accuracy of resonances in ion clouds is limited by the 2nd-order Doppler effect which can be eliminated by means of optical cooling. Unfortunately, cooling of large ion clouds (i.e. more than a few ten particles) confined in an rf trap has turned out impossible due to an effect commonly known as rf heating [8]. The ionic micromotion induced by the trapping fields in an rf trap (which leads to the ponderomotive confining forces) is strong for ions far off the trap's center and may collisionally couple to slow ions. This energy transfer is avoided by experimenting with a single trapped ion, however, at the cost of signal/noise ratio. It has been shown that the velocity and spatial distributions of ion clouds in a Paul trap are quasistationary. Comparison of a recent model [9] with spectroscopic data shows that the kinetic energy of an ion cloud can be calculated from the trap parameters, and its dependence on the driving frequency can be derived. Thus we are able to determine the mean kinetic energy of an ion cloud with an overall uncertainty of less than 5% [10] leading to a resulting uncertainty of $\Delta\nu/\nu = 6\cdot 10^{-14}$ (mean ion cloud energy $E \simeq 0.4$ eV) in the determination of the $^{171}Yb^+$ h.f.s. splitting due to 2nd-order Doppler effect. With the signal/noise ratio obtained in ref. [7], the estimated performance of a frequency standard based on the $^{171}Yb^+$ h.f.s. splitting (using trapped clouds) is comparable to that of currently used Cs clocks. Improvements can result from using a second laser at 2.44 μm to reduce the population of the D state.

In order to obtain higher accuracy, laser cooling is necessary to get rid of the ion motion. This approach requires an experiment with single trapped ions, and it is possible with a single $^{171}Yb^+$ ion even in the presence of h.f.s. splitting. Figure 2 shows the energy levels involved in this cooling scheme. The 369-nm laser is tuned to the $|S, F=1\rangle \leftrightarrow |P, F=0\rangle$ resonance; no transitions to the $|S, F=0\rangle$ ground-state are dipole-allowed. In the same way, the 2.44-μm radiation is tuned to the $|P, F=0\rangle \leftrightarrow |D, F=1\rangle$ transition. Since the lifetime of the $D_{3/2}$ state is 52 ms [11] and laser cooling takes only about 10 ms, the Yb^+ ion can be cooled in a similar way as is a trapped Ba^+ ion [3]. To prevent optical pumping to the $|S, F=0\rangle$ ground state (either by pumping in the wing of the transition or by spontaneous decay from the D state) the 369-nm radiation needs a sideband at 12.6 GHz by electro-optic modulation. After cooling is

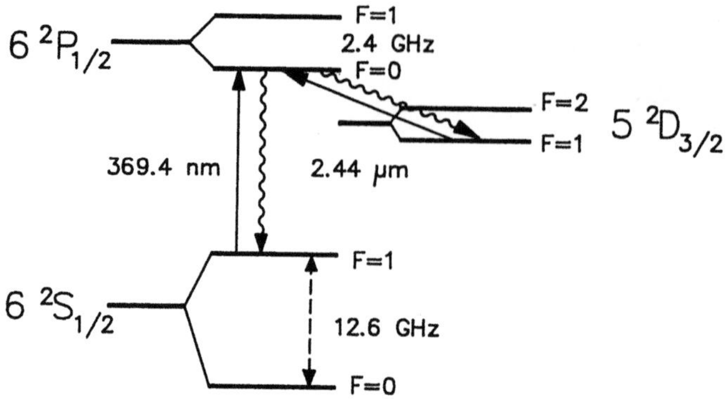

Fig. 2 Cooling scheme for a single $^{171}Yb^+$ ion.

achieved, the sideband is turned off, and microwave radiation at 12.6 GHz is applied. Resulting transitions to the $|S, F=0>$ ground state are detected by the absence of fluorescence light at 369 nm. This "electron shelving technique" [12] allows us to observe the absorption of single microwave quanta as has been pointed out in recent work on "quantum jumps" [13-15]. Counting the number of jumps as a function of the microwave detuning [16] will reveal resonances of high Q-values $Q = \nu_0/\Delta\nu$, where $\Delta\nu$ is the observed linewidth and ν_0 denotes its center frequency. Frequency standards are usually characterized by the (approximate) measurement imprecision

$$S_e \equiv \frac{\delta\nu_{error}}{\nu_0} \simeq (Q \cdot S/N)^{-1}$$

where S/N is the signal/noise ratio. In a passive frequency standard the measurement imprecision S_e decreases as a function of time, i.e. $S_e \sim 1/\sqrt{t}$, since the S/N ratio for shot-noise-limited detection increases as $\sqrt{t}$. Application of the electron-shelving technique allows the observation of every individual microwave transition, hence the signal/noise ratio is given by $S/N = \sqrt{n}$, where n denotes the number of observed quantum jumps during the measurement period. For detection by counting quantum jumps the S/N ratio can be kept constant by increasing the average interaction time and simultaneously reducing the power of the radiation at the clock transition, i.e. the number of observed jumps is constant. Thus the Q value of the transition increases linearly with time, and correspondingly we get $S_e \sim 1/t$. Since the interaction time can be made very long (550 sec have been reported [17]), indeed ultranarrow lines can be observed. Of course, this procedure requires very stable local oscillators as well. A combination of measurements on ion clouds (for short term stability of the local oscillator) and on a single cooled $^{171}Yb^+$ ion (for the long term stability and accuracy) is proposed as a promising frequency standard.

3. Yb^+ Optical Frequency Standards

Further improvement in the performance of frequency standards can be achieved when utilizing higher clock frequency, i.e. with laser radiation. It has been shown that frequency measurements can even be done in the optical domain [18]. The required narrow reference frequencies also can be derived from trapped Yb^+ ions. As has been suggested for trapped Ba^+ ions [19], the S – P – D Raman resonance in Yb^+ ions could be utilized for an optical frequency standard (cf. figure 3). The width of the two-photon resonance is determined by the natural linewidth of the D state and would give $Q \simeq 2 \cdot 10^{14}$. If the lasers are equally detuned from the intermediate P level, its occupation probability and hence the observed fluorescence rate drops and indicates the Raman resonance condition. The observation of this resonance is possible also by means of 4-wave-mixing as has been shown by previous measurements on trapped Ba^+ ions [20]. Stabilizing one of the lasers to the Raman resonance condition yields a stable difference frequency at 435 nm generated in a $LiIO_3$ crystal by mixing 369 nm and 2.44 μm. However, realization of such a standard frequency requires the detailed investigation of light shifts and a thorough understanding of heating and cooling processes at the Raman resonance frequencies [21].

A different scheme for an optical frequency standard based on Yb^+ ions is shown in figure 4. A narrow optical resonance can be obtained by observing quantum jumps to and from the metastable $D_{5/2}$ level induced either by a single-photon E2 transition at 411 nm or by a two-photon transition at 822 nm. The lifetime of the $D_{5/2}$ state has not been measured yet; however, it is expected that it is on the same order of magnitude as or larger than the $D_{3/2}$ lifetime, 52 ms.

In addition to the metastable D states (i.e. $(4f^{14})\ 5d\ ^2D_{3/2}$ and $(4f^{14})\ 5d\ ^2D_{5/2}$) in Yb^+ ions

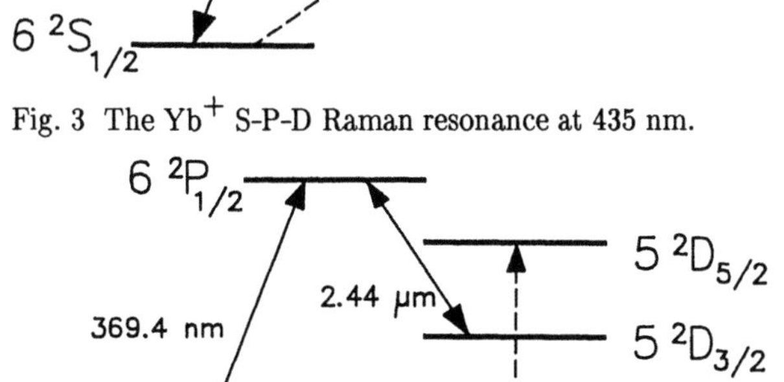

Fig. 3 The Yb^+ S-P-D Raman resonance at 435 nm.

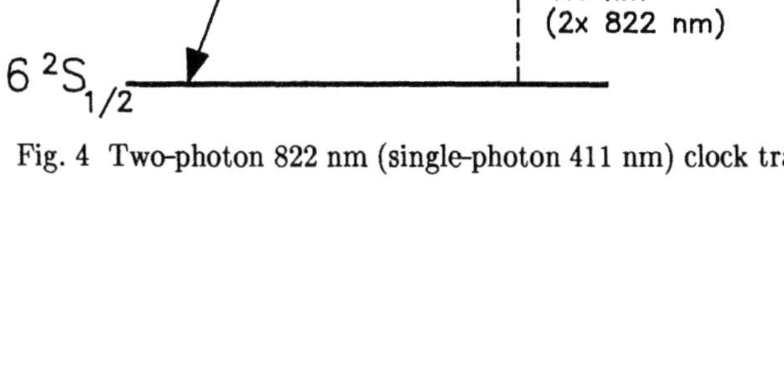

Fig. 4 Two-photon 822 nm (single-photon 411 nm) clock transition in Yb^+.

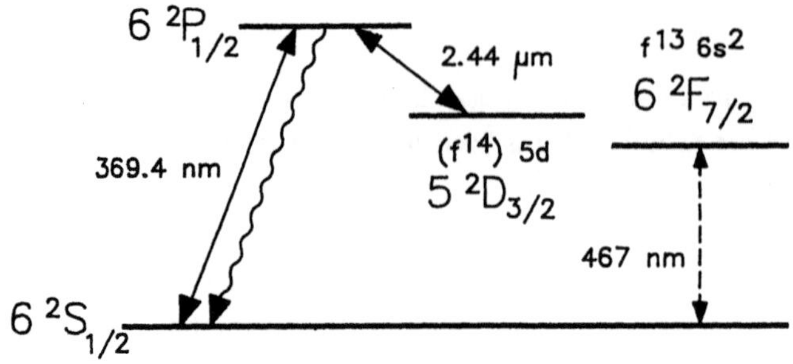

Fig. 5 Single-photon E3 clock transition in Yb^+.

there is a lower lying $^2F^o_{7/2}$ state resulting from a $4f^{13}\,6s^2$ configuration (cf. figure 5). The $D_{3/2}$ state and the $F_{7/2}$ state have different parity with $\Delta J = 2$, and the configuration requires two electrons to undergo a transition between $D_{3/2}$ and $F_{7/2}$. Hence no leakage from $D_{3/2}$ to $F_{7/2}$ is expected. However, in preliminary measurements occupation of the $F_{7/2}$ has been observed and was proved by deexciting the Yb^+ ions in the $F_{7/2}$ state by neighboring transitions at 369.11 nm and 2x 369.37 nm [22]. It has been suggested [23] that configuration mixing might be responsible for the finite probability of populating the $F_{7/2}$ level.

Application of the aforementioned clock transitions at 435 nm and 411 nm (2x 822 nm) hence requires further investigation of the leakage probabilities to the $F_{7/2}$ state. If leakage turns out to be a serious problem, deexcitation of the $F_{7/2}$ level via the observed transitions at 369.11 nm or 369.37 nm (via higher lying states which cascade to the ground state) will be necessary to avoid population trapping. Deexcitation could be achieved by laser diodes (doubling 739 nm) which relieve the need for expensive and complex light sources.

The lifetime of the $F_{7/2}$ state has been measured to exceed 7 days [22]. This long lifetime in turn allows the observation of an ultranarrow transition (E3) $S \leftrightarrow F$ at 467 nm with a Q value of $2 \cdot 10^{21}$ which can be obtained in two ways: i) Since cooling takes place on a time scale of 10 ms, the ion is already cold when the F state will be occupied, i.e. a single ion undergoes a quantum jump, and the fluorescence stops. Illumination by narrow-band radiation at 467 nm then results in a quantum jump to the ground state, followed by strong fluorescence again (flop-in-technique). ii) If the leakage rate turns out to be very small, the ion can be readily cooled, and the illumination by narrow-band 467 nm radiation results in quantum jumps to the F state (flop-out-technique). Although the laser sources which are required to exploit such a narrow transition are not yet available, some progress towards this regime has been made [24], and it seems not unlikely that the 467 nm transition can be employed for future frequency standards.

The realization of frequency standards based on the narrow transitions to the F and D states in Yb^+ certainly requires the detailed investigation of a multitude of systematic effects. Even after laser cooling, the ion experiences the small confining forces of the ion trap, i.e. the quadrupole potential. For example, Stark shifts have to be evaluated, and the electric quadrupole interaction has to be accounted for particularly with the quadrupole transitions to the D states. Eventually, experimental determination of these perturbations will reveal the detection scheme best suited to overcome systematic effects.

4. Conclusion

Several narrow transitions in stored Yb^+ ions have been proposed for frequency standard applications. Since the radiation required for excitation and optical pumping can be derived from diode lasers, it is expected that such devices will operate for long times between failures. With the help of the electron shelving technique very high Q values may be obtained. At present, preliminary experiments have shown that one can match the performance of Cs clocks when using the ion storage and h.f.s. lines as clock transitions. Application of laser cooling and observing narrow optical resonances in single ions promises frequency standards of unprecedented stability and accuracy.

References

1. G. Werth, Metrologia **22**, 190 (1986)
2. D.J. Wineland, Proc. of the IEEE **74**, 147 (1986)
3. W. Neuhauser, M. Hohenstatt, P.E. Toschek, and H. Dehmelt, Appl. Phys. **17**, 123 (1978), Phys. Rev. Lett. **41**, 233 (1978)
4. G. Werth, see these proceedings
5. L.S. Cutler et al., see these proceedings
6. F. Strumia, Proc. 2nd ann. freq. control symp., p. 444, Atlantic City 1978
7. R. Blatt, H. Schnatz, and G. Werth, Phys. Rev. Lett. **48**, 1601 (1982)
8. H. Dehmelt, Adv. At. Molec. Physics **3**, 53 (1967)
9. R. Blatt, P. Zoller, G. Holzmüller, and I. Siemers, Z. Phys. **D 4**, 121 (1986)
10. I. Siemers, R. Blatt, Th. Sauter, and W. Neuhauser, Phys. Rev. **A38**, (November 1988)
11. Ch. Gerz, J. Roths, F. Vedel, and G. Werth, Z. Phys. **D 8**, 235 (1988)
12. H. Dehmelt, IEEE Trans. Instr. Meas. **IM–31**, 83 (1983)
13. W. Nagourney, J. Sandberg, and H. Dehmelt, Phys. Rev. Lett. **56**, 2797 (1986)
14. Th. Sauter, W. Neuhauser, R. Blatt, and P.E. Toschek, Phys. Rev. Lett. **57**, 1696 (1986)
15. J.C. Bergquist, R.G. Hulet, W.M. Itano, and D.J. Wineland, Phys. Rev. Lett. **57**, 1699 (1986)
16. J. C. Bergquist, W.M. Itano, and D.J. Wineland, Phys. Rev. **A36**, 428 (1987)
17. J. Bollinger et al., see these proceedings
18. D.A. Jennings, C.R. Pollock, F.R. Petersen, R.E. Drullinger, K.M. Evenson, J.L. Wells, J.L. Hall, and H.P. Layer, Opt. Lett. **8**, 136 (1983)
19. W. Neuhauser, M. Hohenstatt, P. Toschek, and H. Dehmelt in *Spectral Line Shapes*, B. Wende, ed., Vol. 5, p. 1045; de Gruyter, Hawthorne, New York 1981
20. J. Bialas, R. Blatt, W. Neuhauser, and P.E. Toschek, Opt. Commun. **59**, 27 (1986)
21. H. Gilhaus, Th. Sauter, R. Blatt, W. Neuhauser, and P.E. Toschek, see these proceedings
22. H. Lehmitz, J. Hattendorf-Ledwoch, R. Blatt, and H. Harde, XVI. IQEC Digest of Technical Papers, p. 592, Tokio 1988
23. I. I. Sobel'man, private communication
24. D. Hils and J.L. Hall, see these proceedings

Laser Spectroscopy of a Single Barium Ion Using "Shelving"

W. Nagourney, Nan Yu, and H. Dehmelt

University of Washington, Dept. of Physics FM-15,
Seattle, WA 98195, USA

1.0 Introduction

It has been pointed out elsewhere [1] that a single, laser-cooled atomic ion can be used to perform optical spectroscopy that is free from all conventional shifts and broadenings to the part in 10^{18} level. If one is to construct a practical spectrometer based on a single ion with comparable long term stability, signal to noise considerations make it imperative that virtually every transition between the ground state and metastable excited state be detected. This is particularly true when working with single ions, where conventional techniques which depend upon the observation of a single fluorescent photon from each excitation would detect only a small percentage of the forbidden transitions, resulting in an unacceptably low data rate.

A solution to this problem appeared in 1975 when one of the authors briefly described a scheme [2] in which the strong resonance fluorescence of an individual atomic ion confined in high vacuum serves as a monitor which determines whether or not the ion is in its electronic ground state. The usefulness of this "shelving" scheme as a millionfold amplifier of the very weak *optical* fluorescence from one of the ion's metastable levels was emphasized. This technique would allow one to detect virtually every forbidden transition, even when such transitions are made infrequently to avoid power broadening. When combined with a spectrally narrow laser, it would make it possible to construct an optical spectrometer which closely approaches the "ideal".

We report here the application of the shelving scheme to the detection of forbidden optical transitions in a single trapped barium ion.

2.0 Quantum Jumps

The apparatus used in the single barium ion experiments has been described elsewhere [3,4]. The ion trap has a cap-to-cap spacing of

Frequency Standards and Metrology Editor: A. De Marchi

.5 mm and a ring diameter of .7 mm. The trap drive is at 25 MHz; a drive voltage of about 250 V will result in the observed 2 MHz secular frequency. The trap is enclosed in a stainless steel vacuum envelope maintained at a pressure of about 2×10^{-11} Torr by an ion pump.

The lasers responsible for cooling are made collinear and sent along one body diagonal of the trap. A photomultiplier observes the fluorescence at an angle of about 70° from the incident beam. The light is collected by a f/1.4 camera lens and imaged onto a pinhole placed before the PMT; the pinhole helps discriminate against instrumental scattering. A second lens (f/2.8) on the opposite side of the trap together with a ×20 microscope eyepiece allows one to easily observe the single ions visually.

The level structure of Ba^+ appears in Fig. 1. Two lasers are needed for cooling: one causes $6^2S_{1/2} - 6^2P_{1/2}$ transitions at 493 nm and the other excites $5^2D_{3/2} - 6^2P_{1/2}$ transitions at 650 nm, thus "cleaning out" the $5^2D_{3/2}$ level into which the ion will occasionally fall. To prevent optical pumping by the linearly polarized lasers into the $m_J = \pm 3/2$ sublevels of the $D_{3/2}$ level, a ≈6 Gauss magnetic field is oriented along the direction of the laser beams. The lasers both cool the single barium ion (confining it to <1 μm) and monitor its presence in the trap via strong fluorescence (at 493 nm) from the $6^2P_{1/2} - 6^2S_{1/2}$ transition. If the ion is excited to the $5^2D_{5/2}$ state, it will remain in this level for the ≈32 sec lifetime of that state. The presence of the ion in the metastable level can now be determined with unity "quantum efficiency" by the easily observed *suppression* of the 493 nm fluorescence.

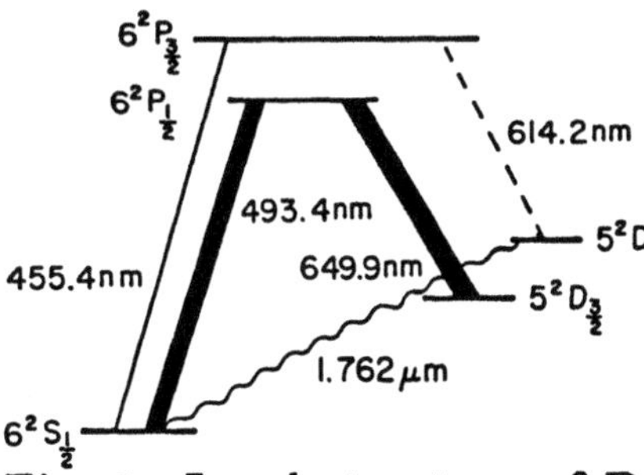

Fig. 1. Level structure of Ba^+. The red and green laser excitation is shown by the bold lines; the lamp excitation is indicated by the light solid line while the subsequent decay into the $5^2D_{5/2}$ level is indicated by the dotted line. The direct excitation of the $5^2D_{5/2}$ level by the color center laser is shown by the wavy line.

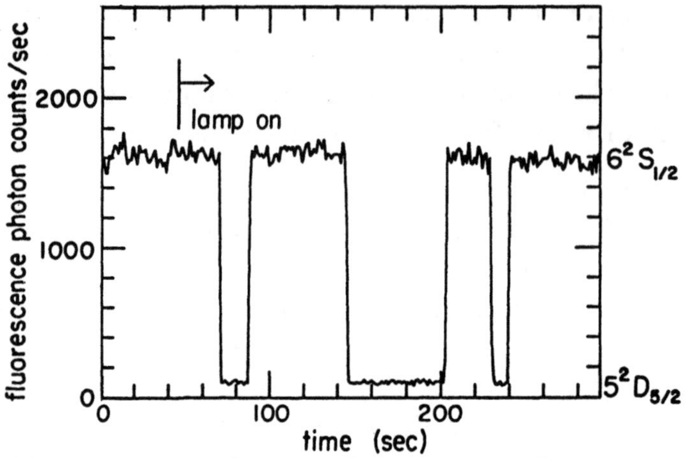

Fig. 2. A typical trace of the 493 nm fluorescence from the $6^2P_{1/2}$ level showing the quantum jumps after the barium resonance lamp is turned on. The atom is definitely known to be in the $5^2D_{5/2}$ level during the low fluorescence periods.

To demonstrate "quantum jumps", a filtered barium resonance lamp was used to make relatively infrequent (roughly 1 every 10 seconds) transitions to the $6^2P_{3/2}$ state. Approximately one third of the subsequent decays are to the metastable "shelf" ($5^2D_{5/2}$) level. A plot of the 493 nm fluorescence versus time appears in Fig. 2. The "dwell times" in the $5^2D_{5/2}$ level can be easily shown to be distributed exponentially, with an average value equal to the metastable lifetime. From a sample of 203 dwell times, we have verified the exponential distribution and measured the metastable lifetime. The result is 32±5 sec, which includes a roughly 10% measured correction due to collisional quenching from the background gas. This is within one standard deviation of the previously measured [5] lifetime of 47±16 sec.

3.0 Laser Spectroscopy

Direct excitation of the $5^2D_{5/2}$ metastable level via a quadrupole transition at 1.762 μm was performed with a tunable color center laser. The color center laser was locked to a Fabry-Perot interferometer. A computer derived analogue voltage scanned the Fabry-Perot, thus scanning the laser. At each frequency step in the scan, the number of abrupt changes in the strong fluorescence from the $6^2P_{1/2}$ level was counted; this number is equal to the number of transitions that the ion makes between the ground and metastable states. It was assumed that virtually all of these transitions were due to stimulated events; this assumption is valid because the spontaneous lifetime (32 sec) is so much greater than the average time ($\approx$.1 sec) between transitions at resonance. To prevent ac Stark broadening of the ground state by the blue laser, the blue laser and the color center laser were alternated using mechanical choppers controlled by the computer. For a typical run each laser was turned on for 15 ms and the infrared laser frequency was advanced after 74 such cycles. Thus, about 2 seconds were spent in each "frequency channel"; a complete run consisted of 256 frequency channels and took about 8 minutes.

A block diagram of the laser stabilization scheme appears in Fig. 3. The color center laser frequency is electronically controlled in two ways. A slow control is provided by a piezo-driven fold mirror which allows about 3 GHz of frequency control; its speed is limited by the roughly 3.5 kHz resonant frequency of the piezo element. Since most of the laser noise is due to mechanical resonances at about 500 Hz, a stable feedback system using the fold mirror alone would provide only a marginal reduction of the laser's frequency noise. Therefore, a faster frequency control is needed. This is provided by a $\approx$80 MHz acousto-optical modulator and VCO combination whose speed limitation is due to a $\approx$.5 μs acoustical delay between the application of the signal and the desired frequency shift. The AO modulator provides a $\pm$10 MHz shift of the laser at speeds of up to 100 kHz. Together, these inputs allow both long term and short term stabilization of the laser.

The correction signal (discriminant) needed for the stabilization is provided by an external EO modulator which frequency modulates the laser beam at 3.6 MHz with a modulation index of about .8. The beam then passes through a 50 cm confocal Fabry-Perot interferometer and is detected by a Ge photodiode. The linewidth of the reference cavity is about 1.5 MHz. The modification of the amplitude and phase of the FM sidebands by the Fabry-Perot filter will cause the detected DC signal to be amplitude modulated at 3.6 MHz. The desired discriminant will appear when the amplified 3.6 MHz signal is mixed with a 90 degree phase shifted sample of the 3.6 MHz modulation voltage.

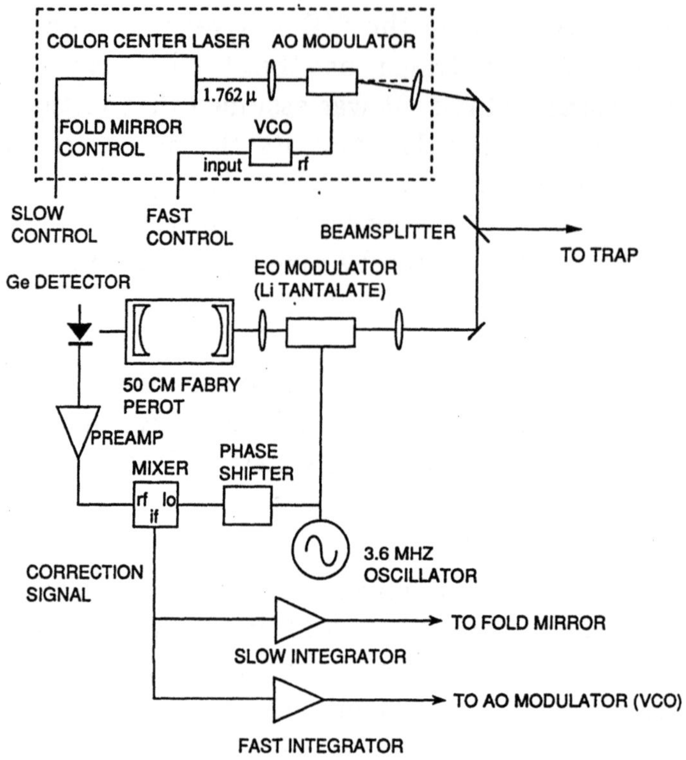

Fig. 3 Block diagram of color center laser stabilization scheme.

The correction signal passes through two integrators: one drives the fold mirror PZT and the other drives the acousto-optical modulator-VCO. The integrator for the slow loop has a single pole while that for the fast loop has two poles in order to obtain the maximum servo gain at the frequency ($\approx$500 Hz) where most of the laser noise is found. The measured closed loop bandwidth was 20 kHz. From the reference cavity signal, a laser linewidth of about 30 kHz was inferred. At this level of stability, the drift of the commercial, temperature stabilized reference cavity (measured to be about 1 MHz/hour) would be the principal limitation in the spectral resolution and might account for the slightly larger linewidth found for the ion resonance (50 kHz).

4.0 Results

A typical recently obtained spectrum appears in Fig. 4. A fit to a Gaussian resulted in a 50 kHz linewidth. The sweep range was insuf-

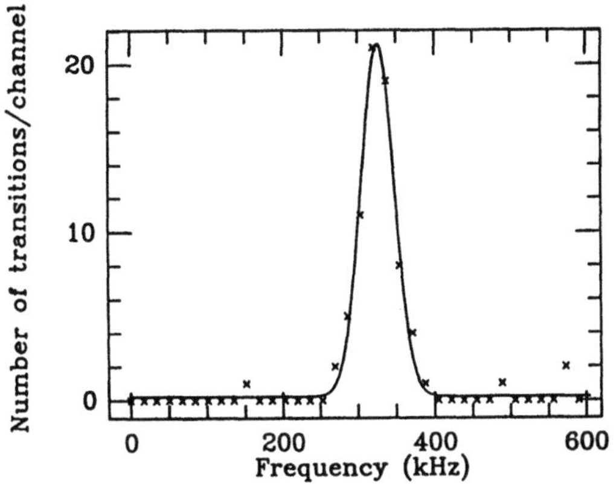

Fig. 4. Spectrum of $6^2S_{1/2}$ to $5^2D_{5/2}$ quadrupole transition at 1.762 μm of a single barium ion. The laser spent 2 seconds per frequency channel; 36 channels are shown. The experimental points are the x's and the solid line is the fitted gaussian. The fitted linewidth is 50 kHz.

ficient to observe sidebands at the axial (2 MHz) or radial (1 MHz) secular frequencies. Earlier data with a wider sweep range (3 MHz) and greater linewidth (330 kHz) displayed no radial sidebands that were above the noise, which was about 5 times less than the signal. This implies an average radial rms amplitude, x_0, given by $(\frac{\pi x_0}{\lambda_0})^2 \leq \frac{1}{5}$ or $x_0 \leq \frac{\lambda_0}{2\pi} \approx 0.3$ μm which in turn yields a radial ion temperature $\leq$20 mK and confirms earlier results [4]. Our 50 kHz linewidth may be compared with a $\approx$30 kHz width recently published for an ultraviolet transition in a single trapped Hg^+ ion [6].

5.0 Conclusion

Our experiment demonstrates the use of the "shelving" technique to provide an essentially noiseless "front end" of a high resolution optical spectrometer. Spectroscopy of the $5^2D_{5/2}$ level of Ba^+ could be done with potentially 5 mHz resolution; virtually every transition to that level would be detected. Such spectroscopy would be free from Doppler shifts to all orders and would not be subject to collisional or transit time broadening; power broadening would be avoided by alternating the lasers. In fact, with adequate magnetic shielding, the principal broadening mechanism for the foreseeable future would be the breadth of the laser.

We would like to thank Clifford R. Pollock for kindly providing the OH-doped NaCl crystal used in the color center laser and Thomas Gallagher for kindly providing the Li Tantalate EO modulator. The work was supported by the National Science Foundation under the SAPARIS PROJECT (PHY-8506028).

6.0 References

1. H. Dehmelt, IEEE Trans. Instrum. Meas. *IM – 31*, 83 (1982).
2. H. Dehmelt, Bull. Am. Phys. Soc. 20, 60 (1975).
3. W. Nagourney, J. Sandberg, and H. Dehmelt, Phys. Rev. Lett. 56, 2797 (1986).
4. G. Janik, W. Nagourney, and H. Dehmelt, J. Opt. Soc. Am. B 2, 1251 (1985).
5. F. Plumelle, M. Desaintfuscien, J. L. Duchene, and C. Audoin, Opt. Commun. 34, 71 (1980).
6. J. C. Bergquist, W. M. Itano, D. J. Wineland, Phys. Rev. A 36, 428 (1986).

Frequency Standards Utilizing Penning Traps *

J.J. Bollinger, S.L. Gilbert, W.M. Itano, and D.J. Wineland

Time and Frequency Div., National Institute of Standards and Technology, Boulder, CO 80303, USA

1. Introduction

Ion traps have a number of characteristics that make them attractive candidates for frequency standards [1]. Ion confinement times from a few hours to a few days permit long interrogation times and therefore narrow linewidths and large line Q's. In addition, ion traps typically do not suffer from the usual perturbations associated with confinement. For example, collisions of hydrogen atoms with the confining walls of a hydrogen maser can produce fractional frequency shifts greater than 1×10^{-11} on the masing transition [2]. Stark shifts due to the confining electrostatic field of a Penning trap can, in many cases, be made less than 1×10^{-15} [3,4].

The largest systematic shift and uncertainty in a frequency standard using many stored ions is typically the second order Doppler (time dilation) frequency shift $\Delta\nu_{D2}$. For a clock transition of frequency ν_0, $\Delta\nu_{D2}$ is given by

$$\langle \Delta\nu_{D2}/\nu_0 \rangle = - \langle \vec{v}^2 \rangle/(2c^2) \qquad (1)$$

where c is the speed of light, $\vec{v}$ is the ion velocity, and < > denotes an average over the ions in the trap. Without a means of achieving low ion temperatures, this shift can be unacceptably large. For example, a 300 K $^9Be^+$ ion has a second order Doppler shift of -4.6×10^{-12}. Laser cooling [5,6] can be used to reduce the ion temperature to much less than 1 K. For Be^+ ions, this results in second order Doppler shifts smaller than -1.5×10^{-14}. In our laboratory, up to 10^5 Be^+ ions stored in a Penning trap have been laser cooled to temperatures less than 1 K. The Penning trap is therefore a suitable candidate for a laboratory microwave frequency standard where the Dicke confinement criterion [7] is easily satisfied and where it is desirable to store many ions in a trap in order to obtain a signal-to-noise ratio that gives a useful fractional frequency stability. The number of ions used in a Penning trap frequency standard is determined from a tradeoff of signal-to-noise considerations and the second order Doppler frequency shift due to the $\vec{E}\times\vec{B}$ rotation of the ions[3,8]. We note that in an rf (Paul)

Frequency Standards and Metrology Editor: A. De Marchi

trap the number of ions that can be laser cooled has been typically small (< 50) due to the problem of rf heating[9].

At the previous (Third) Symposium on Frequency Standards and Metrology, the results of an initial experiment done at the National Bureau of Standards (NBS) on $^{25}Mg^+$ ions stored in a Penning trap were reported[10]. In this experiment a 12 mHz linewidth on a 292 MHz magnetic field-independent hyperfine transition in the ground state of $^{25}Mg^+$ was obtained[11]. Below we summarize the Penning trap frequency metrology work done in our laboratory since the previous symposium. In the next section the results of a frequency standard based on up to 2000 laser cooled $^9Be^+$ ions stored in a Penning trap are discussed. The inaccuracy of this frequency standard was on the order of 1×10^{-13}[4]. It was limited by the second order Doppler shift of the ions due to the ion heating during the clock interrogation period when the cooling laser was turned off. The final section of this paper discusses our current work using sympathetic cooling[12] to maintain a cold Be^+ cloud and obtain a 0.9 mHz linewidth on the Be^+ clock transition. With sympathetic cooling we project an order of magnitude improvement in the Be^+ frequency standard. In addition to the work done at NBS discussed here, frequency metrology work with ions stored in Penning traps is being conducted by groups at Orsay, France[13] and Teddington, England[14].

2. Laser-Cooled $^9Be^+$ Frequency Standard

The Penning trap uses a uniform magnetic field to confine ions in a direction perpendicular to the magnetic field. The ions are prevented from leaving the trap along the magnetic field direction by a static electric field, typically provided by three hyperbolic electrodes[1]. Figure 1 shows the hyperfine structure of the ground state of $^9Be^+$ as a function of magnetic field. At a magnetic field of 0.8194 T the hyperfine transition ν_1 (see Fig. 1) depends only quadratically on the magnetic field deviation ΔB according to $\Delta\nu_1/\nu_1 = -0.017(\Delta B/B)^2$. In spite of its relatively low 303 MHz transition frequency, ν_1 provides the basis for a frequency standard. (We have chosen to work with Be^+, because it is easy to trap and laser cool. Heavier ions such as Hg^+ offer large microwave transition frequencies but are technically much harder to work with.) A narrow-band radiation source (power $\cong$ 20-60 μW) tuned to the low frequency side of the 2s $^2S_{\frac{1}{2}}(m_I = -3/2,\ m_J = -1/2) \rightarrow$ 2p $^2P_{3/2}(-3/2,\ -3/2)$ (λ = 313 nm) transition of $^9Be^+$ was used to cool and spatially compress the ions and optically pump them into the (-3/2, -1/2) ground state[11,15]. The resonance fluorescence induced by this cooling laser was used to detect the ions. The size, density, and temperature of the ion clouds were determined by use of a similar radiation source as a probe[15]. Typical clouds ranged from a few hundred to 2000 ions with cloud diameters from 300 to 500 μm and densities of about 3×10^7 ions/cm^3. Cyclotron and axial temperatures of less than 100 mK and a second order Doppler

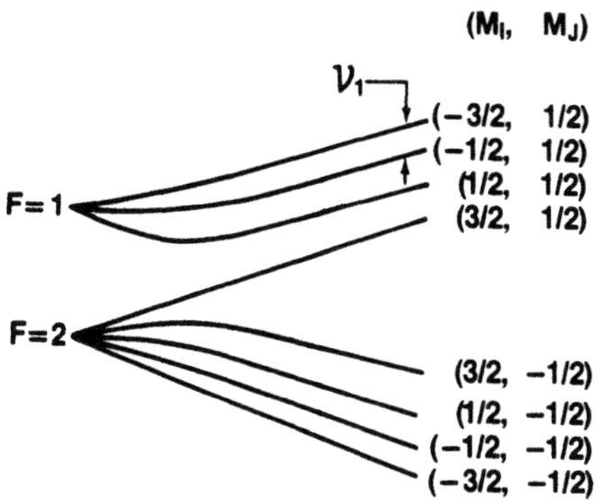

Figure 1. Hyperfine structure (not drawn to scale) of the $^9Be^+$ 2s $^2S_{\frac{1}{2}}$ ground state as a function of magnetic field. ν_1 is independent of magnetic field to first order at B = 0.8194 T.

shift due to the cloud rotation of less than 3×10^{-14} were obtained with the cooling laser applied continuously.

The ν_1 transition was detected by optical-microwave-rf triple resonance. Microwave radiation tuned to the electron spin-flip resonance transferred half of the ion population from the optically pumped (-3/2, -1/2) state to the (-3/2, +1/2) state. Some of the (-3/2, +1/2) state population was transferred to the (-1/2, +1/2) state by application of rf near the 303 MHz ν_1 transition frequency. As a result of the microwave mixing this resulted in an additional decrease in the (-3/2, -1/2) state population and therefore a decrease in the observed fluorescence. The cooling laser and mixing microwaves were typically on for a period of 3 s, during which the $^9Be^+$ ions were prepared in the (-3/2, -1/2) and (-3/2, +1/2) states. The cooling laser and mixing microwaves were then turned off in order to avoid light and ac Zeeman shifts during the ν_1 interrogation period. Ramsey's method of separated oscillatory fields was used to interrogate ν_1. Typically an rf pulse of duration t = 0.5 s was applied, followed by a free precession interval of duration T = 19 s and a second rf pulse of duration t = 0.5 s coherent with the first pulse. After the interrogation period, the laser and microwaves were turned back on, and the signal was obtained from the absence of fluorescence during the first 0.3 s of this time interval. The (t,T) = (0.5 s, 19 s) interrogation resulted in a linewidth $\Delta\nu_1$ = 25 mHz and a $Q \equiv \nu_1/\Delta\nu_1$ of 1.2×10^{10} on the 303 MHz ν_1 transition frequency.

A synthesized rf source was locked to the ν_1 transition. A description of the servosystem is given in reference [4]. With about 1000 ions in the trap, a fractional frequency stability of $\sigma_y(\tau) \cong 2\times10^{-11}\tau^{-\frac{1}{2}}$ for measurement time τ in seconds was obtained with the (0.5 s, 19 s) Ramsey interrogation. The systematic uncertainty of 1×10^{-13} was dominated by the second order Doppler frequency shift [4]. During the 20 s interrogation period when the cooling laser was turned off, the ion temperature increased from less than 1 K to 35 K. This heating produced a second order Doppler shift of 4×10^{-13} on the ν_1 transition and was measured with about 25% uncertainty. The rapid heating that occurred during the interrogation period was probably caused by asymmetry-induced transport [16]. With the cooling laser off, axial asymmetries of the trap can increase the total canonical angular momentum of the ions, resulting in an increase in the ion cloud radius. As the ion cloud expands, electrostatic potential energy of the ions due to the space-charge and trap electric fields is converted into thermal energy of the ions.

3. Sympathetic Cooling

An improvement in Be^+ frequency standard performance requires a reduction in the observed heating. This can possibly be done by improvement of the trap axial symmetry. With this in mind we have constructed a Penning trap with care taken to minimize axial asymmetries. Rather than the usual hyperbolic electrodes, the trap was constructed with hollow cylindrical electrodes [17]. In addition, in order to avoid asymmetric contact potentials due to plating of the trap electrodes from the Be^+ oven when ions were loaded into the trap, a second trap, mounted below the experimental trap, was used to load the ions. The ions were then transferred to the clean experimental trap. The expansion of the Be^+ ion cloud in the cylindrical Penning trap was less than in the previous Be^+ Penning trap. However, the expansion was eliminated by using a second ion species to sympathetically cool and compress [12] the Be^+ ions. Specifically, $^{26}Mg^+$ ions were loaded into the same trap with the $^9Be^+$ ions. The Mg^+ ions were continuously laser-cooled and spatially compressed by a 280 nm (power ≅ 50-80 μW) radiation source. Through the Coulomb interaction with the Mg^+ ions, the Be^+ ions were then cooled and compressed. Due to the ion cloud rotation, the Mg^+ ions were observed to centrifugally separate and form a doughnut surrounding the Be^+ ions.

The 313 nm source was used to optically pump and detect the Be^+ ground state. It had only a small effect on the Be^+ ion temperature and cloud size and, as in the previous experiment, was turned off during the Ramsey interrogation period. Because the 280 nm Mg^+ cooling laser beam is nonresonant with any Be^+ transitions, it was left on continuously during the Ramsey interrogation period without seriously perturbing the ν_1 transition frequency. In this way a cold, steady state cloud of Be^+ ions was obtained independently of the 313 nm radiation source, which permitted the use of very long interrogation times.

In the previous triple resonance experiment (optical-microwave-rf) only half of the stored Be^+ ions were used to measure the ν_1 transition due to the microwave mixing. The stable field of the superconducting magnet used in the present experiment permitted the following quadruple resonance procedure which used all of the stored ions to measure the ν_1 transition. The Be^+ ions were optically pumped into the (3/2, 1/2) ground state by tuning the 313 nm source to the 2s $^2S_{\frac{1}{2}}(3/2,1/2) \rightarrow$ 2p $^2P_{3/2}(3/2,3/2)$ transition. The 313 nm source was turned off and the ions were transferred to the (1/2,1/2) and then the (-1/2,1/2) state by successive 0.2 s Rabi π pulses. After Ramsey's method of separated oscillatory fields was used to interrogate ν_1, the ions remaining in the (-1/2,1/2) state were transferred back to the (3/2,1/2) state by reversing the order of the previous two Rabi π pulses. The 313 nm source was turned back on, and the signal was obtained from the absence of the Be^+ fluorescence during the first few seconds after the 313 nm source was turned on. In addition to increasing the number of ions used to measure the ν_1 transition, this procedure increased the signal by a factor of 4. The long (-3/2, +1/2) → (3/2, 1/2) laser repumping time also allowed the signal counting time after the 313 nm source was turned back on to be increased. This increased the signal-to-noise ratio by increasing the change in the number of detected photons for each ion that made the clock transition.

With sympathetic cooling, long interrogation times were possible. Figure 2 shows the signal obtained with a (1 s, 550 s) Ramsey interrogation. The 0.9 mHz linewidth gives $Q = 3.3\times10^{11}$ on the ν_1 clock transition. If used to servo a local oscillator, this linewidth along with the estimated signal-to-noise ratio should result in a fractional frequency stability of $\sigma_y(\tau) = 3.8\times10^{-12}\tau^{-\frac{1}{2}}$ ($\tau > 1100$ s). This stability is consistent with the stability of our reference oscillator and is about a factor of 8 or 9 larger than the stability estimate based on the number of stored ions [8]. Shorter (1 s, 5 s) Ramsey interrogations resulted in a signal-to-noise ratio within a factor of 2 of the estimate based on the number of stored ions.

Preliminary estimates of all known systematic uncertainties for the transition frequency of Fig. 2 are less than 1×10^{-14}. With the use of an imaging tube, the cloud rotation frequency and radial distribution of the Be^+ ions were determined. The second order Doppler shift due to the cloud rotation was calculated to be

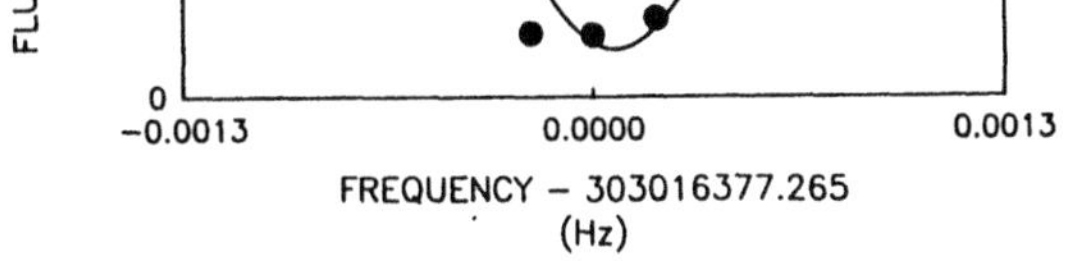

Figure 2. Signal obtained with 3200 Be^+ ions on the ν_1 field independent transition with a 550 s Ramsey free precession period. The data are the result of one sweep. The sweep width is 2.4 mHz and the frequency interval between points is 0.2 mHz. The dots are experimental and the curve is a least squares fit. A signal-to-noise ratio of 12 is estimated from the fit.

-4×10^{-15}. From previous observations, we believe the sympathetic cooling provided Be^+ temperatures $\ll$ 1 K and second order Doppler shifts less than 1×10^{-14}. The ac Stark shift due to the 280 nm source is estimated to be less than 1×10^{-15}. Due to the rotation of the earth, there is a systematic shift of $2.46(0.05)\times10^{-14}$ [4]. All other known systematic shifts are estimated to be less than 1×10^{-15}.

In the future we plan to evaluate the use of sympathetically cooled Be^+ ions to steer a local oscillator. With a passive hydrogen maser [2] as a source for the local oscillator, we may be able to obtain a fractional frequency stability of $\sigma_y(\tau) = 2\times10^{-12}\tau^{-\frac{1}{2}}$ ($\tau > 1000$ s), limited by the stability of the passive hydrogen maser. An inaccuracy of less than 1×10^{-14}, limited by the second order Doppler shift, appears feasible.

We gratefully acknowledge the support of the U.S. Air Force Office of Scientific Research and the U.S. Office of Naval Research. We thank C. Weimer and D. Heinzen for carefully reading the manuscript.

References

1. See for example D.J. Wineland, W.M. Itano, and R.S. Van Dyck, Jr.: Adv. At. Mol. Phys. 19, 135 (1983) and references therein
2. F.L. Walls: Proc. of the IEEE 74, 142 (1986)
3. D.J. Wineland, W.M. Itano, J.C. Bergquist, and F.L. Walls: in Proc. of 35th Ann. Freq. Control Symposium (USAERADCOM, Ft. Monmouth, NJ 07703) p. 602 (1981); D.J. Wineland in Precision Measurement and Fundamental Constants II, ed. by B.N. Taylor and W.D. Phillips (Natl. Bur. Stands. (U.S) Spec. Publ. 617, 1984), p. 83
4. J.J. Bollinger, J.D. Prestage, W.M Itano, and D.J. Wineland: Phys. Rev. Lett. 54, 1000 (1985)
5. S. Stenholm: Rev. Mod. Phys. 58, 699 (1986)
6. D.J. Wineland and W.M Itano: Physics Today 40, 34 (June 1987)
7. R.H. Dicke: Phys. Rev. 89, 472 (1953)
8. D.J. Wineland et al.: these proceedings
9. D.A. Church and H.G. Dehmelt: J. Appl. Phys. 40, 3421 (1969)
10. D.J. Wineland, J.C. Bergquist, R.E. Drullinger, H. Hemmati, W.M. Itano, and F.L. Walls: J. Phys. (Paris) Colloq. 42, C8-307 (1981)
11. W.M. Itano and D.J. Wineland: Phys. Rev. A24, 1364 (1981)
12. D.J. Larson, J.C. Bergquist, J.J. Bollinger, W.M. Itano, and D.J. Wineland: Phys. Rev. Lett. 57 70 (1986)
13. F. Plumelle, M. Desaintfuscien, M. Jardino, and P. Petit: Appl. Phys. B41, 183 (1986)
14. R.C. Thompson, G.P. Barwood, and P. Gill: Appl. Phys. B46, 87 (1988)

15. L.R. Brewer, J.D. Prestage, J.J. Bollinger, W.M. Itano, D.J. Larson, and D.J. Wineland: Phys. Rev. A38, 859 (1988)
16. C.F. Driscoll, K.S. Fine, and J.H. Malmberg: Phys. Fluids 29, 2015 (1986)
17. S.L. Gilbert, J.J. Bollinger, and D.J. Wineland: Phys. Rev. Lett. 60, 2022 (1988)

The Interferometric Detection of Gravitational Waves – a Status Report

A. Brillet

Groupe de Recherches sur les Ondes de Gravitation,
Bât. 104, Campus, F-91405 Orsay, France, and
Laboratoire de l'Horloge Atomique,
Equipe de Recherche du CNRS associée à l'Université Paris-Sud,
Bât. 221, Université Paris-Sud, F-91405 Orsay Cedex, France, and
Laboratoire de Gravitation et Cosmologie Relativistes,
Université Pierre et Marie Curie, Paris, France

There are presently five groups in the world working towards the construction of a large interferometer for the detection of gravitational waves. They are:

- the Max Planck Institüt fur Quantenoptik in Garching (FRG), operating a 30m prototype, and planning to build a 3 km antenna
- the University of Glasgow (Scotland), operating a 10m prototype, and planning to build a 1 km antenna
- the "LIGO" project of the NSF (USA), a collaboration between M.I.T. and Caltech, operating a 40m prototype, and planning to build two 4 km antennas
- the VIRGO italo-french collaboration involving laboratories in Pisa, Orsay-Paris, and Napoli, and preparing the construction of a 3km antenna, funded by the Istituto Nazionale de Fisica Nucleare.
- the ISAS group in Japan, working at a 10 m prototype and planning to build soon a 200 m system.

The VIRGO project has been approved and the construction of the 3 km antenna will start in 1991. There is now a good probability that, within 5 to 7 years, there will be an array of 4 to 5 large interferometric gravitational wave antennas in the world. Not only this would allow to detect gravitational waves of astrophysical origin, but it will also constitute an observatory, able to measure the parameters of the gravitational wave signal, and to bring a new kind of informations on the universe, especially from "dark" objects.

Frequency Standards and Metrology Editor: A. De Marchi

1. SOURCES OF GRAVITATIONAL RADIATION

Among the conceivable sources which have been modelized, the most interesting ones seem to be the following:

1.1. Stellar collapses

When the core of a star collapes to a $1 M_\odot$ neutron star, wether or not producing a supernova, one expects the generation of a burst of gravitational waves. This burst will have a typical duration of 1 ms; its spectrum will then extend from a few hundred Hz to a few kHz. The probability of such events is relatively well known (about 0.1 event per year per galaxy), but the conversion efficiency (from the gravitational energy of the star to gravitational wave energy) can only be evaluated through a difficult modelisation using numerical relativity. This work is not completed, but preliminary results suggest that an efficiency of 0.1% is possible, leading to a gravitational wave amplitude:

$$h = 1.5\ 10^{-21}.\ (\frac{15\ \mathrm{MPC}}{R})$$

where R is the distance to the source.

An array of two or more antennas with a sensitivity $h = 3.\ 10^{-23}\ \mathrm{Hz}^{-1/2}$ (or 10^{-21} in a 1 kHz bandwidth) should be able to detect at least a few events per year. Beyond that, the number of observations will increase proportionnally to the cube of the inverse sensitivity.

1.2. Coalescing binary systems

Binary systems involving two neutron stars, such as PSR 1913+16, are known to loose energy by emission of gravitational waves. Both the amplitude of the gravitational wave and the evolution of the rotation frequency can be calculated very precisely until the very last moment before the coalescence, when the frequency can reach a few hundred hertz. The observed amplitude for $1.4\ M_\odot$ stars, will be:

$$h = 5\ 10^{-23}.\ (\frac{100\ \mathrm{Mpc}}{R}) \cdot (\frac{\nu}{100\ \mathrm{Hz}})^{2/3}$$

while the frequency will chirp, with a doubling time of approximately 5 s. The fact that the time dependance of the signal is known allows one to use matched filters for the detection. Depending on the low frequency cut-off of the antenna, which determines the number of observable cycles, it may be possible to improve the S/N ration by a factor of 40 to100.

The unknown parameter is here the probability of an event, which is believed to lie between 0.01 and 100 events per year within the observation range of the first generation antennas (a radius of 100 Mpc).

1.3. Continuous sources

Any rotating asymetric object should emit gravitational waves. A pulsar of mass asymetry δ will emit a wave of amplitude:

$$h = 10^{-21}.\ \delta.\ (\frac{\nu}{100\ \mathrm{Hz}})^{2}.\ \frac{10\ \mathrm{Kpc}}{R}$$

at twice its rotation frequency. There are no reliable estimations of the effective asymetry, just upper limits, which correspond to $h<10^{-24}$ at 60 Hz for the Crab pulsar and to $h< 5.10^{-25}$ at 30 Hz for Vela. The VIRGO antenna should reach these sensitivities after a month of integration time.

1.4. Stochastic background

This gravitational radiation "noise" is expected to result from phase transitions in the early universe, binaries, primordial nucleosynthesis, cosmic strings.... It could be detected by noise correlation measurements between two neighbouring antennas. Its amplitude is widely uknown.

2. PRESENT STATUS

The first three of the prototypes have reached sensitivities of the order of 10^{-18}, comparable with the best Weber bars. The Garching prototype, in particular, is very well understood, and a simple extension of its arm length *must* produce a proportionnal improvement of sensitivity, down to 3.10^{-20}. All the large projects are aiming for an initial sensitivity better than 10^{-21}, which seems to be the threshold for having a chance to detect a few events per year. All these interferometers rely on the same detection principle: the anisotropic strain ($h=\delta L/L$) produced by the gravitational wave is transformed in the displacement of an interference fringe in a Michelson interferometer. To enhance the effect, the effective length L_{eff} of the arms is made much larger than the mechanical length L, either by a multipass or through the use of a Fabry-Perot cavity.

3. THE SENSITIVITY LIMITATIONS

The main intrinsic limitations to the sensitivity of an interferometric antenna operating in the 10 Hz-1 kHz range are shot-noise, thermal noise, and seismic noise.

3.1. Shot-noise

It defines the minimum detectable strain, $h = \frac{1}{L_{eff}} \cdot \sqrt{\frac{h.c}{\lambda.P}}$ $Hz^{-1/2}$, for a laser of wavelength λ and power P, and an interferometer of effective length L_{eff}. For a typical L_{eff} of 100km, a power of about 1 kW is required. This white noise will limit the sensitivity at high frequencies.

3.2. Thermal Noise

The resonances of the mirrors themselves, and of their suspensions are thermally excited, producing a displacement noise. In particular, at the frequency ω, the thermal noise of the suspension gives an equivalent strain

$$h = \frac{1}{L}\sqrt{\frac{8.kT.\omega_0}{M.Q.\omega^4}}$$

where M is the mass of the mirror, T the temperature, Q and ω_0 the suspension surtension coefficient and the pulsation of the suspension, and L the mechanical length of the interferometer. Even if Q reaches 10^6, this will limit the sensitivity in the 100 Hz range.

3.3. Seismic Noise

Many measurements performed in various places show an average seismic noise of the order of $\delta x \approx 10^{-6}/\omega^2$ m.Hz$^{-1/2}$ above 10 Hz. A mirror suspended from a chain of n (ideal) low frequency pendulums would simulate a strain

$$h \approx 10^{-6}.(\frac{\omega_o}{\omega})^{2n}/(\omega^2.\ L).$$

If n=1 would suffice at 1 kHz, a multiple pendulum is required for operation at very low frequency. Anyway, this noise source will dominate at low frequency. The strain noise due to the last two sources decreases with increasing length L

Below a few Hz, it becomes useless to improve the seismic isolation, because gravitation gradients from moving objects (vehicles, clouds) and even animals cannot be avoided: space experiments will be the only possibility in the very low frequency domain.

Most of the early work on the prototypes (1975-1985) consisted in trying to understand and eliminate many other technical noise sources, in order to reach the sensitivity limits established by the above noise sources. A detailed study of these problems is given in SHOEMAKER et al. [1] .

4. FIGHTING THE SHOT NOISE AND THE SEISMIC NOISE

In all the groups, most of the activity is now oriented towards these "fundamental" noise sources (and towards some engineering problems which are specific of the the large antennas).

4.1. Shot-noise

The problem of the light source (a 1 kW, ultrastable, short wavelength, CW laser) would be hopeless without the idea of "recycling" the light, which was proposed by R.Drever in 1982: it consists in tuning the transmission of the Michelson interferometer to a dark fringe, to collect with a mirror the reflected beam, and to send it again in the interferometer. If the phase of the beam is properly controlled, it interferes constructively with the original beam: the interferometer is now inside a resonant cavity and the power reaching the beamsplitter, which defines the shot-noise limited sensitivity, can be increased by a large factor if the losses are low. This idea has been developped, and a variety of recycling schemes have been studied, VINET et al. [2], which bring a large sensitivity improvement for periodical or impulsive sources. In a experiment recently performed in Orsay, MAN et al. [3], we tested a simple recycling scheme on a rigid Michelson interferometer. To avoid losses in the interferometer, we replaced the standard phase modulation detection technique by an heterodyne detection scheme where the (weak) signal beam beats with a modulated (or frequency shifted) reference beam. With an incident power of only 300 mW, this device reaches an equivalent shot-noise limit of 6 W. The *power* is limited by the appearance of optical damage on the beamsplitter, but it must be emphasized that the light *intensity* is already close to the value it will have in the large VIRGO interferometer. Furthermore, better mirrors are now available, which give a higher power handling capacity and lower losses, so that we are now fully confident that it is possible to store a kW of light in the interferometer with an initial laser power of 20 to 30 W.

The realization of a suitable light source becomes possible, and two directions are investigated: the coherent superposition of a few Argon lasers, whose feasibility was demonstrated in Orsay a few years ago, MAN et al.[4], and the realisation of a high power single-frequency Nd:Yag laser, phase-locked by injection to a low power, ultrastable, diode-pumped Nd:Yag, SHOEMAKER et al. [5]. A future solution may be to use an array of high power phase-locked laser diodes, since their frequency control is now possible with a good precision, HOLLBREG et al.[6].

4.2. Seismic noise

The obvious way to provide the necessary seismic isolation is to suspend each optical component of the interferometer as a pendulum; the double pendulum used in all the prototypes does not provide sufficient isolation for operating a full scale antenna at frequencies lower than 200 Hz, but the Pisa group recently tested a sophisticated 7 stage system, whose attenuation (10^{12} at 10 Hz) is sufficient, DEL FABBRO et al. [8]. One difficulty is to generate a suitable frequency dependent surtension coefficient: the Q factor must be very low around the resonance frequencies in order to avoid large oscillations, and very large at higher frequencies, to keep the thermal noise low. This has already been satisfactorily realized on the double pendulum systems, [1], using a servo loop which reacts on the suspended mass through coil and magnet actuators. The implementation of a similar system on the 7 stage pendulum is underway.

5. Some remaining difficulties

Most physical difficulties seem to be mastered now, except for the thermal noise, which has not been properly studied yet, and for the problem of the scattered light.

There remains also a lot of engineering difficulties, concerning the vacuum, and the manufacture of large high quality optics, but these are questions of cost rather than of feasibility.

Except in the search for periodic sources with very long averaging times, the data acquisition and the data analysis problems are not especially difficult.

6. Conclusion

Most of the technical noise sources which could impair the sensitivity of gravitational wave interferometers have been identified and mastered by the efforts of the pioneering groups. Even the shot noise and the seismic isolation problems seem to be well in hand now. Tight collaboration is already taking place, particularly within the EEC, to solve the engineering problems related to the construction of the large scale antennas. The funding of the Virgo project is now secured, and we expect some other positive decisions in the coming year.

REFERENCES

1. D. Shoemaker, R. Schilling, L. Schnupp, W. Winkler, K. Maischberger, A. Rüdiger, Phys. Rev. D, 38, 423 (1988)

2. J.Y. Vinet, B. Meers, C.N. Man, A. Brillet, Phys. Rev. D, 38, 433 (1988)

3. the VIRGO group, in Proceedings of the 5th Marcel Grossman Meeting, in press (World Scientific)

4. C.N. Man, A. Brillet, Optics Lett., 9, 333 (1984)

5. D.Shoemaker, A. Brillet, C.N. Man, O. Crégut, submitted to Optics Lett.

6. L. Hollberg, to be published in this volume

Relativity, Pulsar Time, and Atomic Time

J.H. Taylor

Joseph Henry Laboratories and Physics Department,
Princeton University, Princeton, NJ 08544, USA

Imagine a solar mass of self-gravitating nuclear matter in free space, with a radius of 10 km and a frozen-in magnetic field of 3×10^8 gauss, spinning at hundreds of revolutions per second and emitting a highly directive beam of broadband radio noise easily detectable over interstellar distances. Eight such millisecond pulsars have now been detected in our Galaxy, and many more undoubtedly exist. The remnants of supernovae which probably exploded 10^8 y or more ago, these pulsars are fascinating objects for a host of astrophysical reasons. In this paper I will focus on one particular characteristic which they all appear to share: they are natural clocks of astonishing stability, apparently exceeding that of the very best man-made frequency standards. This trait makes them exquisite tools for use in diverse experiments in cosmology, relativity, fundamental astrometry, and metrology.

Experiments in which "pulsar time" is compared with terrestrial atomic time are sensitive to a number of relativistic nuances that can arise when widely separated clocks are compared. Motion of the Earth at varying speeds through the complex gravitational potential of the solar system causes small but detectable special- and general-relativistic effects on terrestrial clocks. Many of the fastest pulsars are themselves members of binary star systems, and some have orbital characteristics such that higher order relativistic effects, including gravitational radiation, are dynamically important. Millisecond pulsars as a class appear to be "clean" systems, free of complicating, hard-to-evaluate astrophysical effects. They are, in short, the answer to an experimental relativist's prayer.

Timing observations of millisecond pulsars are invariably signal-to-noise limited, so they are carried out with the largest available radio telescopes. Steep RF spectra make the sources hard to observe much above 2000 MHz, and interstellar propagation conditions and galactic background noise become annoyingly troublesome below 300 MHz, so observations are generally made somewhere within this frequency range. Since the received signals have a detected envelope that is periodic at the neutron star's rotation

Frequency Standards and Metrology Editor: A. De Marchi

Table 1. Analysis of pulsar timing data.

1. Determine topocentric TOA's by template matching.
2. Apply clock corrections to yield TDT for each datum.
3. Transform to TDB at solar system barycenter.
4. Solve for α, δ, ϕ_0, P, $\dot{P}$; possibly also $\dot{\alpha}$, $\dot{\delta}$, orbital parameters.
5. Analyze residuals for signs of clock errors, gravitational waves, ...

frequency, synchronous signal averaging is used to accumulate an average waveform over convenient intervals of a few minutes.

Analysis of the data begins by following a procedure similar to that outlined in Table 1. The first step involves "template matching" to determine the phase offset of an average profile relative to its accurately recorded starting time. The offset is added to the start time to yield a topocentric pulse time of arrival, or TOA, according to the observatory's master clock. GPS common-view observations are used to convert these TOA's to an atomic timescale based on a more stable standard, such as UTC(NBS), and published tables of offsets between the timescales kept at major timekeeping laboratories around the world can be used to transform to other standards as well. A weighted average of the best independent timescales—such as the TT(BIPM88) scale computed at the *Bureau International des Poids et Measures*—represents the best currently available approximation to an ideal implementation of terrestrial dynamical time, or TDT. This is the reference standard of choice for the most precise comparisons between pulsar time and atomic time.

The observed TOA's must be analyzed relative to an inertial reference frame, for which the solar system barycenter is an adequate approximation. Transformation from TDT to barycentric dynamical time, or TDB, must be done with full general relativistic rigor [3,4]. An iterative least-squares solution is performed to obtain optimized estimates of the pulsar's celestial coordinates, α and δ, as well as the period P, spin-down rate $\dot{P}$, and reference rotational phase ϕ_0. If the pulsar's angular proper motion is significant, one must also solve for the rates of change of α and δ. If the pulsar is a member of an orbiting binary system, the parameters of the orbit are extracted from the timing data as well.

The analysis procedure is straightforward in principle, although complicated in implementation, and can yield remarkable precisions for the measured quantities. A principal reason for the high precision is that the rotational phase of the pulsar can be measured to 10^{-3} of a turn or better, and predicted with comparable accuracies over intervals as long as years. Therefore, phase-coherent solutions are possible spanning many years of observations and perhaps 10^8 to 10^{11} rotations of the neutron star. In such solutions the integer pulse number corresponding to each observed TOA is determined unambiguously. Consequently the fractional precision for the pulsar rotation frequency, for example, can be one part in 10^{14} or better.

It is worth emphasizing that such levels of precision exceed the accuracy of available standards implementing the definition of the second. In addition, the precise numbers obtained depend on such things as the convention used to define a constant gravitational-potential platform for a standard solar system clock, and on the details of the model solar system underlying the numerical ephemeris used to describe the Earth's motion. In due course, the pulsar data will probably be used to improve various solar-system parameters and thus the resulting ephemeris.

For the topics being discussed at this symposium, the most interesting results of pulsar timing observations are those derived from a series of measurements of the millisecond pulsars PSR 1855+09, 1937+21, 1953+29, and 1957+20 carried out with the 305 m radio telescope at the Arecibo Observatory in Puerto Rico [1,2,6]. State-of-the-art measurements at intervals of about two weeks have yielded arrival-time accuracies of 1.2, 0.3, 8.6, and 1.0 μs, respectively, for these four pulsars. Over the available data spans, which now range from 5 months for PSR 1957+20 to 5.8 y for PSR 1937+21, no intrinsic instabilities have yet been detected in any of the pulsar clocks.

Two general relativistic effects that are readily detected when comparing pulsar time with terrestrial atomic time are illustrated using portions of the PSR 1937+21 data in Fig. 1. At the top I show the effect of intentionally neglecting the "Shapiro time delay" of the signals as they propagate through the solar gravitational potential [7,8]. The effect has a ne year period but but is non-sinusoidal, and thus is readily distinguishable from variations caused by a small error in the pulsar's celestial coordinates. From these data the "parametrized post-Newtonian" quantity γ, which has the value 1.0 if general relativity is correct, is experimentally found to be 1.03 ± 0.05.

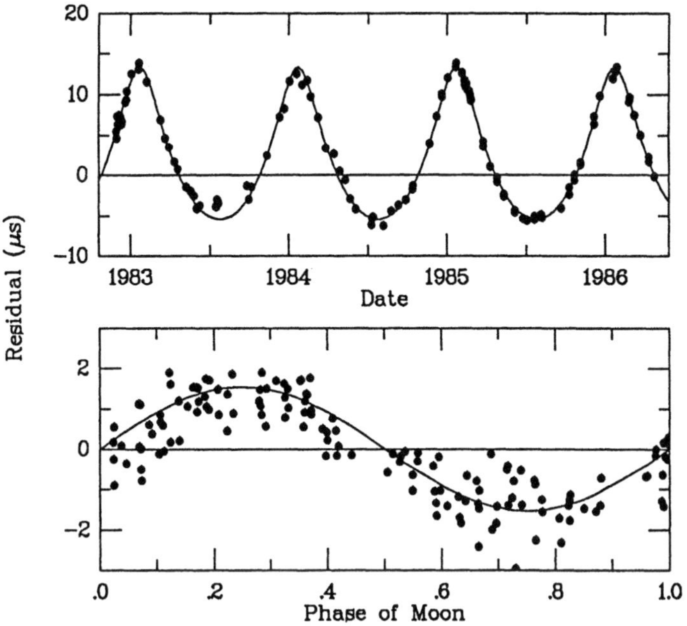

Figure 1: Top: the Shapiro time delay for PSR 1937+21; bottom: monthly part of the variable gravitational redshift of terrestrial clocks relative to PSR 1937+21. Smooth curves illustrate the expected effects in general relativity.

In a similar manner, the bottom panel of Fig. 1 illustrates the difference between pulsar time and terrestrial atomic time when the monthly portion of the variation in gravitational redshift is intentionally omitted. Near the time of full moon the Earth is closer to the sun, and hence deeper in its gravitational potential, so that all clocks on Earth run more slowly. With PSR 1937+21 used as a standard, the observed effect has an amplitude of about 1.5 μs, and agrees with general relativistic expectations [5] to within a factor 1.04 ± 0.06.

Since an equipment upgrade in late 1984, the Arecibo timing data have been of uniform quality. Post-fit residuals for the two best-studied millisecond pulsars are plotted *vs.* date in Fig. 2. For sake of comparison the data for PSR 1937+21 are plotted twice: at the top with respect to UTC(NBS), and in the center with respect to TT(BIPM88), which arguably represents the "world's best clock" over the range of dates in question. The effects of small frequency changes in the steered timescale UTC(NBS) are easily seen. As shown by Rawley *et al.* [1], over averaging intervals greater than about six months some of the best available man-made time standards are demonstrably less stable than millisecond pulsars.

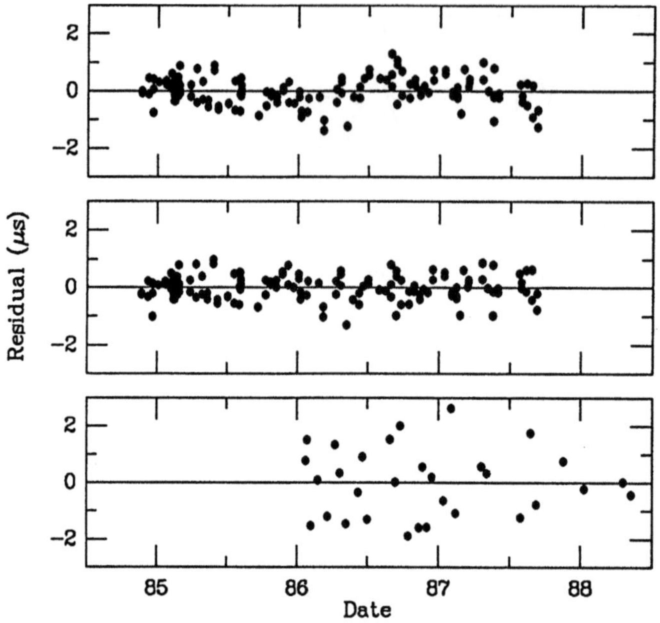

Figure 2: Post-fit arrival time residuals for PSR 1937+21 relative to UTC(NBS) (top) and TT(BIPM88) (center); and for PSR 1855+09 (bottom).

Some important applications of millisecond pulsar timing data, such as its use for placing limits on the energy density of a cosmic background of gravitational radiation [1], are in danger of being limited by inadequate long-term stability of available frequency standards. I take advantage of this opportunity to urge the timekeeping community to redouble its efforts toward further progress in building more stable clocks. Let us all hope that some of the fascinating work discussed at this Symposium will help toward rectifying the present inadequacy!

References

[1] L. A. Rawley, J. H. Taylor, M. M. Davis, D. W. Allen: *Science*, **238**, 761 (1987)

[2] L. A. Rawley, J. H. Taylor, M. M. Davis: *Astrophys. J.*, **326**, 947 (1988)

[3] R. N. Manchester, J. H. Taylor, *Pulsars* (Freeman, San Francisco 1977), chapter 6

[4] D. C. Backer, R. W. Hellings: *Ann. Rev. Astron. Astrophys.*, **24**, 537 (1986)

[5] B. Y. Gelfand: unpublished Princeton University Junior Paper (1988)

[6] A. S. Fruchter, D. R. Stinebring, and J. H. Taylor: *Nature*, **333**, 327 (1988)

[7] I. I. Shapiro, *Phys. Rev. Letters*, **13**, 789 (1964)

[8] J. H. Taylor, in *General Relativity and Gravitation*, ed. M. A. H. MacCallum (Cambridge, Cambridge University Press), p. 209

Testing Spatial Isotropy Using Optically-Pumped ^{199}Hg and ^{201}Hg

S.K. Lamoreaux

Department of Physics, University of Washington, Seattle, WA 98195, USA

Using ^{201}Hg ($^1S_0$, $I = 3/2$) atoms, we have developed a technique to studying static quadrupole interactions of nuclei in $^1S_0$ atoms/1/. Using this technique, we have set limits on how the ^{201}Hg nuclear Zeeman levels depend on the orientation of the quantization axis in space with the null result $\Delta f < 5\times10^{-7}$Hz, which reduces previous limits on possible local Lorentz invariance (LLI) violating interactions by over three orders of magnitude.

According to the Einstein equivalence principle (EEP), a measurement made in a reference frame which is freely falling and is shielded from external non-gravitational interactions is not sensitive to gravitational fields generated by sources outside that reference frame (except for tidal effects which can be made neglible by making the apparatus small enough; a "local" experiment). The EEP is the foundation on which all metric theories of gravity are based/2/ and requires, for example, that the outcome of any local non-gravitational experiment is independent of the velocity and orientation of the freely falling measurement apparatus (LLI). LLI implies that two different atomic clocks located at the same point in space-time will have relative rates that are independent of the orientation and velocity of the local Lorentz frame.

A possible LLI violation which has been studied extensively is whether the mass-energy of a nucleus depends on its orientation in space. For example, in the $TH\epsilon\mu$ model/3/, massive (material) objects and photons propagate with different limiting velocities c_0 and c respectively; a measure of the breakdown of LLI is the parameter $\eta = |1 - (c_0/c))^2|$. This leads to a coupling between the internal structure of the nucleus and its center-of-mass motion, generating a quadrupole energy shift/4/

$$\delta E = \eta \frac{(Z-1)e^2 Q}{6R^3} \frac{V^2}{c^2} P_2(\cos\beta) \tag{1}$$

where Z is the atomic number, Q the nuclear quadrupole moment, R the nuclear charge radius, V the velocity relative to some preferred frame, and β is the angle of the quantization axis relative to that preferred frame. Possible dipole interactions include coupling of the nuclear spin to a gravitational field/5,6/

Frequency Standards and Metrology Editor: A. De Marchi

$$U_1 = \alpha_1(gh/2c)\vec{I}\cdot\vec{R}/R, \quad U_2 = \alpha_2(gh/2c)\vec{I}\cdot\vec{V}/c \tag{2}$$

where g is the gravitational acceleration of the local frame due to a source labeled by position and velocity $\vec{R}$ and $\vec{V}$. These lead to a direction-dependent energy shifts of the form $P_1(\cos\beta)$.

Early tests of nuclear spin coupled spatial anisotropy were carried out by Hughes and collaborators/7/ and Drever/8/, who searched for a variation in quadrupole splitting among the Zeeman levels of ^{7}Li (I= 3/2) as the quantization axis rotated with the earth, and obtained an upper limit of 2×10^{-2}Hz. More recently, a measurement on trapped $^9Be^+$ ions lowered the experimental limits on anisotropy to approximately 10^{-4}Hz /4/.

In our experiment we compare the nuclear spin-precession frequencies of optically pumped ^{201}Hg (I= 3/2) and ^{199}Hg (I= 1/2) isotopes contained in the same cell, and look for a relative shift as the orientation of the precession axis relative to the fixed stars changes with the earth's rotation (our two "clocks"). We expand the frequency shift of a nuclear Zeeman level due to small spatially anisotropic coupling as:

$$\delta\nu(m,I) = \sum_{k=1}^{2I} \nu_k(I) T_0^k(I)_{mm} P_k(\cos\beta) \tag{3}$$

where m is the projection of the spin I along the axis of quantization, $T_0^k(I)$ are multipole spin operators/9/ normalized to $\mathrm{Tr}[T_0^k, T_0^{k\dagger}] = 1$; $T_0^1(\frac{1}{2})_{mm} = \sqrt{2}m$, $T_0^1(\frac{3}{2})_{mm} = m/\sqrt{5}$ and $T_0^2(\frac{3}{2})_{mm} = (3m^2 - \mathrm{I}^2)/6$ and where $P_k(\cos\beta)$ are Legendre polynomials with β the angle of quantization relative to some assumed preferred direction in space. The quantization axis of our apparatus rotates with the earth parallel to the equatorial plane.

Our results for the dipole and quadrupole shifts from (3) are:

$$\delta\nu_1 \equiv [\nu_1(1/2) + .85\nu_1(3/2)]\cos\theta = 1.03 \pm .82\mu\mathrm{Hz}$$

$$\delta\nu_2 \equiv \nu_2(3/2)\cos^2\theta = .17 \pm .19\mu\mathrm{Hz} \tag{4}$$

where θ is the latitude of an assumed anisotropy axis which we leave unspecified since we did not resolve any effect.

Our experiment uses two optically-pumped light absorbtion oscillators based on the nuclear spin Larmor frequencies of the two Hg isotopes/10/. Each oscillator operates with the geometry shown in Fig. 1. Circularly polarized resonance light propagates along the z-axis. At an angle of 45° is a static magnetic field B_0 (the quantization axis) which lies in the yz plane, (the earth's equitorial plane.) Along the x-axis is an oscillating magnetic field whose frequency, ω, is close to the Larmor frequency ω_0. Through optical pumping, the nuclear spins become polarized; the ex-

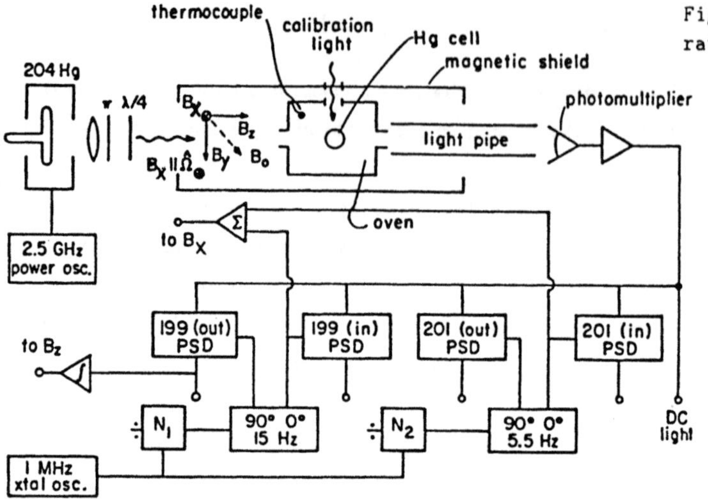

Fig 1 Schematic of the apparatus

pectation value of the spin $\langle\vec{I}\rangle$ precesses at frequency ω in a cone about $\vec{B}_0$. The projection of $\langle\vec{I}\rangle$ on the z-axis has a static and an oscillating component. Since the absorption of the light depends on $\langle I_z\rangle$ the light will acquire a modulation at ω. For $\omega = \omega_0$ (and in the absence of additional phase shifts), the modulation will be exactly in phase with B_x. For small deviations of ω from ω_0 there will be a phase shift between B_x and the light modulation of magnitude $\phi \approx T_2\Delta\omega$, where T_2 is the transverse spin relaxation time.

Averaging the shift in (3) over all Zeeman levels, it is evident that the dipole sensitivity is proportional to $\langle T_0^1\rangle$, the vector spin polarization of the atoms, while the quadrupole sensitivity is proportional to $\langle T_0^2\rangle$, the spin alignment of the atoms. These sensitivities can be calibrated under operating conditions by applying known dipole and quadrupole perturbations as described below. An important conceptual difference between our technique to detect external quadrupole interactions and the techniques of previous attempts is that in our case neither the Zeeman transitions nor the quadrupole shifts need be resolved. Such sub-linewidth quadrupole sensitivity arises from the presence of a non-zero spin alignment.

As seen in Fig. 1, the pumping light for both isotopes comes from a stabilized ^{204}Hg electrodeless microwave discharge lamp. The single 253.7 nm line of the ^{204}Hg overlaps the F= 1/2 to F= 1/2 transition in ^{199}Hg and the F= 3/2 to F= 5/2 transition in ^{201}Hg. The cell containing the mercury is a 2 cm diameter sphere made of high purity fused silica (Suprasil) which was filled with mercury vapor (16% ^{199}Hg, 13% ^{201}Hg) at the 0°C vapor pressure. To achieve long spin relaxation times, the cell is maintained at 300°C in a temperature regulated oven in the center of three Helmholtz coils which provide the static and oscillating magnetic fields. The static field has a magnitude of about 20 mG, corresponding to Larmor frequencies for ^{201}Hg and ^{199}Hg of 5.5 Hz and 15 Hz respectively.

The vector spin polarization lifetime (T_2) for the ^{199}Hg is 1120±40 seconds and for the ^{201}Hg is 80±5 seconds. The ^{201}Hg spin alignment lifetime is 40±5 seconds. With the resonance light optically pumping the ensemble of atoms the ^{199}Hg lifetime is shortened to 80 seconds and the ^{201}Hg lifetime (vector) is shortened to 50 seconds. The amplitude of the oscillating field for each isotope was set so that the in-phase components were 0.7 times their respective limiting values. For the ^{201}Hg this corresponds to about 19 μG and for the ^{199}Hg about 6μG.

The ^{199}Hg quadrature signal is used to stabilize the magnetic field by adjusting the current in the z-axis Helmholtz coil to hold the ^{199}Hg quadrature signal at zero. The ^{201}Hg quadrature signal is then insensitive to fluctuations in either the magnetic field or the reference oscillator. Thus the ^{201}Hg quadrature signal V_{201}^{out} becomes a measure of any shift between the Zeeman energy levels of the two isotopes that does not scale as the ratio of the nuclear magnetic moments. In particular, V_{201}^{out} is sensitive to a quadrupole energy shift in ^{201}Hg since a quadrupole cannot affect the energy levels of ^{199}Hg. Because the ^{199}Hg servo also forces $\nu_1(\frac{1}{2})$ in (3) to be zero, V_{201}^{out} is sensitive to an anisotropic dipole energy shift given by $\nu_1(\frac{3}{2}) - \sqrt{10}(\gamma_{201}/\gamma_{199})\nu_1(\frac{1}{2})$ where $\gamma_{201}/\gamma_{199} = -.37$ is the ratio of the nuclear magnetic moments.

The sensitivity of the apparatus to dipole and quadrupole energy shifts is calibrated experimentally. The dipole sensitivity is measured by changing the ^{201}Hg drive frequency by a small amount ($\Delta\nu$=120μHz) and measuring the resulting change in V_{201}^{out}. The quadrupole sensitivity, is determined by applying a known quadrupole light shift using light from a second lamp containing enriched ^{200}Hg. This isotope emits a single 253.7 nm. line far enough from resonance with either pumped isotope to induce readily measurable quadrupole and dipole light shifts without seriously affecting the pump rates. The ratio of these virtual light shifts for ^{201}Hg can be calculated accurately/9/, and thus a measurement of them determines the ratio of quadrupole sensitivity to the measured dipole sensitivity which is typically 0.7.

A computer records the in-phase and quadrature components of the transmitted light modulation for both isotopes, as well as the lamp intensity, magnetic field correction signal, oven drive current and room temperature. We have taken six runs, each of 2.5 days continuous duration. The direction of $\vec{B}_o$ was changed by 90° for half of the data to discriminate against systematic effects associated with the time of day. By multiple regression we then fit the ^{201}Hg phase, $(V_{out}^{201}/V_{in}^{201})$ to sine and cosine components at $\omega_\oplus$ and $2\omega_\oplus$ where $\omega_\oplus$ is the earth's rotation frequency in side-

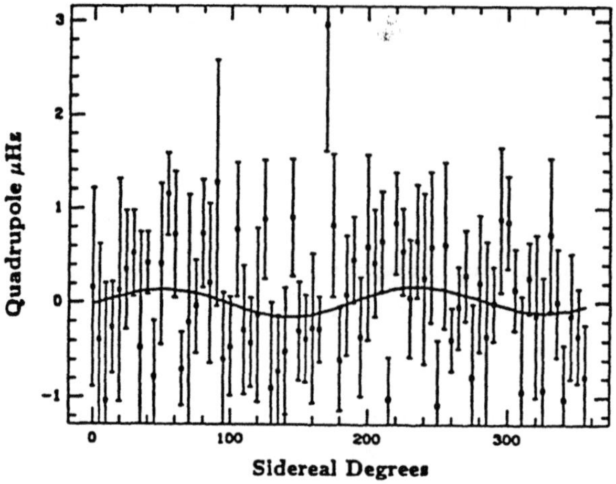

Fig. 2 Results of the entire data set averaged into 5° bins modulo one sidereal day (after subtracting instrumental drifts)

real time. We also remove various instrumental drifts by use of a simple linear model which takes into account, for example, changes in the light intensity. Figure 2 is the result for all of the runs (after subtracting instrumental drifts) averaged into 5° bins modulo one sidereal day.

We have achieved a sensitivity of better than 10^{-6}Hz to Zeeman frequency shifts. Using (1), (2), and (3), we can set the following limits. Our null result, combined with the relatively large quadrupole moment of ^{201}Hg, further improves limits on anisotropic quadrupolar coupling by a factor of 2500, constraining, for example, $\eta \leq 10^{-21}$. We can also set limits of $\alpha_1 < 3 \times 10^5$ and $\alpha_2 < 3 \times 10^9$. Although we expect the α's to be of order unity, these limits are nonetheless currently the best for such couplings to intrinsic spin.

I gratefully acknowledge the assistance of B.R. Heckel and J.P. Jacobs, and of E.N. Fortson who suggested this work and provided guidance throughout.

1. S.K. Lamoreaux et al., Phys. Rev. Lett. 57, 3125 (1986).
2. C. M. Will, Theory and Experiment in Gravitational Physics, Cambridge University Press, Cambridge, 1981.
3. M. P. Haugan, Ann. Phys. 118, 156, (1979), A. P. Lightman and D. L. Lee, Phys. Rev. D 8, 364, (1973).
4. J. D. Prestage, J. J. Bollinger, W. M. Itano, and D. J. Wineland, Phys. Rev. Lett. 54, 2387, (1985).
5. J. Leitner and S. Okubo, Phys. Rev. 136, B1542, (1964), N. D. Hari Dass, Phys. Rev. Lett. 36, 393, (1976).
6. P. R. Phillips and D. Woolum, Nuovo Cimento 64B, 28, (1969).
7. V. W. Hughes, H. G. Robinson, and V. Beltran-Lopez, Phys. Rev. Lett. 4, 342, (1960).
8. R. W. P. Drever, Philos. Mag. 6, 683, (1961).
9. W. Happer, Rev. Mod. Phys. 44, 169, (1972).
10. H. G. Dehmelt, Phys. Rev. 105, 1924, (1957), and A. L. Bloom, Appl. Opt. 1, 61, (1962).

Squeezed States, Interferometric Limits and Back-Action Evasion

R.E. Slusher, A. La Porta, B. Yurke, and P. Grangier [1]

AT&T Bell Laboratories, Murray Hill, NJ 07974, USA
[1] Permanent address: Institut d'Optique Théorique, France

Quantum fluctuations in the electromagnetic field are described as limits to fundamental physical measurements. Squeezed light experiments now allow interferometric measurements beyond the shot-noise limit. A dark fringe polarization interferometer experiment in progress is reported. A simple scheme to obtain back-action evasion in quantum metrology is also described based on a single pass parametric downconverter.

I. Introduction

At the limit of all physical measurement is the quantum uncertainty of the measurement process. It appears that at present none of the measurements of the fundamental constants or standards are at this quantum metrology limit. However, this limit is being approached in experiments such as laser stabilization where locking loops operate near the shot-noise limit. Shot noise can be interpreted as originating from the "zero-point" fluctuations in harmonic oscillator model of the electromagnetic field. These zero-point or vacuum fluctuations can be amplified *and de-amplified* by parametric processes to form squeezed states of the field [1,2]. One phase quadrature of a squeezed vacuum field has a variance or fluctuation level less than the normal vacuum while the other phase quadrature has a correspondingly larger variance than the vacuum in order to satisfy the uncertainty principle. Large squeezed noise reduction, greater than a factor of two, can now be obtained routinely in the laboratory [3]. This new type of light allows measurement precision beyond the shot-noise limit [4,5]. For example, it could be used to enhance interferometric gravitation wave antennae or for laser stabilization. Variations of the parametric amplification process used to generate squeezed light also allow back-action evasion [6] of noise inherent in almost all measurement processes. Both interferometric enhancement and a simple back-action evasion experiment are briefly discussed here.

It is also interesting to consider the quantum limits to simultaneous measurements of a set of fundamental constants or standards. For example, it is possible to decrease the variance for a quantum limited measurement of the period of a light source by using squeezing techniques.

Frequency Standards and Metrology Editor: A. De Marchi

However, the squeezing process in this case would increase the measurement error in the amplitude of the light. It would be interesting to study similar constraints for measurements of the complete set of standards and fundamental constants.

II. Interferometric Limits

As described by Caves [7], interferometric measurement is limited by vacuum fluctuations entering the normally ignored "dark" input port of the interferometer. These vacuum fluctuations limit the measurement of the phase ϕ to a variance $\Delta\phi$ of $1/\sqrt{\eta N}$ where η is the net efficiently of the optical system including optical losses and detector efficiency and N is the number of photon detection events in the measurement time interval. This is typically referred to as the shot-noise limit. If squeezed vacuum is injected at the dark input port with a reduction in one phase-quadrature field variance, $\Delta E_1^2 < 1$ (where one is the vacuum level), the phase variance is reduced by a factor $(1-\eta+\eta\Delta E_1^2)^{1/2}$. Realistic values of $\eta = 0.9$ and $\Delta E_1^2 = 0.1$ result in a reduction in the phase variance by a factor of 2.3 at fixed input power and measurement interval. This is equivalent to reducing the input power or measurement interval by a factor of $(1-\eta+\eta\Delta E_1^2)$ or nearly 5 for the assumed parameters. This factor may be very important in experiments where the ultimate sensitivity is required (e.g. gravitation wave detectors) or where there are limits on power and time interval (e.g. studies of dynamics in biological systems).

Squeezed state enhancement of interferometric measurement has been demonstrated for both Mach-Zehnder [4] and polarization [5] interferometers by using balanced detection schemes and operating at the side of a fringe. In these experiments the input laser power is limited by the maximum power the detectors can withstand before burning out, typically less than a few milliwatts for an efficient photodiode. Clearly one should operate at a dark fringe so that the detector only sees the small imbalance in light due to the measured phase shift and the exciting laser passes directly through the interferometer. This arrangement also has the advantage that the exciting laser light, having passed through the interferometer, can be recycled [8] in a loop to further enhanced the interferometer's sensitivity. Can the ultimate squeeze enhanced interferometer sensitivity $\Delta\phi = [(1-\eta+\eta\Delta E_1^2)/\eta N]^{1/2}$ be obtained in this dark fringe case ($N = P\tau$ where P is the power including the recycled light)? The answer is a qualified "yes". Gea-Banacloche and Leuchs [9,10] have analyzed the dark fringe case in detail. They point out that imperfections in the interferometer, e.g. wave front distortions or imbalance between the two arms, quickly destroy the squeezed state enhancement.

One might ask if there is any point to using squeezed states for interferometric enhancement. Why not use a many pass Fabry-Perot system where the light simply bounces back and forth until it is lost due to mirror imperfections or coating losses? The sensitivity is very high because of the high finesse. In this case squeezed states won't help because the net loss is very high. However, the time response of this system is slow, dominated by the many passes of light through the system. It is possible to regain the sensitivity *and* decrease response time by using a lossless system with fewer passes and squeezed-light at the dark input port. This combination of improved response time and high sensitivity is crucial in some experiments, such as the gravitational-wave detector.

We are presently investigating the squeeze state enhanced dark fringe case experimentally. As in our previous studies [5,11] we are using a polarization interferometer and squeezed vacuum generated at a wavelength of 1.064 μm in a cavity containing a KTP crystal used as a parametric downconverter. Squeezed noise reduction in one phase quadrature as large as 3dB or 50% in power has been obtained as shown in Fig. 1. The reduction in the measured noise level is limited in this case by detector efficiency and optical losses.

The ultimate sensitivity can be obtained in the dark fringe case by using a phase modulator [9,10] in the interferometer as shown in Fig. 2. We use an electro-optic modulator driven at 400kHz to produce a peak phase modulation near 0.15 radians. With 100 mW laser input into the bright port of the interferometer this phase modulation around the dark fringe results in a power near 1 mW at the detector, enough for it to operate in the shot-noise limited regime. In this configuration the interferometer operates near the shot-noise limit as shown in Fig. 3. With 100 mW at the input and averaging times near 1 ms this interferometer is extremely sensitive to effective polarization rotations in the nanoradian region. We have had great difficulty in actually obtaining the shot-noise limit in preparation for studying the squeezed light enhanced operation beyond this limit. In Fig. 3 the peak above the shot-noise level is due to angular motion of the incident beam. The peak shifts in frequency with the length of the laser rod indicating it is probably due to weakly excited mechanical bending modes of the laser rod. Over a wide frequency range, the fluctuations above the shot-noise level occur intermittently with a characteristic time near one second. We think this noise can be suppressed by a low frequency feedback loop to the voltage on the electrooptic modulator. If this is successful we will proceed to study the squeezed light enhancement.

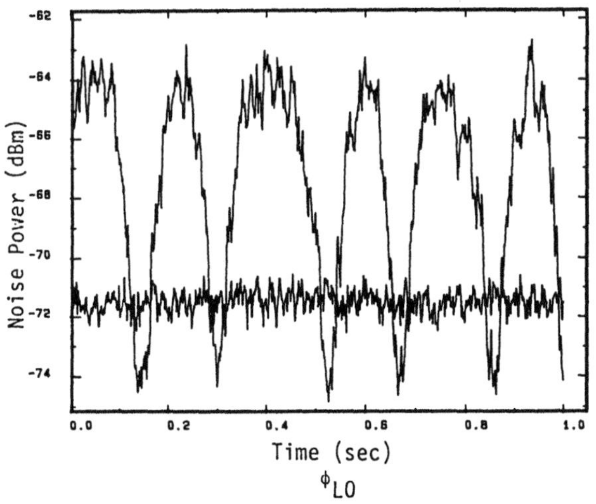

Fig. 1 - Squeezed noise, varying from +8 dBm to -3 dBm above and below the shot-noise level (flat trace), from a CW cavity squeezer. Both time and local oscillator phase (ϕ_{LO}) are swept along the x-axis. The phase change between adjacent noise minima is π. Sweep time is 1s, rf bandwidth is 30 kHz, video bandwidth is 100 Hz and the center frequency is 486 kHz.

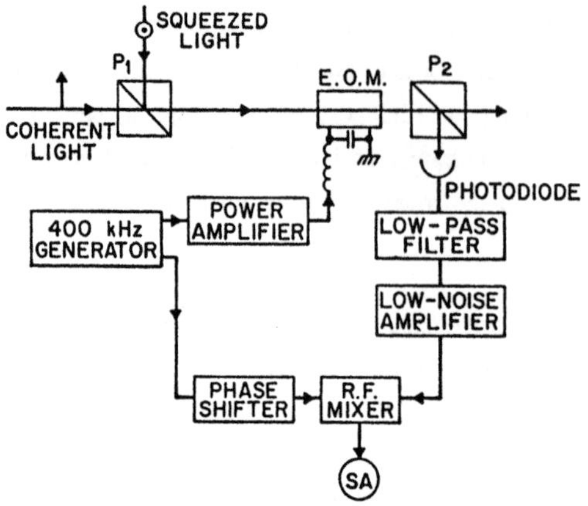

Fig. 2 - A schematic diagram of the apparatus for a shot-noise limited polarization interferometer operating at a dark fringe. Coherent light (~100 mW) enters one port of the interferomcter formed by polarization beamsplitters P_1 and P_2 while squeezed light enters the normally dark port. An electrooptic modulator (EOM) produces a 1 mW signal near 400 kHz at the "dark" output port photodiode. The photocurrent is filtered to remove the strong spectral component near 800 kHz, amplified and mixed with a phase shifted component of the 400 kHz generator. The noise out of the mixer is analyzed on a spectrum analyzer (SA).

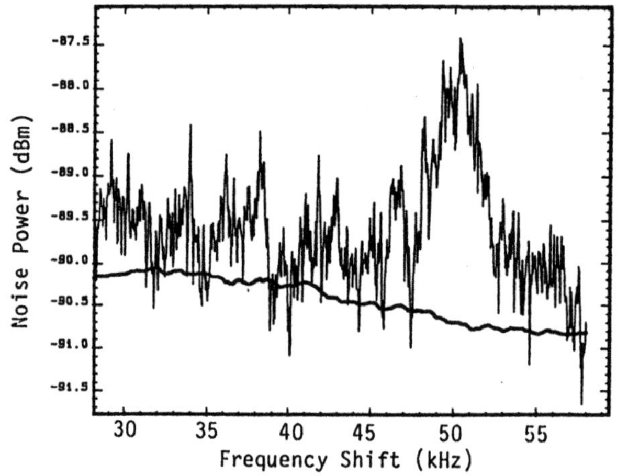

Fig. 3 - Noise spectrum at the dark fringe compared to the shot-noise level (smoothed trace). The sweep time is 20s. Random excursions above the shot-noise level occur with a characteristic time of a few seconds. The rf bandwidth is 1 kHz and the video bandwidth is 3 Hz. Frequencies are shifts from the EOM driving frequency at 560 kHz.

III. Back-Action Evasion

In most measurement processes the measured system is perturbed so that a second measurement of the same parameter will have an increased variance. A simple example is an optical fiber tap. Consider a simple fiber tap where two fibers are mode coupled over a small distance so that a fraction of the input in one fiber is transferred to the second fiber by exchange between the modes in the interaction region. The amplitude or phase of the incident signal is measured by tapping out a fraction of it into the second fiber. However, in the exchange process vacuum fluctuation noise from the second fiber input ‘‘contaminants’’ the signal transmitted in the first fiber after a portion is tapped out for measurement. The back-action in this case is the vacuum fluctuation contamination of the measured and transmitted signals. Shapiro [12] has suggested that this contamination can be reduced by using squeezed vacuum at the input of the second fiber. In this case it is easy to show a remarkable result; as the squeezing is increased the signal/noise ratio for both the transmitted *and* measurement beams approaches that of the input. Of course, the transmitted signal is reduced by the fraction of the tap, but in the limit that the tapping fraction is small this is a back-action evasion or quantum nondemolition measurement. Back-action evasion is the term used for quantum nondemolition when phase quadrature measurement is involved.

Yurke [6] has suggested a simple method to achieve back-action evasion measurements. A parametric amplifier replaces the passive optical tap in the example above. The paramp is preceded and followed by optical converters. We have implemented a Yurke evader by using half-wave plate polarization rotators as the converters and a single-pass KTP crystal operating as a downconverting amplifier as the paramp. The pump wavelength is 0.53 μm and the signal and idler are orthogonally polarized (type II phase matching) at a wavelengths near 1.064 μm. Sufficient single pass gain (3-4 dB) is obtained by using a pulsed pump. The experimental arrangement is shown in Fig. 4. A portion of the input beam is converted by the first rotator (a half-wave plate) R1 to the orthogonal polarization. Both polarizations are parametrically amplified in the KTP crystal. A quarter-wave plate CP after the KTP crystal compensates for birefringence in the crystal. The second converter R2 is also a half-wave plate which recombines the amplified signal and idler beams so as to recreate the input beam with no noise contamination in one phase quadrature. The orthogonal polarization exiting the second converter is the readout beam which contains the measured information from the input beam. Both the readout beam and the recreated incident beam are monitored by balanced homodyne detectors. A correlation of these two detector signals demonstrates the back-action evading measurement. Preliminary experimental results show that the correlations predicted for back-action evasion are obtained. A detailed analysis of this experiment will be published elsewhere.

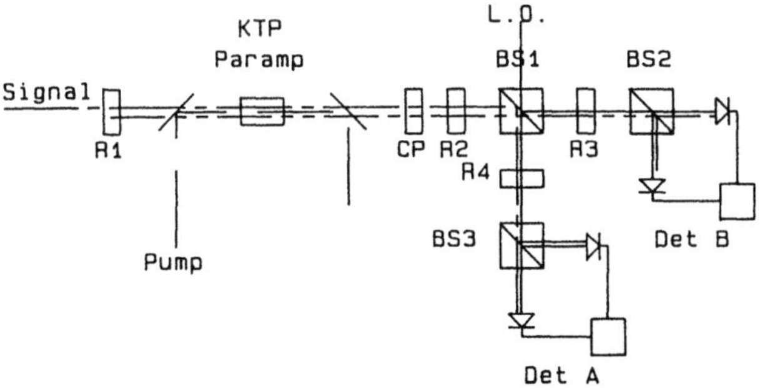

Fig. 4 - Vertically polarized light enters the signal port and horizontally polarized vacuum fluctuations enter the readout port. R1 rotates the polarization vectors at an angle θ, mixing the inputs. After the action of the KTP parametric amplifier a compensator CP cancels phase shifts that result from the birefringence of the KTP crystal, and R2 again rotates the polarizations by θ. Polarization beamsplitter BS1 separates the vertically polarized signal beam from the horizontally polarized readout beam. The field quadratures of signal and readout are detected by introducing a local oscillator field at BS1. R3 and R4 rotate the polarizations by 45 degrees and BS2 and BS3 mix the LO field with the signal and readout fields on detectors A and B, respectively. The quadratures are selected by adjusting the relative phase of the LO field.

References

1. "Squeezed States of the Electromagnetic Field," J. Opt. Soc. Am. B*4*, 1450-1741 (1987).
2. R. E. Slusher and B. Yurke, Sci. Am. (May 1988).
3. L. A. Wu, H. J. Kimble, J. L. Hall, H. Wu, Phys. Rev. Lett. *57*, 2520 (1986).
4. Min Xiao, Ling-An Wu and H. J. Kimble, Phys. Rev. Lett. *59*, 278 (1987).
5. P. Grangier, R. E. Slusher, B. Yurke and A. LaPorta, Phys. Rev. Lett. *59*, 2153 (1987).
6. B. Yurke, J. Opt. Soc. Am. B*2*, 732 (1985).
7. C. M. Caves, Phys. Rev. D*23*, 1693 (1981).
8. R. W. P. Drever et. al., in *Quantum Optics, Experimental Gravitation, and Measurement Theory*, P. Meystre and M. O. Scully, eds. (Plenum, New York, London, 1983).
9. J. Gea-Banacloche and G. Leuchs, J. Opt. Soc. Am. B*4*, 1667 (1987).
10. J. Gea-Banacloche and G. Leuchs, J. Mod. Opt. *34*, 793 (1987).
11. P. Grangier, R. E. Slusher, B. Yurke and A. LaPorta, Proc. Advanced Research Workshop on Squeezed and Nonclassical Light, Cortina, Italy, 1988; NATO ASI Series.
12. J. H. Shapiro, Opt. Lett. *5*, 351 (1980).

Mass Ratio Spectroscopy and the Proton's Atomic Mass

R.S. Van Dyck, Jr., F.L. Moore, D.L. Farnham, and P.B. Schwinberg

Department of Physics, FM-15, University of Washington,
Seattle, WA 98195, USA

1. Introduction

The traditional compensated Penning trap has been used in the past to precisely measure mass ratios of positrons[1] and protons[2] relative to the electron. Present investigations may allow one to indirectly determine the neutron/proton mass ratio as well as the $^3H - {}^3He$ mass difference to sub-ppb precision. The latter measurement may even have direct impact on the neutrino rest mass experiments[3]. It is also probable that variations of this trap will be used to make precision measurements of the antiproton relative to the proton[4]. Because of the present interest in such mass spectroscopy, a review of the basic method is in order with an emphasis on the use of multiply-charged ions for these measurements with an application to C^{4+}.

As a spectrometer, one has the potential to measure any ion mass relative to a proton or ^{12}C with potential sub-ppb accuracies or in absolute terms, measure electron binding energies with a resolution approaching 1-eV. The advantages of this device are inherent in the very small ion samples required for the measurements, the very long containment times and ionization periods which are possible, the lack of interactions with background gases, and the easily controlled electric and magnetic fields whose effects can be accurately taken into account for ion cyclotron measurements.

The basic compensated Penning trap[5] is a 5-electrode device that consists of two end caps and a ring electrode as hyperboloids of revolution, with guard rings (between each end cap and the main ring electrode) used to compensate for the non-quadratic even-order terms in the trapping potential. In addition, this device is modified by splitting the ring electrode into four equal parts in order to efficiently drive and cool the cyclotron motion as well as to detect the cyclotron fre-

Frequency Standards and Metrology Editor: A. De Marchi

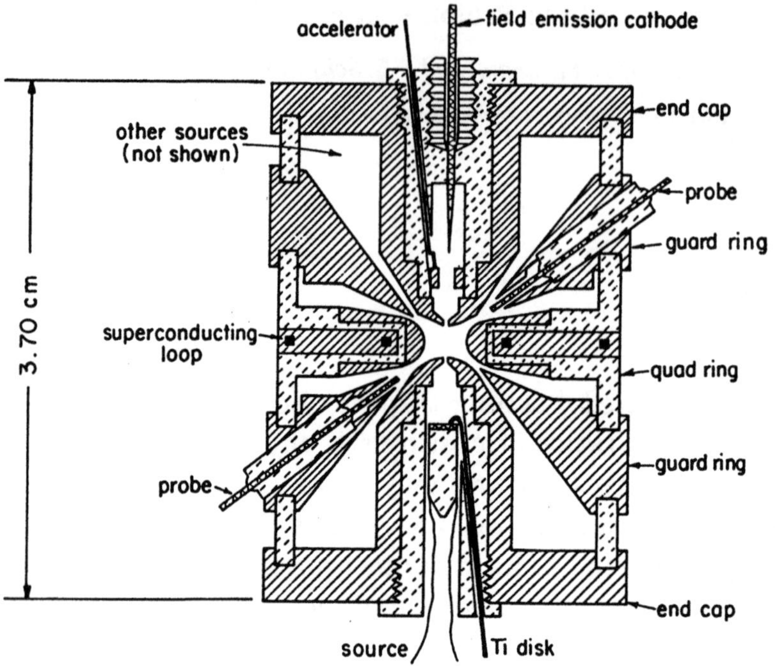

Fig. 1. Detailed trap schematic. In addition to the usual electrodes, the actual trap includes probes for sideband cooling, a superconducting loop for magnetic field diagnostics, a field emission cathode to generate an axial electron beam, and several sources of hydrogen isotopes.

quency of the ion motion directly. A detailed example of a trap with sources and probes is shown in Fig. 1.

2. Ionic Mass ratios

The set of normal mode frequencies for a charged particle in an ideal Penning trap are as follows:

$$\nu_z = \frac{1}{2\pi}\sqrt{\frac{qV_0}{mZ_0^2}}, \qquad \nu_m = \frac{\nu_z^2}{2\nu_c'}, \qquad \nu_c' = \frac{qB_0}{2\pi mc} - \nu_m \tag{1}$$

where B_0 is the applied magnetic field, V_0 is the applied ring-end cap potential, q/m is the characteristic charge-to-mass ratio of a single trapped ion, and $2Z_0$ is the trap's characteristic minimum end cap spacing. As is evident from (1), in principal only the perturbed cyclotron frequency, ν_c', and the axial frequency, ν_z, need be measured in order to obtain the free space cyclotron frequency $\nu_c = \nu_c' + \nu_z^2/2\nu_c'$. Both frequencies can be directly measured using tuned LC circuits as in past work[6]. However, recently we have found it convenient to detect ν_c' through axial frequency shifts (as described in the next section). In practice, all three normal modes are measured in order to obtain the free space cyclotron frequency according to the quadrature invariance equation:

$$\nu_c^2 = \nu_m^2 + \nu_z^2 + {\nu_c'}^2 \tag{2}$$

which is valid for small misalignments of the electronic and magnetic axes or certain asymmetries of the trap electrodes[7]. It thus follows that an ion with charge $q_1 \equiv ne$ (where n is the number of elementary charges) has a mass ratio (relative to the proton) given by

$$\frac{M(ion)}{m_p} = \frac{\nu_c(p^+)}{\nu_c(ion)} n \tag{3}$$

to the extent that both particles are effectively located within the same magnetic field.

3. Axial Resonance Spectrometer

The primary detection system utilizes the fact that ν_z is principally determined by the trapping potential. Thus, any ion can be tuned into "resonance" with an external tuned circuit by adjusting the trapping voltage V_0, assuming $2\nu_z^2 < \nu_c^2$. The bound motion then becomes equivalent to a series lc tuned circuit in parallel with the detection tuned circuit, where $l_1 = (4mZ_0^2)/(\kappa q)^2$ represents the basic single ion inductance which then determines its coupling to the external world, $l = l_1/N$. The quantity N is the number of trapped ions and κ is a geometry factor on the order of unity. The corresponding axial linewidth is subsequently given by

$$\gamma_z = \frac{R}{l} = N \frac{(\kappa q_1)^2 R}{4 m_1 Z_0^2} = N \gamma_1 \tag{4}$$

where R is the effective damping resistor of the tuned circuit resonant at ν_z. Therefore, the axial linewidth scales as the number of trapped ions and as q_1^2/m_1. By driving the axial motion and mixing the driven signal with the appropriately phased drive signal, an error signal is produced which is integrated and fed back to the ring to keep the ion's axial resonance "locked" to the drive oscillator. This correction voltage then represents the frequency shift signal which, in the present ion work, represents the particle's non-centered position via the residual anharmonicity of the trap. In general, the driven ion current, $I = (\kappa q/2Z_0)dz/dt$, generates a signal given by the voltage drop across the tuned circuit. It can also be shown that under ideal conditions and on resonance, this driven signal voltage must always be equal to the amplitude of the drive voltage V_D applied from drive end cap to ring electrode. Since the motion is simply that of a driven harmonic oscillator, it follows that the spacial drive amplitude Z_D will be given by

$$\frac{Z_D}{Z_0} = \frac{2V_D}{(N\kappa q_1)R\omega_z} \tag{5}$$

where $R\omega_z = Q/C \approx$ constant with Q the tuned circuit quality factor and C its total capacitance. This last expression (5) is crucial since typically we drive the ion(s) as hard as possible before anharmonic terms in the potential destroy one's resolution in frequency space. Thus, for each trap, there is some characteristic allowed $(Z_D/2Z_0)_c$ (determined by geometry and the desired frequency resolution) for which:

$$V_{sig} = V_D = (Z_D/2Z_0)_c \cdot (Q/C) \cdot N\kappa q_1 \; . \tag{6}$$

Thus, the maximum undistorted signal increases with N, the number of particles trapped, and with q_1, their state of ionization! Also in practice, a greater anharmonic distortion appears to be tolerable due to the greater axial width as q_1 increases.

4. Proton Atomic Mass

To illustrate the potential precision of this ion mass-ratio spectrometer, we recently completed two runs in which C^{4+} and H^+ ions were alternately stored within the trap. In the first run, a single C^{4+} ion was used and in the second, we had two such ions. In all cases, great care is taken to clean the trap of all extraneous ions which may have been trapped during the loading process. These "garbage" ions tend to make the primary ion cloud unstable in configuration, leading to possible shifts in the normal mode frequencies via the space charge interaction. The preferred method for removing them is a hard (noise broadened) axial drive, swept slowly over their respective axial frequencies.

Figure 2 illustrates the excellent signal-to-noise (S/N) which is possible for the axial resonance of such multiply charged ions. When used in the frequency shift detector, we have adequate sensitivity to measure the magnetron frequency by applying two different end cap drives. The preferred method utilizes the axial sideband at $\nu_z - \nu_m$ since its application reduces the magnetron orbit while yielding ν_m (with typical resolution of $\approx$ 50 ppb). The more sensitive method uses a direct drive at ν_m and for a single ion, a beating is observed between the free magnetron motion and the drive field, yielding a precision better than 10 ppb, but with the disadvantage of increasing orbit size. Cyclotron resonances can be excited by driving two quadrants of the split ring and likewise detected using the locked axial resonance. Figure 3 illustrates

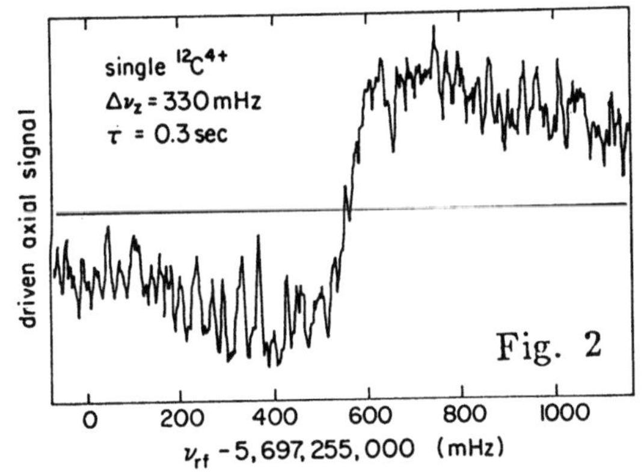

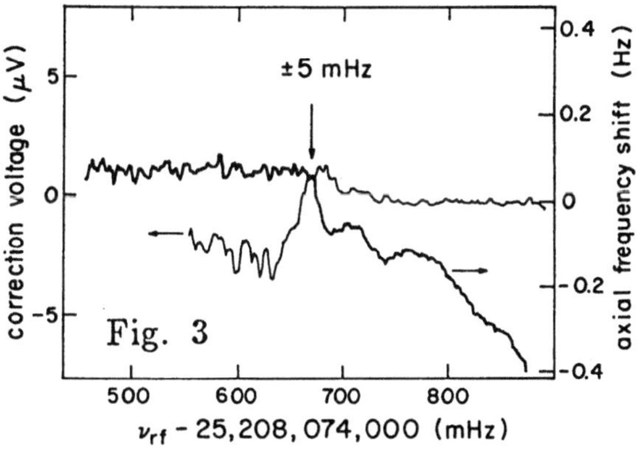

Fig. 2. Single C^{4+} axial resonance. The signal's phase is appropriate for the lock loop to generate a correction voltage which keeps ν_z fixed. S/N is adequate to yield frequency resolution better than 10 ppb.

Fig. 3. Single C^{4+} cyclotron resonance observed, after excitation, via anharmonicity induced axial frequency shifts. Even though the ion was cooled via a $\nu_z + \nu_m$ sideband drive prior to each excitation, S/N of the heavy (second) trace is not as good as the thin (first) trace due to decaying cyclotron energy [τ(decay) $\approx$ 10 m]. (Note 0.2 ppb precision.)

a typical example of such a resonance, but there have been many occasions when the cyclotron resonance's precision has reached the limit of our measurement ability at 0.1 ppb.

Field drift represents the major systematic in the present measurements. To illustrate the problem, we show in Fig. 4 the residual scatter of cyclotron resonances with time (relative to a constant drift) for protons and C^{4+} ions. Another small systematic is associated with ion-number and is produced by the space-charge repulsion of the like-ions. For protons, this effect has been measured to be 0.2 ppb/ion. As for the run with two C^{4+} ions, we estimated a dependence equivalent to 8 protons. Thus, using the quadrature invariance equation (2), we obtain free space cyclotron resonances for each ion and the corresponding mass ratio of $M(C^{4+})/4m_p = 2.977,783,714(10)$.

Since ^{12}C is the standard for the atomic mass tables, one can arrive at the proton's atomic mass through

$$\frac{M(C^0)}{m_p} = \frac{M(C^{4+})}{m_p} - \frac{E_B}{m_p} + 4\frac{m_e}{m_p} \tag{7}$$

where $E_B = 148.019$ eV is the sum of all binding energies for the four liberated electrons. Thus, we conclude that $M_p = 1.007,276,468(3)$ amu compared to the accepted value[8,9] of 1.007,276,470(12) amu. This

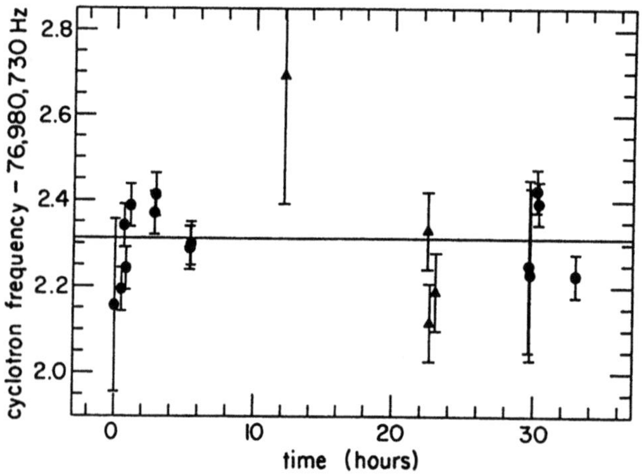

Fig. 4. Free space cyclotron frequency ν_c versus time for protons (circles), corrected for a drift of +3 ppb/h. The C^{4+} resonances (triangles) are also plotted after being scaled by the fitted mass ratio. The solid line represents the weighted average for all data, corrected for drift.

agreement is encouraging for future comparisons at the ppb level, especially when we note that the two runs yielded exactly the same result under widely different conditions.

Acknowledgements

This work is supported by the National Science Foundation under the "Mono-Ion Research" grant.

References

1. P.B. Schwinberg, H.G. Dehmelt, and R.S. Van Dyck, Jr., Phys. Rev. 81A, 119 (1981).
2. R.S. Van Dyck, Jr., F.L. Moore, D.L. Farnham, and P.B. Schwinberg, Bull. Am. Phys. Soc. 31, 244 (1986).
3. J.J. Simpson and P. Vogel, in Low Energy Test of Conservation laws in Particle Physics, 1983, AIP Conference Proceedings, No. 114 (AIP, New York 1984) p. 220.
4. G. Gabrielse, X. Fei, K. Helmerson, S.L. Rolston, R. Tjoelker, T.A. Trainor, H. Kalinowsky, J. Haas, W. Kells, Phys. Rev. Lett. 57, 2504 (1986).
5. R.S. Van Dyck, Jr., D.J. Wineland, P.A. Ekstrom, and H.G. Dehmelt, Appl. Phys. Letters 28, 446 (1976); also G. Gabrielse, Phys. Rev. A 27, 2277 (1983).
6. R.S. Van Dyck, Jr., P.B. Schwinberg, and H.G. Dehmelt, Phys. Rev. D 34, 722 (1986), and New Frontiers in High Energy Physics (Plenum, N.Y., 1978) p. 159.
7. L. Brown and G. Gabrielse, Rapid Communications of Phys. Rev. A 25, 2423 (1982).
8. A.H. Wapstra and G. Audi, Nucl. Phys. A432, 1, (1985).
9. E.R. Cohen and B.N. Taylor, CODATA Bull. No. 63 (Pergamon Press, Oxford 1987) p. 23.

The Physics of Circular Atoms and the Measurement of the Rydberg Constant

M. Gross, J. Hare, P. Goy, and S. Haroche

Laboratoire de Spectroscopie Hertzienne de l'E.N.S.,
24, rue Lhomond, F-75231 Paris Cedex 05, France

INTRODUCTION

Circular atoms are highly excited atoms, i.e. Rydberg atoms (with a large principal quantum number, $n \geq 10$) with a maximum value of the angular momentum l and of its projection m along a quantization axis ($l = |m| = n-1$). These atoms are quasi-classical : the wave function of the valence electron has the shape of a thin torus lying around the classical circular orbit of the Bohr theory, whose radius is $a_0 n^2$ (where a_0 is the Bohr radius). These atoms fit the Bohr model of Hydrogen and are well adapted for measuring the Rydberg constant R. This metrology is presently performed by optical laser spectroscopy of atomic hydrogen (1,2,3). The present accuracy ($1.6\ 10^{-10}$) is limited by the laser wavelength measurement.

A natural direction to overcome this limit is to attempt an experiment at longer wavelengths, in a domain where the frequency of the radiation can be linked directly to the atomic time standard. This can be done by measuring R on microwave transitions between adjacent circular levels. The corresponding lifetime is very long (1–2 ms for $n \sim 25$–30) and yields a transition resonance quality factor of about 10^9. This method offers the following other advantages compatible with frequency determinations with a 10^{-11} or better precision :

- In circular atoms, the Rydberg electron always remains very far from the nucleus. Contact and QED corrections are thus much smaller than in the lower angular momentum states involved in optical R measurements.
- At the 10^{-11} precision level, it is not even necessary to perform the Rydberg measurement on a Hydrogen circular state : in alkali atoms, the circular electron sees the core made of the nucleus and the Z–1 remaining electrons mainly as a point-like charge. The small correction due to the core polarizability can be expressed as a very small quantum defect (4).

Frequency Standards and Metrology Editor: A. De Marchi

Actual knowledge of this polarizability in Lithium corresponds to uncertainties in the 10^{-11} range for R determination.

- Circular Rydberg atoms have no permanent electric dipole moment and hence no linear Stark effect. Stray electric field shifts and broadenings, which are usually very important in the spectroscopy of the low angular momentum Rydberg levels, can thus be minimized and much better controlled.

External electric and magnetic fields play a fundamental role in the circular atoms physics and will be discussed first. The preliminary results of an experiment performed on Lithium quasi-hydrogenic circular Rydberg states will then be reported. We have observed transit time limited resonances on the n=24 to n=25 circular transition around 447GHz (5).

I. CIRCULAR ATOMS IN ELECTRIC AND MAGNETIC FIELDS

In zero fields, the circular state is quasi degenerate with the other high angular momentum states. In low electric and/or magnetic fields, this degeneracy is removed, and the structure of the eigenspaces is related to the fields. The Hamiltonian is :

$$H = H_0(n) - \vec{d} \cdot \vec{E} + \mu_B \vec{B}.\vec{L} + 2\mu_B \vec{B}.\vec{S} + A_{(n,\ell)}\vec{L}.\vec{S} \qquad (1)$$

For typical $\vec{E}$ = 1V/cm and $\vec{B}$ = 1 Gauss fields and for n = 25 the electric coupling $\vec{d}.\vec{E}$ overcomes all the other terms.

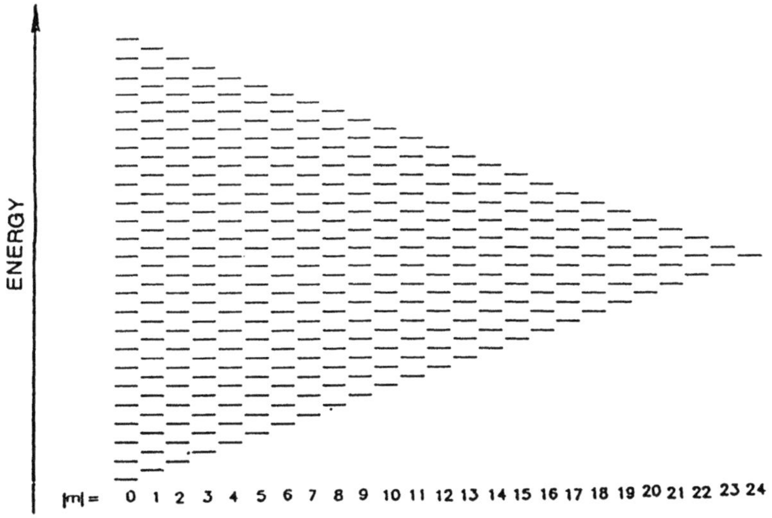

Fig. 1 : *Stark level diagram for n = 25*

Figure 1 shows the n=25 Stark level diagram. The level spacing ν is proportional to E (linear Stark effect), and is about 48 MHz for 1V/cm. The right-hand-side eigenstate ($|m|=24$) corresponds to the circular one, with its angular momentum aligned along $\vec{E}$. Its electric dipole $\vec{d}$ is zero to first order and its Stark shift is quadratic ($7.11 kHz/(V/cm)^2$ on the 24 to 25 circular-circular transition).

The orbital motion is driven by $\vec{E}$, and if the $\vec{E}$ direction changes slowly, the circular atoms' orientation will follow it adiabatically (6). In order to control the atomic orientation during the experiment, a low electric field (about 1V/cm) is applied along the path of the atoms.

The magnetic field affects the orbital and spin parts of the atomic state differently. The orbital Zeeman term $\mu_B\vec{B}.\vec{L}$ shifts the levels proportionally to the projection of $\vec{B}$ along $\vec{E}$. The spin Zeeman term $2\mu_B\vec{B}.\vec{S}$, overcomes the spin orbit $A\vec{L}.\vec{S}$ coupling ($A.L \sim 18$ kHz for the n=25 circular state) and drives the spin orientation. The spin is thus fully decoupled with respect to the orbital angular momentum (they are oriented along $\vec{B}$ and $\vec{E}$ respectively) and the spin-orbit interaction induces a fine structure splitting proportional to the cosine of the angle between $\vec{E}$ and $\vec{B}$.

II. THE EXPERIMENT

The circular atom microwave spectroscopy experimental set-up is sketched on Fig. 1. A thermal beam of Li atoms crosses three sections of the apparatus : the excitation region, the microwave interaction region and the detection one. The whole set-up is protected from room temperature thermal radiation by a liquid nitrogen-cooled shield (which can be replaced in a later stage of the experiment by a liquid helium-cooled one).

The preparation of Li Rydberg circular states is performed by pulsed lasers resonant with the 2S-2P (6708Å), 2P-3D (6104Å) and 3D-n=24 (~8320Å) transitions in presence of an $E \sim 190$V/cm field, produced by the stack of plates seen in Fig. 2. A 9GHz microwave, resonant with the Stark level spacing, induces a succession of adiabatic transitions and brings to the atoms the needed angular momentum (7).

The circular atoms then cross a Fabry-Perot semi confocal cavity, which sustains a microwave around 447 GHz produced by a phase-locked backwave oscillator (Thomson-CSF carcinotron) driven by a microcomputer. Its frequency can be stabilized, swept and measured continuously with reference to a time standard. After crossing the microwave cavity, the atoms enter the ionization zone, where they are selectively detected by the field ionization method (8). The circular n=24 to circular n=25 transfer rate is then plotted as a function of the microwave frequency (see fig. 3).

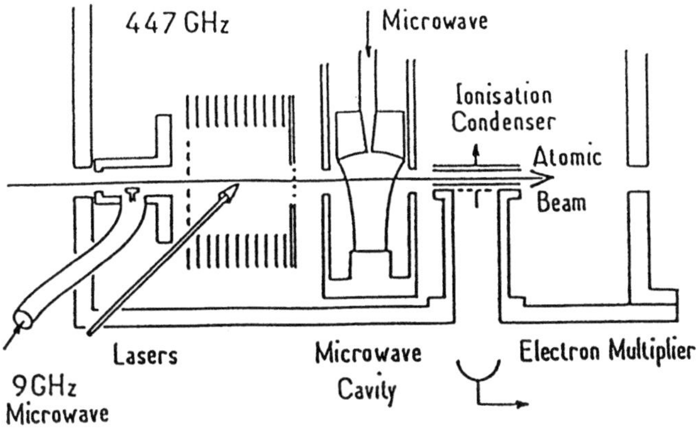

Fig. 2 : *Schematic of the experimental set-up.*

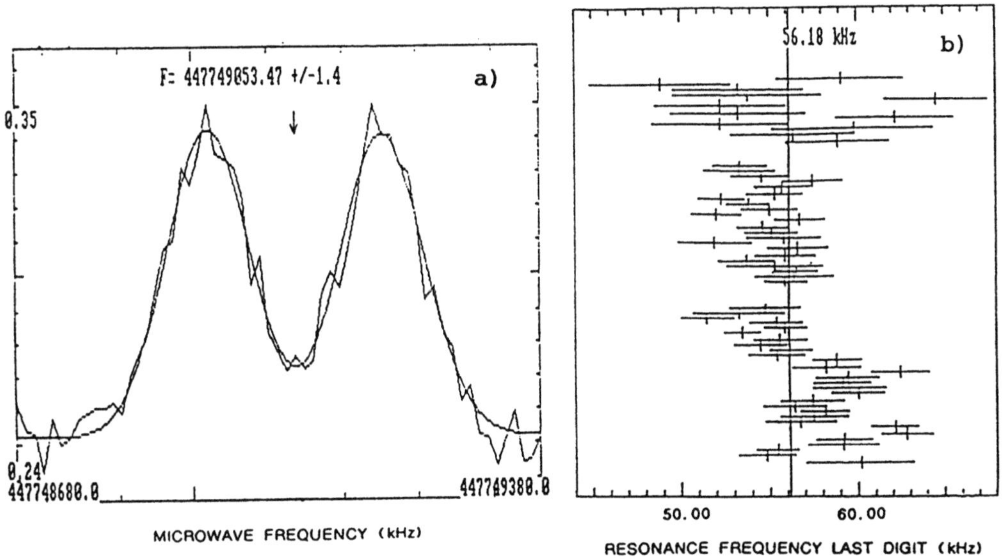

Fig. 3-a) : *n=24 to n=25 circular-circular transition microwave spectrum. The population transfer rate is plotted as a function of the microwave frequency. The smooth curve corresponds to the Gaussian fit. F is the fitted central frequency (arrow)*

Fig. 3-b) : *Plot of the resonance frequency (447 749OXX kHZ) of the n=24 to n=25 circular-circular transition of Li^7. The vertical straight line corresponds to the data weighted average.*

The Fig. 3 spectrum has been recorded with the atoms interacting with controlled fields, E = 1V/cm and B = 0.1 Gauss, E and B both being oriented along the cavity axis. The B field splits the line into two Zeeman components seen in Fig. 3, corresponding to both circular orbit helicities. Each component is shifted by about 7 kHz by the E field, and split into two fine structure components (~ 4 kHz splitting not resolved).

The Fig. 3 experimental points result from the averaging of 20 pulses recorded for each frequency. The frequency scan is performed randomly to avoid systematic drift effects. The width at half maximum of the resonance, 120 kHz, corresponds to the theoretical limit due to the finite transit time across the waist of the microwave Gaussian beam.

We have performed a precise measurement of the n=24 to n=25 circular-circular transition in Lithium based on 63 atomic resonance spectra recorded in about the same conditions as the spectrum seen in Fig. 3. Using at least squares iterative method, each spectrum is fitted with two Gaussian lines (smooth curve seen on Fig. 3). This procedure provides the frequency of the center of the two Zeeman lines, and after correction of the Stark shift (inferred from a precise measurement of $\vec{E}$), the unperturbed transition frequency, F_{exp}.

Fig. 3-b summarizes the results. F_{exp} is given for each spectrum with the error bar related to the experimental noise on the spectrum. Data are seen in time order in three groups separated by a small vertical blank space, corresponding to three different days. Two cavities have been used : in the first one the interaction time is shorter, making the linewidth typically twice as large as the Fig. 3 spectrum. This corresponds to the first group of data, where the error bars are therefore twice as large. The two following groups have been recorded with the second cavity.

The averaged value of F_{exp} (weighted by the inverse square of the uncertainties) is :

$$F_{exp} = 447\ 749\ 056.2 \text{ kHz}$$

The theoretical predicted value F_o using the current best measured Rydberg (3) is :

$$F_o = 447\ 749\ 056.43 \text{ kHz}$$

The relative uncertainty on F_o is $1.7\ 10^{-10}$ due to the uncertainty in the Rydberg constant and 4.10^{-11} related to the actual one percent precision on the dipolar polarizability of Li^+ $(1s)^2$ ground state (9). The uncertainty on the other corrective terms is about 10^{-12}.

The $\overline{F}_{exp}$ mean value is equal to F_0. However Fig. 3 exhibits a dispersion of 2.7 kHz standard deviation, which is related to fluctuating systematic shifts. The miscontrolled horizontal component of the magnetic field affects the fine structure splitting and unbalances the weight of the two fine structure components. It results in a time-dependent drift within ±500Hz. The mistuning of the cavity shifts the line and yields a systematic ±500Hz. For large atomic populations (about 1000 atoms detected), we have observed shifts of a few kHz, attributed to collective effects. Although all the reported measurements have been performed with less than 300 atoms, we are actually unable to exclude the possibility of a residual shift of about ±500 Hz.

Finally, this preliminary experiment yields an F_0 measurement within 2.5kHz, which can be reinterpreted as an R measurement with a ±6 10^{-9} accuracy.

This accuracy is limited mainly by the interaction time which is 2 orders of magnitude smaller than the atomic life-time, and can be improved by slowing down the atoms or by using a Ramsey design split cavity. This experiment demonstrates the feasibility of a competitive R measurement in frequency units on microwave transitions involving circular states.

REFERENCES

[1] P. ZHAO, W. LICHTEN, H.P. LAYER, J.C. BERGQUIST, Phys. Rev. A34, 5138 (1986)
[2] E.A. HILDUM, U. BOESL, D.H. McINTYRE, R.G. BEAUSOLEIL and T.W. HANSCH, Phys. Rev. Lett. 56, 576 (1986)
[3] . F. BIRABEN, J.C. GARREAU and L. JULIEN, Europhys. Lett. 2, 925 (1986)
. F. BIRABEN et al., Proceedings of the Symposium on *"The Hydrogen Atom"*. Pisa 1988.
[4] R.R. FREEMAN, D. KLEPPNER, Phys. Rev. A14, 1614 (1976)
[5] J. HARE, M. GROSS, P. GOY, S. HAROCHE, Post deadline paper in XVth International Quantum Electronics Conference, Baltimore (1987)
[6] J. LIANG, M. GROSS, P. GOY and S. HAROCHE, Phys. Rev. A33, 4437 (1986)
[7] R.G. HULET, D. KLEPPNER, Phys. Rev. Lett. 51, 1430 (1983).
[8] T. DUCAS, M. LITTMAN, R. FREEMAN, D. KLEPPNER, Phys. Rev. Lett. 35, 366 (1975)
[9] W.E. COOKE, T.F. GALLAGHER, R.M. HILL, and S.A. EDELSTINE, Phys. Rev. A 16, 1141 (1977).

Determination of the Rydberg Constant by Doppler-Free Two-Photon Spectroscopy of Atomic Hydrogen Rydberg States

M. Allegrini[1], F. Biraben, B. Cagnac, J.C. Garreau, and L. Julien

Laboratoire de Spectroscopie Hertzienne de l'E.N.S.,
4, Place Jussieu, Tour 12 E01, F-75252 Paris Cedex 05, France
[1]Permanent address: Istituto di Fisica Atomica e Molecolare del C.N.R., Via del Giardino 7, I-56100 Pisa, Italy

1. Introduction

During the last ten years several laser measurements of the Rydberg constant [1-8] have improved its accuracy from 3 parts in 10^9 to few parts in 10^{10} and the discrepancies among absolute values are now removed. Indeed recent measurements of R_∞ (see table 1), obtained from the study of different transitions in atomic hydrogen and deuterium, agree very well each with other . The accuracy of all these measurements is approaching the precision (1.6×10^{-10}) of the I_2-stabilized He-Ne laser which serves as primary frequency/wavelength standard in the visible region.

Table 1. Recent values obtained for the Rydberg constant by laser spectroscopy of atomic hydrogen and deuterium

Transition	R_∞ ($-109737 cm^{-1}$)	Relative(*) precision (rms error)	Group	Ref.
H,D: 2S- 8D, 2S-10D	.31569(6)	5×10^{-10}	PARIS, 1986	3
H,D: 2S- 3P	.31569(7)	6×10^{-10}	YALE, 1986	4
H,D: 2S- 4P	.31573(3)	3×10^{-10}	YALE, 1987	5
H : 1S- 2S	.31571(7)	6×10^{-10}	STANFORD, 1987	6
H,D: 1S- 2S	.31573(3)	3×10^{-10}	OXFORD, 1987	7
H,D: 2S- 8D, 2S-10D, 2S-12D	.315714(19)	1.7×10^{-10}	PARIS, 1988	this work

(*) The absolute frequency calibration in the experiments of ref.[6,7] is done through an intermediate standard ($^{130}Te_2$), whose accuracy is 4×10^{-10} [6] and 2.7×10^{-10} [7].

In the last line of table 1 we report also our most recent measurement of R_∞ [8], whose precision with respect to the frequency/wavelength standard is 4.4×10^{-11}. Thus our measurement of the Rydberg constant is not limited by the uncertainties inherent in the experiment, but rather by the precision with which

Frequency Standards and Metrology Editor: A. De Marchi

absolute frequencies are measured in the optical domain. In other words our result shows that the present realization of the metre is no more satisfactory in the optical domain and that better optical frequency standards are required.

To perform this measurement of R_∞ we have used the same scheme of the Doppler-free two-photon laser spectroscopy experiment of ref.[3], improving several parts of the apparatus.

2. Measurement Method and Apparatus

We measure the Rydberg constant by comparing the 2S-nD transitions frequencies (n = 8, 10, 12) in hydrogen and deuterium with the one of a home-made I_2-stabilized He-Ne laser. This laser has been calibrated directly with respect to one of the Institut National de Métrologie which had been previously compared to the standard lasers of the Bureau International des Poids et Mesures.

In order to obtain narrow signals, we use a metastable atomic beam collinear with two counter-propagating laser beams. Thus the line broadening due to the finite transit time of the atoms into the lasers beams is small. The experimental set-up is made of three major parts: i) metastable beam apparatus, ii) dye laser apparatus iii) frequency measurement apparatus.

i) The metastable beam apparatus (Fig. 1) is divided in three vacuum chambers. In the first one a ground state effusive atomic beam, obtained from molecular hydrogen (or deuterium) through r.f.discharge, is excited to the metastable 2S state by electron bombardment. Because of the inelastic collisions with electrons, the atomic beam is deviated by about 20° and, after the collision, the metastable atomic beam is collinear with the laser beams. In the second chamber where electric and magnetic stray fields are minimized, (<2mV, <50mG), the 2S-atoms are excited to the nD Rydberg states by two counter-propagating laser beams. The whole metastable beam apparatus is placed inside a Fabry-Perot cavity mounted to enhance the two-photon absorption. In a third chamber an applied electric field quenches the 2S-level and the metastable atoms are detected through the resulting Lyman-α fluorescence. The two-photon transitions are observed as a

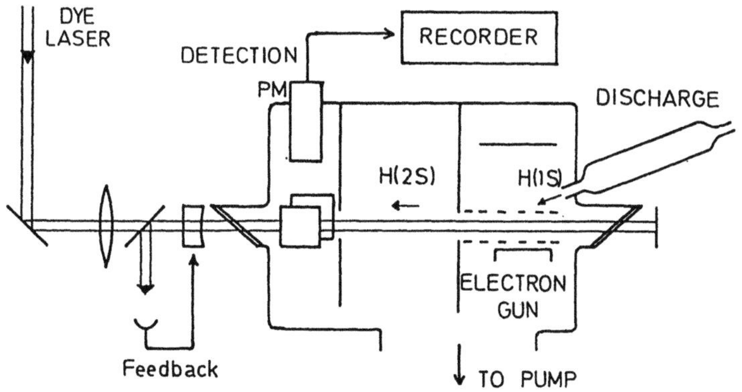

Figure 1 : Metastable beam apparatus

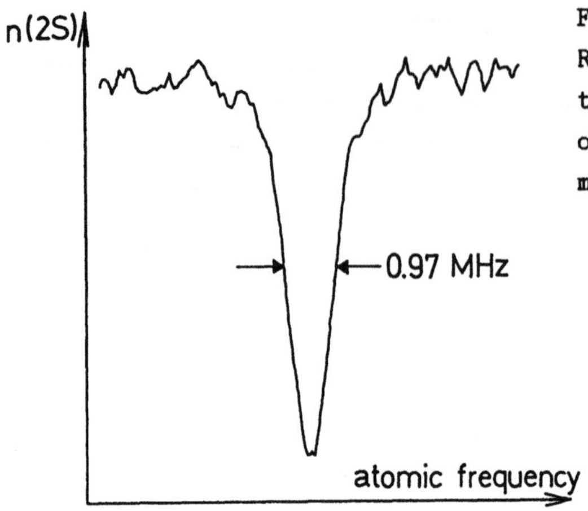

Figure 2 :
Recording of the $2S_{1/2}$ - $12D_{5/2}$ two-photon transition in hydrogen observed as a decrease of the metastable beam intensity n(2S)

decrease of the metastable beam intensity (see Fig. 2). Typical values are ≈ 10^6 atoms/s metastable atom flux (or a metastable density of few atoms per c.c.), ≈ 10^4 W/cm^2 laser power density in the excitation chamber, up to 15% quenching of the metastables and a signal-to-noise ratio up to 30 in the detection chamber.

ii) The laser used to excite the 2S-nD transitions is a home made, cw, single-mode, ring, LD 700 dye laser, pumped by a Kr^+ laser. The frequency stabilization is made by active locking to an external auxiliary Fabry-Perot cavity while the scanning is done through an acousto-optics device (see Fig. 3). The modulation frequency of the acousto-optics crystal is provided by a computer-controlled frequency sinthesizer and the result is a very precise control of the dye laser frequency. The overall characteristics of the dye laser are ≈ 1 W power in the 730-780 nm region of our interest, ≈ 50 kHz linewidth and ≈ 250 MHz tunability without mode-hopping, centered at any desired frequency.

iii) The absolute frequency position measurement of the two-photon transitions is done by comparing the wavelengths of the i.r. dye laser and of the red He-Ne reference laser. This comparison is done inside a nonconfocal Fabry-Perot cavity of high finesse and kept under high vacuum. The construction of this cavity requires the use of optical adhesion technique and metallic mirrors of good quality. An auxiliary He-Ne laser as well as the dye laser are mode-matched and locked to this optical cavity while the beat frequency between the two He-Ne lasers is measured. Two lengths of the Fabry-Perot cavity are used (10 cm and 50 cm) so that the phase shift due to the metallic mirror reflection is eliminated by the method of the virtual mirrors.

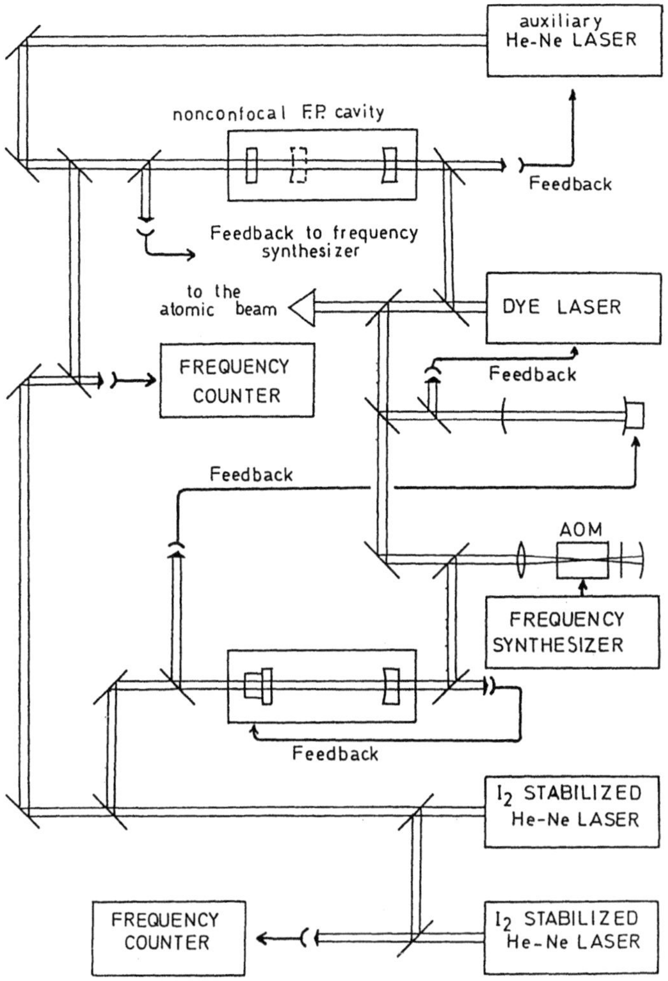

Figure 3 : Experimental set-up for the control and the measurement of the dye laser frequency

3. Results and conclusion

We have measured the frequency of the three transitions $2S_{1/2}$-$8D_{5/2}$, $2S_{1/2}$-$10D_{5/2}$ and $2S_{1/2}$-$12D_{5/2}$ in both atomic hydrogen and deuterium and we have corrected them for h.f.s. splitting and second-order Doppler effect. In order to eliminate the light shift we have extrapolated the line position at null light power. Using Erickson values for the 2P-nD interval [9], the experimental value of H-2S Lamb-shift [10] and the theoretical value of the D-2S Lamb-shift [11], we have deduced six independent values of the Rydberg constant which give as final result:

$$R_\infty = 109\ 737.315\ 714\ (19)\ \mathrm{cm}^{-1}$$

In figure 4, this result is compared with those recently obtained by other groups.

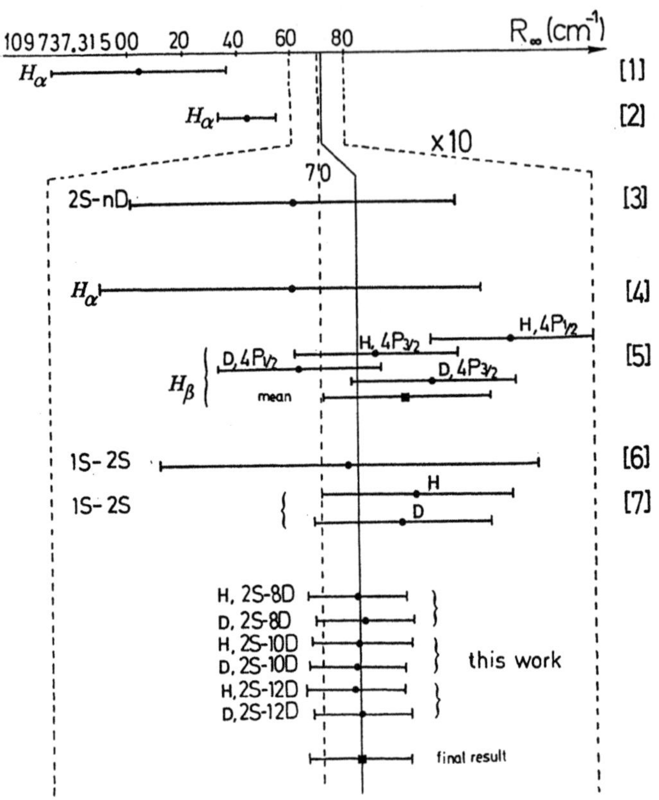

Figure 4 : Comparison of recent Rydberg constant measurements

From the analysis of the error sources we find that the determination of the frequency of the standard laser gives the limit to the precision of our measurement of R_∞. In fact our precision with respect to the standard laser is less than 5×10^{-11} while the reliability of the standard laser is 1.6 parts in 10^{10} [12]. Work is in progress in order to reach a more refined analysis of our data, including the effects of the velocity distributions inside the atomic beam.

This work is partially supported by the Bureau National de Métrologie under grant BNM-87 2 46 0017 and by the European Economic Community under grant EEC-ST2*0423.

References

1. J.E.M.Goldsmith, E.W.Weber and T.W.Hänsch, Phys. Rev. Lett. **41**, 1525 (1978)
2. S.R.Amin, C.D.Caldwell and W.Lichten, Phys. Rev. Lett. **47**, 1234 (1981)
3. F.Biraben, J.C.Garreau and L.Julien, Europhys.Lett. **2**, 925 (1986)
4. P.Zhao, W.Lichten, H.P.Layer and J.C.Bergquist, Phys. Rev. **A34**, 5138 (1986)

5. P.Zhao, W.Lichten, H.P.Layer and J.C.Bergquist, Phys. Rev.Lett. 58, 1293 (1987)
6. R.G.Beausoleil, D.H.McIntyre, C.J.Foot, E.A.Hildum, B.Couillaud and T.W.Hänsch, Phys. Rev.A35, 4878 (1987)
7. M.G.Boshier, P.E.G.Baird, C.J.Foot, E.A.Hinds, M.D.Plimmer, D.N.Stacey, J.B.Swan, D.A.Tate, D.M.Warrington and G.K.Woodgate, Nature 330, 463 (1987)
8. This result has been presented also to : Intern. Symposium on "The Hydrogen Atom", Pisa, June 1988; XIth Intern. Conference on Atomic Physics, Paris,July 1988; XXth Conference European Group of Atomic Spectroscopy, Graz, July 1988
9. G.W.Erickson, J. Phys. Chem. Ref. Data 6, 831 (1977)
10. V.G.Pal'chikov, Yu.L.Sokolov and V.P.Yakovlev, JEPT Lett 38, 418 (1983); Metrologia 21, 99 (1985)
11. G.W.Erickson and H.Grotch, Phys. Rev. Lett. 60, 2611 (1988) Using the theoretical expression of the Lamb-shift given in this reference, we obtain 1059.229 MHz for the 2S Lamb-shift in deuterium.
12. D.A.Jennings, C.R.Pollock, F.R.Petersen, R.E.Drullinger, K.M.Evenson, J.S.Wells, J.L.Hall and H.P.Layer, Opt. Lett. 8, 136 (1983)

Part XIII

Poster Sessions

The Primary Cesium Atomic Clocks of the PTB

A. Bauch and T. Heindorff

Physikalisch-Technische Bundesanstalt, Bundesallee 100,
D-3300 Braunschweig, Fed. Rep. of Germany

1. Introduction

Almost 20 years ago, in 1969, the well-known primary atomic clock CS1 of the PTB was put into operation, and it has been in continuous operation since 1978. In 1985, a second device, CS2, was completed. The intercomparison data of the CS2 versus CS1 will be presented in section 2. Research work at the PTB is mainly directed at reducing the total uncertainty of its primary clocks. One aspect, the determination of the cavity phase difference, is addressed in section 3. A few days ago (August 1988), the first beam signal was recorded from the newest primary clock, CS3. A few remarks on this are to be found in section 4.

2. Long-Term Comparison of CS2 Versus CS1

The CS2 was introduced for the first time at the CPEM'86 /1/. A detailed description has recently been given and the uncertainty budget discussed in /2/. The essentials of the CS1 and CS2 are as follows: A narrow cylindrical atomic beam runs along the symmetry axis of the device. Magnetic lenses are used for the state selection and velocity filtering. At a mean atomic velocity of 90 m/s, the velocity interval is only 8 m/s (FWHM). The interaction length of the Ramsey-type microwave cavity is 79 cm, which results in a linewidth of 60 Hz (FWHM). The calculated S/N of the clock transition line is 550 and 1000 for CS1 and CS2, respectively (1 Hz bandwidth).

The total uncertainty (1σ) is estimated to be $30 \cdot 10^{-15}$ (CS1) and $15 \cdot 10^{-15}$ (CS2). 900 one-day mean values of frequency intercomparisons (mean value $25 \cdot 10^{-15}$) are now available. The results are shown in Fig. 1. The difference in time-scale reading is now 2.1 µs. From the σ-τ diagram, the existence of non-white noise components is assumed, and these determine the frequency instability at sample times longer than 10 days. The flicker floor appears at a level of $1 \cdot 10^{-14}$.

Frequency Standards and Metrology Editor: A. De Marchi

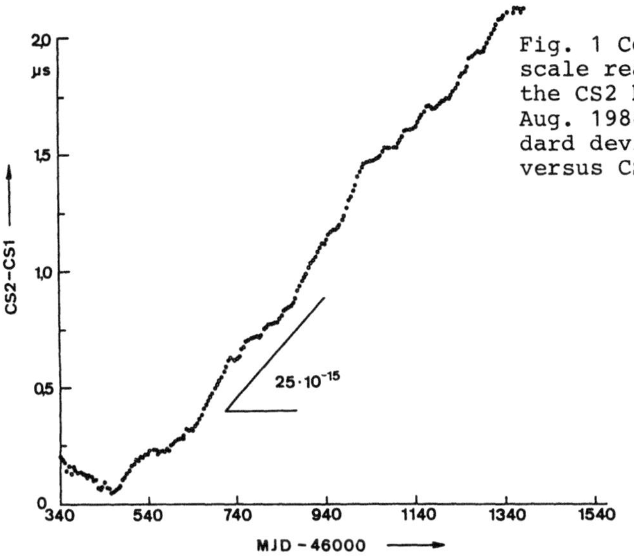

Fig. 1 Comparison of the time scale readings of the CS1 and the CS2 between Oct. 1985 and Aug. 1988 (top), 2-sample standard deviation $\sigma_A(\tau)$ of CS2 versus CS1 (bottom)

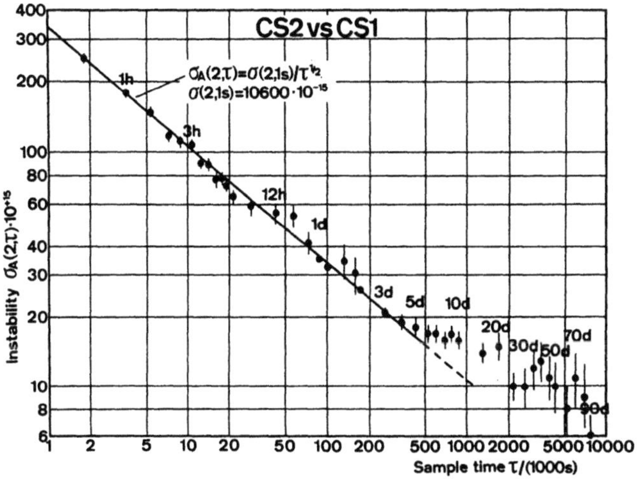

3. Reduction of the Total Uncertainty: Test of a Ring Cavity

The main source of the uncertainties of the PTB's primary clocks is connected with the determination of the phase difference φ of the Ramsey cavity. In short, an ideal standing wave cannot be sustained in a cavity with walls of finite conductivity. This is why in the conventional design, a small net energy flux is present at the places of interaction between atoms and field which causes a spatial distribution of the phase as seen by the atoms. This limits the uncertainty of measuring φ. A. DE MARCHI /3/ proposed ring-shaped end sections for the cavity. The net energy flux in the interaction region should then

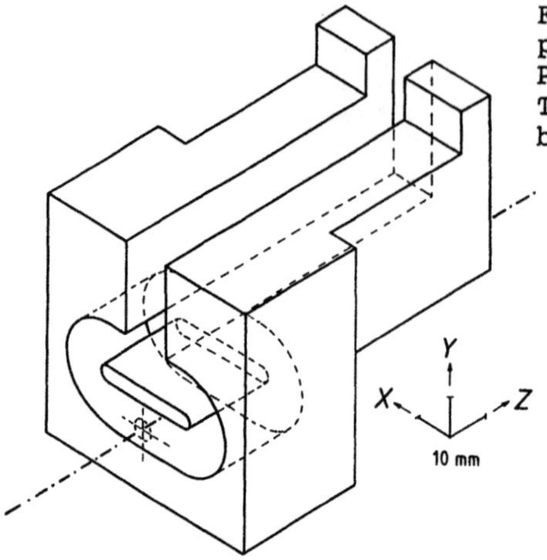

Fig. 2 Ring-type cavity end part, proposed by De Marchi /3/, tested in the PTB's experimental atomic clock, CSX. The left and top covering plates have been omitted

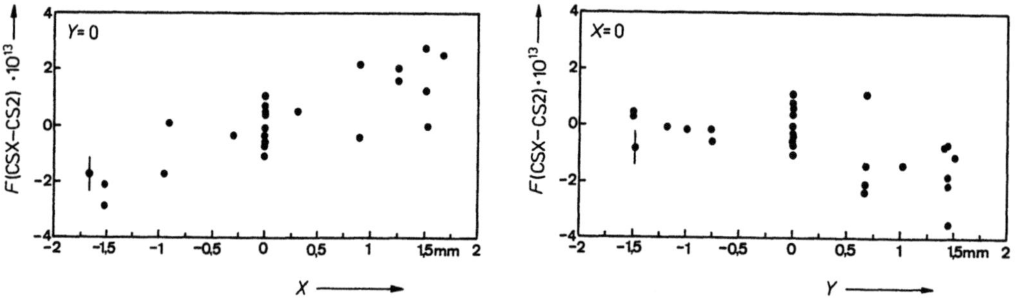

Fig. 3 Relative frequency difference F(CSX-CS2) as a function of the cavity position. X=0,Y=0 is the on-axis position of the cavity

vanish due to the symmetry. A prototype of the cavity end section depicted in Fig. 2 was tested in the CSX, the PTB's experimental clock. The results in Fig. 3 show a reduction of the spatial dependence of the phase by a factor of 3, compared with the conventional design. The results, however, are not in perfect agreement with the theory developed by De Marchi. Further studies seem to be inevitable.

4. First Results of the Newest Device, CS3

Construction of the new generation of primary clocks, the CS3 and CS4, has almost been completed. The chief objective has been to use atoms even slower than those used in the CS2. To circumvent a complicated beam geometry caused by the gravitational deflection, the atomic beam is directed vertically. The first recorded Ramsey resonances exhibit a linewidth of 44 Hz, corresponding to a mean beam velocity of 65 m/s. What will now follow is a period of detailed investigations of the performance of the numerous components, and a determination of the frequency deviations and their uncertainties. We hope finally to achieve a total uncertainty similar to that of the CS2.

This unique group of four primary cesium clocks at the PTB will enable interesting research to be done and will considerably influence the international time scale TAI.

5. Literature

1. A. Bauch et al.: IEEE Trans. Instr. Meas. IM-36, 613 (1987)
2. A. Bauch et al.: Proc. 2nd EFTF, 425 (1988)
3. A. De Marchi: Proc. 40th Ann. Symp. Frequ. Contr., 441 (1986)

Metrological Parameters of the Primary Cesium Beam Frequency Standards in the National Time and Frequency Standard of the USSR

Y.G. Abashev, V.N. Baryshev, G.A. Yolkin, and Y.N. Yakovlev

USSR Gosstandart, Moscow, USSR

The design features and some results obtained during researches of cesium-beam frequency standards MCs-1, MCs-2, MCs-3 (Fig.1-3) and their comparisons with hydrogen clocks in the National Time and Frequency Standard are given in [1-4]. The above three machines substantially differ from each other. MCs-1 and MCs-3 are dipole systems with a transverse homogeneous magnetic field. The cavity length is 65 cm and 194 cm respectively. The MCs-2 system has a hexapole optics, an axial C-field and a point-contact beam detector; the length of the cavity is 100 cm. The beam-direction recersals through the cavity can be accomplished in all of three devices; the magnetic shields and the microwave cavity are temperature-controlled.

The resonance amplitude of the cesium beam standards is 5 to 10pA; the spectral linewidth for MCs-1, MCs-2 and MCs-3 is 180 Hz, 110Hz and 50 Hz respectively. The electronic systems of MCs-1 and MCs-2

Fig.1

Frequency Standards and Metrology Editor: A. De Marchi

Fig.2

Fig.3

are identical and permit frequency measurement of hydrogen clocks with regard to the spectral line of Cs atoms [5]. The electronics of MCs-3 allows continuous tuning the hydrogen clock to the top of the Cs beam line (for one beam direction).

The most important metrological parameters of the cesium standard are accuracy, frequency stability and reproducibility. The standards' frequency stability for 1 sec is 3 to 5 in 10^{-12}. As to frequency reproducibility,it should be noted that in practice the important parameters are both the reproducibility of a frequency value averaged for two beam directions and frequency reproducibility for one beam direction. The value for the first parameter is (5 to 10) x 10^{-14} for MCs-1 and MCs-2 (a single measurement lasts 5-6 hours) and 3 x 10^{-14} for MCs-3 (the time of a single measurement is (15-20) hours). When operating with one beam direction the frequency reproducibility for MCs-1 and MCs-2 is (10 to 15) x 10^{-14} and for MCs-3 is 5 x 10^{-14}.

The accuracy of all the three cesium standards is estimated to be 1 x 10^{-13}. The criteria of the exact evaluation of the accuracy is, in particular, the degree of coincidence of frequencies produced by cesium beam standards that significantly differ from each other both in their design and principle of operation.

For the period of 1981-1988 the mean difference in frequencies of MCs-1 and MCs-2 did not exceed 1 x 10^{-13}. In 1985-1988 this value was less than 5 x 10^{-14}. However the MCs-3 frequency is less than the mean frequency value of MCs-1 and MCs-2 by a factor of 1,5 x 10^{-13}. The preliminary analysis of the data of long term researches of frequency dependences on the microwave power showed that such difference in frequencies reproduced by cesium standards could be due to spurious spectral components available in the microwave signal. One should keep in mind that the final value of the MCs-3 frequency was determined by extrapolation of the results obtained to a zero power, while all the measurements on MCs-1 and MCs-2 were made at an optimum power.

Such difference in the procedure of determination of the frequency of the unperturbed transition of the cesium 133 atom with spectral impurities in the exciting signal could apparently result in frequency divergence. However, additional investigations should be performed to make a final conclusion.

In 1985-1986 MCs-4 was developed. The apparatus is planned to be used for investigations in the efficiency of optical pumping of the cesium beam with the aim of construction the primary optically pumped cesium beam frequency standard. The atomic beam MCs-4 system is designed in such a way that it can employ both the dipole and hexapole selecting systems. For optically pumped experiments the magnets are removed and the systems for optical pumping and detecting are symmetrically arranged at each end of the instrument (Fig.4).

A small set of frequency comparisons between MCs-4 (using magnetic selectin of the atoms) and hydrogen clocks comprising the National

Fig.4

standard showed that the frequency produced by MCs-4 is lower than the mean frequency value of MCs-1 and MCs-2 by 8×10^{-14}.

At present our efforts are concentrated on putting into operation the optically pumped system for MCs-4 for making frequency intercomparisons; the frequencies being produced by one and the same machine employing the traditional magnetic selection and optical pumping of a cesium beam. Optical pumping seems to significantly improve metrological parameters both of commercial and laboratory cesium beam frequency standard [6,7].

References.

1. Yu.G.Abashev, G.A.Yolkin and S.B.Pushkin, Izmeritelynaya technika, 1983, N 12,pp. 36-38.
2. Yu.G.Abashev, V.N.Baryshev, G.A.Yolkin and Yu.V.Timofeev, Izmeritelynaya technika, 1986, N 2, pp.18-19.
3. Yu.G.Abashev, B.N.Baryshev, G.A.Yolkin et.al., "Investigations in the atomic time and frequency standards", VNIIFTRI Transactions, 1987, pp. 73-79.
4. Yu.G.Abashev, L.P.Yolkina, S.B.Pushkin, IEEE Transactions on Instrumentation and Measurement, V.IM-36, No.2, 1987.

5. G.A.Yolkin, V.I.Purtov, Izmeritelynaya technika, 1976, N 10,p.47.
6. R.E.Drullinger, Jon Shirley, D.J.Glaze, L.W.Hollberg, A.De Marchi, "Proc. 40th Annual Freq. Control Symp.", 1986, pp. 428-431.
7. McClelland T., Pascari I., Zacharski J., Tran N.H., Meris M. "Proc. 41 Annual Frequency Control Symp.", Philadelphia, Pa, 1987, N.Y., pp.59-65.

Work on Cesium Standards at NRC

J.-S. Boulanger and R.J. Douglas

Division of Physics, National Research Council of Canada,
Ottawa, Ontario K1A 0R6, Canada

NRC was the first laboratory to operate a primary laboratory cesium frequency standard as a clock[1,2,3], starting in 1975. Since then we have been emulated by a few other laboratories. Our clocks are among the oldest primary frequency standards still in operation. We are in a position to see the effects of aging: better stability of the phase of the cavity; better understanding of various processes; deterioration of some components; mishaps; etc.

Among the various improvements that have been made recently are the better determination of the second order Doppler effect[4], the Rabi pulling[5] and the microwave instabilities. Among the disappointments: the recent loss of CsVIB (one of the three one-meter clocks, due to a leak), and aging of various electronic components.

Good Side of Aging

The first obvious side effect of aging is the stabilization of the microwave cavity. The phase difference has relaxed with a time constant of less than two years. After fourteen years of operation the cavity still has an offset, but its value appears quite constant; at least within the accuracy of the measurements (of 2 to 5 parts in 10^{14}). But we cannot take it for granted: in 1984, when we had to open the clock to refill the ovens with cesium (every nine years in principle) we observed a change in the frequency bias due to the cavity phase shift of 6×10^{-14}. Fortunately, there was no drift apparent thereafter.

The determination of the velocity distribution by the variation of the power at resonance[4] has proven a very efficient tool to analyze the behaviour of our clocks. It is easy to determine the effect of the Rabi pulling knowing all the neighboring transitions. The worst case is less than 1 in 10^{14} at optimum power (maximum transition: P_{opt}) on our one-meter clocks (CsVI's). In the case of CsV, the effect is smaller still, by more than a factor of two, over the operating range of microwave power (P_{opt}:P_{opt}-6dB). The velocity distribution is constant up to 24 hours before the oven is empty of cesium. This means that the beam is stable as long as the oven and detectors are at the same positions and as long as the temperature of the oven is kept constant.

We have redesigned the servos so they can skip a cycle (or more) and we will use this time slot to modify a parameter and measure its effect. It will then be possible to have a continuously updated record of parameters such as the C field and power shift.

We are also studying the possibility of monitoring directly the cavity phase shift through the power shift without having to reverse the beam direction. This could be made possible by having the servo operated as described above. Every minute, the loop is opened to take a measurement at power level different than the normal one. A fictitious "servo" uses this data to calculate the offset in frequency compared to the normal operation.

Frequency Standards and Metrology Editor: A. De Marchi

Due to the stability of the beam's velocity distribution, such a measurement can be carried over long period of time (perhaps months). With the help of the theory of moments of velocity distribution[6] it is easy to calculate the various contribution to the power shift due to the second order Doppler effect. We also know the Rabi pulling. These two shifts being removed from the offset measured by the pseudo servo, the cavity phase shift can be evaluated. A simple calculation shows that if a clock is stable to 1 in 10^{14} at one day, a duty cycle of 10% for the pseudo servo should give an equivalent accuracy in less than ten days, with only about a 5% degradation in shorter term stability for the clock.

This mode of operation will probably be tested first on the experimental clock (CsVIB). The assumption behind this theory is that the phase shift experienced by the atoms is independent of the velocity. This would be true if all the atoms were going through the same trajectory. Unfortunately, in our clocks the paths are strongly dependent on the velocity: it is of no importance if the phase distribution is orthogonal to the velocity spreading in space. Thus we will have to first study the phase distribution across the opening of the microwave cavity. Even if the orthogonality condition is verified there is the possibility of a change in the phase value itself at different power levels. Since our clocks can attain stabilities of better than 1 in 10^{14} in a few hours at high beam currents it will be possible to use a hydrogen maser and the experimental clock to conduct these tests.

Bad Side of Aging

One problem we faced recently was the degradation of the short term stability of our clocks. We found that this was related to the parasitic surface conduction on the glass and ceramic insulators of the detectors. In air, the resistance measured between the collector and the body of the detector could be less than two megohms! Even under vacuum, the noise added to the electrometer caused a significant loss of frequency stability. The only cure we have so far is to open the clock and wipe the insulators. Obviously a better cure is needed although the problem appears only after many years of operation at normal beam current.

Another source of noise are the mechanical vibrations inducing oscillations in the hot wire of the detector. These oscillations (ranging from 40 to 300 Hz) can induce signals as large as a few picoamperes through the oscillation of the capacitance. These parasitic signals add noise and can even saturate the amplifiers and low pass filters of the servo system. They present the largest difficulty if the oscillations are near an odd harmonic of the square-wave modulation frequency. Sometimes the problem can be solved by adjusting the temperature of the hot wire and the potential of acceleration. Unfortunately it does not totally eliminate these oscillations for every detector. We are considering redesigning these detectors.

One aspect of our clocks that particularly needs monitoring is the frequency shift due to spectral impurities in the microwave signal. A few years ago we observed a gradual shift in frequency in CsV. We suspected the microwaves and/or the servo. Unfortunately, diagnosing the problem was complicated by the fact that there was an intermittent parasitic signal. We did replace many cables and connectors; the silver on these had blackened over the years. They were replaced by gold plated connectors. We also replaced the servo and the local oscillator. The problem seemed to be aggravated by the lack of humidity in the building ($<5\%$). We tried to measure the effect of humidity but could not find consistent results. Now the humidity in the building is under control (40%-55%) and the clock is more stable.

We still suspect the microwave impurities to be the source of the problem. We have tried replacing part of the frequency multiplying chain by a commercial unit claiming good phase noise; it was a disaster: the frequency of the clock would change by many parts in 10^{13} with a 3 dB change in the power level. We found diffuse side bands around 10 Hz from the carrier in that unit. We are still investigating.

In our clocks, square-wave FM at a frequency of 0.8 or 0.08 Hz is used for the main frequency control servo. We lock a Gunn oscillator to the difference of the multiplied 9180 MHz signal and a switching 12.6... MHz synthesizer. The 12.6...MHz coherent switching synthesizer can generate side bands out to a few kilohertz. In our oldest design, using the 4044 type phase-lock IC, variable single side bands can develop as the unit ages. We have designed a completely different unit which will allow the use of any (reasonable) C field. It uses direct frequency synthesis for the agile loop. It will receive extended testing before we replace the old unit by the new one. Its main operational attraction, aside from its expected long-term stability, is that there is no need to change the C-field anymore to compensate for the various offsets. The synthesizer frequency can be adjusted, with ample resolution, to the needed value: under computer control if necessary.

Conclusion

With a little bit of care, in good environment, our clocks should be able to deliver another ten years of reliable data. The conversion of CsVIB into an experimental clock will provide us with a tool to better understand the behaviour of classic type of cesium tube. The construction of CsVII, a cesium clock using optical pumping, and the construction of two hydrogen masers will give us even better opportunity to investigate cesium standards.

1 A.G. Mungall et al, *Metrologia*, **9**, 113 (1973)

2 A.G. Mungall et al, *Metrologia*, **13**, 105 (1977)

3 A.G. Mungall et al, *Metrologia*, **17**,123 (1981)

4 J.-S. Boulanger, *Metrologia*, **23**, 37 (1986)

5 A. DeMarchi, *Metrologia*, **20**, 37 (1984)

6 C. Audouin, *IEEE Trans. Instr. Meas.*, **IM-23**, 501 (1974)

Optical Pumping with Finite-Linewidth Lasers: Predictions for the Optically Pumped Cesium Beam Frequency Standard

P. Tremblay[1] *and C. Jacques*[2]

[1]Laboratoire de recherches sur les oscillateurs et systèmes, Université Laval, Québec G1K 7P4, Canada
[2]Laboratoire des étalons fondamentaux, Conseil national de recherches, Ottawa K1A 0R6, Canada

1. Introduction

Many laboratories are presently working on the project of an optically pumped cesium beam frequency standard using semiconductor laser(s). Particularly interesting pumping schemes using two lasers have been proposed by CUTLER [1] and have been tried experimentally [2],[3],[4]. It is expected that these schemes, where one of the lasers must be π-polarized, will totally populate one of the two ground state $m_F=0$ Zeeman sublevels, based on the fact that the transition $\Delta F=\Delta m_F=0$ is forbidden. However restrictions imposed by the fundamental properties of the cesium atoms and by experimental conditions will prevent the achievement of a perfect pumping. Precise knowledge of the pumping effects on the populations of the various Zeeman sublevels will help to choose appropriate conditions to insure an efficient optical pumping.

2. Theory

We have theoretically studied the problem of optical pumping by means of two lasers having lorentzian profiles. We developed a new model using the density matrix formalism and fluctuating laser field [5],[6]. This model represents more closely the optical pumping experiment on the cesium beam since it takes into account the effects of the atomic coherences, those of the neighboring transitions ("wings effect") and the fact that the semiconductor lasers used for the pumping have lorentzian profiles with intermediate linewidths (of the order of the natural linewidth).

In this model, we are interested in the mean values of the populations of each Zeeman sublevel. In order to find the expected value of the populations difference between the F=4, $m_F=0$ and the F=3, $m_F=0$ sublevels, we must also evaluate the mean values of the Zeeman coherences, the correlations between optical coherences and field fluctuations of the corresponding laser, and the correlations between ground state hyperfine coherences and field fluctuations of both lasers.

3. Results

The model we developed was used to predict the efficiency of the optical pumping process in creating a population difference between the two $m_F=0$ ground state Zeeman sublevels. Firstly, in Fig. 1, we compare the efficiency of various pumping schemes for typical experimental conditions. Under these conditions, the $(3\to4',\pi);(4\to4',\pi)$ pumping scheme is much more efficient than the six other possible pumping schemes. This conclusion differs from that obtained using a model neglecting atomic coherences and neighboring transitions [3]. This shows that, under typical conditions, the atomic coherences induced by laser radiation are present, and they greatly reduce the efficiency when σ-polarization is used.

Because of those last results, our calculations were concentrated on the efficiency of the $(3\to4',\pi);(4\to4',\pi)$ pumping scheme, since it appears to be the most promising one. Figure 2 presents the asymptotic pumping efficiency achievable with this scheme. For each intensity, there exists an optimum linewidth which increases as the intensity increases. With narrow laser linewidth, the creation of correlations between ground state hyperfine coherences and field fluctuations decreases the efficiency. The wings effect does the same very rapidly for broad lasers (line-

Frequency Standards and Metrology Editor: A. De Marchi

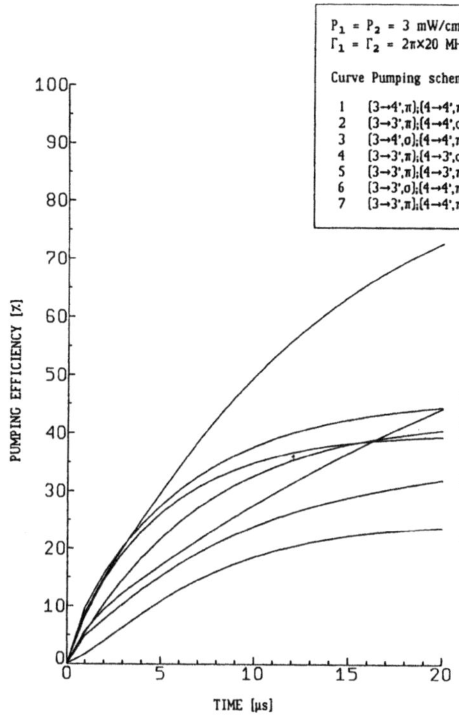

Fig. 1. Pumping efficiency of the seven possible two lasers pumping schemes for an interaction time of 20 μs [5]

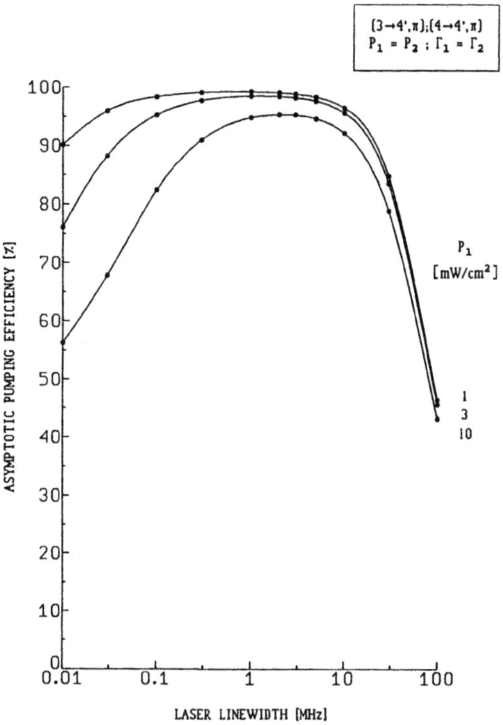

Fig. 2. Asymptotic pumping efficiency as a function of the linewidth of the lasers for various intensities [5]

widths greater than 10 MHz), since the Zeeman sublevel F=4, m_F=0 is depleted as more and more atoms are excited to the $\Delta F=\pm 1$, $\Delta m_F=0$ Zeeman sublevels, even if those transitions are 200 MHz apart with natural linewidths of 5 MHz. For lower intensities, the optimum efficiency is very close to 100%, but it decreases as the intensity go beyond 10 mW/cm^2. It is interesting to note that the increase in intensity corresponds not only to an increase in coherence effect at narrow linewidths, but also to an increase in wings effect at larger linewidths, thus to less optical pumping efficiency.

We calculate the time needed to achieve 95% of the asymptotic efficiency (illustrated in Fig. 3). The pumping time decreases as the intensity is increased. However there are various values for the laser linewidth that minimize the pumping time for each intensity. The pumping times obtained with narrow linewidth lasers are quite large due to the creation of correlation between ground state hyperfine coherences and lasers fluctuations. At the opposite, for much wider laser linewidths, the correlations between optical coherences and corresponding field fluctuations decreases, which increases the pumping time. It is interesting to note that good values for the laser linewidth minimizing the pumping time also give excellent pumping efficiency, for a given intensity. However, as the intensity increases, the pumping time and the asymptotic efficiency decrease. Thus, since experimentally the interaction time is limited, there will be an optimum intensity.

For a given available laser power, one must make a compromise between intensity and interaction time, depending on the beam pattern. For a finite interaction time, we have to consider the real transient which takes into account the asymptotic efficiency as well as the pumping time. The pumping efficiency achievable after 20 μs

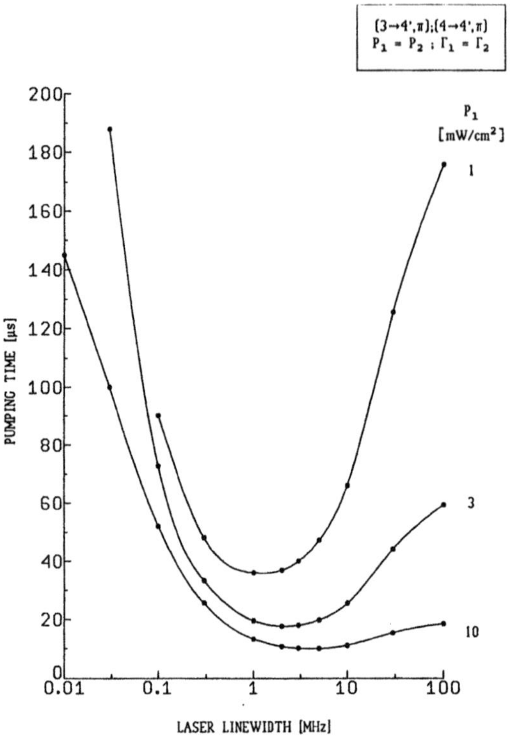

Fig. 3. Pumping time as a function of the linewidth of the lasers for various intensities

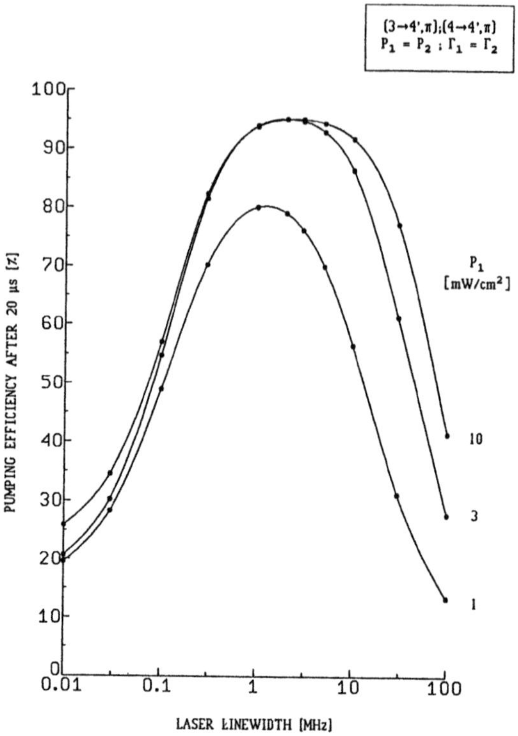

Fig. 4. Pumping efficiency after an interaction time of 20 µs as a function of the linewidth of the lasers for various intensities

is given in Fig. 4. We see that the pumping efficiency is reduced for low intensities since the pumping time is rather long. The optimum efficiency is around 95% and it is obtained for intensities around 3 to 10 mW/cm^2 and for linewidths around 1 to 2 MHz. These optimum conditions will obviously vary with the interaction time.

Finally, we have calculated the variation of the pumping efficiency when the laser 2 is slightly detuned from the center of the (4→4') transition. This is illustrated in Fig. 5. The first curve corresponds to the monochromatic case. We see the appearance of a "black line" as has been observed and verified using a three-level model [7]. The pumping efficiency decreases from 97% for a detuning of ±3 MHz to 31% for a detuning of +101 KHz; when both lasers are tuned, the pumping efficiency is around 45%. This result is quite different from that obtained considering the coherence effect, but neglecting the wings effect (14%) [8]. Thus, the neighboring transitions have an important contribution that must be accounted for. The width of this "black line" is around 400 kHz. Its depth decreases rapidly as the linewidth of the lasers is increased, and it is already negligible for laser linewidths of 2 MHz.

4. Conclusions

We have introduced a new model to evaluate the pumping efficiency of two lasers pumping schemes taking into account the effect of atomic coherences, those of neighboring transitions, and fluctuating laser fields. This model has been used to calculate the efficiency of two lasers pumping schemes for the cesium beam frequency standard. It has been shown that the (3→4',π);(4→4',π) pumping scheme seems to be

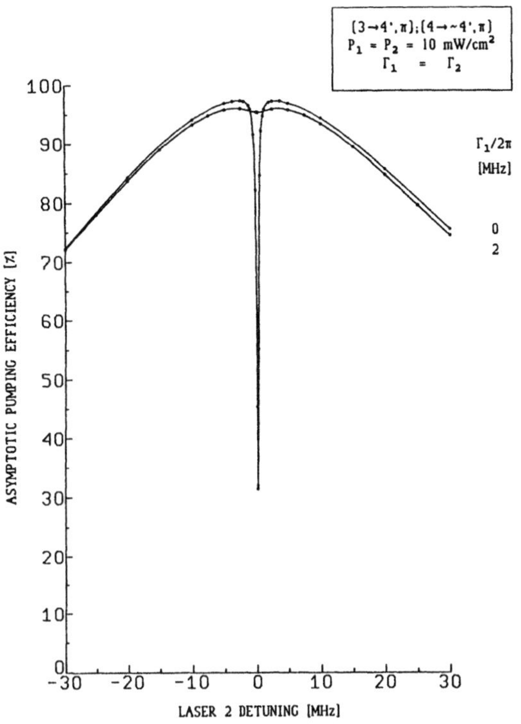

Fig. 5. Asymptotic pumping efficiency as a function of the laser 2 frequency detuning for various linewidths

the most efficient for typical conditions. Using this scheme, the best efficiencies are obtained with an intensity around 3 mW/cm^2 coming from lasers having a linewidth around 2 MHz. Finally, it has been shown that the "black line" seen with monochromatic lasers disappears rapidly when the linewidth of the lasers is increased, and will not be perceptible for typical intensities.

Acknowledgements

We wish to thank Dr. Jacques Vanier from C.N.R.C. and Alain Michaud from Université Laval for their numerous and helpful comments. One of the authors wish to thank the Conseil de recherches en sciences naturelles et en génie and the Conseil national de recherches for their financial support.

References

1. L.S. Cutler: U.S. patent 4,425,653
2. G. Avila, E. de Clercq, M. de Labachellerie, P. Cérez: IEEE Trans. Instrum. Meas. IM-34, 2 (1985)
3. G. Avila, V. Giordano, V. Candelier, E. de Clercq, G. Théobald, P. Cérez: Phys. Rev. A 36, 8 (1987)
4. S.-I. Ohshima, Y. Nakadan, Y. Koga: IEEE Trans. Instrum. Meas. IM-37, 3 (1988)
5. C. Jacques and P. Tremblay: in Proc. of the 42nd Annual Frequency Control Symposium, Baltimore, Md, U.S.A., 1988
6. P. Tremblay and C. Jacques: to be published
7. G. Orriols: Il Nuovo Cimento 53B, 1 (1979)
8. E. de Clercq, M. de Labachellerie, G. Avila, P. Cérez, M. Têtu: J. Physique 45, 2 (1984)

Work on Primary Cesium Beam Frequency Standard at CRL

K. Nakagiri, M. Shibuki, H. Okazawa, M. Aida, and N. Kotake

Communications Research Laboratory, Ministry of Posts and Telecommunications, Nukui-Kitamachi 4-2-1, Koganei, Tokyo 184, Japan

1. Beam Tube

Initially, beam tube experiments adopted an H-plane-bend Ramsey cavity using an E-plane feed and a solenoid coil to make a uniform C-field [1], [2]. This system is similar to the Cs 1 and 2 atomic clocks at PTB. Thereafter a conventional type E-bend Ramsey cavity with a Helmholz coil replaced the initial equipment, in order to decrease large and unreasonable characteristics in microwave power dependent frequency shift (power shift in abbreviation). However, the power shift was not effectively improved until 1984 when slow-velocity ($\sim$110[m/s]) beam optics were realized by using 3cm-long hexapole magnets [3]. Moreover, the position-adjustable beam-stoppers were installed and slightly but significantly improved beam uniformity and beam direction reversal reproducibility [4].

The improvements to date achieved 4[pA] peak-to-valley current in a 100[Hz] linewidth Ramsey pattern at 112[°C], and its linewidth is 100[Hz]. Experimental characteristics and theoretical calculations agree well [5],[8].

2. Electronic Servo System

We formerly adopted a new squarewave frequency modulation. It is performed directly by changing the control voltage of a 5.00688 [MHz] crystal oscillator [2],[6]. To control the C-field automatically by using ($F=4$, $m_F=\pm1$)$\rightarrow$($F=3$, $m_F=\pm1$) transitions sequentially, the system was replaced with the conventioal one [3].

To avoid spurious sidebands effect due to ground-loop noise, three types of dc-cut isolation couplers were installed between electronic circuits in the servo system [4]. The single sideband around 9.2[GHz] microwave are -40, -50 and -60 [dB] at offset frequencies of 1, 10 and 100 [Hz], respectively.

3. Accuracy Evaluation

Since 1984, accuracy evaluation data has been compiled once or twice a year for delivery to the International Bureau of Weights and Measures (BIPM) to calibrate the International Atomic Time (TAI).

Frequency Standards and Metrology Editor: A. De Marchi

Table 1 shows a typical uncertainty list from CRL Cs 1 accuracy evaluation. Primary uncertainty evaluation methods are given as follows.

The large power shift was formerly assumed to be caused by distributed cavity phase [2]. However, the calculated results on this supposition can not explain all the phenomena of past experiments, and due consideration of the analysis in [7] suggests that this physical interpretation is incorrect [4]. Therefore we assume that the sources of travelling waves are derived from TE_{11} mode radiation leakage in the cylindrical cut-off waveguide tube attached to the Ramsey cavity, and from insufficient short at the cavity end [8]. The calculated results should physically explain the power shift. These calculations also consider the effects of gravity and modulation frequency. Additionally, the calculated weights of distributed cavity phase difference can be used to obtain the mean value for first-order Doppler shifts in beam direction reversal.

The uniformity was roughly one order better in a C-field made by a solenoid coil than in that by a helmholz coil. In the latter C-field the measured mean value of the drift region before assembling was 1% higher than that of either interaction region. The effect of this on the power shift of ($F=4$, $m_F=\pm1$)$\rightarrow$($F=3$, $m_F=\pm1$) and clock transition is estimated by measuring the frequency difference between the Rabi and Ramsey peaks of the neiboring field-dependent transitions [3]. As the results obtained are in good agreement with the experiments, these field-dependent transitions are used to control C-field intensity and compensate for its distortion [4].

Figure 1 shows two examples of pulling by neighboring transitions in comparison with the calculated ones. The microwave field shape of interaction assumed to be squarewave, because the results obtained by using the smoothed shape like B in the reference [4] do not agree with the

Table 1. CRL Cs 1 Uncertainties, 1986

Primary Source	Uncertainty($\times10^{-13}$)
1. 1st-order Doppler and cavity phase shift	0.42
2. 2nd-order Doppler shift	0.02
3. Magnetic field	
1) Offset due to finite field	0.2
2) Field inhomogeneity	0.1
4. Pulling by neighboring transitions	0.25
5. Servo system offsets	0.25
6. Measurement error	0.43
Total (*rms*)	0.7

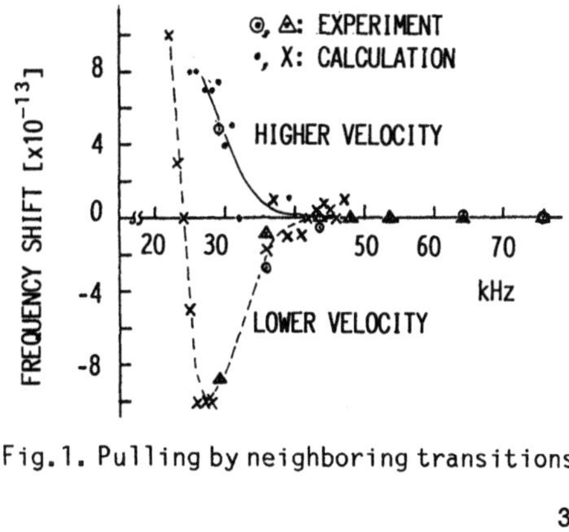

Fig.1. Pulling by neighboring transitions

experimental results of the lower velocity beam in Fig. 1. To confirm this physical phenomenon of the TE_{11} mode in cutoff waveguides, the theoretical calculations of π-transition are carried out. As in CRL Cs 1 such π-transitions cannot be observed around the σ-transitions, the TE_{11} cutoff mode does not physically contribute to the transition, though its leakage should induce four low-intensity traveling waves along the beam path, as described above.

The frequency shift due to microwave spectrum impurities is calculated by the method in the reference [9] and evaluated by measuring the power shifts in beam direction reversal. To estimate the effect of impurities more precisely, it is necessary to investigate them in detail experimentally and theoretically.

4. Future Work

A new cesium beam primary frequency standard is under construction at CRL. It will start operation in 1989 as primary clock Cs 2. Its design and characteristics are the same as those of Cs 1. The joint goal for Cs 1 and 2 is to constantly attain 10^{-14} accuracy as is obtained by the present standard.

References

1. M.Kobayashi, K.Nakagiri, S.Urabe, M.Shibuki, and Y.Saburi: IEEE Trans. Meas. IM-27(1978)447.
2. K.Nakagiri, M.Shibuki, S.Urabe, R.Hayashi and Y.Saburi: 3rd Symp. Frequency Standard and Metrology, Aussoi(France), 1981, J. Phys. (Paris)42(1981)Suppl. 12, c-8-253.
3. K.Nakagiri, M.Shibuki, S.Urabe, Y.Ohta, T.Morikawa, and Y.Saburi: Proc. 38th Ann. Symp. Frequency Control, Philadelphia, 1984(IEEE, Piscataway, 1985)p.447.
4. K.Nakagiri, M.Shibuki, H.OKazawa, J.Umezu, Y.Ohta, and H. Saitoh: IEEE Trans. Instrum. Meas. IM-36(1987)617.
5. K.Nakagiri, H.Okazawa, and S.Urabe: Proc. 18th Ann. Precise Time and Time Interval Applications and Planning Meeting, Washington, D.C., 1986(Naval Observatory, Washington, D.C.,1987)p.503.
6. K.Nakagiri, S.Urabe, M.Shibuki, M.Kobayashi, and Y.Saburi: J. Jpn. Soc. Appl. Phys. 50(1981)1221 [in Japanese].
7. J. Vanier, A.G.Mungall, and J.-S. Boulanger: Metrologia, 20(1984)101.
8. K.Nakagiri, H.Okazawa, M.Shibuki, and S.Urabe: submitted to Jpn. J. Appl. Phys.
9. C.Audoin, M.Jardino, L.S.Cutler and R.F.Lacey: IEEE Trans. Instrum. Meas. IM-27(1978)325 .

Improvement and Reevaluation of Chinese Primary Frequency Standard

Yang Xiaoren, Wu Changhua, and Li Mingshou

National Institute of Metrology (NIM), Beijing, People's Republic of China

I. Introduction

The Chinese primary cesium beam frequendy standard CSIII has been reported at the previous Symposium [1]. This paper describes briefly its major improvements and its reevaluation by the Time/Frequency Division, NIM in the recent years.

II. New Dipole Beam Optics System

Since 1982, a series of improvements to CSIII has been performed [2,3] and the most important one is the use of a new dipole beam optics system.

The traditional dipole state selection magnets are modified so that the distribution of the deflection field H in the magnetic gaps approaches:

$$H = H_0 + kz^2$$

and the force F exerted on the atoms in the gap approaches:

$$F = \pm 2k\mu z$$

where H_0 is a constant, k is about 0.4T/mm, z is the distance from the beam optical axis and μ is the effective magnetic moment of the atoms. The new beam optics disposition is shown in Fig.1 and the hand swept Ramsey pattern for the new optics system is shown in Fig.2.

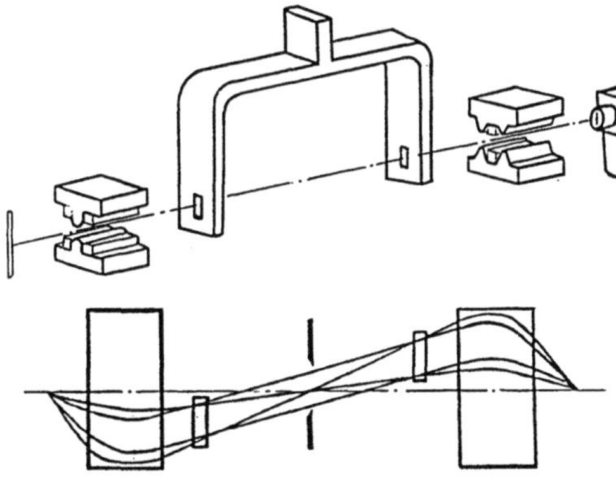

Fig.1 New Beam Optics Disposition

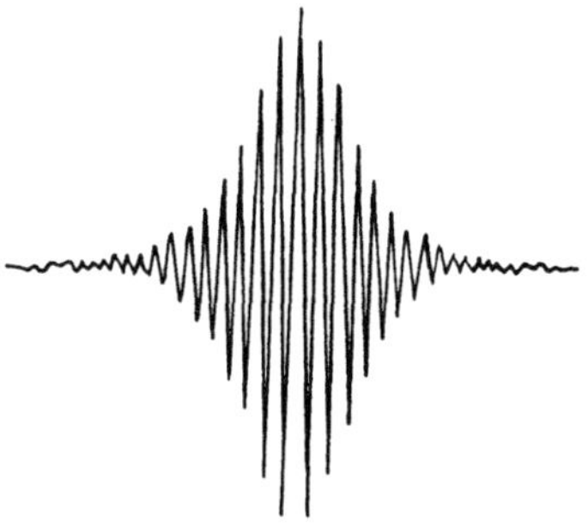

Fig.2 Hand Swept Ramsey Pattern for New Optics System

The advantages of the new beam optics are as follows:

- .Narrower velocity range of transition atoms;
- .Better focusing capability and better symmetry;
- .Stronger transition signal;
- .Easier selection of slow atoms;
- .Lower background noise by using flop-in detection.

The problems induced are trimming the shape of the magnet pole to keep $k\mu$ constant at different points and regulating the direction change of the magnetic field at the region between the state selection magnets and the C field to eliminate the Majorana effect. For the latter, mumetal rings and an armco iron tube have been added to the cover shield of each magnet as shown in Fig.3.

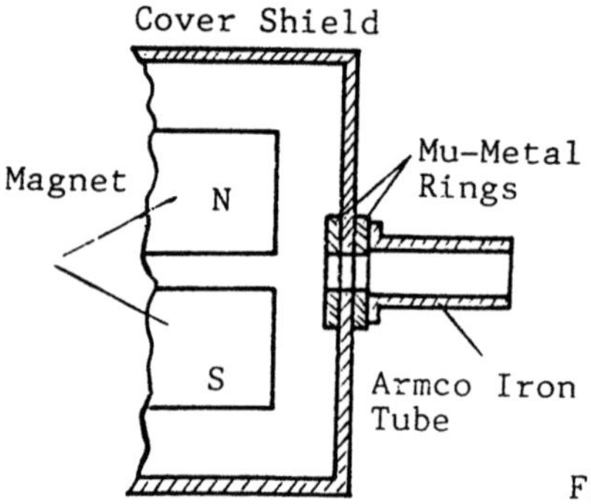

Fig.3 Additional Shielding of State Selection Magnet

III. Effective Measurement of Frequency Shift Due to Cavity Phase Difference [4]

In primary cesium beam frequency standard it is rather difficult to determine accurately the effective velocity of atoms V_P in the formula of cavity phase shift bias and V_D in the formula of second order Doppler effect, since they are related to the practical velocity distribution of atoms. By defining an effective velocity of atoms V_W:

$$V_W = 2LW$$

where L is the distance between the two interaction regions and W is the linewidth of the central resonant line in the Ramsey pattern, we find that V_P and V_D can be replaced by V_W in the evaluation of cavity phase shift bias and second order Doppler effect with an uncertainty in the order of 10^{-15} for various existing velocity distributions.

IV. Reevaluation of CSIII

After all the improvements performed, CSIII has been reevaluated in the period from July 1986 to February 1987. The resonance linewidth decreased to 34 Hz. The error budget of the modified CSIII is shown in Table 1.

Table 1. Error budget of modified CSIII

Bias source	Fractional frequency bias ($\times 10^{-13}$)	Uncertainty ($\times 10^{-13}$)
Cavity phase shift	-9.3 (E-W) +9.3 (W-E)	1.2
Rabi pulling & spectrum impurities	+4.4	2.0
Second order Doppler effect	-5.7	0.4
C field	+3.2	0.5
Cavity pulling	—	0.1
Servo system	—	0.6
Total	-7.4 (E-W) +11.2(W-E)	2.5 (1σ)

V.Subsidiary Ramsey Pattern [5]

For a quasi-rectangular velocity distribution of atoms with a narrow velocity range ΔV and centered at the velocity V_0, the transition probability P can be derived as:

$$P \propto \cos(2\pi f/2^{n+1}/B/T_f)\cos(2\pi f/2^{n}/B/T_f)\cos(2\pi f/2^{n-1}/B/T_f)\ldots$$
$$\ldots\cos(2\pi f/2^{3}/B/T_f)\cos(2\pi f/2^{2}/B/T_f)\cos(2\pi f/T_f)$$

where f is the frequency offset from the center of the resonance, $B=V_0/\Delta V$ and $T_f=V_0/L$. This formula predicts that all nodes of the decaying sinusoidal envelopes occur at:

$$f = KBT_f\ , \qquad K=1,2,3,\ldots$$

and between two nodes there is a group of resonant lines, which is called the sub-Ramsey pattern. Figure 4 shows the Ramsey and sub-Ramsey patterns observed at L=368 cm, V_0=270m/s and ΔV=44.2 m/s.

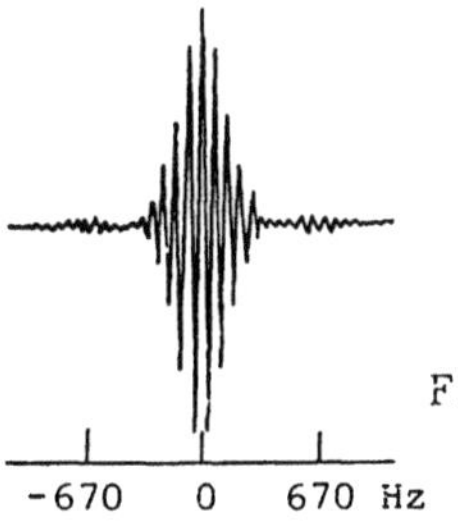

Fig.4 Ramsey and Sub-Ramsey Patterns Observed

VI.References

1. Yang Xiaoren, Journal de Physique, supplement, Tome 42 (1981) C8-257.
2. Chuanyun Xue, IEEE Trans. IM-29 (1980) 301.
3. Li Mingshou, CPEM'88 Digest (1988) 386.
4. Xue Chuanyun, Acta Metrologia Sinica (Chinese), Vol.5 (1984) 91.
5. Li Enxian and Xue Chuanyun, IEEE Trans. IM-36 (1987) 620.

Preliminary Results on Small Optically Pumped Cesium Resonator

P. Thomann, H. Schweda, and G. Busca

Oscilloquartz S.A., CH-2000 Neuchâtel, Switzerland

We have constructed a commercial size optically pumped cesium resonator. Optical pumping of the atomic beam and detection of the Ramsey signal are both achieved with a single laser. The linewidth of the Ramsey signal is 950 Hz and the present signal-to-noise ratio is 1500 $Hz^{-1/2}$. Based on these measured values, the calculated short-term frequency stability is given by $\sigma_y(\tau) \cong 5\ 10^{-11}\ \tau^{-1/2}$. The main present limitation is the noise contribution of the scattered light. A tenfold reduction of scattered light, recently achieved, should improve the S/N ratio by the same amount.

Description of the Resonator

The principle of operation of an optically pumped tube is illustrated in Fig. 1. The atomic beam produced by the oven first intersects the pump laser beam in interaction region A. The laser frequency is tuned to one of the hyperfine components of the D_2 resonance line. A combination of hyperfine and/or Zeeman pumping leads to a population difference between the two clock states (F = 3, m = 0 and F = 4, m = 0) The fluorescent light

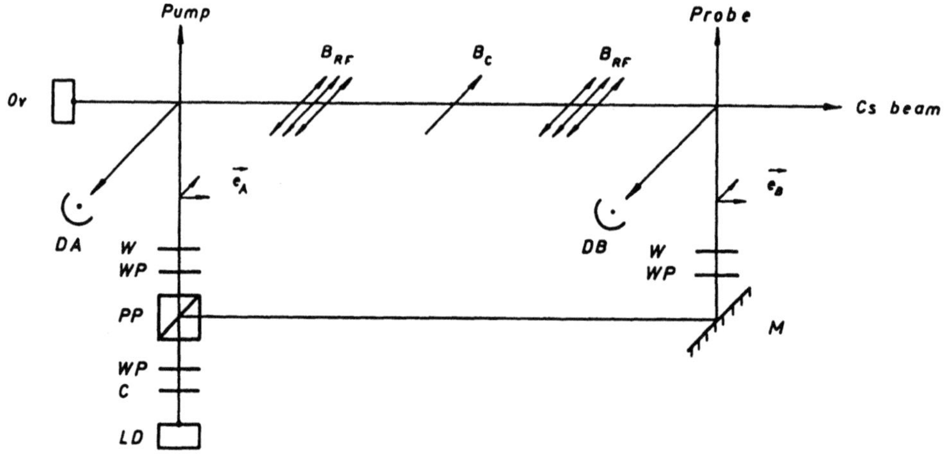

Ov	: Cs oven	$e_{A,B}$	: possible polarization of laser beams	M	: mirror
B_{RF}	: microwave interaction regions	W	: windows in vacuum envelope	C	: collimating optics
B_C	: C-field	WP	: waveplate	LD	: laser diode
DA,DB	: photodetectors	PP	: polarizing prism		

FIGURE 1 : Schematic diagram of the resonator

Frequency Standards and Metrology Editor: A. De Marchi

emitted by the atoms in the optical pumping region is directed by the "A" collecting optics towards the A detector. This detector can be used to lock the laser frequency to the suitable hyperfine transition. A photocell collecting the fluorescent light from the B interaction region detects the Ramsey signal.

Several pumping and probing schemes can be envisaged involving the use of one, two or even three lasers [1-4]. For reasons of simplicity and reliability, we consider here a single-laser set-up, where an unmodified commercial GaAlAs laser with a spectral width of 50 MHz is used.

The light collection system was designed to maximize the solid angle of light collection. It is made of two concave mirrors, one elliptical and one spherical, facing each other. The interaction region is centered at one focus of the elliptical mirror while the hole in the spherical mirror and in the magnetic shield are at the second focus.

Preliminary Results

A recording of the B photodetector signal as a function of RF frequency (Ramsey signal) is shown in Fig. 2. The signal-to-noise ratio, S/N, measured at the center of the Ramsey resonance in a 1 Hz bandwidth around 500 Hz, is S/N = 1500 in both cases. The linewidth (FWHM) is 950Hz, as expected from the length of the Ramsey cavity. From these measured values, we calculate the Allan variance $\sigma_y(\tau)$ of relative frequency fluctuations of an oscillator locked to such a resonator :

$$\sigma_y(\tau) = \frac{W}{\sqrt{2}\,(S/N)\,\nu_o}\,\tau^{-1/2} = 5\;10^{-11}\tau^{-1/2}$$

(ν_o = 9.2 GHz). This is comparable to the short-term stability of conventional resonators of the same size where, however, both linewidth and S/N ratio are smaller by a factor of two.

We have carried out an analysis of the contributions to the S/N ratio. The Ramsey signal amplitude is 0.3 nA (photodetector current, responsivity 0.6

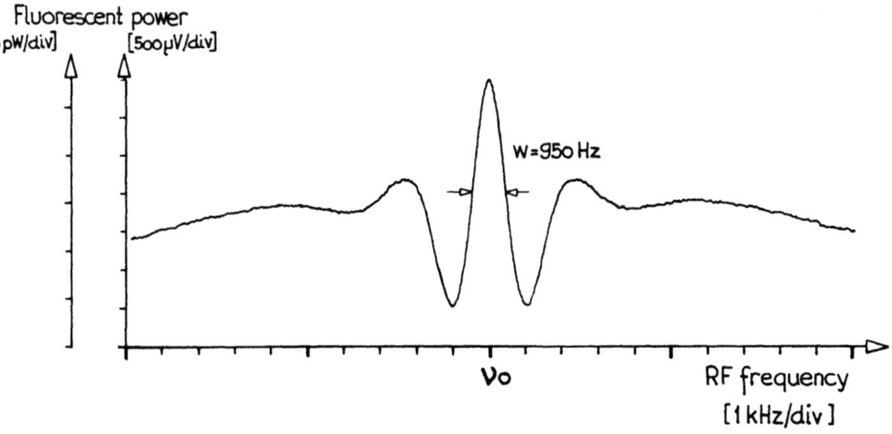

Fig. 2 : Ramsey pattern (4 - 4 σ transition, detector UV250B/10Ω)

A/W); the background signal due to unwanted scattered light is 16 nA. Among the various contributions to the total noise, laser intensity noise associated with the background light dominate largely. By reducing the amount of background light, which we have already achieved in separate experiments by a factor of 10, and by moving the detector closer to its ideal position, we expect to improve the S/N ratio by an order of magnitude.

A similar scan to that of Fig. 2 but over a larger RF frequency range (Fig.3) shows the Ramsey patterns corresponding to all the field dependent transitions (m_F = ± 1, ± 2, ± 3). The symmetry of the overall spectrum contrasts with the generally asymmetric spectra of magnetic deflection resonators; this symmetry is a consequence of optical pumping with linearly polarized light. It is a desirable property since Rabi pulling and Majorana effects are reduced. This spectrum was recorded with a single magnetic shield. The good visibility of the Ramsey patterns on the outermost resonances is proof of the C-field homogeneity (B = 53 mG, homogeneity 1%).

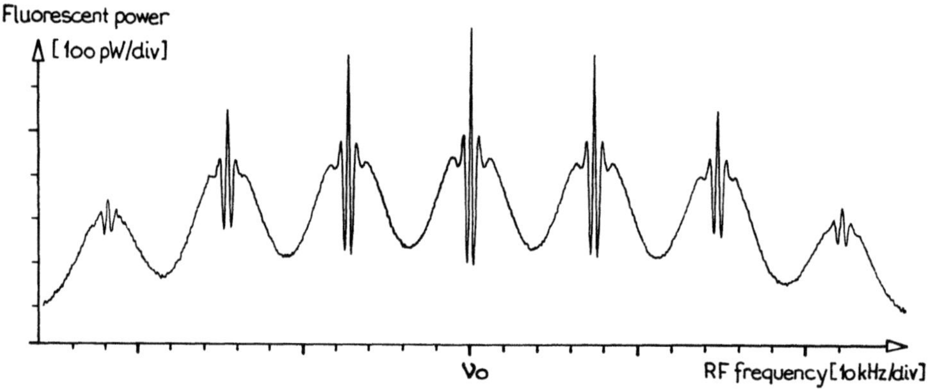

Fig. 3 : RF spectrum 4 - 4 σ transition, detector UV250B/10Ω)

Conclusion

Preliminary measurements on a prototype of commercial-size optically pumped cesium beam tube show that the short-term stability, already equivalent to that of a conventional resonator of same size, may be improved further by a factor of ten.

References

1. V. Candelier, V. Giordano, A. Hamel, G. Theobald, P. Cerez, C. Audoin : proceedings of the second European Frequency and Time Forum (March 1988, Neuchâtel), p. 483
2. SHANG Song-quan, WU Xin-Xin, YAO Shu-Tong, XIE Lin-Zhen, WANG Yi-qiu, Chinese Phys. Lett. 2, 12(1985) 557-560
3. T.M. Clelland et al., Proc. of the 41st Annual frequency Control Symposium, 1987, p. 59
4. G. Avila, V. Giordano, V. Candelier, E. de Clercq, G. Theobald, P. Cerez : Phays. Rev. A 36, 8(1987) 3719-28.

Determining the Velocity Distribution in Cesium Beams from the Response to a Microwave Pulse

G.D. Rovera[1], *A. Premoli*[2], *and A. De Marchi*[3]

[1]Theo-tek, Paesana

[2]Dip. Elettronica, Elettrotecnica ed Informatica, Università di Trieste, Italy

[3]Dip. Elettronica ed Automatica, Università di Ancona, Italy

The determination of the velocity or time-of-flight (TOF) distribution of the atoms in atomic beam frequency standards is important in order to evaluate to a good approximation some physical phenomena determining their accuracy.

Optical methods like the analysis of the laser doppler profile or of laser pulse response could be used in experimental and/or optically pumped beams. For sealed Ramsey tubes proposed methods range from Fizeau-type pulsed microwave measurement to Fourier analysis of Ramsey curves or beam intensity vs microwave power curves.

It is proposed here to use the double-bump time response to a microwave pulse. This method is thought to offer fast acquisition and good before-analysis feeling for the distribution. The measurement is performed by pulsing the excitation microwave signal in the Ramsey cavity and observing the signal produced by atoms that reach the detector having been excited in either Rabi cavity. In the ideal case of short pulses, short Rabi cavities, and fast detector this signal yields directly the sum of the two TOF distributions originated from the two interaction zones, weighted by the TOF; it is then very easy to obtain with simple calculations the desired distribution. In reality this signal is distorted by several effects, the most important of which are the duration of the microwave pulse, the finite length of the Rabi cavities, and the detector's dynamic response. The first gives a velocity selection because not all atoms can see the full pulse before they exit the interaction zone, the second gives a spread of the geometrical origin for the TOF length base, the third gives a distortion of the observed response with respect to the incoming atomic flux. Unfolding the third is in principle the most difficult task because the dynamic response of the detector cannot be measured directly and must be inferred. This can be done by comparing the shape of the two bumps of the observed response, provided other distortion effects can be supposed unimportant. Unfolding the first two is in principle easier because the geometry of the tube is known. However trying to correct for all these effects at the same time may result in a too undefined problem and yield inaccurate aswers. It is best to identify in each case the biggest problem and correct for that as if it were the only one.

In long tubes, because of low beam flux, a high detector sensitivity is required, which may then be paid with a narrow band in its circuitry. Because of this, and because the aspect ratio ℓ/L of the Ramsey cavity is favorable, the biggest effect in this case is usually the detector response. The output signal y(t) is given by the convolution of the number z(t) of atoms per second reaching the detector and the impulse time domain response $h_d(t)$ of the latter (for example $h_d(t)$ can be a decreasing exponential or a damped sinewave). On turn z(t) is given by the sum of two terms z1(t) and z2(t), coming from the two interaction zones. These have equal length ℓ and differ only by the different distances L_1 and L_2 from the detector. If the aspect ratio of the cavity were not favorable, as in short tubes, one should also take into account that both $z_1(t)$ and $z_2(t)$ depend on the integral of the TOF distribution $\rho(\tau)$ over ℓ (τ it the TOF through a given base length).

Frequency Standards and Metrology Editor: A. De Marchi

In short tubes instead the important effect is usually the finite length ℓ of the interactions, which is a relevant fraction of distances L_1 and L_2. This gives an integration over ℓ which must be undone for the best results.

The procedure for the calculation follows these lines: the unknown distribution $\rho(\tau)$ is expressed as a linear combination with unknown coefficients of known bell shaped functions (Fig.1). The unknown coefficients ρ_j which describe $\rho(\tau)$ are determined by a linear mean squares algorithm to best fit the observed response after the distortion introduced by the considered effect.

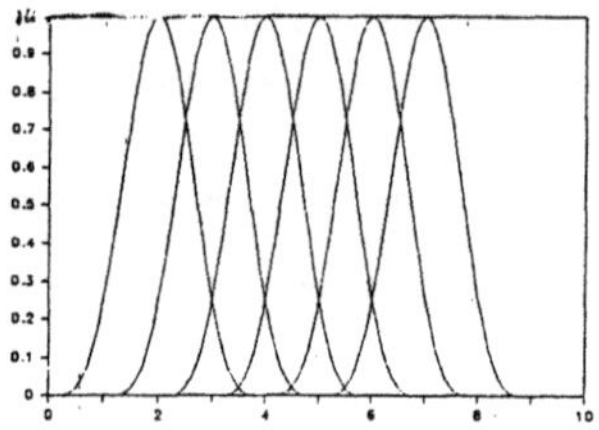

Fig. 1: Bell shaped functions.

In the case of the long tube we assume the detector to be an undetermined linear dynamic system of order not exceeding two, then $h_d(t)$ is characterized by the frequency and Q-factor of the complex pole pair. These parameters are supposed determined for the calculation of the coefficients ρ_j. On turn, their optimal values are estimated by minimization of the average square residuals. A simulation of this process has been done by building up a reasonable $\rho(\tau)$ by the use of the bell functions, computing the double-bump pulse response, passing it through a second order transfer function to simulate the detector, and adding noise. The ρ_j and the detector characteristics are then ricomputed by the described optimization process and compared in Fig.2 and 3.

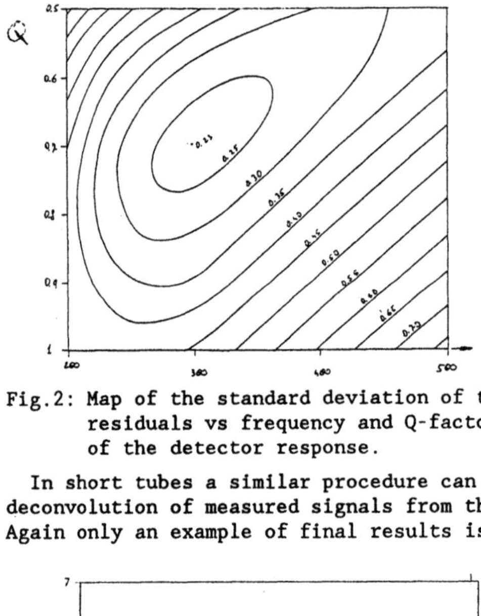

Fig.2: Map of the standard deviation of the residuals vs frequency and Q-factor of the detector response.

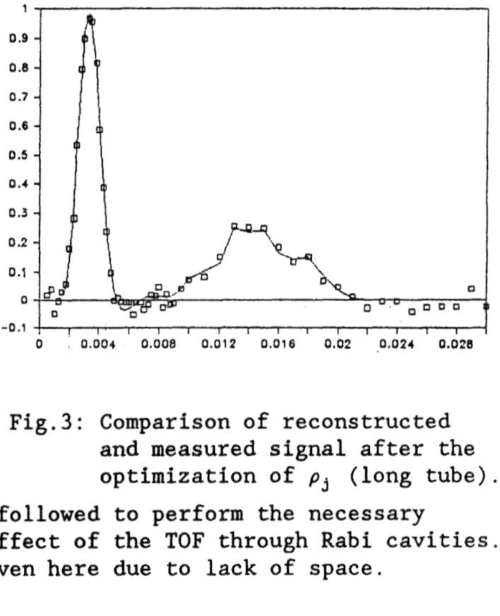

Fig.3: Comparison of reconstructed and measured signal after the optimization of ρ_j (long tube).

In short tubes a similar procedure can be followed to perform the necessary deconvolution of measured signals from the effect of the TOF through Rabi cavities. Again only an example of final results is given here due to lack of space.

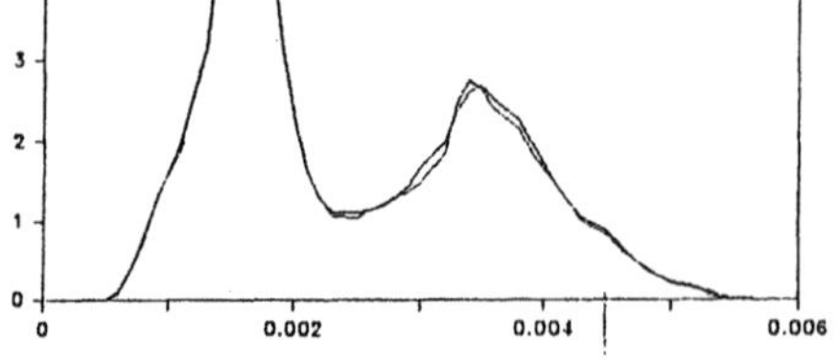

Fig.4: Comparison as in Fig.3, (short tube).

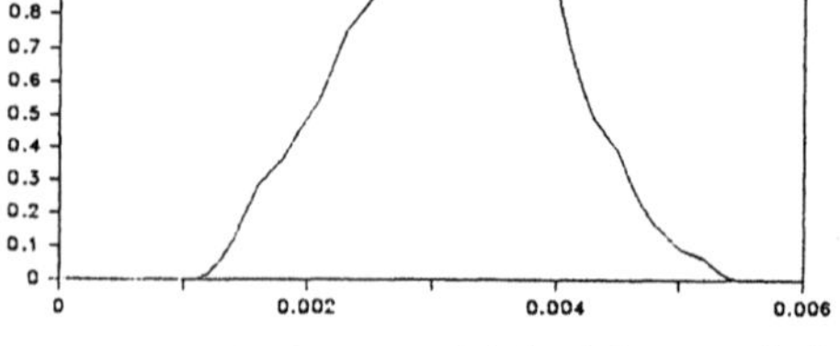

Fig.5: Calculated $\rho(\tau)/\tau$ (short tube).

In both cases it is assumed that enough power is available to saturate the transition with a single pulse of about 10 μs, in this way all velocity groups are equally excited to a good approximation and the first problem mentioned above is considered unimportant. Of course the C field in this case must be consequently increased so that neighboring transitions don't alter the data.

The method presented here can give good results if properly applied. Although the computation time may be longer than in other methods, the measurement time on the tube is very short. This is of great value because of the limited time usually available for testing on these devices.

Physical Problems Determining the Reliability of the Classical Microwave Frequency Standards

G. Busca[1], *H. Schweda*[1], *P. Thomann*[2], *and L.G. Bernier*[2]

[1]Observatoire Cantonal de Neuchâtel, CH-2000 Neuchâtel, Switzerland
[2]Oscilloquartz S.A., CH-2000 Neuchâtel, Switzerland

Even if the microwave frequency standards are widely used in our modern life, their reliability has not reached the level which is expected for a "mature" technical product, the type of the market requiring such a standards preventing, up to now, the "purification stage" which is typical of "consumer products".

In this situation the reliability can only be improved by a better understanding of basic physical problems which are not yet totally understood or that are still out of our control. TABLE 1 gives a list of such problems and illustrate the associated research (to the best knowledge of the authors).

TABLE 1 : Primary and Commercial Cs Beam

Problems	Cause	Research effort
1. Signal decay V/S time	Elect. Multiplier gain deterioration V/S time	Hewlett-Packard [1] Oscilloquartz SA [2]
2. Frequency shifts associated with non-homogeneous static magnetic field	Magnetic field homogeneity depending upon the design	P.T.B. [3]
3. Ramsay cavity interrogation	Cavity phase shift and Rabi pulling	N.B.S. [4] Un. of Ancona [5] P.T.B. [in this proceedings]
	H Maser	
4. Maser amplitude decay V/S time	Aging of the dissociator efficiency	N.B.S. [6] In progress at Oscilloquartz SA. William College [in this proceedings]
5. Aging of Teflon coating	unknown	SAO [7] William College [in this proceedings]

Frequency Standards and Metrology Editor: A. De Marchi

Passive Rubidium

Problems	Cause	Research effort
6. Lamp lifetime	Rubidium Diffusion into the glass	Aerospace Corp [8]
7. Light-shift	Distributed light-shift	Aerospace Corp [9] Oscilloquartz SA [10]
8. Frequency drift associated with the cell	unknown	In progress at Observatory of Neuchâtel.
9. Compensation of light-shift, buffer gas shift spin-exchange shift	Mechanism unknown at the level of few parts in 10^{-10}. Value of the spin-exchange shift unknown.	Laval Un. [11] In progress at Observatory of Neuchâtel.

Positive solutions are in view for the problems 1, 3, 6, 8. The other problems need still a basic and extensive experimental investigation.

References

1. E.R. Straka: Proc. 36th Ann. Symp. Frequency Control, 1982, p. 236
2. H.S. Schweda, P. Thomann, G. Busca: Proc. 2nd European Time and Frequency Forum, 1988, p. 435
3. A. Bauch, T. Heindorff: Proc. 37th Ann. Symp. Frequency Control, 1985, p. 8
4. A. de Marchi, J. Shirley, R.E. Drullinger: IEEE Trans. on Instr. and Measurement, 1988
5. A. de Marchi: IEEE Trans. UFFC-34, 598 (1987)
6. K.B. Person, F.L. Walls: NBS Techn. Note, to be published
7. E. Mattison, R.F.C. Vessot: Proc. 44th Ann. Symp. Frequency Control, 1987, p. 95
8. C.H. Volk et al.: Proc. 38th Ann. Symp. Frequency Control, 1984, p. 387 and 41st ASFC, 1987, p. 36
9. J.C. Camparo, R.P. Frueholz: J. Appl. Phys. 59, 301 (1986)
10. G. Busca, L. Pan: Proc. 1st European Time and Frequency Forum, 1987, p. 331
11. J. Vanier et al.: Proc. 3rd Symp. on Frequency Standards and Metrology, J. de Physique C-8, 139 (1981)

Difference Frequency in Two Isotopes Atomic Beam Magnetic Resonance Apparatus as Candidate for Frequency Standard

K.W. Radecki

Institute of Radioelectronics, Warsaw University of Technology, Nowowiejska 15/19, PL-00-665 Warsaw, Poland

An idea of the difference frequency measurement in two isotopes atomic beam magnetic resonance apparatus is presented and the main advantages are briefly described.

1. Introduction

In an atomic beam magnetic resonance apparatus using two isotopes it is possible to measure a simultaneous resonance of both isotopes. For practical applications to frequency standards two modes of operation can be considered:

1. A so-colled superposition line is measured [1]. The useful pairs of isotopes are for example: ^{107}Ag and ^{109}Ag, ^{63}Cu and ^{65}Cu or ^{203}Tl and ^{205}Tl. A silver beam magnetic resonance apparatus was built some time ago by HAHN and RADECKI [2]. Also the feasibility to build a similar device with laser optical pumping was described [3].
2. The difference frequency between resonant frequencies of both isotopes is measured. It is possible to use the same isotopes as previous and also combination of ^{87}Rb and ^{133}Cs. Some properties of silver, rubiduim and cesium isotopes are presented in Table 1.

A block diagram of the measurement circuit is shown in Fig.1. In a conventional Ramsey type apparatus exciting rf signals are applied to the mixer inputs. The difference frequency is equal:

$$f_d = f_{r1} - f_{r2} = \frac{f_1}{n_1} - \frac{f_2}{n_2} ,$$

where f_1 and f_2 are the microwave frequencies exciting both isotopes, n_1 and n_2 are the frequency multiplication factors.

2. Factors affecting reproducibility

The results of several calculations of frequency offsets due to magnetic C-field δf_C and 2-nd order Doppler effect δf_D in silver and cesium-rubidium atomic beam apparatus are given in Table 2. The following assumptions were made: the beam has Maxwellian velocity distribution (oven temperature is 1300 K for Ag and 373 K for Cs-Rb), C-field

Frequency Standards and Metrology Editor: A. De Marchi

intensity is 1 A/m for Ag and 5 A/m for Cs-Rb, $n_1=n_2$. If one Ramsey cavity with lenght 1 m is used, the offset due to excitation of both isotopes is $7.8\cdot10^{-12}$ for Ag and $3\cdot10^{-14}$ for Cs-Rb.

3. Frequency stability

It seems that some fluctuations of mixed signals may be correlated because of using the same oven, detector and similar electronics for both isotopes. The theoretical analysis of influence of various noises to the different frequency is under investigation.

Table 1. Some properties of candidate isotopes.

isotope	melting point [K]	nucl. spin I	hyperfine splitting [MHz]
^{107}Ag	1233.8	1/2	1712.512111
^{109}Ag			1976.932075
^{87}Rb	312	3/2	6834.682614
^{133}Cs	301.5	7/2	9192.631776

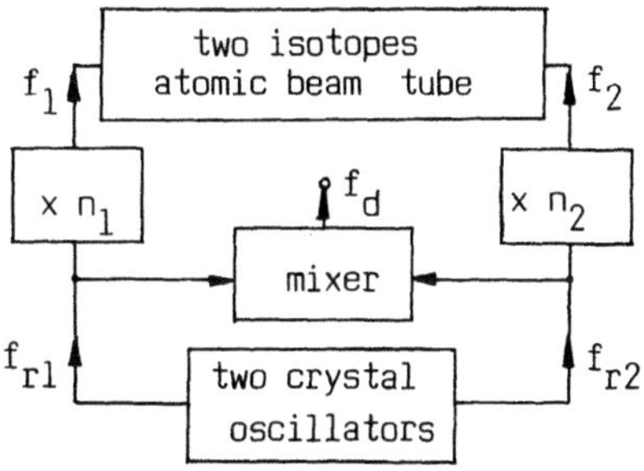

Fig.1. Block diagram of measuring circuit

Table 2. Factors affecting the reproducibility

	one isotope		two isotopes		
	δf_C	δf_D	f_1-f_2 [MHz]	δf_C	δf_D
^{107}Ag	$2.11\cdot10^{-10}$	$1.12\cdot10^{-12}$	264.4	$1.83\cdot10^{-10}$	$9.69\cdot10^{-13}$
^{109}Ag	$1.58\cdot10^{-10}$	$1.10\cdot10^{-12}$			
^{87}Rb	$3.11\cdot10^{-10}$	$3.96\cdot10^{-13}$	2357.9	$2.46\cdot10^{-10}$	$1.38\cdot10^{-13}$
^{133}Cs	$1.85\cdot10^{-10}$	$2.59\cdot10^{-13}$			

4. Conclusions

In comparison to one isotope tube similar frequency offsets due to C-field and Doppler effect are in two isotopes tube. Both isotopes should have similar physical and microwave properties. In optically pumped apparatus , because of many common elements for both isotopes, the influence of some perturbations to measured frequency may be reduced.

References

1. S.L. Hahn: Electron. Lett. 9, 609 (1973)
2. S.L. Hahn, K.W. Radecki: IEEE Trans. IM-27, 335 (1978)
3. S.L. Hahn, K.W. Radecki: IEEE Trans. IM-36, 639 (1987)

Studies and Developments of H-Masers at LHA

R. Barillet, P. Lesage, P. Petit, J. Viennet, and C. Audoin

Laboratoire de l'Horloge Atomique,
Equipe de Recherche du CNRS associée à l'Université Paris-Sud,
Bât. 221, Université Paris-Sud, F-91405 Orsay Cedex, France

1. Compact Active H-Masers for VLBI Applications

Three active H-masers have been manufactured at LHA for french radioastronomers and geophysicists. They are transportable and compact. Including electronics, power supplies and batteries, their volume is about 0.5 m^3 and their mass is 280 kg. They are operated from 220 Vac of 24 Vdc power supplies, and their internal battery makes it possible to operate them for 2 hours without external supply. The signal receiver and the P.L.L. have been designed to optimize the frequency stability transfer. Operating conditions are microprocessor controlled and monitored.

Frequency stability is about 1.10^{-15} for τ = 1 000 s /1/ and measurements of systematic effects are presented.

2. Fast Cavity Auto-tuning Systems

The long term frequency stability of hydrogen masers is degraded by the drifts of the resonant cavity.

We have realized an electronic apparatus able to tune the resonant cavity according to the method proposed in /2/. A square wave frequency modulated microwave signal, without carrier, is transmitted through the cavity. The amplitude of the transmitted signal is modulated when the mean value of the cavity resonant frequency is different from the reference frequency. A synchronous detector provides an error signal. After being filtered by a convenient transfer function, the voltage obtained is applied to the varactor that controls the cavity resonant frequency.

In these conditions, a tight control of the cavity resonant frequency is insured. The experimental results show that the apparatus does not degrade the short and medium term frequency stability, and allows a long term frequency stability of few part in 10^{-15}.

3. Studies of Dielectrically Loaded Cavities for Small Size Hydrogen Masers

A dielectrically loaded cavity made of a hollow alumina cylinder, used in order to reduce the hydrogen maser size,has been investigated /3/. A theoretical study has been performed which shows the influence of the cavity geometry on the frequency stability capability of the maser. The active mode as well as the passive mode of operation have been considered /4/, /5/. The analysis shows that achievable frequency stabilities do not differ very much for both modes of operation with,however, a slight advantage to the active mode.

Frequency Standards and Metrology Editor: A. De Marchi

References

1. R. Barillet, P. Petit, J. Viennet, C. Audoin : Proc. of the 40th Annual Frequency Control Symposium, Philadelphia (USA) pp. 410-412 (May 1986)
2. C. Audoin : Rev: Phys. Appl. 16, pp. 125-130 (1981) and 17, p. 273 (1982)
3. N. Yahyabey, P. Lesage, C. Audoin : Proc. of the Second European Frequency and Time Forum, Neuchâtel (Switzerland), to be published
4. P. Lesage, C. Audoin, M. Têtu : Proc. of the 33rd Annual Symposium on Frequency Control, Atlantic City, N.J. (USA), pp. 515-535 (1979)
5. C. Audoin, J. Viennet, P. Lesage : Journal de Physique 12, Supplément C8, pp.159-170 (1981)

Hydrogen Maser Auto-Tuning Systems

D. Morris

Division of Physics, National Research Council of Canada,
Ottawa, Ontario K1A 0R6, Canada

The active hydrogen maser is one of the most stable frequency standards for averaging times up to a few thousand seconds. In the longer term the frequency stability deteriorates, mainly as the result of frequency pulling due to changes in dimensions (and hence, resonant frequency) of the resonant cavity in the maser. In order to reduce this effect an auto-tuner (a device which causes the resonant cavity to be maintained automatically at the correct frequency) is employed. Two auto-tuning methods are considered in this paper. They are the beam flux modulation method [1] and the signal injection method [2].

Using two masers, H1 and H2, at the National Research Council of Canada (NRC) some comparison results of frequency stability were obtained using the two systems. The first system has been described in [1]. Both masers were auto-tuned, each being used alternately as a reference. The frequency comparison was carried out at the maser frequency. The 5 MHz signal used for supplying the frequency multipliers in the receivers was obtained from a quartz crystal oscillator locked to one of the NRC primary cesium clocks (CsV).

The second auto-tuning system used fast cavity tuning, employing injected signals. The output of a commercial synthesizer was switched between frequencies of approximately 0.390 MHz and 0.422 MHz at a 10 Hz rate. By up-conversion switched frequencies of 1420.390 MHz and 1420.422 MHz were applied to one coupling loop of the maser resonant cavity. Amplitude modulation at 10 Hz on the maser output (from a second coupling loop) was detected by a synchronous detector and used to provide a correction signal to set the cavity to the correct frequency. Using this system on both masers, the frequencies of H1 and H2 were compared with each other and with one of the NRC primary cesium clocks over periods of several days. The longest continuous comparison was made over a period of 14 days. The results of the maser-maser comparison are shown in Fig. 1. In this figure the data are compared with those obtained at an earlier date [1] both for the free-running masers and for the masers under the control of the beam flux modulation auto-tuner.

It can be seen that the results obtained with the two different methods of automatic control are fairly similar. At short averaging times the signal injection method appears to cause somewhat greater degradation of the frequency stability than the other method. However, for τ > 1000s, the results are comparable using the two systems, within the error limits.

Frequency Standards and Metrology Editor: A. De Marchi

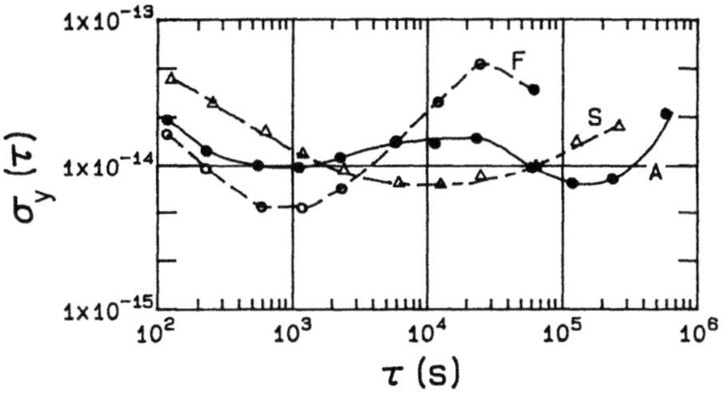

Fig. 1 Comparison of the maser frequency stability when both masers were free-running (F), under control of the beam flux modulation auto-tuner (A), and under control of the signal injection auto-tuner (S).

Phase comparisons between a 5 MHz crystal oscillator locked to H2 and cesium clock CsV1A were also carried out, with the maser under control of the signal injection auto-tuner. Fig. 2 shows the residuals (in ns) from a linear fit to the phase difference between the maser and the cesium clock over a period of 8 days, the data being recorded at intervals of 2 hours. The residuals remained within ± 2 ns over the 8 day period. The equivalent $\sigma_y(\tau)$ derived from these measurements indicates a stability of 3.2×10^{-14} over 2 hours and 6.8×10^{-15} over 24 hours.

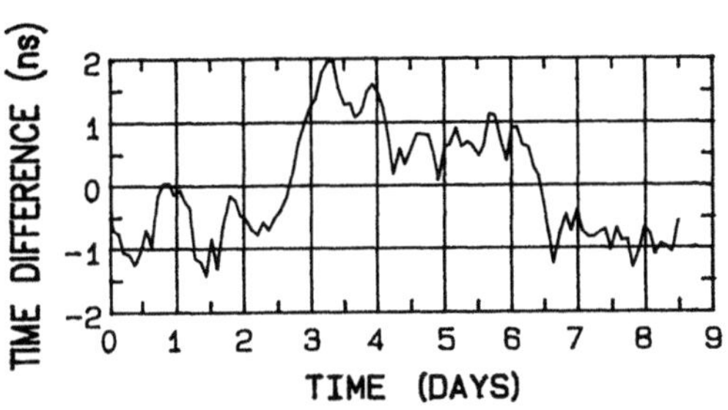

Fig. 2 Residuals from a linear fit to the phase difference between maser H2 under control of the signal injection auto-tuner and NRC cesium clock CsV1A over a period of 8 days.

It is planned to carry out further comparison measurements when two new hydrogen masers H3 and H4, now under construction at NRC, are in operation.

1. References

1. D. Morris and K. Nakagiri, Metrologia 12, 1 (1976)
2. C. Audoin, Rev. Phys. Appl. 16, 125 (1981)

Work on Hydrogen Maser

Cheng Yumin

Time & Frequency Department, Institute of Testing Technology,
Shanghai Bureau of Metrology, 1226 Chang Le Rd.,
Shanghai, People's Republic of China

Abstract, Hydrogen masers have been developed at Shanghai Bureau of Metrology since 1973. This article summarizes design consideration, construction, and performance of masers.

1. Introduction

The research work of hydrogen masers started at Shanghai Bureau of Metrology since 1973. The maser 1 and maser 2 was constructed in 1975. The testing results shown that they are subject to environment mainly due to two causes, i.e., for magnetic shield insufficient and for the temperature control imperfect. Great attention was paid to redesign the new masers[1] . From 1975 to 1980 we have developed eight hydrogen masers. Among these, three were supplied to Shanxi Astronomical Observatory, two were in Department of Aerospace Industry, and one was installed in National Institute of Metrology of China. The remainders oprate continuously in our institute. An improved hydrogen maser has been built in 1983[2]. The physics package and electronic system are housed in a aluminium fram measuring 600mm X 720mm X 1410mm. Its size is only 1/3 of our older masers, and can be moved easily by four small wheels.

2. Design and Construction

In our masers, a LaNi5-H storage have been developed as the moleculor hydrogen supply. The purifier to be used consists of a thermally controlled palladium silver alloy tube with quick response, and can adjuct the hydrogen flex by direct heating method. The RF discharge chamber with collimator is made of pyrex glass. The state selector employs a hexapole magnet construction. The storage bulb coats with local made F4 coating. The TE011 mode cavity is construceted of cervit with approximately a diameter of 276mm and a length 270mm, b=0.22 and Q=35000. In order to reduce the temperature effect, the cover plate of the cavity is made of bulk aluminum plate. It is necessary to consider the thermal compersation of aluminum plate and cavity. The temperature coefficient of the cavity-bulb can be reduced to 100Hz/℃. We also employ inner and outer as well as neckzone temperature control whith 40±0. 01℃ . The mechanical stability of the cavity have been concerned, barometrically induced stresses on the vacuum tank are isolated from the cavity resonator by a double-bass structure. The C field coil is a solenoid assembly with a magnetic field compensation. The magnetic shielding uses four shieldas which are made by permalloy. In order to provide a high vacuum, we have developed a 350 l/s pump.

Frequency Standards and Metrology Editor: A. De Marchi

3. Electronics System

The phase locked loop for the maser is shown in fig.1. As far as the noise figure is concerned, the first stage of amplifiers determing the performance of the system, in order to reduce the noise from the electronics a 405,7xxHz cystel filter has been used in maser reciever, and a preamplifier at 1.4 GHz with a noise figure 2.5db is available. The frequency synthesizer 405,7xx,xxx mHz is directly synthesized from 5 MHz VCXO, and its resolution is 1 mHz. Some analysis of the locked loop has been made[3]. The resolts shown that considering the attation noise of cercuits and the inner noise of VCXO, the loop filter presents different function, for the former as a low pass filter, for the latter as a high pass filter. We can choose the loop band to make the totel phase fluctuation as low as possible, in our case the loop band is equel to 65 Hz[3].

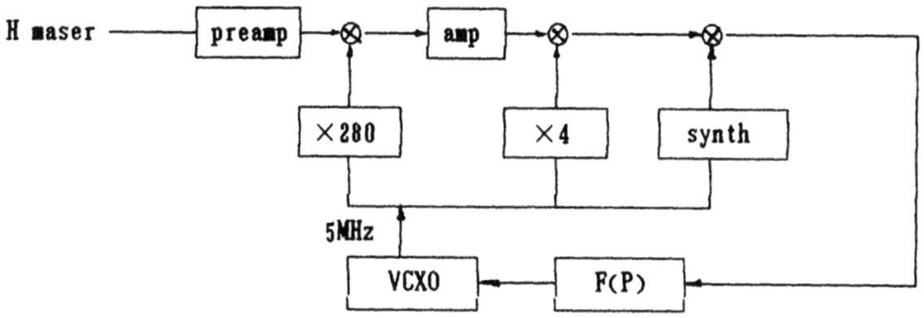

Fig.1.The block diagram of the control electronics

4. Performances

The maser wall shift has been determination, with local made F4 coating we got the value, K(40°C)=−293±17 mHz·cm; a(40°C)=(−17±2)x10^{-3}°C^{-1}, the unperturbated hyperfine frequency of the ground state of hydrogen atom is 1420,405,751.768±0.002 Hz[4]. The frequency stability are 3x10^{-13}/s;6x10^{-14}/10s;9x10^{-15}/100s. The improvede maser have almost the same performance. Since the cavity temperature control imperfect and room temperature variant, the long term stability is unsatisfied, therefore a microcomputer autotunning system of masers has been developed, it may improve the masers performance[5].

5. Reference

1. Cheng Yumin, Proceedings of The First Atomic Frequency Standards Symposium, China, 1976.
2. Chen zhonba, Cheng Yumin et al, Proceedings of Han Zhao Symposium of Time and Frequency, China, 1983.
3. Cheng Yumin , Proceedings of Metrology Society of Shanghai, 1986.
4. Cheng Yumin, IEEE Trans.Instrum.Meas., Vol.IM-29 pp.316-319.Dec.1980.
5. Wei Jinxin, Cheng Yumin et al, Proceedings of Metrology Society of China, 1988.

Advanced Atomic Hydrogen Masers

H. Peters, B. Owings, and P. Koppang

Sigma Tau Standards Corporation, P.O. Box 1877, Tuscaloosa, AL 35403, USA

Atomic hydrogen masers presently being delivered by Sigma Tau Standards Corporation have several features which make them particularly useful for remote field station operation as well as for long term basic time and frequency standards applications. Most important is a self-contained automatic cavity frequency tuning system which makes the maser a stand-alone standard not requiring frequent spin-exchange cavity tuning against a secondary reference to achieve precise frequency reproducibility and accurate timing.

Another feature is a magnetic field servo system which counteracts the effect of external magnetic field variations. A low noise detector senses variations in the Z-axis field between two of the maser magnetic shields; the signal is then amplified and drives a correcting coil which nulls the intershield field. Experimental results indicate that external field variations of 1 gauss result in less than 1 x 10E-14 variation in maser frequency.

The Sigma Tau Standards Corporation hydrogen masers are active oscillators, characterized by exceptional stability as shown in Fig. 1. The masers are relatively light, compact (107cm x 46cm x 76cm) and include on board batteries which supply up to eight hours of operation without external power applied. An additional feature is a convenient monitor and control interface system which provides for either local or remote observation and recording of 32 channels of instrumental data as well as either local or remote control of the maser receiver-synthesizer which gives an output frequency resolution of 5 x 10E-17. Remote transmission or reception of data is via a full duplex RS232C serial port.

In the advanced hydrogen maser design attention has been given to electronic packaging, placement of sub-systems and controls in convenient, replaceable modules and other details to facilitate maintenance and repair. There are two hydrogen ion pumps, each with an estimated life over 10 years, and a separate ion pump for regions surrounding the source and storage bulb which isolates the "inner"

Frequency Standards and Metrology Editor: A. De Marchi

MODEL VLBA-112 HYDROGEN MASER STABILITY

Sigma Tau Standards Corporation MASER V-6 Vs MASER V-5 Data of May 15, 1988

NRAO ADIOS FST PROGRAM OF 05/5/87

BATCH START 13:20.6 05/14/88
BATCH END 03:22.3 05/15/88

EQUAL REFERENCE ASSUMMED
TEMP=23.39C ST, PHASE=33.2PS
TEMP=23.16C END PHASE=-31PS
TLOW=22.61C, THIGH=24.13C

TAU SEC	SAMPLES	SIGMA *EXP-15
1	50499	80.8
2	25249	48.2
5	10099	26.3
10	5048	17.0
20	2523	11.2
50	1008	6.9
100	503	4.8
200	251	3.5
500	99	1.7
1000	49	1.6
2000	24	1.0
5000	9	0.8
10000	4	0.6

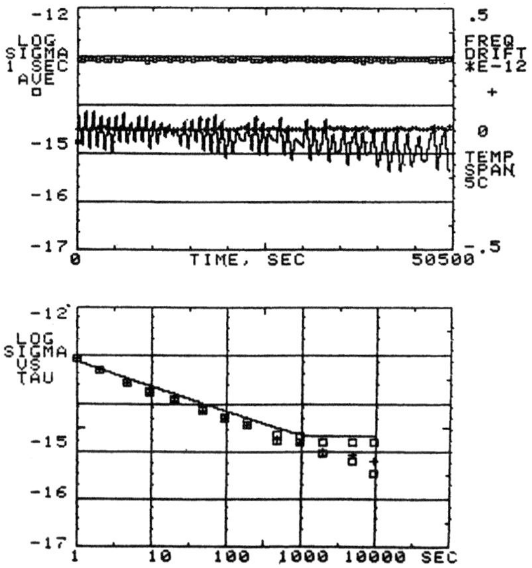

Fig. 1 Stability Data, 1 Hz Band Width, No Drift Removed

(hydrogen pump) system from vacuum background contaminants. Each pump may be changed in the field in a few hours without serious interruption of operation if required. Other features incorporated in the maser design include a very low noise receiver preamplifier having a noise figure under 1 dB, a heavy copper cavity providing very good thermal homogeniety and large thermal capacity to reduce susceptibility to ambient temperature variations, a separate DC-DC converter for each of the three ion pumps and temperature controlled multipliers and dividers for the 100 MHz and 5 MHz outputs.

Recent Progress in Oscillating Compact Hydrogen Masers [1]

H.T.M. Wang and R.R. Hayes

Hughes Research Laboratories, Malibu, CA 90265, USA

Several oscillating compact hydrogen masers in three progressively smaller sizes [2] (Table I), including an engineering model for spaceborne applications, have been built. A common feature of the masers is the use of electronic cavity Q enhancement to obtain sustained maser oscillation, and an electronic servo system to control the frequency of the cavity. The masers employ the same basic electronic design and have comparable electronics-limited long term stability. As shown in Table I, the approximately 30% sacrifice in short term stability for each 3-fold reduction in the volume of the physics unit is a very favorable trade-off. The subcompact CHYMNS-III physics unit is housed in a hermetically sealed isothermal enclosure with thermoelectric coolers for bi-directional temperature control. This assures a stable operating environment over a wide range of ambient conditions.

Table I. Characteristics of oscillating compact hydrogen masers

	CHYMNS-I	CHYMNS-II	EDM SPACE MASER	CHYMNS-IIIb	CHYMNS-III-1
Cavity Dimension, cm	15.2D x 15.2L	10.4D x 15.2L	10.7D x 15.8L	7.6D x 7.6L	7.6D x 7.6L
Physics Unit Vol, L	65.5	22.1	28.5	8.8	8.8
System Dimension, cm	Experimental	Experimental	21.8x21.8x59.7 10.7x32.8x39.1	17.8x30.5x43.2	26.0x61.0x43.2
Weight, kg			29.1	19.5	40.0
Power Consumption, W			64	63	100
$\sigma(\tau$=400s), Theory	1.9×10^{-14}	3.1×10^{-14}	3.1×10^{-14}	3.9×10^{-14}	3.9×10^{-14}
Measured	2.4	3.0	6.9	4.1	4.1

The stability of the oscillating compact masers has been evaluated by comparing the masers with each other, with a full-size hydrogen maser, and with MC-USNO via GPS. A stability comparison of the subcompact maser CHYMNS-IIIb versus CHYMNS-I is shown in Fig. 1. A fitted drift rate of several parts in 10^{-16} per day is generally obtained in a comparison between two oscillating compact masers in a stable laboratory environment. Confirmation of the low drift rate was provided by a comparison of CHYMNS-II versus MC-USNO via GPS common view where, over a 35-day interval, a drift rate of -8.55×10^{-16} per day was obtained. The EDM space maser,

Frequency Standards and Metrology Editor: A. De Marchi

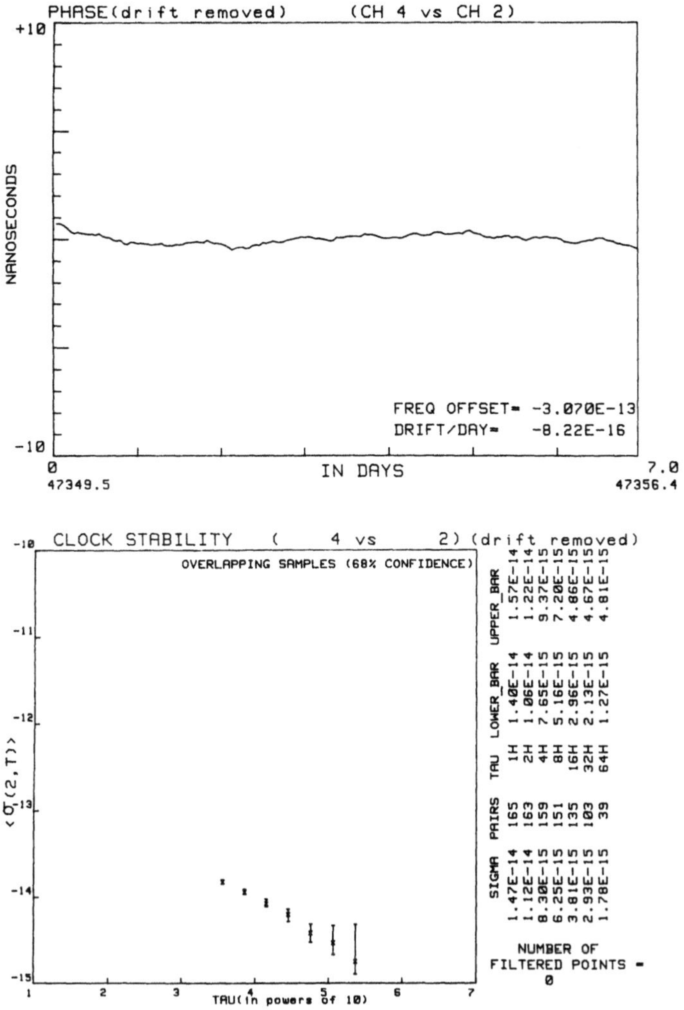

Fig. 1. Stability of CHYMNS-IIIb versus CHYMNS-I

operating in vacuum, has shown a remarkable insensitivity to temperature variations. During tests, the maser frequency was unperturbed by a 10°C, 12-h-period sinusoidal temperature modulation and a 30° to 37°C temperature step, both applied to the baseplate of the maser. The result is significant, since temperature variation is the dominant perturbation on a spaceborne atomic clock. Low thermal sensitivity is essential to long term stability.

REFERENCES

1. This work has been supported by US Naval Research Laboratory under contracts N00014-83-C-2023, N00014-83-C-2120, N00014-83-C-2144, and N00014-88-C-2116
2. R.R. Hayes and H.T.M. Wang, Proc. 18th Annual PTTI (1986), p. 611

Shapes of Storage Bulbs for Hydrogen Masers for Maximum Filling Factor

J.E.B. Ponsonby

University of Manchester, N.R.A.L., Jodrell Bank, Cheshire, UK

The shape of the starage bulb in the H-maser cavity determines the value of the filling factor η' defined as

$$\eta' = \frac{\langle H_z \rangle_B^2 \cdot V_B}{\langle H^2 \rangle_C \cdot V_C}$$

Where symbols have their usual meanings and the z direction is that of the static bias field.

Quite modest fractional increases in η' can dramatically improve the maser's ability to oscillate and the greater the value of η' the greater the possible radiated power, the greater the range of atomic flux over which oscillation can be achieved and the better the ultimate frequency stability: see AUDOIN et al (1981). [1]

Masers have been built with bulbs of various shapes such as spherical, cylindrical and cylindrical with hemispherical ends, but despite the importance of maximizing η' the way to achieve this seems not hitherto to have been reported. The shape depends on the chosen cavity mode and on the chosen z direction, but in all cases it may be found by application of the following theorem which is presented without proof.

THEOREM The filling factor η' is maximized by making the bulb that surface of constant H_z such that the average of H_z within the bulb is twice its value on it.

i.e. $$H_{zs} = \frac{1}{2} \langle H_z \rangle_B$$

Here H_{zs} is the value of H_z over the entire surface of the bulb.

The shape of bulbs thus prescribed have been computed for both the traditional TE_{011} mode and for the TE_{111} mode, both in cylindrical cavities. The corresponding maximum attainable values of η' for the two modes have been computed as functions of cavity length/radius ratio. The results are presented in Fig. 1. For the TE_{011} mode it is found the maximum η' is only marginally greater than what is achieved in existing designs, but for the TE_{111} mode it is found that the optimum configuration is for two equal "blister" bulbs. With these much larger values of η' are attainable than with a single bulb in the TE_{011} mode. This can go some way towards making up for the lower Q factor of the TE_{111} mode.

Frequency Standards and Metrology Editor: A. De Marchi

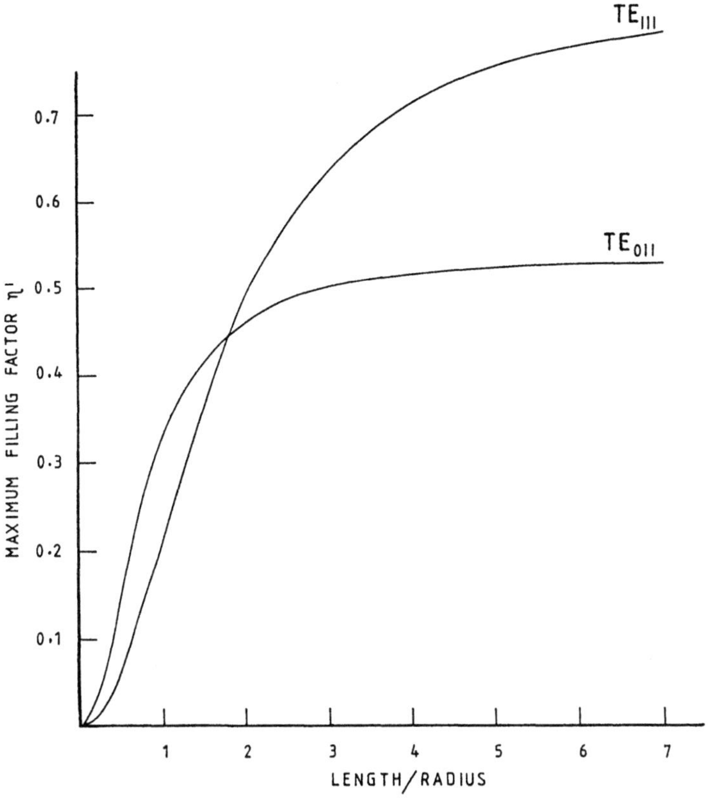

Fig. 1. Maximum filling factor η' versus cavity length/radius ratio for cylindrical cavities in the TE_{011} and TE_{111} modes. (No dielectric loading).

Computations which allow for the dielectric loss introduced by the bulbs and for a realistic external coupling factor (β=0.25) suggest that for a copper cavity with L/a = 4 the spin exchange parameter q can be made one half of its threshold value of 0.172 required for oscillation in this twin "blister" bulb configuration. This margin is sufficient to make this configuration, which requires two atomic beams, in principle a viable miniature active maser with an ultimate short term stability nearly as good as for a large active maser of conventional design, and very much better than for a passive miniature H-maser.

Reference

1. C. Audoin, J.Viennet and P.Lesage. Hydrogen maser: Active or Passive? Troisieme Symposium sur les Etalons de Frequence et la Metrologie.

H Maser and Rb Work in Romania

O. Gheorghiu, L. Giurgiu, M. Dinca, C. Szekely, and C. Mandache
Central Institute of Physics, Bucharest-Magurele, POB MG-7, Romania

1. Hydrogen Maser

In the last generation of H-masers, developed in our institute, the hydrogen is stored in FeTi hydrure. The hydrogen flow regulator has a very fast response (3 sec) (Fig. 1) and over-shoots less than 1% /1/. In the hydrogen source, the discharge tube is placed between the plates of the terminal condenser of a coaxial cavity, working around 400 MHz. We are obtained that over a certain critical power, the maser signal level does not depend on the absorbed power in the discharge (Fig. 2) and a power of about 3.5 W is enough to obtain the maximum degree of dissociation /2/.

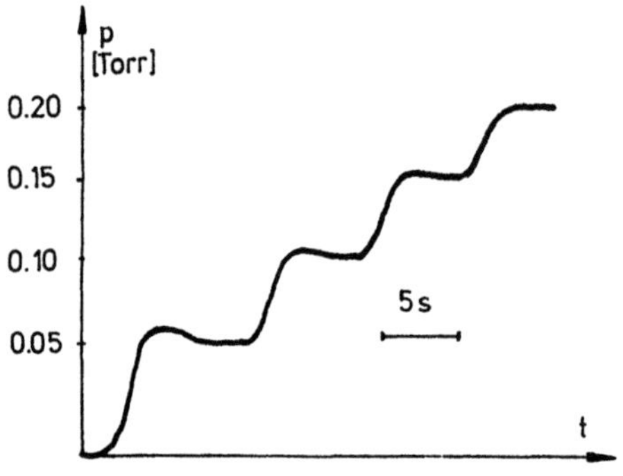

Fig. 1

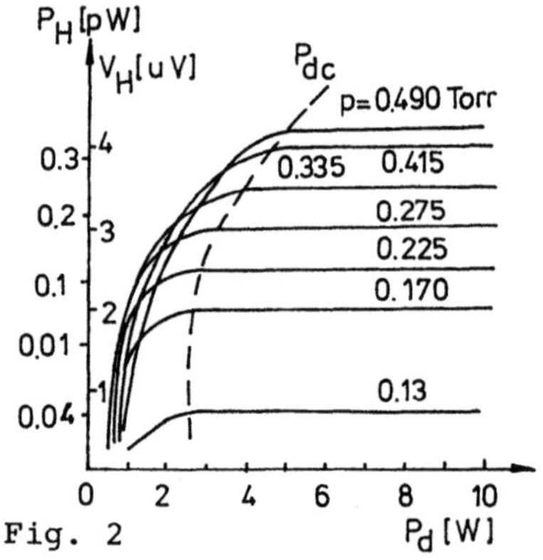

Fig. 2

The thermal regulation of the maser cavity (made of aluminium) consists of five independent ovens. Around the cavity a thin Al box provides additional r.f. shielding and thermal insulation. The controler of each oven has a PID transfer function in order to provide a high DC gain and a good stability. The temperature stability is better than 10^{-5} K/hour /3/. Five magnetic screens assure a shielding factor of about 35 000 in the axial direction and 400 000 in the horizontal one. The fluctuations of the residual magnetic field are below 2 μG. To measure the bias magnetic field a set of four rods are placed inside the

Frequency Standards and Metrology Editor: A. De Marchi

cavity. In the receiver, the phase comparison is realized at 62 Hz. The masers are mains-operated, 24 V battery buffered, with dc-dc converters and the overall electrical consumption of 200 W.

Our experimental data are consistent with the published values: for the spin exchange cross section we are obtained 2.42×10^{-15} [cm^2] /4/. The preliminary measurements gives a stability of about 10^{-14} for τ = 1000 sec.

2. Rubidium Standard

A passive rubidium standard, using a separated filtre approach and alternativey a composite lamp-filter system with natural rubidium is building. The first measurements will be made at the end of this year.

References

1. L.C. Giurgiu, M.P. Dinca, O.C. Gheorghiu, to be published
2. O.C. Gheorghiu, C. Mandache, C.R. Acad. Sci. Paris, 7, 295, 131 (1982)
3. L. Giurgiu, Thesis, Bucarest (1987)
4. C. Mandache, Thesis, Bucarest (1982)

Collisions of Atoms in Microwave and Laser Fields

B.J. Verhaar and H.T.C. Stoof
Eindhoven University of Technology, The Netherlands

In order to Bose condense spin-polarized atomic hydrogen, a few types of traps have been proposed to isolate the atoms from the walls where recombination takes place. Recently, Agosta and Silvera [1] proposed a microwave trap operating on the b-c electron spin-flip transition at ~ 5 tesla. They made a preliminary calculation of the collisional relaxation loss of atoms in such a trap using dressed atom states. In this contribution the relaxation rates are worked out more completely. Although we mention only the microwave trap explicitly, our theoretical techniques apply also to laser traps.

The dressed states are superpositions of bare levels b and c with admixture depending on detuning δ:

$$|1(N)\rangle = \cos\Theta\ |c,N\rangle + \sin\Theta\ |b,N+1\rangle = \text{unbound state}$$
$$|2(N)\rangle = -\sin\Theta\ |c,N\rangle + \cos\Theta\ |b,N+1\rangle = \text{bound state}$$

with $\cot an2\Theta = -\delta/\omega_1$, where $\omega_1(\underline{r})$ is the local Rabi frequency. The energy splitting of the uncoupled $|c,N\rangle$ and $|b,N+1\rangle$ states is $\hbar\delta$, that of the dressed states is $\hbar\Omega$ with $\Omega = \sqrt{(\delta^2 + \omega_1^2)}$. The 2(N) atoms are trapped in a potential $-½\hbar\Omega(\underline{r})$. Our problem is to calculate the decay of such a gas due to spin-exchange and magnetic dipole interactions. It turns out to be possible to express all relevant rate constants in terms of collision amplitudes for bare states, for which rigorous values are available [2]. The values of the latter at about 5 T indicate that we may confine ourselves to collisions with only b and c states as initial and final states, i.e. only dressed initial and final states are to be taken into account [1]. A total number of 15 transitions are then to be considered: from 2(N)2(N) to 2(N)1(N), 1(N)1(N), 2(N)1(N-1), 1(N)1(N-1), ...

Our method to calculate these rates is to use dressed-atom states as asymptotic states in two-body collisions. A typical S-matrix element is then $S_{fi} = S_{2(N)1(N),2(N)2(N)}$. We apply the degenerate-internal-states (DIS) approximation, i.e. we neglect the splitting $\hbar\Omega$ of levels within each of the manifolds. As a consequence we have

Frequency Standards and Metrology Editor: A. De Marchi

$$S_{2(N)1(N),2(N)2(N)} = \sum_{h_i h_i'} \langle 2(N)1(N) | h_1'h_2' \rangle \, S_{h_1'h_2',h_1h_2} \, \langle h_1h_2 | 2(N)2(N) \rangle$$

In view of the low temperatures involved, the initial ℓ value is 0 and the final ℓ is 2 (0) for dipolar (exchange) transitions. To simplify the notation we have suppressed these quantum numbers as well as the symmetrization of the spin states. The rate constants then follow from

$$G_{2(N)2(N)\to 2(N)1(N)} = \langle \frac{2\pi\hbar}{m_H k} \left| S_{2(N)1(N),2(N)2(N)} \right|^2 \rangle_{thermal}$$

Combining the foregoing equations it is easy to find the total rates.

All relaxation processes, except those within the same N dressed-states manifold, are <u>inelastic dipolar transitions</u>. They were calculated previously by Agosta and Silvera. In that calculation cross-terms between bare S-matrix elements were left out. This does not change the order of magnitude. Relaxation processes within the same N-manifold are <u>elastic exchange collisions</u>. We find

$$G_{2(N)2(N)\to 2(N)1(N)} = \frac{\pi\sqrt{8}\,\hbar}{m_H \lambda_{th}} e^{-\hbar\Omega/k_BT} \sin^4\theta \sin^2 2\theta \, \left| (S_{cc,cc} - S_{trip})/k \right|^2_{k\to 0}$$

$$= 2.3 \times 10^{-13} \sqrt{T} \, e^{-\hbar\Omega/k_BT} \sin^4 2\theta_B \sin^4\theta \sin^2 2\theta \ \mathrm{cm^3 K^{-1/2}}$$

with $\cot 2\theta_B = B/0.0569T$, and a similar expression for $G_{2(N)2(N)\to 1(N)1(N)}$.

1. C.C. Agosta and I.F. Silvera, Contribution to Conference on Spin Polarized Quantum Systems, Torino 1988.
2. A. Lagendijk, I.F. Silvera, and B.J. Verhaar, Phys. Rev. B33, 626 (1986); H.T.C. Stoof, J.M.V.A. Koelman, and B.J. Verhaar, to be published (Phys. Rev. B38).

Temperature of Mercury Ions Stored in an R.F. Cylindrical Trap

C. Meis, M. Jardino, B. Gely, and M. Desaintfuscien

**Laboratoire de l'Horloge Atomique,
Equipe de Recherche du CNRS associée à l'Université Paris-Sud,
Bât. 221, Université Paris-Sud, F-91405 Orsay Cedex, France**

Stored ions temperature is deduced using an analytical model with experimentally measured physical parameters.

Assuming a Maxwellian charge density distribution of N ions at thermal equilibrium at temperature T, in a spherical effective potential caracterized by a theoretical secular frequency $\omega_s/2\pi$, it can be shown /1/ :

$$T = N^{2/3}\left[\frac{m\omega_m^2}{2\pi k_B}\right]\left[\frac{|\mu|q^2}{m\varepsilon_0(\omega_s^2-\omega_m^2)}\right]^{2/3}$$

where m and q are respectively the mass and charge of stored ions, $\omega_m/2\pi$ the experimentally observed secular frequency reduced by space charge effects and μ =-0.224...

We show here the cooling of a spherical ion cloud of about 2.10^6 ions by helium buffer gas at different pressures (fig (1)). The ions are stored in an R.F.cylindrical trap in a 40 kHz effective potential. The theoretical secular frequency in this case has been calculated numerically /2/.

The value of $\omega_m/2\pi$ is deduced from the first order Doppler spectrum of the microwave hyperfine transition at 40.5 GHz of the fundamental level $^2S_{1/2}$ of 199 mercury ions /3/. We control the total number of stored ions by measuring image currents created on the trap electrodes when a modulated signal at $\omega_m/2\pi$ is applied on the endcaps electrodes /4/.

We have also tested the influence of trap irregularities (holes etc) on the sphericity of the pseudopotential comparing the measured values of the secular frequency along (x,y) and z axes of the cylindrical trap.

In fact, the microwave excitation for the hyperfine transition is a running wave along y axe while the ion number signal is supplied by an electronic excitation of the ion motion along z axe.The precision on the

Frequency Standards and Metrology Editor: A. De Marchi

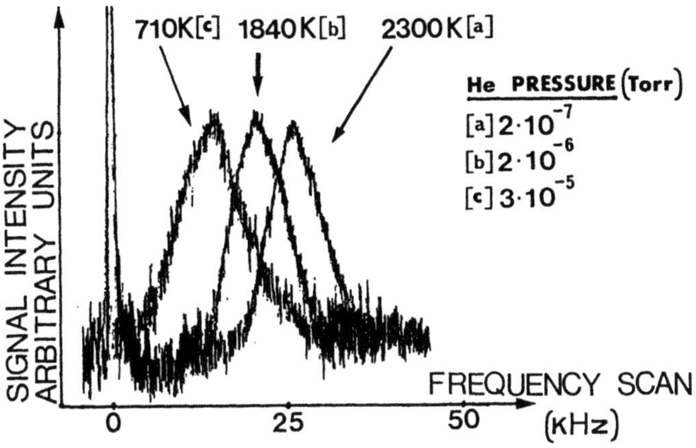

Fig.1 : Effect of buffer gas cooling on first order Doppler spectrum.

secular frequencies $\omega_{mz}/2\pi$ and $\omega_{my}/2\pi$ detected by image currents and the first order Doppler hyperfine spectrum respectively is of about 1%. We have, thus, concluded that the particular form of the effective potential created in a cylindrical R.F. trap compensates well the trap asymmetry /5/.

/1/.C. Meis, M. Desaintfuscien and M. Jardino
Appl. Phys. B45,59-64 (1988).

/2/.H. Lagadec,C. Meis and M. Jardino.
Accepted to Int. Jour. of Mass Spec. and Ion Pr.

/3/.M. Jardino, F. Plumelle, M. Desaintfuscien and J.L. Duchêne.
Proc. 38th Ann. Freq. Control Symposium, IEEE Press 84 CH 2062-8 P-431 (1984).

/4/.M.N. Gaboriaud, M. Desaintfuscien and F.G. Major.
Int. Jour.of Mass Spec. and Ion Phys. 41, 109-123 (1981).

/5/.C. Meis, M. Jardino, B. Gely and M. Desaintfuscien.
Accepted to Appl. Phys. B.

A High Stability Oscillator Based on a Sapphire Loaded Superconducting Cavity Resonator

S.K. Jones, D.G. Blair, and A. Giles

Physics Department, University of Western Australia,
Nedlands, Western Australia, 6008, Australia

1. Introduction

This paper describes, in simple terms, a new ultrastable oscillator which features the following properties; low maintenance, high reliability, simple low cost cryogenic system, high power levels which result in short term stability superior to all other frequency standards and performance comparable to a hydrogen maser at a fraction of the cost. The heart of the oscillator is a high Q sapphire loaded superconducting cavity (SLOSC) resonator which provides a frequency reference with relatively high immunity to environmental perturbations. Using a novel combination of active and passive frequency control we can achieve fractional frequency stabilities of around $9x10^{-15}$. This result compares favourably with the state-of-the-art hydrogen masers and, in fact, for measurement intervals up to a few tens of seconds, our oscillator is superior to these systems. It is expected that several improvements should lead to fractional frequency stabilities around the 10^{-16} level. The door is now open for this oscillator to be useful in applications such as precision tests of general relativity and very long baseline interferometry (VLBI) radio astronomy.

2. The High Stability Oscillator

At the heart of any accurate clock is a high stability oscillator. The better the oscillator, the better the clock. We have developed an ultra stable oscillator which derives its frequency stability from a sapphire loaded superconducting cavity resonator, described in the next section. The frequency of this oscillator is 9.733 GHz. The oscillator configuration, shown schematically in figure 1, is novel in that it uses the same resonator to act both as the frequency determining element in a loop oscillator and as the dispersive element in a discriminator circuit which forms part of a servo to control unwanted phase shifts in the loop oscillator circuit. The loop oscillator consists essentially of a microwave amplifier operated in positive feedback with the SLOSC resonator acting as a very narrow band transmission filter. By carefully adjusting the electrical length of the feedback path and

Frequency Standards and Metrology Editor: A. De Marchi

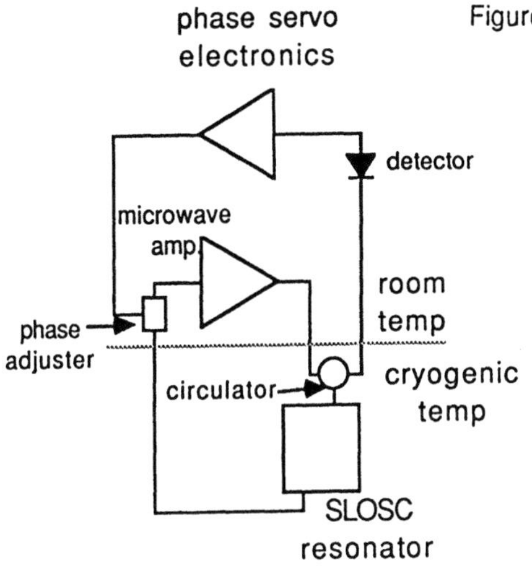

Figure 1: A schematic of the high stability oscillator showing the partially cryogenic loop oscillator and the phase stabilization servo.

tuning the transmission characteristics of the SLOSC it is possible to select the desired oscillation frequency. The phase adjusting servo is necessary because of unavoidable changes of the electrical path length around the loop. Without the servo these changes would give rise to frequency jitter at the oscillator output. The servo effectively detects the SLOSC resonant frequency and adjusts the loop circuit so that the oscillator frequency is the same as the SLOSC frequency. While it is of utmost importance that the SLOSC be kept in a highly stabilized environment, the rest of the circuit is relatively immune to environmental disturbance. Thus, all the electronics for this oscillator are located at room temperature except for the SLOSC itself which needs to be cooled to liquid helium temperatures (see next section). Perturbations such as thermally induced length changes in the microwave lines which connect the cold SLOSC to the room temperature electronics are controlled by the servo system.

Figure 2 compares the recently measured stability of this stabilized loop oscillator against that of some other systems. The vertical scale, labelled Square Root Allan Variance (SRAV), is equivalent to the average fractional frequency fluctuation, $\Delta f/f$, averaged over the measurement interval indicated by the horizontal scale. Using a pair of similar oscillators we obtain an SRAV of 9×10^{-15} over a 100 sec measurement time. Already this performance compares favourably with state of the art hydrogen masers, especially for times less than 100 seconds, and we are confidently predicting around in order of magnitude improvement in the foreseable future. A low noise synthesizer is currently under construction so that the 9.73GHz output from the oscillator is converted to a 5MHz signal. This will make our high stability oscillator compatible with other electronic timing systems.

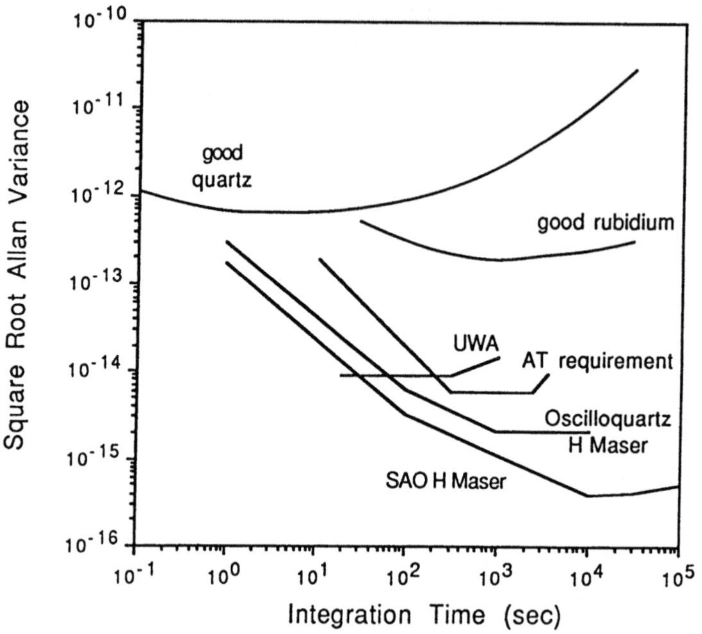

Figure 2: Graph showing the stability of the high stability oscillator at UWA compared against some other systems.

3. The Sapphire Loaded Superconducting Cavity

The heart of the high stability oscillator is a Sapphire Loaded Superconducting Cavity (SLOSC) resonator. As the name suggests, the SLOSC consists of a single crystal of pure artificial sapphire, 30mm high and with 30mm diameter, mounted inside a cylindrical can with inside diameter 50mm and 50mm inside height, made from niobium metal. The main cylindrical body of sapphire is fabricated with a small spindle at each end so that it can be supported in the middle of the niobium cavity (see fig 3). Superconducting probes are used to couple energy into and out of the SLOSC. This configuration effectively forms a tuned circuit which has many resonances at microwave frequencies.

The high dielectric constant of sapphire ($\varepsilon \sim 10$) results in nearly all the microwave energy of a resonant mode being confined to the sapphire . This means that important characteristics such as the frequency and bandwidth of each resonant mode and their responses to environmental perturbations are primarily determined by the properties of the sapphire. One of the major advantages that the SLOSC has over other types of microwave resonator is that some of the resonant modes have extremely high quality factors (Q's), or very narrow linewidths, which means that the SLOSC can be used as a very precise

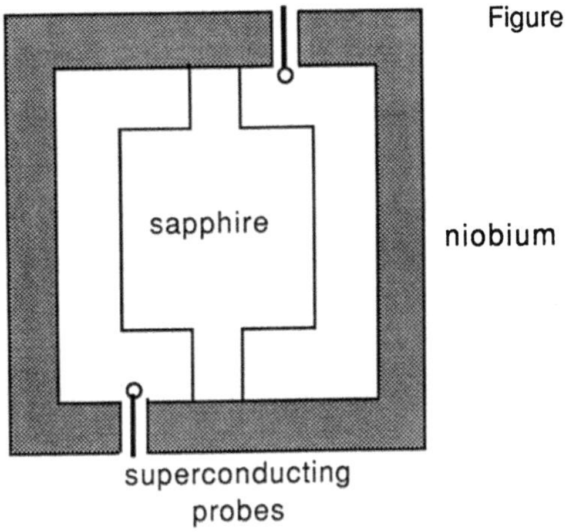

Figure 3: A cross section through the SLOSC.

frequency reference. We have measured a record Q of $4.2x10^9$ for a mode at 9.733 GHz (i.e. a linewidth of about 2Hz). This is the highest Q yet reported for any resonator system based on a dielectric material. To achieve these results it is necessary to cool the SLOSC to liquid helium temperatures (4.2K or -269°C) because at these temperatures the intrinsic losses in the sapphire, which contribute to increased linewidth or decreased Q, are reduced to a minimum. In addition, the niobium cavity which is necessary to prevent radiative loss, becomes superconducting so that any small currents induced in it by the electromagnetic fields outside the sapphire see zero electrical resistance and hence do not contribute any further loss.

It should be noted that in the past, empty superconducting niobium cavities have themselves been successfully used as extremely precise frequency references. The SLOSC is superior to conventional superconducting cavities for the following reasons:-

a)The Q of the SLOSC (discussed above) is relatively weakly dependent on temperature and so it is not necessary to try to operate the SLOSC at the same very low temperatures at which conventional superconducting cavities must operate.

b)The SLOSC frequency has a maximum at about 6K which provides an ideal operating point and leads to more than a thousand fold reduction in the level of temperature control necessary to achieve a high level of frequency stability.

c)The mechanical rigidity of sapphire is considerably more than that of niobium which results in the frequency of the SLOSC being far more immune to the effects of mechanical vibration and long term sag. Because of the high degree of confinement of the microwave energy to the sapphire the niobium part of the SLOSC plays only a minor role in determining the resonant frequency.

During operation the SLOSC is mounted inside a small stainless steel vacuum pot which has been designed to enable precise temperature control and to minimize the level of mechanical strain placed on the SLOSC. This pot, once evacuated and sealed, forms a small self contained unit which can be stored at room temperature for months without any degradation of performance. A high capacity, long hold time dewar is being built to keep the SLOSC cold for up to 100 days of continuous, non-interrupted operation.

References

D.G. Blair & S.K. Jones, "A High Q Sapphire Loaded Superconducting Cavity Resonator", J. Phys. D: Appl. Phys., 20 (1987), pp. 1559 - 1566.

S.K. Jones & D.G. Blair, "High Quality Factors for a Sapphire Loaded Superconducting Cavity Resonator", Electron. Lett., 23, No. 16 (1987), pp. 817 - 818.

S. K. Jones, D. G. Blair & M. J. Buckingham, "Effect of Paramagnetic Impurities on the Frequency of Sapphire Loaded Superconducting Cavity Resonators", Electron. Lett., 24, No. 6 (1988), pp. 346 - 347.

Recent Scientific Results at the Istituto di Radioastronomia

R. Ambrosini

Istituto di Radioastronomia, (Consiglio Nationale delle Ricerche), Bologna, Italy

- The Medicina VLBI parabola partecipates for more than 100 days/year to the radioastronomical campaigns of the European and Global VLBI Networks at the 1.4 GHz, 1.7 GHz, 5.0 GHz, 10.7 GHz and 22 GHz bands.

- With a dedicated, prime focus, S/X receiver other 100 days/year are dedicated to Geodynamic observations within the IRIS and the Crustal Dynamic projects, and to study short term fluctuations of UT1.

- The CALC & SOLVE programs have been modified to run on the HP-1000F of the Observatory to reduce GEO-VLBI data.
With this software the geocentric coordinates of our antenna have been obtained with decimetric uncertainity.

- LASER ranging and GPS measurement campaigns have been performed at the site in cooperation with other institutions.

- An autocorrelator with 1024 channels for observations of H2O-lines has been implemented at the radiotelescope by the Florence group of Arcetri. Recentely, with a Digital Tone Extractor it has been succesfully received at X-band the coherent transponder of the VOYAGER II spacecraft, now at 25 a.u., (>3 G Km).
This test has shown that our receiving equipments and atomic frequency reference are suitable in the field of detection of gravitational waves perturbing the Earth- Spacecraft system.

- The BLOCK 0, MKII type, correlator has been given indefinitely on loan to our Observatory by Caltech.
It needs some repairing before coming back operational as a full 5 stations processor.

- The opening cerimonia of our second VLBI parabola at Noto, in the southernest part of the isle of Sicily, is scheduled for the 28th of October. At first it will be equipped with an H-Maser frequency standard, a 10.7 GHz receiver, and an MKII data acquisition terminal. Full equipment of the station is still to be funded.

Frequency Standards and Metrology Editor: A. De Marchi

- The Northen Cross radiotelescope with its 10, 000 square meters of effective collecting area has been recentely updated with a new, PC controlled, data acquisition system.

Apart from the routine observations of variable and galactic radiosources, now, in cooperation with the University of Palermo, a new scientific project has started to observe Pulsars.

Fig. 1 Air-view of the Medicina Observatory

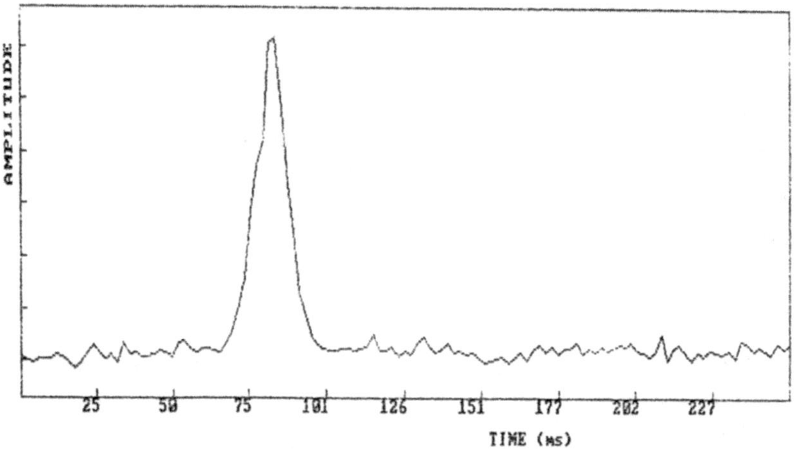

Fig. 2 An example of a Pulsar profile received with the East-West arm of the Northern cross

The Time and Frequency System at the Matera Space Geodesy Center

A. Cenci[1], *L. Garramone*[1], *G. Bianco*[2], *B. Pernice*[2], *and G. Sylos Labini*[2]

[1]Telespazio SpA, Via A. Bergamini 50, I-00159 Rome, Italy
[2]CNR, Space Geodesy Center, Matera, Italy

1. Introduction

The Matera Space Geodesy Center started its activity in 1983 with the installation of a Satellite Laser Ranging System (SLR), under an agreement between National Aeronautics and Space Administration (NASA), Italian National Space Plan (CNR/PSN) and Regione Basilicata. The Matera Space Geodesy Center, operated by Telespazio under CNR/PSN management, is devoted to Geodesy and Geodynamics, such as study of crustal deformations of the Earth, plate tectonics, polar motion and Earth rotation, as well as to improve the knowledge of the Earth gravity field and Earth and ocean tides. A new Very Long Baseline Interferometry station (VLBI) is now under construction. The new buildings have been already completed, while the 20-meter antenna will be completed within 1988. The electronic subsystems (such as the MARK III terminal) will be installed by Telespazio by the end of 1989, together with a new Time and Frequency system, in order to meet the requirements for both the SLR and VLBI systems. This system could be also used in the future as prime reference for a time and frequency distribution station, to be possibly installed near Matera. An overview of the configuration of the system is given, together with a description of the special room dedicated to host the frequency standards.

2. The New Time and Frequency System

The Time and Frequency system at Matera was based up to fall 1984 on the use of two Rubidium vapour standards to generate the local time scale (average stability $\simeq$ 3x10-12, T $>$1 day) and a LORAN-C receiver to keep it synchronized to UTC at microsecond level. Since 1985, the main oscillator has been replaced with a Cesium beam HP 5061A standard (average stability $\simeq 3\times10^{-13}$, T $>$ 1 day) and since 1986, a GPS FTS 8400 receiver has been installed /1/. For SLR, timing of the epoch of observation must be of an accuracy that is comparable with the range measurement /2/. The average velocity of a satellite (e.g. LAGEOS) is of the order of 5 Km/s, so that absolute timing of better than 1 microsecond is required to fix the satellite in its orbit with, say, 5 mm accuracy. For VLBI /3/, let us suppose a duration of 1000 seconds in reception of signals from a given radio source at, say, 10 GHz frequency. To avoid ambiguity errors, the maximum phase rotation must not exceede 90°, i.e. 25 picoseconds, corresponding to a maximum variation of the frequency used for the synthesis $\Delta f/f= \Delta t/t=2.5\times10^{-14}$. The stability of each local oscillator must be therefore in the order of 1x10-14 for periods of 1000 seconds. In order to meet these requirements, a new time and

Frequency Standards and Metrology Editor: A. De Marchi

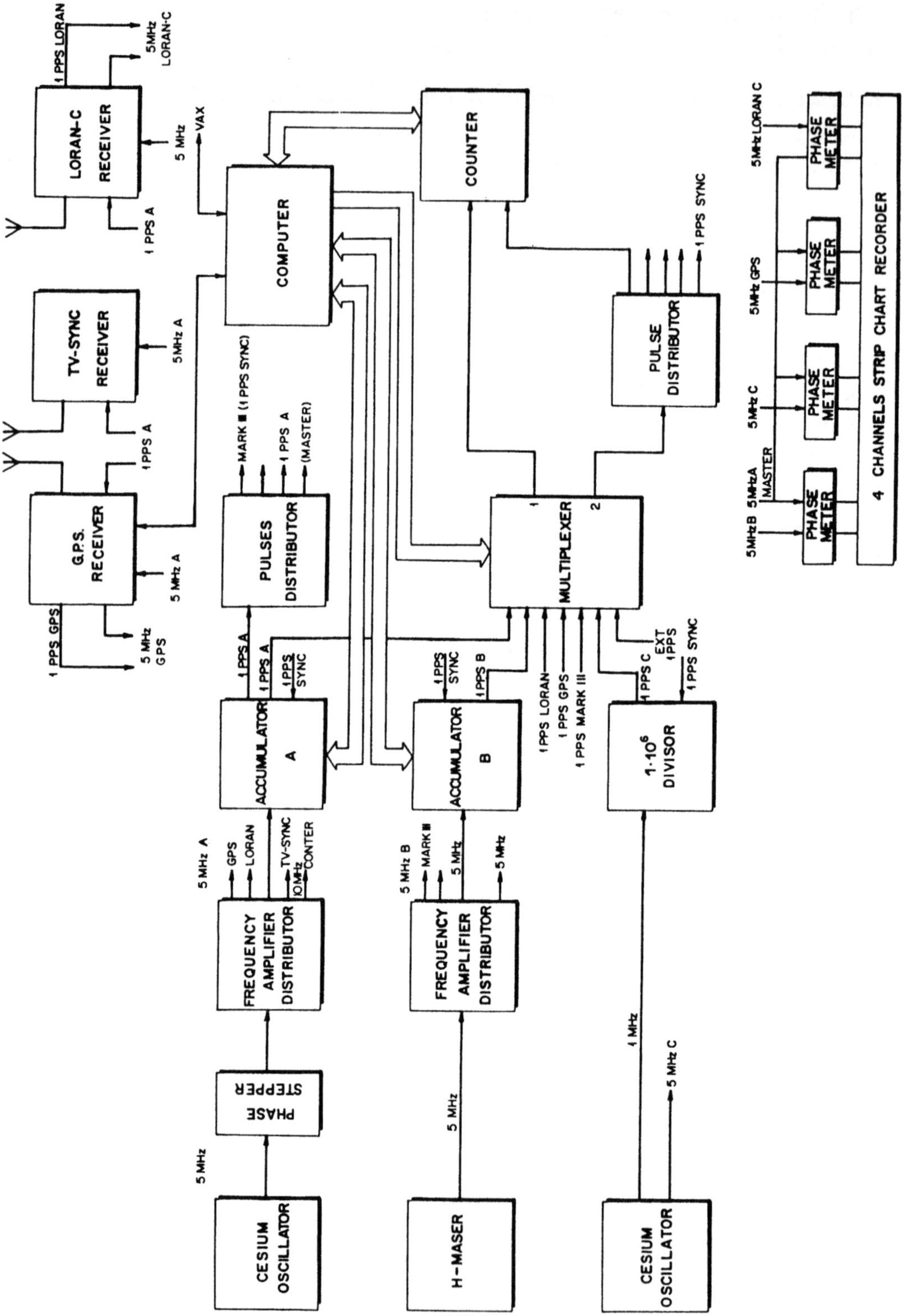

Fig.1 The new Frequency and Time system to be installed at Matera.

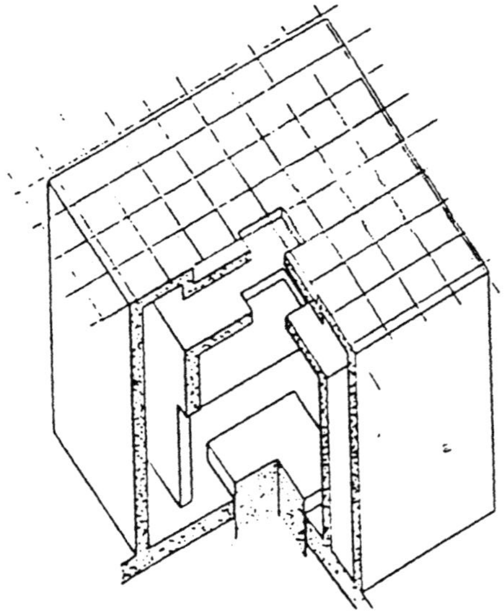

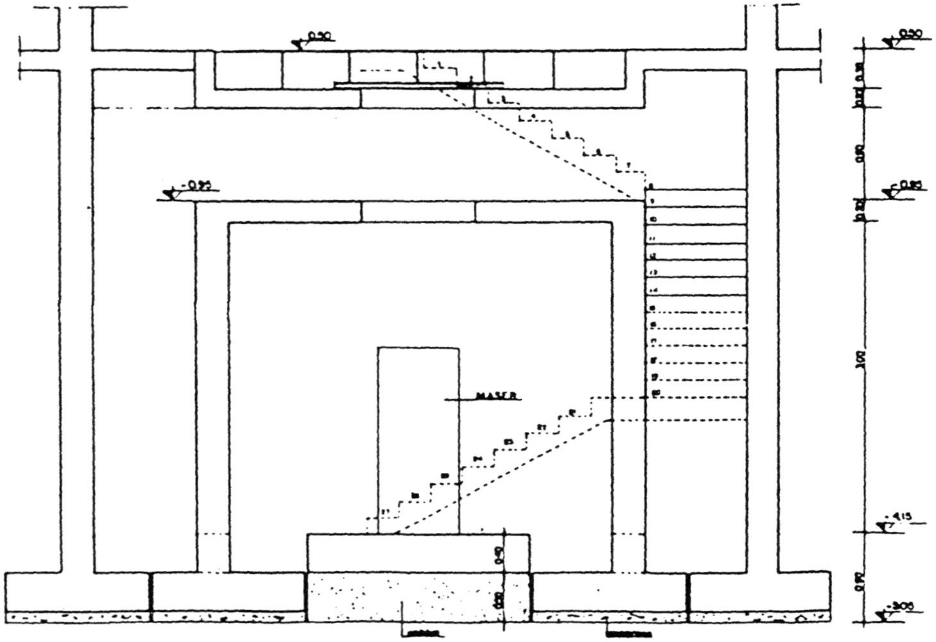

Fig.2 Axonometry and cross-section of the frequency standards room.

frequency system will be installed by the end of 1989 at Matera. The overall block diagram of this system is shown in fig. 1. It consists of three frequency sources and three independent clock chains. The OSA EFOS-8 H-MASER will be used as frequency source for VLBI. This system, already available, showed excellent performance in terms of stability ($\Delta f/f < 2\times10^{-15}$, $1000 > \tau > 10000$ seconds) spectral purity (>110 dB, 1 Hz), temperature sensitivity ($< 2\times10^{-14}/^{\circ}C$). The prime time scale is generated by a Cesium standard (a second Cesium is intended as back-up). Synchronization to UTC to better than 100 nanoseconds is guaranteed by the use of a GPS time receiver. A dedicated computer system manages all the control and monitoring operations, such as clock comparisons, time and/or frequency adjustement, synchronization, environmental control, etc. Absolute continuity of operations is guaranteed by the main power inverter. An additional low power inverter battery back-up will be dedicated to the frequency standards. To minimize the effect of environmental conditions, in particular for the H-MASER, the frequency standards will be hosted in a dedicated room, as described in the next section.

3. The Frequency Standard Room

A special underground room has been built into rock to host the frequency standards (see Fig. 2). It consist of a 4x4x3.4 meter "box" located 5 meters (floor) below the equipment room. Particular criteria were adopted in the construction of this room, following the suggestions of Prof. S. Leschiutta. An 1-meter air space (aisle) around the room and above the ceiling, with double access door, has been designed to improve the thermal insulation. The installation of the H-MASER will be facilitated by the use of a double 1.3x1.3 meter porthole in the ceiling (the room is also reachable by stairway in the aisle for maintenance). A 2x2x0.4 meter concrete pedestal on the floor, insulated from the rest of the building to reduce vibrations, will host the standards, allowing also easy maintenance. Active air conditioning is present in the above equipment room. With no active air conditioning neither in the room nor in the air space, temperature variations less than 2°C/month and 1°C/week have been measured (with the room empty) from January to July 1988. Diurnal temperature variations are very well absorbed at a level not measurable with the chart instrument used ($\sim$0.1°C). The room will be equipped with temperature and humidity sensors and monitored by the computer.

4. References

1. A. Cenci: Matera Laser Station Performance Analysis from July 1986 to June 1987, Telespazio, CNR/PSN contract, 1987
2. Proceedings of the 6th International Workshop on Laser Ranging Instrumentation, Antibes, 1986
3. J.B. Thomas: An analysis of Long Baseline Interferometry, JPL Technical Report 32-1526, 1972 (Vol. VII, VIII, XVI)

Optical Pumping of Rubidium Vapor with Laser-Diodes: Application to Coherent Optical Communications

M. Têtu, N. Cyr, B. Villeneuve, V. Giordano, S. Thériault, and M. Breton

Laboratoire de recherches sur les oscillateurs et systèmes,
Département de génie électrique, Université Laval,
Ste-Foy, Québec, G1K 7P4, Canada

1. Introduction

In recent years, much work has been done on the near-infrared spectroscopy of atoms and molecules with laser-diodes [1,2]. These optical sources are easy to use, can be tuned over a wide range and most are inexpensive. Some of this work have used the techniques developed with optical frequency standards [3] to frequency-lock a laser-diode to an atomic or molecular transition. The absolute value of the laser frequency is then known with high precision and its stability is greatly improved.

We present here some of the results obtained in the use of rubidium as a frequency reference to lock AlGaAs laser-diodes in the 0.8 μm region. Both linear absorption and saturated absorption resonances are considered and a theoretical model allowing the computation of the light intensity transmitted through the cell has been developed. The profile features are well explained so that the resonance frequencies are precisely established. The application of such locked lasers to multiwavelength coherent optical communications is studied [4].

2. Theoretical Model

The physical situation is typical of a saturated absorption experiment [5]. A cell containing a low density rubidium vapor is exposed to two optical beams of the same frequency, propagating colinearly in opposite directions with their linear polarizations crossed. We assume that the light intensity of each beam has a uniform profile and that the pump intensity is constant over the cell length. One beam (pump beam) is much more intense than the other (probe beam) so that high order mixing effects are negligible.

The rate equations formalism is applied to evaluate the population of each energy level. We assume that the hyperfine splitting of the ground state is much larger than the Doppler width while the hyperfine structure of the excited state is not resolved. In order to simplify the writing, we postulate that the pump light and the probe light have an even distribution of π, σ^+ and σ^- polarization so we can avoid specifying each Zeeman sublevel involved in the transitions. The main non-radiative relaxation rate of each state comes from the transit time of the atoms in the light beam. From this model, we get an analytical solution for the light intensity transmitted through the cell.

3. Experimental Results

We have observed the saturated absorption of the D_2-line (780.0 nm) of ^{87}Rb. Taking simple system parameters like cell temperature, laser power and linewidth, and geometrical dimensions, we have determined the hyperfine transition frequencies involved through a nonlinear fit technique. Although the data reduction is only partly completed, we give the preliminary values for the hyperfine structure in Fig.1. The hyperfine constants [6] we found are A=84.92(06) MHz and B=12.31(30) MHz. These values were used to calculate the saturated absorption profile and the results, for the high frequency part, 87A, are compared to the measured profile in Fig. 2. Their agreement is very satisfactory.

The linear absorption profile reveals many resonance peaks. Figure 3 compare the measured and the calculated frequency offsets between each peak value for the D_1-line (794.7 nm) and D_2-line of a mixture of Rb isotopes. The measured values were obtained through an optical heterodyne technique (beat note) with the laser-diodes

Frequency Standards and Metrology Editor: A. De Marchi

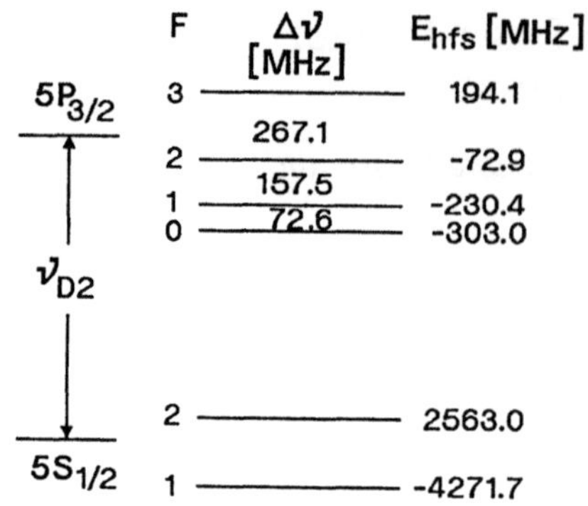

Fig. 1: Hyperfine structure of ^{87}Rb, $5P_{3/2}$ (preliminary)

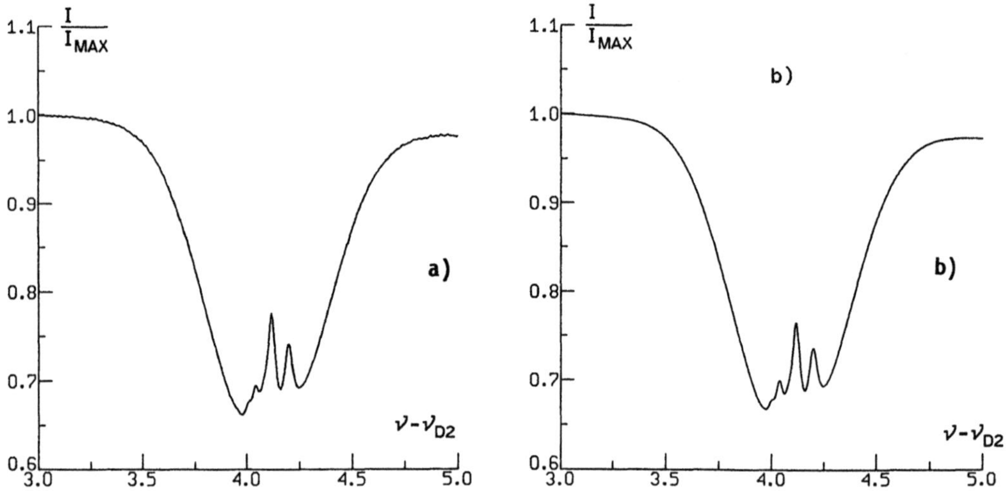

Fig. 2: ^{87}Rb saturated absorption profile of D_2-line, F=1-F'=0,1,2 a)measured, b)calculated

frequency-locked to the various resonances. We note that our model allows a precise determination of these offsets.

4. Application to Coherent Optical Communications

Multi-wavelength coherent optical communications require a great number of closely spaced, stable, narrow-band laser sources [7]. Laser-diodes frequency-locked to different atomic or molecular resonances appear suitable for such systems since the channel selection can be done at precisely defined IF frequencies.

For a low cost system using 0.8 µm laser-diodes, the many resonances shown in the linear absorption profile of rubidium atoms are directly applicable within a simple setup. In order to increase the number of reference lines, saturated absorption offer some possibilities although the setup may become cumbersome. To go further while keeping the predictability, we consider the use of buffer gas in the absorption cells to set the frequency-locked lasers at desired offsets. Using the optical heterodyne technique, we have measured the frequency shift of the ^{87}Rb D_2-line under the influence of N_2 pressure. The coefficient for the high frequency component 87A is -5.8 MHz/Torr and for the low frequency component 87B it is -6.3 MHz/Torr. Further work is in progress to fully characterize this approach.

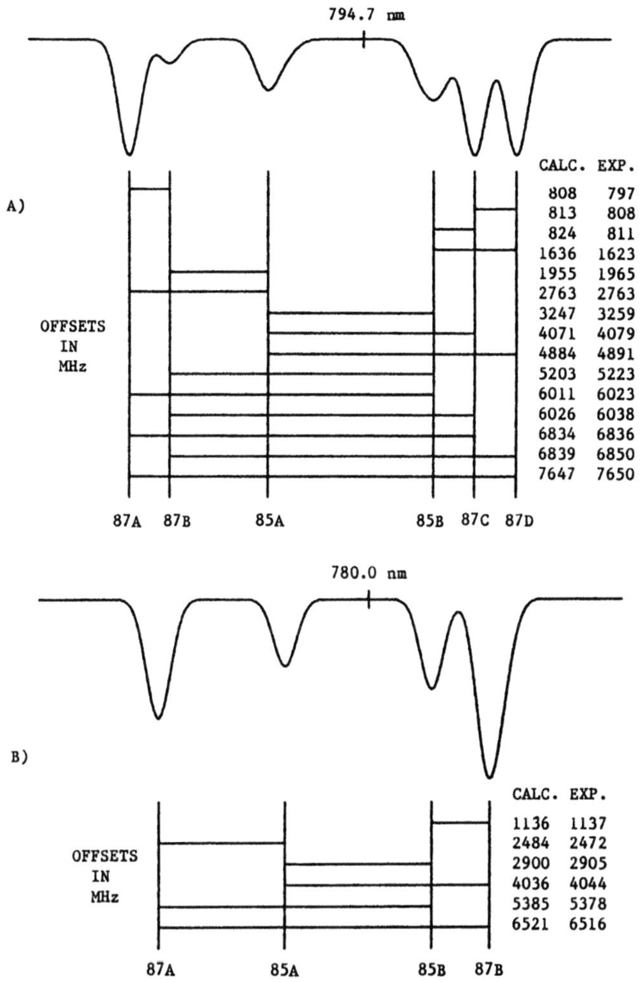

Fig. 3: Possible frequency offsets from the linear absorption profile of a) D_1-line and b) D_2-line for a mixture of Rb isotopes

5. References

1. J. C. Camparo: Contemp. Phys. 26, 5 (1985)
2. M. Ohtsu, T. Tako: Progress in Optics XXV, ed. E. Wolf, Elsevier Science Pub. B. V., 1988, p.191
3. J. L. Hall: Prec. Meas. Fund. Const., eds. B. N. Taylor, W. D. Phillips, N.B.S. (U.S.A.), Spec. Pub. 617 (1984) p.43
4. B. Villeneuve, N. Cyr, M. Têtu, Electron. Lett. 24, 12 (1988)
5. A. Brillet, P. Cérez: J. Physique, Coll. C8, 42, 12 (1981)
6. E. Arimondo, M. Inguscio, P. Violino: Rev. Mod. Phys. 49, 1 (1977)
7. K. Nosu, H. Toba, K. Iwashita: J. Lightwave Technol. LT-5, 9 (1987)

Improvements in Short- and Long-Term Frequency Stabilities of Diode Laser Pumped Rubidium Atomic Clock

M. Hashimoto and M. Ohtsu

Graduate School at Nagatsuta, Tokyo Institute of Technology,
4259 Nagatsuta, Midori-ku, Yokohama, Kanagawa 227, Japan

1. Introduction

Highly stable rubidium (Rb) atomic clocks have been used popularly for global positioning system, microwave communication system, and so on. We have carried out several experiments for the developing a diode laser (DL) pumped Rb atomic clock, and have demonstrated for the first time that a sensitive detection of the optical-microwave double resonance signal in Rb atoms was possible by using a technique of frequency modulation (FM) spectroscopy[1].

In the present paper, we report results of experiments to improve the short- and long-term frequency stabilities of a DL pumped Rb atomic clock.

2. Improvement in Short-Term Frequency Stability

The square root of the Allan variance for the short-term stability of the microwave frequency is expressed as[2]

$$\sigma_y(\tau) = \frac{0.11}{Q\,(S/N)}\tau^{-1/2}, \qquad (1)$$

where τ is the integration time, S/N is the signal-to-noise ratio of resonance signal. The quantity Q is a quality factor of frequency discriminator which is defined by $Q = (\omega_{12}/2\pi)/\Delta\nu$, where $\Delta\nu$ is the linewidth of frequency discriminator, and $\omega_{12}/2\pi$ is the microwave transition frequency (6.8GHz). Therefore, the short-term frequency stability can be improved by increasing the values of Q and of S/N. For this purpose, we employed a technique of FM spectroscopy by using a coherent optical pumping by DL. The optimum values of S/N and $\Delta\nu$ obtained by this technique were 73 dB and 220 Hz, respectively. Substituting these values into eq.(1), the value of $\sigma_y(\tau)$ of a DL pumped Rb atomic clock was estimated as $\sigma_y(\tau) = 7.9\times 10^{-13}\tau^{-1/2}$ which is about 1/10 times that of conventional Rb atomic clocks[3]. Further improvements of the short-term frequency stability of a DL pumped Rb atomic clock are possible by increasing the S/N value by applying a technique of the electrical negative feedback to DL[4].

3. Improvement in Long-Term Frequency Stability

The optical Stark effect by the electric field of pumping light would reduce light shift, i.e., the shift of the microwave transition frequency[5]. If the light shift is induced, the profile of double resonance spectral shape is deformed and become asymmetric, which results from inhomogeneous distribution of laser intensity in the Rb gas cell[6]. The light shift as well as buffer gases would limit the frequency accuracy of Rb atomic clock. Furthermore, the light shift would induce microwave frequency drifts, which have not been reduced yet in conventional Rb atomic clocks. We propose here a new technique to improve long-term frequency

Frequency Standards and Metrology Editor: A. De Marchi

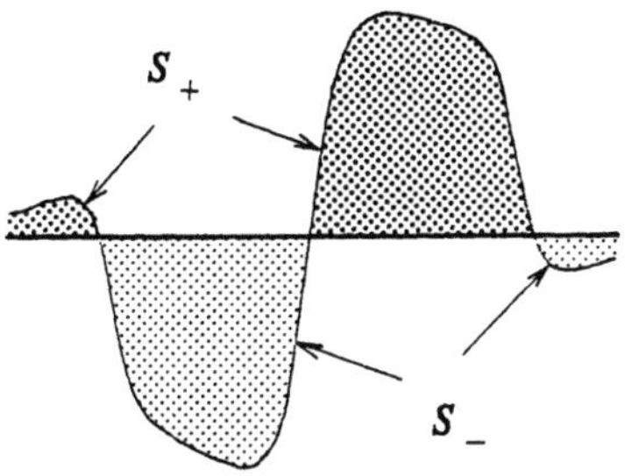

Fig.1 The definition of the asymmetrical factor ΔS.

$$\Delta S = S_+ - S_-$$

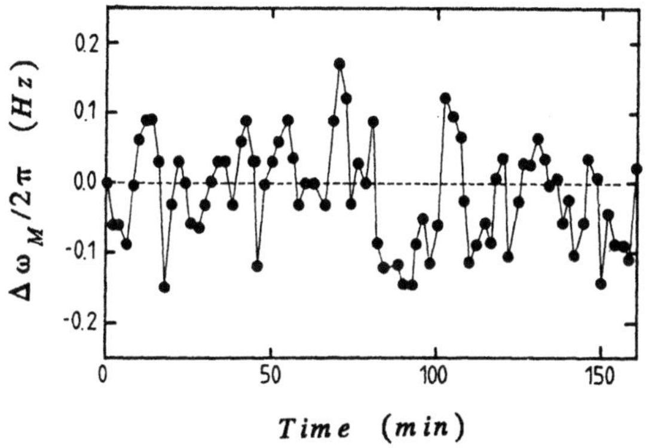

Fig.2 Reduction of microwave frequency drift by a self-tuning system.
$\Delta\omega_M/2\pi$: Microwave frequency drift.

stability. We defined the asymmetical factor ΔS as a difference between the areas S_+ and S_- of the double-resonance spectral profile, which are schematically given in Fig.1. Since the value of ΔS is proportional to the magnitude of the light shift, the microwave frequency drift caused by the light shift can be reduced by controlling the laser frequency so that the measured value of ΔS can be maintained zero. It should be noted that this system does not require any additional reference frequency sources. Therefore, this system can be referred as a self-tuning system.

In order to reduce the microwave frequency drift caused by the light shift, the laser frequency was controlled by this self-tuning system at every 120s. The result is shown by Fig.2. The value of $\sigma_y(\tau)$ at τ =120s was 1.2×10^{-11}, which is 1/10 times that of conventional Rb atomic clocks[3]. This result confirms that the present method is feasible for the improvement of the long-term stability.

4. Summary

The short-term frequency stability of Rb atomic clock was improved to as high as $\sigma_y(\tau) = 7.9\times 10^{-13}\tau^{-1/2}$ by using a technique of FM spectroscopy. A self-tuning system was developed to reduce the frequency drift caused by the light shift. The long-term frequency stability obtained by using this system was $\sigma_y(\tau = 120s) = 1.2\times 10^{-11}$.

REFERENCE

1. M.Hashimoto and M.Ohtsu:IEEE J.Quantum Electron.,QE-23,446(1987)
2. J.Vanier and L.G.Bernier:IEEE Trans.Instrum.Meas.,IM-13,221(1981)
3. H.Hellwig:Proc.28th Annu.Freq.Control Symposium,612(1974)
4. M.Ohtsu:J.Lightwave Technol.,6,245(1988)
5. W.Happer and B.S.Mathur:Phys.Rev.,163,12(1967)
6. J.C.Camparo,R.P.Frueholz, and C.H.Volk:Phys.Rev.,27 ,1914(1983)

Efficient Frequency Noise Reduction of GaAlAs Laser Diodes by Negative Electronic Feedback

H.R. Telle and B. Lipphardt

Physikalisch-Technische Bundesanstalt, Bundesallee 100, D-3300 Braunschweig, Fed. Rep. of Germany

Negative electronic feedback (NEF) is a promising technique to improve the coherence properties of semiconductor laser diodes (LDs). The system considered here consists of a LD (HLP 1400), a frequency discriminator (ring cavity, round-trip length 13 cm, linewidth ≈ 80 MHz FWHM) and a servo amplifier. If properly designed, NEF reduces the frequency noise spectral density S of the free running laser (index FR) at Fourier frequencies f below a certain value f_u (≈ servo loop unity gain frequency), following more or less a 20 dB/decade-dependence. We will assume that the 1/f-corner frequency of S_{FR} is smaller than f_u, i.e. S_{FR} is almost constant in this frequency region. Below a lower corner frequency f_1, S_{NEF} becomes almost frequency independent in many practical cases. This noise limited level $S_{NEF}(f<f_1)$ and the total phase noise power $PNP(f_1)$, defined as the integral over all phase noise components at Fourier frequencies above f_1, are the most important characteristics of a NEF set-up and influence sensitively the spectral properties of the LD. $PNP(f_1)$ has to be less than 1 rad^2 in order to keep the total optical sideband power smaller than the central line ("carrier") power. This condition can be fulfilled only with sufficiently short feedback loop delay times. Our computer simulations show that this time should be less than 6 nsec for a HLP 1400. We used 3.2 nsec during the experiments. $S_{NEF}(f<f_1)$, called S_{MIN} in the following, characterizes the width $\Delta\nu$ of the lorentzian central line: $\Delta\nu = \pi \cdot S_{MIN}$. Here we have assumed $S_{MIN} \ll f_1$. S_{MIN} may depend on different noise contributions:

1. Shot-noise (SN) is of minor interest if linewidths of more than 1 kHz are considered. For example the SN contribution to S_{MIN} of a typical set-up (10 µW at the photodetector, discriminator slope of 1/70 MHz, as in our case) is of the order of 100 Hz^2/Hz, as indicated in Fig. 1.

2. Electronic noise contributions to S_{MIN} can be kept below the SN-level by using sufficiently large photocurrents (>10 µA), small area photodiodes and low noise transimpedance preamplifiers.

3. Relative intensity noise of the dominant longitudinal LD-mode, named RIN(DM) in the following, may be a severe problem, at least for LDs of Fabry-Perot type. It is well known, that RIN(DM) is usually several orders of magnitude higher than the all-mode noise RIN(TOT) for these lasers. For example, we found for our CSP GaAlAs LD operating in a nearly single longitudinal mode: $RIN(DM) = 5 \cdot 10^{-10}\ Hz^{-1}$, $RIN(TOT) < 10^{-13}\ Hz^{-1}$ for f>100 kHz. It is obvious, that a conventional two beam slope discrim-

Frequency Standards and Metrology Editor: A. De Marchi

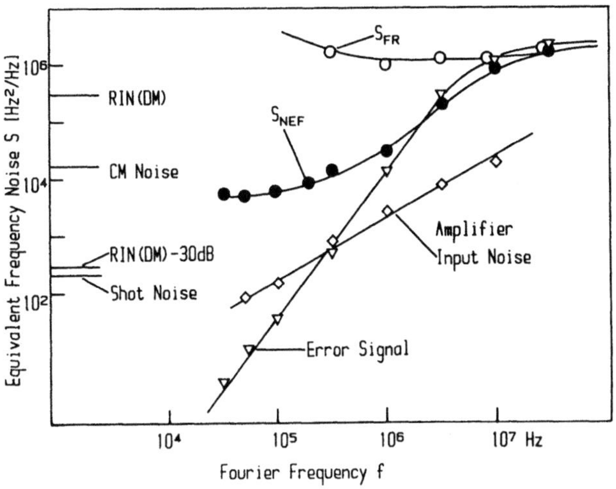

Fig. 1: Frequency noise density under free running, S_{FR}, and electronic feedback conditions, S_{NEF}. Note the >30 dB difference between error signal and S_{MIN} at low Fourier frequencies due to FM/AM cross modulation

inator, where the unfiltered LD output signal is used as the reference for AM noise suppression by electronic substraction from the Fabry-Perot transmission signal, is inadequate for this application because a substantial part of RIN(DM) is converted to FM noise by the control loop. The resulting equivalent S_{MIN}-level of about $3 \cdot 10^5$ Hz^2/Hz is indicated by RIN(DM) in Fig. 1. We reduced this noise contribution by at least three orders of magnitude by using a resonator with internal birefringent plate. Both sets of orthogonally polarized, frequency shifted, resonator eigenmodes were excited equally by choosing a proper input polarisation. The frequency shift was adjusted to about one full cavity linewidth by applying a transversal force to the intracavity plate (stress birefringence). Then, the photocurrent difference of both orthogonally polarized transmission signals displayed dispersion-like spectral features of odd symmetry, which are needed for servo control. Both transmission signals, which were detected by separate photodiodes, carried almost the same amplitude modulation RIN(DM) while AM contribution from the longitudinal neighbour modes was highly suppressed. This scheme results in an efficient RIN(DM) compensation for the difference signal. We measured a RIN(DM) suppression of better than -30 dB, as indicated in Fig. 1, leading to an equivalent S_{MIN} value of about 300 Hz^2/Hz.

4. AM/FM cross modulation noise (CM-noise) is defined here as the mixing product of AM and FM noise components having a frequency difference of less than f_1. We estimated the equivalent value of S_{MIN} due to this contribution to be of the order of 10^4 Hz^2/Hz under our conditions, see CM-noise level in Fig. 1.

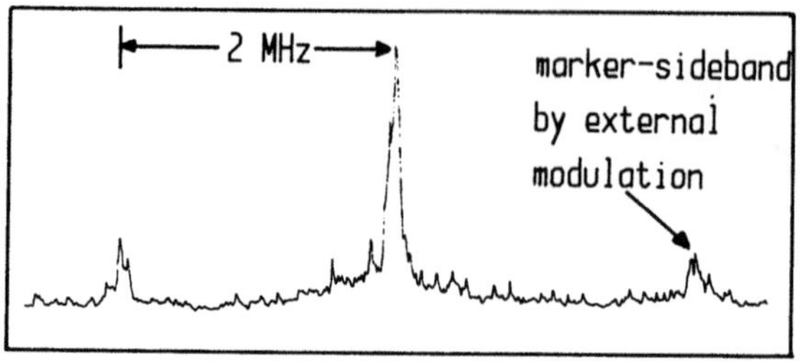

Fig. 2: Fabry-Perot scan of the LD output spectrum under NEF conditions.
F-P length = 1.80 m, Finesse ≃ 700, optical isolation > 100 dB

The experimental $S_{NEF}(f)$ - curve (closed circles in Fig. 1) was measured by using a similar, also RIN(DM)-compensated, monitor discriminator. Obviously, S_{MIN} exceeds the shot-noise level SN, the remaining RIN(DM)-contribution and the preamplifier input noise level by more than 10 dB. Agreement is found between S_{MIN} and the estimated equivalent CM-level within a few dB, which means that CM-noise was the dominant S_{MIN}-limiting noise contribution in our experiment. The experimental S_{NEF}-curve indicates, that the optical power spectrum should consist of an almost lorentzian line of about 30 kHz width (FWHM) containing almost the complete LD output power. In fact, we found by using a 1.8 m long monitor Fabry-Perot resonator a (instrumentally limited) linewidth of less than 120 kHz, see Fig. 2.

New fields of applications, requiring oscillators of high spectral purity, are opened up by these kinds of laser diodes: long-distance communication, high resolution spectroscopy, optical frequency synthesis...

Some Applications of Extended-Cavity Semiconductor Lasers

M. de Labachelerie, K. Diomande, and P. Cérez

Laboratoire de l'Horloge Atomique,
Equipe de Recherche du CNRS associée à l'Université Paris-Sud,
Bât. 221, Université Paris-Sud, F-91405 Orsay Cedex, France

Extended-cavity semiconductor lasers (ECL) have recently been extensively studied to improve the spectral characteristics of laser diodes [1,2] : these lasers have spectral linewidh and tuning properties that are far much better than ordinary laser diodes. The main applications of such lasers are, of course, connected with coherent optical communications, however, they are also of great interest in other fields such as atomic physics, spectroscopy, Cesium frequency standards as well as laser frequency standards. We describe here our latest results concerning these various applications of this new type of tunable laser sources.

1. Laser Description

Extended-cavity lasers are made of an anti-reflection coated laser diode coupled to a long external cavity [1,3] . This external cavity allows to reduce the high linewidth of ordinary laser diodes to values well below 100 kHz. Moreover, a spectrally selective element (usually a diffraction grating) inserted in the external cavity, allows to tune the laser wavelength on wide wavelength bands (100 nm around 1.5 microns). This high tuning capability is of great interest for atomic physics and spectroscopy.

2. Application to Optically Pumped Cesium Beam Frequency Standards

Comparing the optically detected error signals obtained with an ordinary diode or an ECL, we have shown that the low frequency noise of ECLs could lead to an order of magnitude improvement in the stability of optically pumped Cesium beam frequency standards.

3. Lineshape Measurements

Using an heterodyne technique with an ECL as a reference laser, it is possibe to plot the lineshape of ordinary lasers with a great accuracy and a

Frequency Standards and Metrology Editor: A. De Marchi

low sensitivity to optical feedback that usually perturbs such measurements. With this technique, we have recently characterized new DFB lasers that showed a 4 MHz linewidth at 16 mW output power.

4. Spectroscopy

As was already pointed out [4] , it is of great interest to find new absorption lines that may be used for laser frequency standards. in the 1.3-1.6 microns fiber optics band. That's why an extensive spectroscopic work should be made to explore this wavelength region.

Using an ECL we have obtained the first high resolution laser spectrum of Ammonia in the 1.49-1.52 microns range. The 30 strongest lines have been calibrated with a 10^{-7} accuracy using a wavemeter. Other molecules will be studied in the future.

5. LaserLong-term Frequency Stabilization

The strongest line recorded is a doublet with a 1.3 GHz separation between each component; it can be used for a long term frequency stabilization of the ECL. However, as no saturated absorption has been observed, the reference line was too wide to obtain the 10^{-12} stability already recorded [3] with the Cs-D_2 line at 852 nm. We have obtained with NH_3 a 2.10^{-10} stability for 1 to 10 seconds observation time. These results should be improved with saturated absorption lines.

References

1. M.Labachelerie, P.Cerez : Optics comm. 55, (1985), pp. 174-178
2. M.R.Matthews et Al. : Electron. Lett. 21, (1985), pp. 113-115
3. M.Labachelerie, K.Diomande, N.Dimarcq : Proc. EFTF 88, Neuchâtel, march 1988, pp.547-554.
4. DJE Knight : Proc. EFTF 88, Neuchâtel, march1988, pp.23-33

Ultra-Narrow Linewidth of Optically Self-Locked Diode Lasers

C. Bréant[1], *P. Laurent*[1], *and A. Clairon*[2]

[1]Laboratoire Lasers Ultra-Stables, E.T.C.A., 16 bis Av. Prieur de la Côte d'Or, F-94114 Arcueil, France

[2]L.P.T.F., Observatoire de Paris, 61 avenue de l'Observatoire, F-75014 Paris, France

Using diode laser in metrology requires a strong reduction of the important frequency noise of the laser which is due to a low Q cavity and to a strong coupling between the phase and the amplitude of the electric field fluctuations. The high sensitivity of diode lasers to optical feedback can deeply modify the static, dynamic and spectral behaviour of such optically coupled cavities. In our experiment, similar to the one of Dahmani and Hollberg [1], the optical feedback comes from one of the reflected beam of a tilted Confocal Fabry-Perot (CFP) interferometer (fig.1).

A theoretical and experimental analysis [2] shows how the diode laser frequency self-locks to a Fabry-Perot resonance (fig.2). Hence the frequency noise is largely reduced (fig.3) and the relative static reduction of the frequency noise spectral

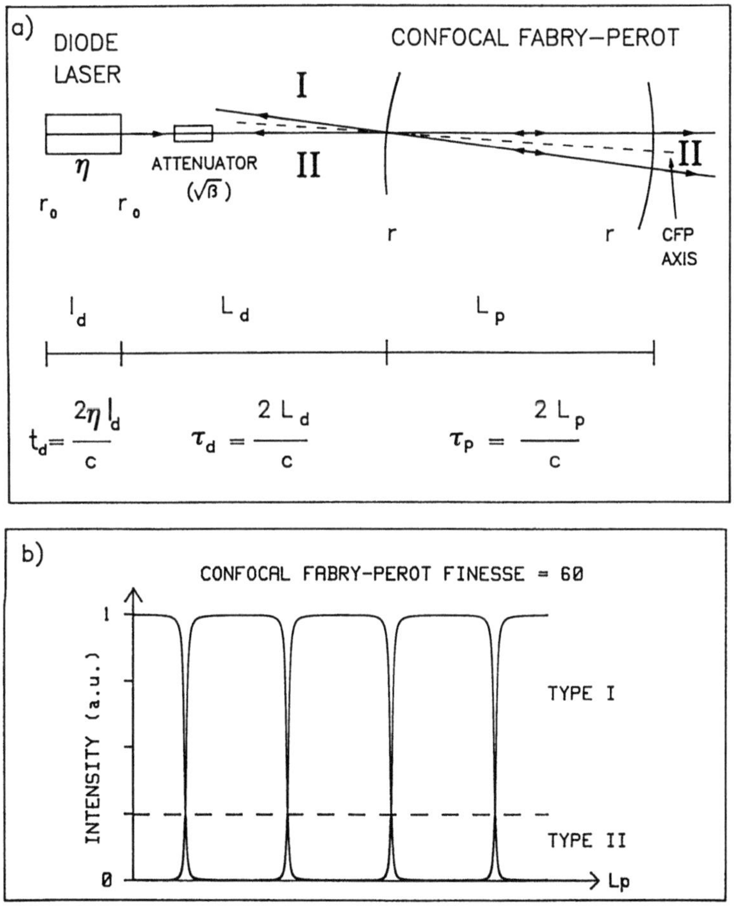

Fig. 1 : Schematic of the optical feedback locking system: a) represents a diagram of the optically self-locked laser cavity coupled to a tilted C.F.P. , b) shows the transmission (II) and reflection (I) fringes obtained with a tilted C.F.P..

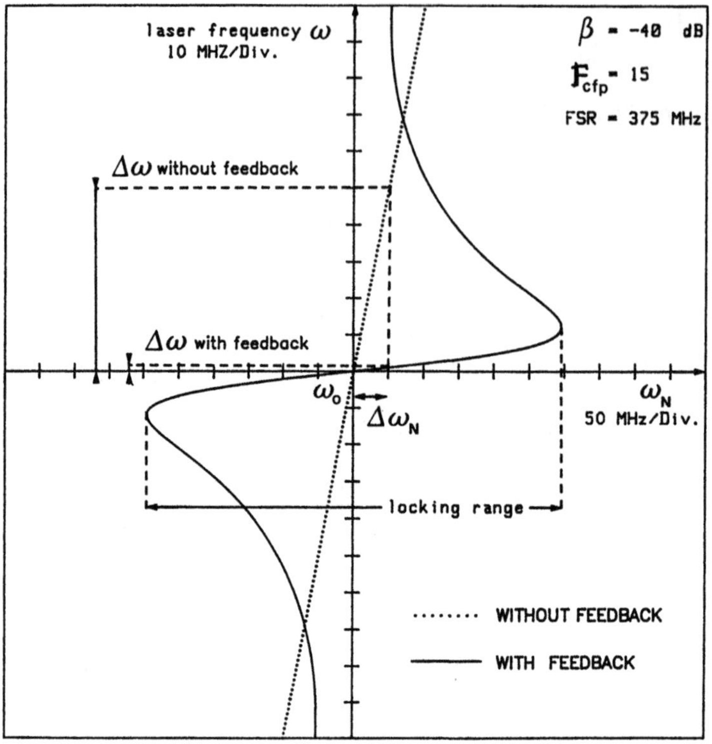

Fig. 2 : Theoretical representation of the laser frequency behaviour ω around frequency ω_0 as a function of the unperturbed laser frequency ω_N. The dotted line represents the free-running laser diode ($\omega = \omega_N$) and the solid line the feedback coupled cavity laser. This figure illustrates the dramatic frequency excursion reduction from Δω without feedback to Δω with feedback, for a given $\Delta\omega_N$.

density between the locked state ($S_{\dot\phi}^{L}(0)$) and the free running state ($S_{\dot\phi}^{F}(0)$) is given by :

$$S_{\dot\phi}^{L}(0)/S_{\dot\phi}^{F}(0) \cong [1+(1+\alpha^2)^{1/2}\sqrt{\beta}\,(L_p F_p/\eta\, l_d F_d)]^{-2}$$

where α is the phase-amplitude coupling factor, β is the power mode coupling factor, η is the real part of the refractive index, L_p is the CFP length, l_d is the diode laser length (≅ 300 μm), F_p is the CFP finesse and F_d is the diode laser finesse (≅ 3).

A full frequency analysis of the frequency noise reduction of such a coupled cavity has been calculated. The response time of the servo system locked to a C.F.P. cavity is a function of both the storage time of the cavity and the feedback level. With a high finesse cavity and a high feedback level, frequency noise reduction of 50 to 60 dB can be achieved and should lead, if white frequency noise is the main source of noise, to a Lorentzian lineshape with a linewidth below 100 Hz given by :

$$\text{HWHM} \cong \pi \, . \, S_{\dot\phi}^{F}(0).$$

For different feedback levels, we recorded and analyzed the RF power spectra of the beat note between two such coupled cavities and found non-Lorentzian lineshapes (fig.4). In fact a careful noise analysis shows that at high feedback level the frequency noise spectrum is mainly Flicker frequency noise ($S_{\dot\phi}^{F}(f) \cong 1/f$) and leads to a linewidth which varies as $(S_{\dot\phi}^{F}(1))^{1/2}$. Hence for noise reduction level of 50-60 dB the HWHM linewidth is much wider, of the order of 4 kHz (fig.5).

Frequency tunability of the coupled cavity laser is achieved by a double servo-loop system so that the frequency is simply determined by the CFP length. Through the modulation of any of the three parameters ω_N, τ_d or τ_p a locking amplifier detects a signal proportional to the derivative of the transmitted power (T) through the CFP and after integration this signal is fed through a PZT controlling the distance between the diode laser and the CFP to ensure maximum transmission through the CFP. In these conditions the coupled cavity laser frequency is determined by the CFP resonance frequency

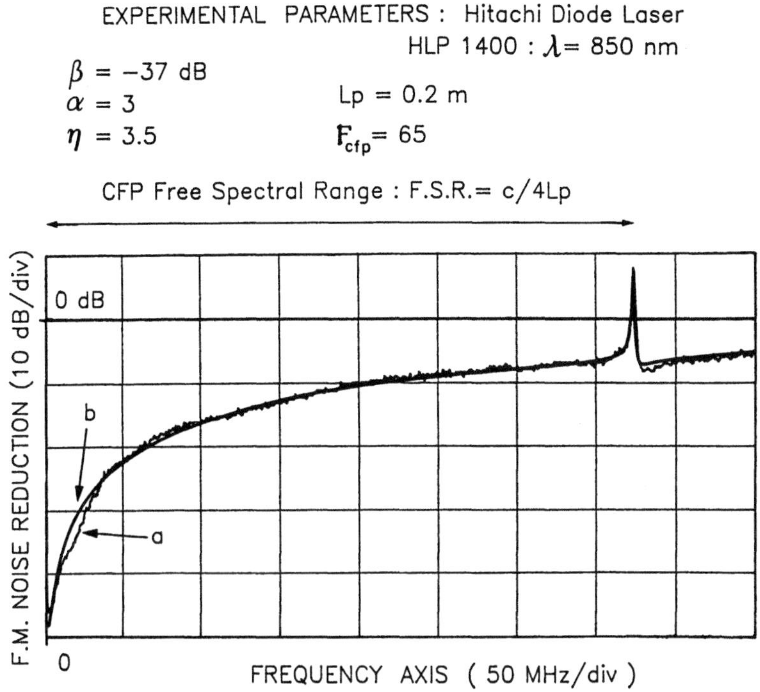

Fig. 3 : Frequency noise reduction of a diode laser coupled to a CFP. Line a) shows the experimental data and line b) represents the theoretical prediction calculated from the experimental parameters. In this case the static noise reduction is of the order of 46 dB. The resolution bandwidth is 300 kHz, the video bandwidth is 1 kHz and the sweeping time is 5 s.

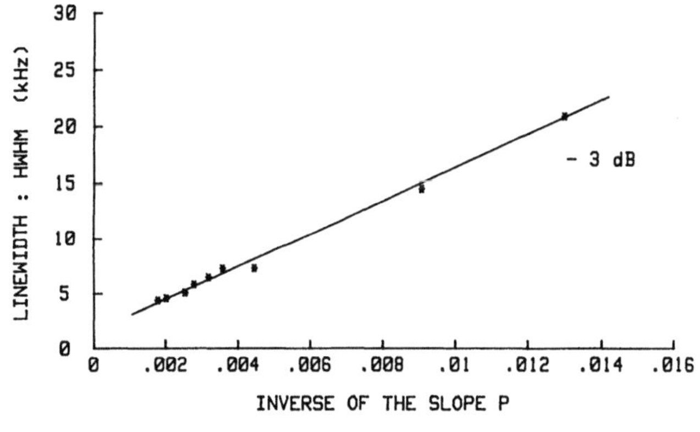

Fig. 4 : Experimental width of the recorded beat-note signal of the two stabilized diode lasers, measured at different feedback levels, as a function of the inverse of the slope P for one of the stabilized system. The other coupled cavity laser was kept locked at its best level. These curves clearly show a linear dependence of the width as a function of 1/P :
$P = d\omega_N / d\omega = 1 + (1+\alpha^2)^{1/2} \sqrt{\beta}\, (L_pF_p / \eta l_d F_d)$.

($\omega\tau_p = k\pi$). Then through the modulation of the feedback level it is possible after detection at the sum frequency of the two modulations to ensure

$$0 = d^2 T / d\beta\, dx = 2\, T F^2 \tau_p \cdot d(\omega \tau_p) / dx \,.\, d(\omega_N - \omega) / d\beta \propto (\omega_N - \omega) / \beta$$

by driving the diode laser current ($x = \omega_N, \tau_d$ or τ_p).

It is interesting to note that the modulation of the feedback level does not induce any frequency modulation of the coupled cavity laser frequency when the two servo-loops are closed.

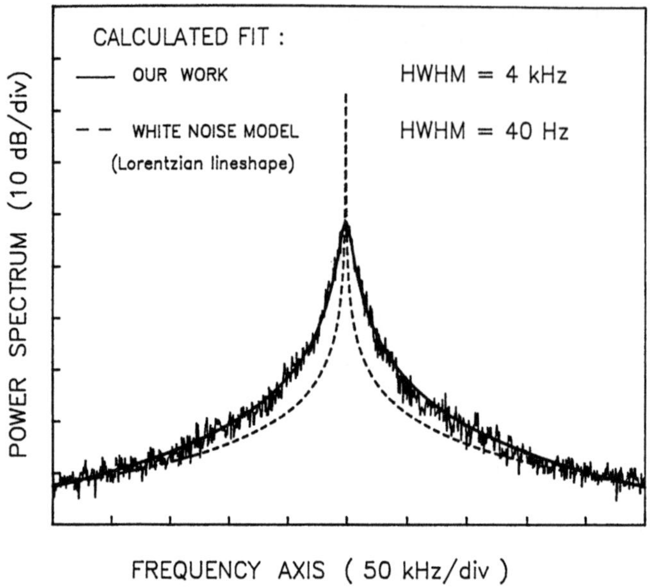

Fig. 5 : Beat-note spectrum of two independently locked diode lasers with frequency noise reduction of 60 dB for diode 1 and 55 dB for diode 2. The dashed line shows the calculated spectrum if we assumed a white frequency noise model. The solid line represents the fit of our best spectrum (Lorentzian lineshape to the power 3/2) and can be interpreted as 1/f frequency noise. The resolution bandwidth is 1 kHz, the video bandwidth is 300 Hz and the sweeping time is 5 s.

This stabilization technique can be easily extended to any kind of single mode diodes, without any modification of the laser chip itself. Hence such narrow linewidth laser systems, available over a large spectral range, will greatly simplify experiments in the research fields of atomic and molecular spectroscopy, optical frequency standards, optical pumping in Cesium clock, cooling and trapping of atoms, for instance.

1. B. Dahmani, L. Hollberg and R. Drullinger, " Frequency stabilization of semiconductor lasers by resonant optical feedback", Optics Lett., vol. 12, n° 11, pp. 876-878, 1987.
2. Ph. Laurent, A. Clairon and Ch. Bréant, " Frequency noise analysis of optically self-locked diode lasers", submitted to I.E.E.E., J. of Quantum Electronics.

Manipulation of Metastable Helium with a LNA Laser

R. Kaiser[1], *N. Vansteenkiste*[1], *E. Arimondo*[2], *A. Aspect*[1], *C. Cohen-Tannoudji*[1], *M. Karrais*[3], *and H. Haberland*[3]

[1]Laboratoire de Spectroscopie Hertzienne de l'E.N.S. et Collège de France, 24, rue Lhomond, F-75231 Paris Cedex 05, France
[2]Permanent address: Dipartimento di Fisica, Università di Pisa, Italy
[3]Fakultät für Physik der Universität Freiburg, D-7800 Freiburg, Fed. Rep. of Germany

It is possible to cool metastable (2^3S_1) Helium by interaction with a laser at 1.08 μm, since the transition 2^3S_1 - 2^3P_2 is a closed transition. Helium is a very attractive system for many reasons : for example it is a simple atom, and progress in its spectroscopic measurements may lead to improved accuracy in fundamental constants (α).

Our source of He* is a supersonic Helium beam, excited by a colinear electron gun [1]. It can be cooled by liquid nitrogen or liquid Helium. We have been thus able to produce a He* beam down to a velocity of 400 m/s (measured by time of flight techniques).

The laser for interacting with (2^3S_1) He* is a home-build single mode ring laser, with a LNA crystal [2] pumped by an Ar^+ laser. The laser linewidth is less than 0.5 MHz, and long term stability is obtained by locking the laser to the atomic transition. This is achieved by a saturated absorption technique in an auxiliary Helium discharge.

We present several experimental results obtained by tranverse action of the laser crossing the atomic beam at right angle. The transverse velocity distribution is studied with a scanable electron multiplier located downstream. Typical results are :

(1) deflection of the He* out of the intense beam of He atom in the ground state with a curved optical wavefront (Fig. 1) ;

(2) increase in the beam intensity by one order of magnitude by one-dimensional optical molasses (Fig. 2) ;

(3) transverse temperature of 53 μK after one-dimensional cooling (by optical molasses) ; a computer simulation shows that the cooling limit (23 μK) is not reached because of too short interaction time.

Frequency Standards and Metrology Editor: A. De Marchi

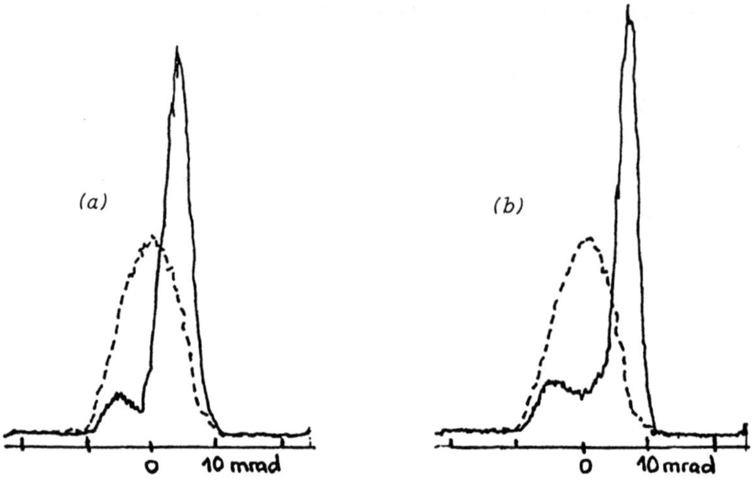

Fig. 1 : *Deflection of an He* beam with a plane wave (a) and with a converging wave of radius of curvature R = 10 m (b). The two figures show the profiles of the atomic flux at 1.6 m from the source, without the laser (dashed line) and with the laser (solid line). Note that there is also a cooling effect in case (b).*

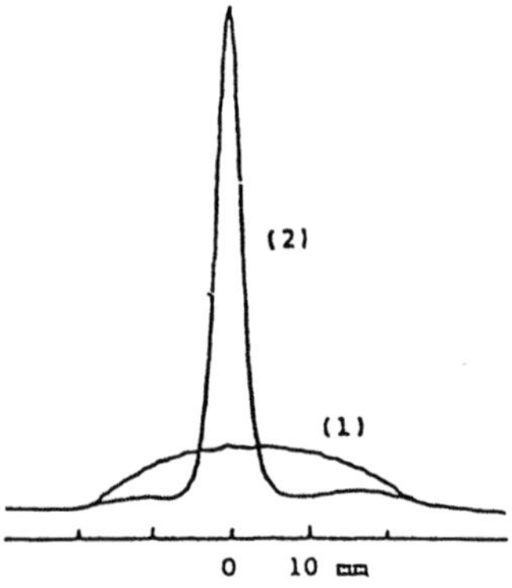

Fig. 2 : *Collimation by one dimensional molasses. The figure shows the profiles of the atomic flux at 2 m from the source, without the laser (1) and with the laser (2). The source width is 2 mm. The temperature corresponding to the final velocity distribution is 53 μK.*

REFERENCES

[1] H. Conrad, G. Ertl, J. Küppers, W. Sesselmann and H. Haberland, Surf. Sci. 121, 161 (1982).

[2] L. Schearer, M. Leduc, D. Vivien, A.M. Lejus and J. Thery, IEEE J. Quantum Electron. 22, 713 (1986).

Quantitative Study of Laser Cooling in a Penning Trap *

W.M. Itano, L.R. Brewer, D.J. Larson, J.J. Bollinger, S.L. Gilbert, and D.J. Wineland

National Institute of Standards and Technology, Boulder, CO 80303, USA

Penning traps are currently being studied at our laboratory [1,2] and elsewhere [3,4] for use in atomic frequency standards. A Penning trap confines ions by a combination of a static, uniform magnetic field and a static, nonuniform electric field. The fact that the fields are static makes it feasible to laser cool more than 10^4 ions to low temperatures [5]. In contrast, the radio-frequency (Paul) ion trap uses oscillating electric fields, which, in combination with the non-linear Coulomb coupling between ions, cause heating and make it difficult to laser cool large numbers of ions [6]. For either trap, large numbers increase the signal-to-noise ratio; low temperatures reduce the second-order Doppler shift [7].

Several experimental groups have reported laser cooling of ions confined in Penning traps to temperatures below 1 K [1-5,8-10]. Typically, the measured temperatures were one or two orders of magnitude higher than the theoretical limit $T_{min} = \hbar\gamma_0/(2k_B)$ (about 1 mK), where γ_0 is the natural linewidth of the optical transition used for laser cooling [9]. The reason for this discrepancy is now understood [10]. The rotation of the ion plasma about the magnetic field axis, due to the presence of crossed electric and magnetic fields, has a great effect on the temperature. In order for the ion plasma to be stable, it is necessary to displace the laser beam radially so that it illuminates the side of the plasma which is receding from the laser. The radiation pressure applies a torque to the plasma which counteracts the torques induced by axially nonsymmetric, static electric and magnetic fields. This radial displacement of the beam, however, results in higher temperatures. The reason stems from the requirement that the work done by the laser on the ions be zero for a steady state. The rate at which work is done on an ion of velocity $\vec{v}$ by a force $\vec{F}$, which, in this case, is parallel to the direction of propagation of the laser beam, is $\vec{F}\cdot\vec{v}$. Because of the radial displacement of the beam and the plasma rotation, $\vec{F}\cdot\vec{v}$ tends to be positive. In order to make the average value $\langle\vec{F}\cdot\vec{v}\rangle$ be zero, the frequency of the laser must be tuned below the optical resonance. Because of the Doppler shift, this makes $\vec{F}$ stronger for ions with $\vec{F}\cdot\vec{v} < 0$ than for ions with $\vec{F}\cdot\vec{v} > 0$. The width of the velocity distribution must have a particular value in order for cancellation between negative and positive contributions to make the average value

Frequency Standards and Metrology Editor: A. De Marchi

be zero. The required width increases with the rotation frequency and with the distance of the laser beam from the axis of rotation of the plasma. This width corresponds to a temperature higher than T_{min}.

Temperatures and rotation frequencies of $^9Be^+$ plasmas were measured for a wide range of experimental parameters [10]. The measured temperatures varied from 40 mK to 2 K. These measured temperatures were in agreement with the calculation, within an rms error of about 10%. For these measurements, the cooling laser beam was in the plane perpendicular to the magnetic field. Lower temperatures have been attained by the use of laser beams whose direction of propagation were not perpendicular to the magnetic field axis, but the calculation in this case is more complicated [5].

For certain ranges of laser frequency detuning and beam position, we have observed strong oscillations in the total ion fluorescence, with periods from less than a second to over a minute. These oscillations have been observed in all Penning traps at our laboratory and are similar to those reported by others [3,11]. Sometimes, a deliberate misalignment of the magnetic field direction with respect to the trap axis was necessary in order for oscillations to occur. This was an indication that external torques were involved. An ion plasma undergoing oscillations was observed with a photon-counting imaging tube. The change in ion fluorescence was accompanied by a change in the radius (which is directly related to the total canonical angular momentum) of the plasma. For the plasma to reach a steady state, the rate of both the energy and angular momentum imparted to the plasma by external forces must be 0. Apparently, these two conditions sometimes cannot be satisfied simultaneously, and this leads to oscillations.

We wish to acknowledge support from the U.S. Office of Naval Research and the U.S. Air Force Office of Scientific Research.

1. J.J. Bollinger et al.: Phys. Rev. Lett. 54, 1000 (1985)
2. J.J. Bollinger et al.: these Proceedings
3. R.C. Thompson et al.: Opt. Acta 33, 535 (1986); Appl. Phys. B 46, 87 (1988)
4. F. Plumelle et al.: Appl. Phys. B 41, 183 (1986)
5. L.R. Brewer et al.: Phys. Rev. A 38, 859 (1988)
6. J. Hoffnagle et al.: Phys. Rev. Lett. 60, 2022 (1988); R. Blümel et al.: Nature, 334, 309 (1988)
7. D.J. Wineland et al.: these Proceedings
8. R.E. Drullinger, D.J. Wineland, and J.C. Bergquist: Appl. Phys. 22, 365 (1980)
9. W.M. Itano and D.J. Wineland: Phys. Rev. A 25, 35 (1982)
10. W.M. Itano et al.: Phys. Rev. A (in press)
11. J. Yu, M. Desaintfuscien, and F. Plumelle: (submitted to Appl. Phys. B)

Laser Cooling and Resonant Transport of MG^+ Ions in a Penning Trap

Jin Yu, F. Plumelle, and M. Desaintfuscien

Laboratoire de l'Horloge Atomique,
Equipe de Recherche du CNRS associée à l'Université Paris-Sud,
Bât. 221, Université Paris-Sud, F-91405 Orsay Cedex, France

We give a simple model to calculate the trajectory of an ion in a Penning trap, submitted to a space charge potential. This model shows the effect of electric field asymmetries and misalignment of the magnetic field : the ion trajectory expands in the cases where the axial frequency ω'_z is equal to some multiples of the magnetron frequency ω'_m (Fig. 1), leading to a limitation of the density. These frequencies are modified by space-charge effect and evaluated in the simple case of a uniform distribution of the ions :

$$\omega'_z = \sqrt{\omega_z^2 - 1/3\ \omega_p^2}$$

$$\omega'_m = \frac{\Omega_c}{2}\left[1 - \sqrt{1 - \frac{4}{\Omega_c^2}\left(\frac{\omega_z^2}{2} + \frac{\omega_p^2}{3}\right)}\right]$$

Where ω_z is the axial frequency of a single ion in the trap, ω_p the plasma frequency and Ω_c the cyclotron frequency.

In a second step, we express the resonant condition :

$\omega'_z = l.\omega'_m$ (l is an integer)

in order to provide a relation between the density for which a resonance occurs and the magnetic field in the trap /1/.

We present an experimental recording of the fluorescence intensity, which is proportional to the density, as a function of the magnetic field. This experiment was realized in the presence of a continuous source of ions, and is in good agreement with the theoretical results.

This enables us to point out that resonance induced by fields asymmetries plays an important role in the thermalisation process of ions in a Penning trap.

We report experimental results about the temperature of a laser cooled Mg^+ ion cloud as a function of the laser beam position in the

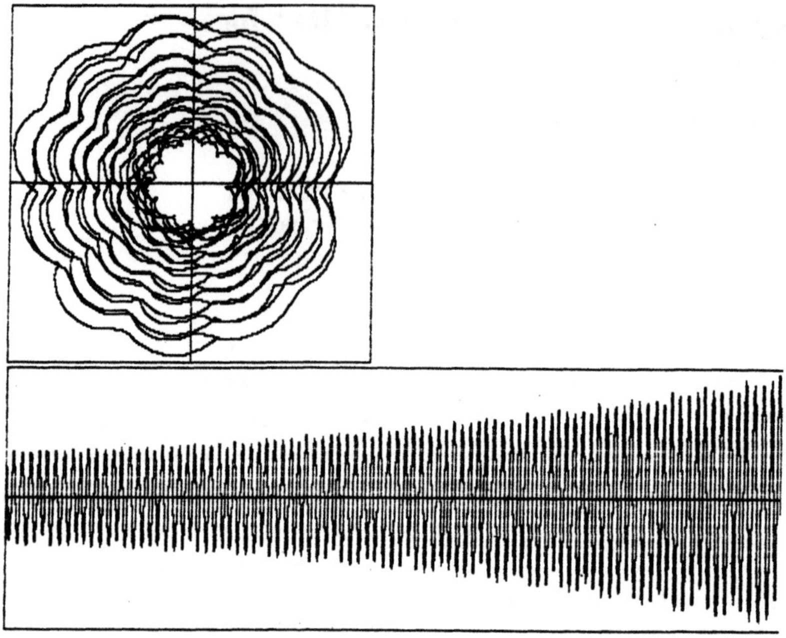

Fig. 1 Ion trajectory for $\omega'_z = 4\omega'_m$ in the xy plane (top) and in the z direction (bottom)

trap and present an absorption line produced by a ten-ion cloud cooled to a temperature of about 60 mK.

REFERENCES.

1. J. Yu, M. Desaintfuscien and F. Plumelle
Submitted to Appl. Phys. B

Trapped Yb^+ as a Potential Optical Frequency Standard

G.P. Barwood, A.S. Bell, P. Gill, and H.A. Klein
National Physical Laboratory, Teddington, Middlesex, TW11 0LW, UK

Trapped ions offer considerable potential for future frequency standards [1]. Confinement by known fields in ultra-high-vacuum conditions can be used to reduce perturbations and allow long interaction times. Doppler shifts can be reduced by laser cooling. At the NPL, following earlier work on cooling of Mg^+ ions in a Penning trap [2], a transition in Yb^+ is being considered as an optical frequency standard [3-5].

A partial term scheme for Yb^+ is shown in Fig. 1. The transition chosen for laser cooling is the $^2S_{1/2}$-$^2P_{1/2}$ resonance transition at 369.5 nm. An alternative optical decay route from the $^2P_{1/2}$ level is the 2.438 μm transition to the $^2D_{3/2}$ metastable level. This could remove ions from the cooling cycle. One solution is to generate radiation at 2.438 μm to drive the metastable population back into the $^2P_{1/2}$ state.

Ytterbium ions have been confined in an r.f. trap of 5 mm internal trap radius. Their presence was originally observed using electronic detection. Background pressures of ~ 10^{-8} mbar allow storage times of order 1 hour. Helium gas at pressures up to 10^{-4} mbar can also be used to provide some collisional cooling of the hot ions.

UV fluorescence from the Yb^+ 369.5 nm transition has been observed by irradiating the ion cloud with UV output at this wavelength from a frequency-doubled ring dye laser. This is a Coherent model 699-21 operating with Rh700 dye at 738 nm, and incorporating an angle-tuned intra-cavity Lithium Iodate crystal for doubling to 369 nm. UV powers of ~ 1 mW are achieved. Reference spectra at 369 nm, observed optogalvanically in a Yb hollow cathode, are shown in Fig. 2. Fluorescence count rates

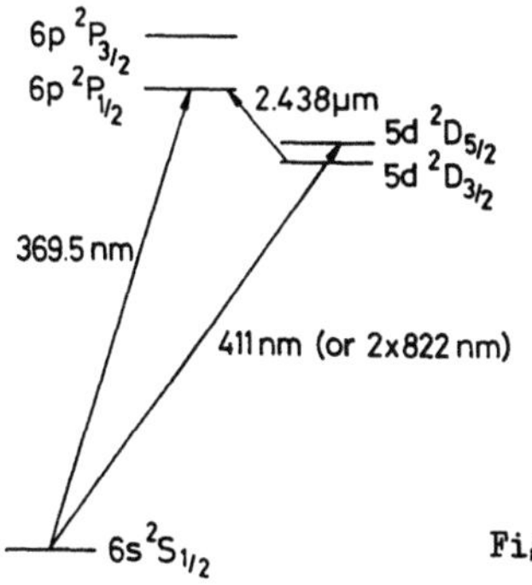

Fig. 1: Yb^+ partial term scheme

Frequency Standards and Metrology Editor: A. De Marchi

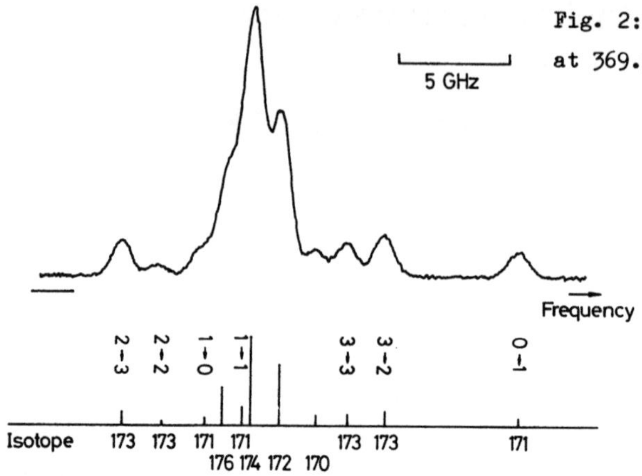

Fig. 2: Yb^+ reference spectra at 369.5 nm

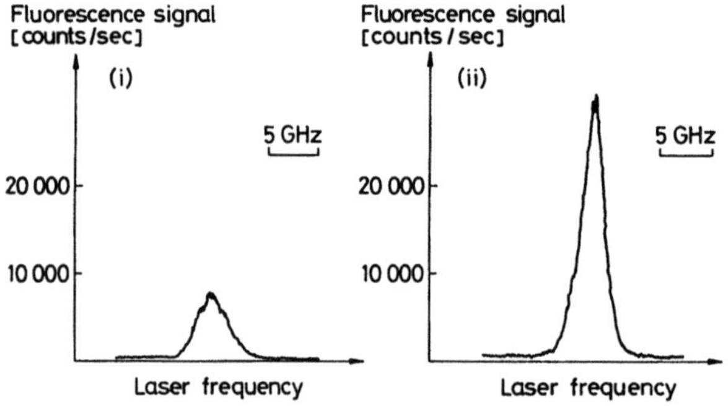

Fig. 3: UV trap fluorescence (i) without and (ii) with infra-red enhancement

from the trapped ion cloud of up to 70 000 s^{-1} are observed. Figure 3(i) shows the UV fluorescence detected as a function of laser frequency. Linewidths of 2-4 GHz are observed, due to isotopic structure and Doppler widths related to r.f. heating. Figure 3(ii) shows enhancement of the detected fluorescence observed when additionally irradiating the trap with an infra-red beam at 2.438 μm to repopulate the $^2P_{1/2}$ level from the $^2D_{3/2}$ metastable level. The infra-red radiation is produced by difference mixing a 477 nm single frequency argon laser output with dye laser output at 592 nm in a crystal of Lithium Niobate.

Work is currently in progress to achieve laser cooling of small numbers of trapped Yb^+ ions. Subsequently, it is planned to interrogate one of the ^2S-^2D clock transitions in the cooled Yb^+, in a single or two-photon arrangement, with a narrow linewidth dye laser.

The authors wish to acknowledge help and advice of R C Thompson, Imperial College, UK.

1. Selected contributions to Physica Scripta T 22 1988 and references therein
2. R.C. Thompson, G.P. Barwood and P. Gill: Appl. Phys. B46, 87 (1988)
3. G. Werth: Metrologia 22, 190 (1986)
4. Ch. Gerz, J. Roths, F. Vedel and G. Werth: Z. Phys. D. 8, 235 (1988)
5. R. Blatt et. al.: these proceedings.

Kinetic Energy and Spatial Distribution of Ion Clouds in Paul Traps

R. Blatt, I. Siemers, M. Schubert, T. Sauter, and W. Neuhauser

I. Institut für Experimentalphysik, Jungiusstr. 9,
D-2000 Hamburg 36, Fed. Rep. of Germany

Ion traps provide a very favourable environment for the conduction of experiments aimed at high spectral resolution: They allow the preparation of ionic samples localized in space free of collisions with walls [1]. In this situation, the second order Doppler effect is the largest systematic error, which can be reduced by laser cooling of the stored ions [2]. However, this merely works with few ions and yields only low signal–to–noise ratio. Another approach is to measure or to calculate the second–order Doppler effect and to hold it constant. In order to do so, an investigation of the dynamics of ion clouds has been performed. The ions' motion is described as Brownian motion in the time–dependent trapping potential. Both damping and diffusion terms due to collisions with the background gas and other ions are introduced on a phenomenological basis. The corresponding Fokker–Planck–equation has been solved analytically and results in the joint distribution of position and velocity as a function of the trap parameters. This Gaussian distribution has time–dependent variances and shows a correlation between position and velocity. With that, the mean kinetic energy can be calculated and a correction for the second–order Doppler effect can be extracted [3].

An experiment has been performed to check these predictions. About 10^7 Ba^+ ions were stored in an rf trap with 40 mm ring diameter. The mean kinetic energy of the ion cloud was derived from the width of the Doppler broadened transition $6^2S_{1/2} \rightleftharpoons 6^2P_{1/2}$ excited by a cw dye laser at 493 nm. In order to avoid optical pumping to the low–lying metastable $5^2D_{3/2}$ state a second laser at 650 nm is required. This configuration – a Λ–type three–level system in interaction with two lasers – should lead to sub–Doppler line widths due to cross saturation. However, due to the oscillatory motion of trapped ions, this cross saturation does not work and the result is a line width equal to the Doppler width [4]. The dependence of the so determined kinetic energies on the trap parameters is in good agreement with the predictions of the model calculations (Fig. 1, [5]).

The spatial distribution of the ions has been measured by observation of the ions' fluorescence by means of an image intensifier and a diode array. They show a dark bar perpendicular to the direction of propagation of the laser fields (Fig. 2, view along the symmetry axis of the trap) with its width increasing with increasing detuning of the light fields. This bar does not have the axial symmetry of the trap. We explain it as spatially localized optical pumping (SLOP), the localization being due to the correlation of position and micromotion–velocity–amplitude encountered in rf traps [5]. If the detunings of both lasers are zero, the

Frequency Standards and Metrology Editor: A. De Marchi

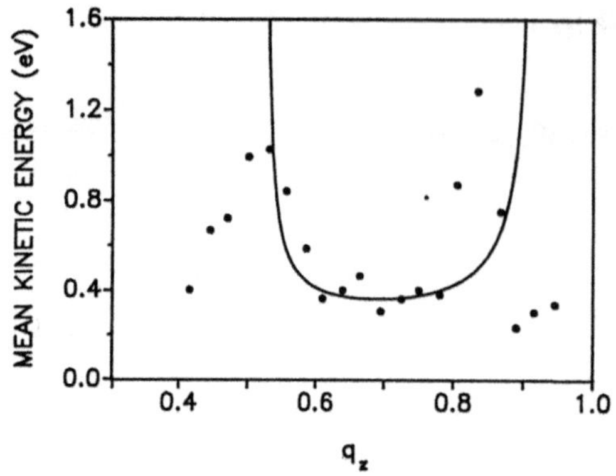

Fig.1: Measured kinetic energy and model prediction (driving frequency 400 kHz, $a_z = -0.022$)

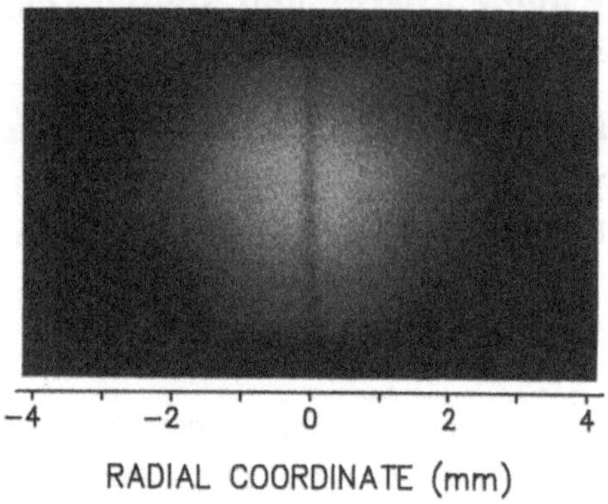

Fig.2: Photograph of the illuminated part of the ion cloud (both lasers have zero detuning)

bar arises from SLOP into a coherent superposition of the levels $6^2S_{1/2}$ and $5^2D_{3/2}$. If one of the lasers is detuned from the ionic resonance, the bar is due to SLOP into an energy eigenstate (S or D state), since the ions near the trap center do not have the velocity required to compensate for the detuning. In both cases the $\vec{k}$–vectors of the laser fields cause the break of the trap's symmetry.

The knowledge of kinetic energies and spatial distributions of the ions in an rf trap, along with the influence of the oscillatory motion on the interaction with the applied light fields, allows to correct for systematic uncertainties in time and frequency standards applications.

1. H. Dehmelt, Adv.At.Mol.Phys. 3, 53 (1967)
2. W. Neuhauser, M. Hohenstatt, P.E. Toschek, and H. Dehmelt Phys.Rev A 22, 1137 (1980)
3. R. Blatt, P. Zoller, G. Holzmüller, and I. Siemers; Z. Phys. D 4, 121 (1986)
4. M. Schubert, I. Siemers, and R. Blatt; to be published
5. I. Siemers, R. Blatt, Th. Sauter, W. Neuhauser: Phys. Rev. A 38, (November 1988)

Tuning Characteristic of 543 nm Internal Mirror He-Ne Laser for Frequency Stabilization

T. Tako, M. Yamashita, and M. Suzuki

Department of Physics, Science University of Tokyo,
2641 Yamazaki, Noda, Chiba 278, Japan

1. Introduction

The 543nm of neon ($3s_2$-$2p_{10}$ transition) is the shortest wavelength in He-Ne laser. This wavelength region is the most sensitive range for human eyes and it is useful for interferometric measurements. The 543nm laser becomes commercially available recently, but its performance is not so stable for the purpose of precise measurement. Tuning characteristics of power, polarization, and intramode beat frequency of 543nm internal-mirror lasers are studied for the practical frequency stabilization.

2. Tuning Curves in Two Modes Operation

Figure 1. shows the tuning curves of mode spectra and powers of two polarized components (perpendicular to each other) divided by a Wollaston prism. It is seen

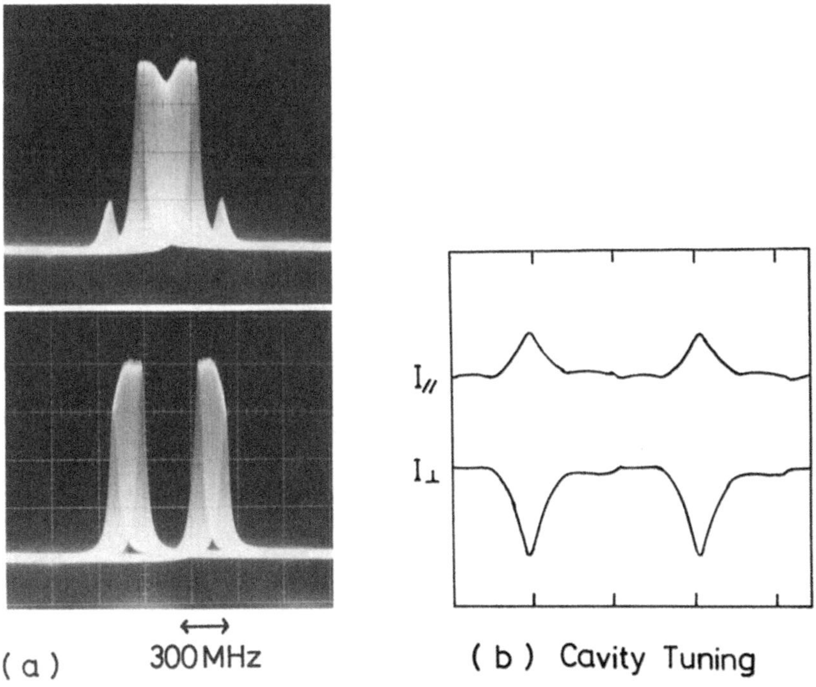

Fig.1. Tuning curves of mode spectra (a) and powers (b) of two polarized components in two modes operation.

Frequency Standards and Metrology Editor: A. De Marchi

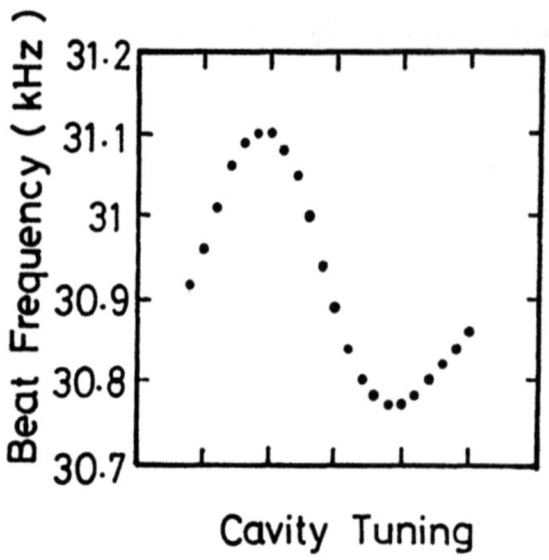

Fig.2. Tuning curve of the intramode beat frequency.

that one mode (upper trace) oscillates at the central part of the gain curve and the other at the two separate ranges. The former has a peak and the latter has a valley in the power tuning curves at the same points. The wing parts of the peak or valley are useful for frequency stabilization as a discriminator.

3. Intramode Beat Frequency

The intramode beat of the 543nm internal-mirror laser under transverse magnetic field observed at the frequency range of 30—100kHz as well as the 633nm laser.[1,2] The tuning curves of beat frequency by cavity tuning has a S-shape as shown in Fig.2. The magnetic field dependences of the center beat frequency and the peak to peak value are rather small compaired with about 200kHz of the 633nm lasers.

These tuning curves are useful for the frequency stabilization better than 10^{-8}. These experiments are in progress.

1. N. Umeda, M. Tsukiji and H. Takasaki: Appl. Opt. 19, 442 (1980)
2. J.B. Ferguson and R.H. Morris: Appl. Opt. 17, 2924 (1978)

Double-Mode CO_2 Laser with Complex Resonator for Frequency Standards and Superhigh Resolution Spectroscopy

V.Y. Kurochkin, V.N. Petrovskiy, E.D. Protsenko, and A.N. Rurukin

Moscow Engineering Physics Institute, Kashirskoe sh., 31, SU-115409 Moscow, USSR

Double-mode gas lasers are used in the laser intracavity saturated-absorption spectroscopy. In these lasers intensity resonances amplitude may be much higher than intensity resonances amplitude in the single-mode lasers and the highly sensitive method of obtaining narrow resonances of saturated dispersion is realised /1/. We report here the results of studying double-mode CO_2/SF_6 laser with three-mirror resonator and telescopic beam expander containing the internal absorption cell for sub-Doppler spectroscopy and laser frequency stabilization /2/. The laser provides stable operation on the two axial modes with orthogonal linear polarizations near the gain curve centre. In order to receive two linearly and orthogonally polarized modes we posed a CdS birefringent wedge into the amplifying section of the three-mirror resonator. Moreover, the wedge makes the caustics of the both modes to separate one from another in space, that results in broad double-mode oscillation range /3/.

The three-mirror resonator and the telescopic beam expander were used for decrease of saturation parameter and transit-time linewidth. The light beam diameter in the absorption cell was 0.02 m, in the active medium it was 0.002 m, the cell length - 0.6 m, the resonator full length - 1.8 m. The laser provides the narrow and high-amplitude resonances both in the output intensity of a separate mode /IR/ and in mode-beat frequency /FR/.

We have measured the dependence of amplitude and width of the resonances on various parameters of active and absorbing media and on the resonator's features. It was found that the widths of both IR and FR are equal and proportional to $/1-R/^{1/2}\ /\ /1+R/^{1/2}$, where R - amplitude reflection factor of the intermediate mirror. The laser light was detected from the amplifying section. The IR amplitude is proportional to /1-R/. Decrease of the resonances' width due to rise of R is caused by absorbing medium saturation fall. The fall of resonances' amplitude is due to light lifetime shortening in the absorbing section of the resonator.

Variation of absorbing gas pressure over 0.1 - 1 Pa range results in nearlinear variation of the resonances amplitudes and width. The increase of the resonances amplitude and width with SF_6 pressure is caused by growing of collisional broadening and linear absorption, respectively.

The less is mode-splitting, the more is amplitude of the IR. This is brought about the modes' interaction strengthening. Meanwhile FR amplitude grows negligibly /1/. We variated the mode-splitting from 2 to 15 MHz.

We have got the maximum contrast of the IR up to 20 % of the total laser power. At SF_6 pressure being 0.1 Pa we have got the IR amplitude of 5 mW, the FR amplitude of 5 kHz and their HWHM 50 kHz. Amplitude of IR may be about two orders of magnitude higher than amplitude of the IR of the single-mode laser with an external SF_6 cell at the same resonance width.

The present construction of double-mode CO_2 laser with an internal absorption cell may be used with OsO_4.

1. N.G.Basov, M.A.Gubin, V.V.Nikitin, E.D.Protsenko: Sov. Quantum Electron., 11, 1084 /1984/.
2. J.B.Cole, C.F.Bruce: Appl. Opt., 14, 1303 /1975/.
3. V.M.Ermachenko, V.N.Petrovskiy, E.D.Protsenko, A.N.Rurukin, R.A.Shananin: Sov. Q#antum Electron., 12, 571 /1985/.

Laser Frequency Measurement at NPL

D.J.E. Knight[1], *K.I. Pharaoh*[1], *and M.J. Padgett*[2]
[1]National Physical Laboratory, Teddington, Middlesex, TW11 0LW, UK
[2]University of Cambridge, Cavendish Laboratory,
Madingley Road, Cambridge, CB3 0HE, UK

In the area of laser frequency measurement at NPL, as elsewhere, there has been a shift of emphasis over recent years, from the measurement of the speed of light to the provision of standards and development of techniques for frequency measurement [1,2].

At present the NPL can measure frequencies up to the 9 to 11 µm band (CO_2 lasers) using two far-infrared optically-pumped lasers and Schottky and MIM mixers. The work on a frequency chain from caesium to 576 nm [3] ceased because of inadequate cw power from a colour-centre laser at 2.3 µm [4].

Recent work has included measurement of far-infrared (FIR) laser frequencies of interest in physical chemistry [5], and development of tunable sideband generation in Schottky diodes for tunable FIR spectroscopy. However, our ordered list of far-infrared cw laser lines (1350) and its pump-ordered version [6] has not been revised and now accounts for only about half the total of known lines [7].

The NPL collaborates with the Cavendish Laboratory, Cambridge on techniques for high-resolution spectroscopy with CO_2 lasers [8]. This is aimed at providing frequency standards based on SF_6, OsO_4 or other gases. Gas cells with large diameter (42 mm diameter beam waist) have been used at low pressures in Fabry Perot and two-beam standing wave configurations to observe saturated absorption spectra [8]. Linewidths down to 12 kHz (FWHM) have been observed in SF_6, and interline frequency-difference measurements using two CO_2 lasers locked to different SF_6 components have been measured to about ± 1 kHz reproducibility.

At higher frequencies, standards for the optical-fibre transmission bands between 1.3 and 1.6 µm are being sought [9]. A 4-m 1.52 µm HeNe laser has been constructed for investigating absorptions in gases. This tunes over about ± 420 MHz single mode at about 3 mW output, and has space for an intracavity gas cell about 760 mm long. A similar, 6 m, laser has been operated as a power oscillator. The frequency of the line centre when operated with ^{20}Ne has been previously measured as 196 780 372(25) MHz [10]. This laser so far gives rather less than 10 mW when running in a single mode. For absolute frequency measurement, experiments on 6th-harmonic mixing with a CO_2 laser in MIM diodes are being attempted. Initial tests of MIM diodes with whiskers of tip radius

Frequency Standards and Metrology Editor: A. De Marchi

about 0.5 to 1 μm [11] have shown intermode beats at 1.52 μm to be about 30 dB below the signal from a photodiode, at offsets of 175 MHz. The signal to noise ratio in 100 kHz bandwidth for these intermode beats was of order 28 dB with dc bias and 14 dB without bias (Z = 150 ohm, and rectified dc = 2 mV (CO_2) and 90 μV (1.52 μm)). The beat signal against the 9.14 μm R(48) CO_2 line harmonic, near 17.4 GHz, is being sought using a spectrum analyser with FET preamplifiers which provide a receiver noise figure of 8 - 10 dB.

For the future there is a continuing need for accurate frequency measurements on stabilised lasers in and near the visible region of the spectrum, e.g. for spectroscopy of the H atom, and to connect ion-trap standards to caesium.

References

1. D A Jennings, K M Evenson and D J E Knight: Proc. IEEE 74, 168-179 (1986)
2. D J E Knight: Metrologia 22, 251-257 (1986)
3. D J E Knight, G J Edwards, P R Pearce, K I Pharaoh, G P Barwood, N R Cross and Fan Xi-Sheng: Sov. J. Quant. Electron. 15, 1374-1377 (1985); Russian translation in: Kvantovaya Elektron.12, 2087-2093 (1985)
4. D J E Knight, K I Pharaoh and M J Padgett: "Laser frequency measurement at NPL" In Proc. IX Vavilov Conference, Novosibirsk, June 1987: Sov. J. Quant. Electron., in the press; Russian translation in: Kvantovaya Elektron., 15, 1185-1193 (1988)
5. I H Davis, K I Pharaoh and D J E Knight: Int. J. Infrared and Millimeter Waves 8, 765-769 (1987)
6. D J E Knight: NPL Report QU-45 1st revision, with pump ordered list (NPL, Teddington, TW11 OlW UK, September 1982)
7. N G Douglas "Ordered list of cw far-infrared laser lines" Kapteyn Instituut preprint, Univ. of Groningen, Netherlands (1987), monograph in preparation.
8. M J Padgett, J A Golby, D D Arnone and R J Butcher: "An ultra-high resolution offset-locked carbon dioxide laser spectrometer" J. Phys. D 21 1352-1358 (1988)
9. D J E Knight: In Proc. 2nd European Frequency and Time Forum, 16 - 18 March 1988, Neuchatel, Switzerland, pp.23-33
10. D A Jennings, F R Petersen and K M Evenson: Optics Lett. 4, 129-130 (1979) see note 10
11. K M Evenson, M Inguscio and D A Jennings: J. Appl. Phys. 57, 956-960 (1985)

Saturation Spectroscopy with the NH_3 Laser

K.J. Siemsen[1], *A.A. Madej*[1], *and J. Reid*[2]

[1]National Research Council of Canada, Ottawa, Ontario K1A 0R6, Canada

[2]Lumonics Research Ltd., Kanata, Ontario K2K 1Y3, Canada

The NH_3 laser, optically pumped by CO_2 laser radiation at 9.2 µm, operates at 12.08 µm wavelength and has been demonstrated to be capable of a short term frequency stability of a few hundred Hertz [1]. Since the laser radiation is strongly absorbed in NH_3 one could gain additional long term stability by locking the laser to a saturated absorption dip in NH_3. The precision of the lock would be considerably improved if the 3 hyperfine structure components of the 12 µm sP(7,0) transition could be resolved. However, using the present published value for the upper state quadrupole coupling constant $(eqQ)\nu_2 = 1,a$ [2] the calculated separation of these components was uncertain by > 100 kHz.

In the present experiments two independent NH_3 lasers were used. One laser served as a frequency reference while the other laser probed the saturated absorption dip of the sP(7,0) transition. The 1m long absorber cell was placed inside an interferometer to enhance the sensitivity. The sP(7,0) saturation dip was 240 kHz wide (HWHM) and showed no hyperfine structure. The sR(5,0) pump transition shares the same upper level as sP(7,0), so it was also investigated. For this purpose one of the CO_2 pump lasers was used as a tunable radiation source. The sR(5,0) transition showed a strong central component (HWHM 150 kHz) and four weaker components. The latter were measured and identified as crossover and $\Delta F = 0$ forbidden transitions. From the frequencies of these components, $(eqQ)\nu_2 = 1,a$ was determined with increased precision as -4106 ± 70 kHz. Consequently the positions of the 3 main hyperfine components of sR(5,0) and sP(7,0) were recalculated as {-23, -12, +20 kHz} and {-19, -5, +33 kHz} respectively, with an uncertainty of approximately 15 kHz for each value. These hyperfine structure components could not be resolved with the present resolution limit of the apparatus of ~ 50 kHz. It was not considered attractive to lock the NH_3 laser to the unresolved saturated absorption dip.

Because of the common upper level another type of saturation experiment can be done which leads to measurements of the gain in the NH_3 laser. One of the CO_2 pump lasers is used to populate the upper $\nu_2 = 1,a$ level while the NH_3 laser is probing the excess population. Two saturation dips are obtained with this technique: a strong narrow dip for the case that the 9 µm and 12 µm beams are propagating in the

Frequency Standards and Metrology Editor: A. De Marchi

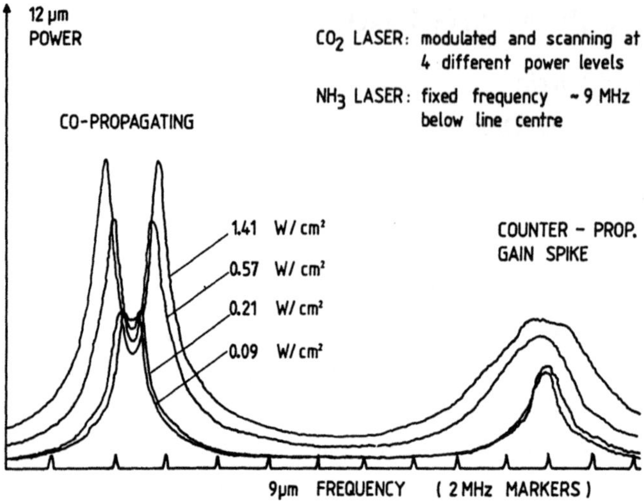

DYNAMIC STARK SPLITTING OF THE sR(5,0) TRANSITION

same direction and a weaker broader dip for the counter-propagating case. At the lowest pump laser power the HWHM of the co-propagating spike was 400 kHz, roughly twice as wide as the HWHM for the self-saturation dips of sR(5,0) or sP(7,0) as expected. At high pump laser power the co-propagating spike begins to split due to the dynamic Stark effect, and the reduced absorption spike turns into a gain spike. The figure below shows the two gain spikes at four different pump power levels. These gain measurements will be used to test our present theory of the NH_3 laser.

References

1. K.J. Siemsen, E. Williams and J. Reid: Opt. Lett 12, 879 (1987)

2. G. Magerl, W. Schupita, J. Frye, W. Kreiner and T. Oka: J. Mol. Spectr. 107, 72 (1984)

OsO_4 Molecular Beam for High Accuracy Frequency Standard

A. Godone[1], *M.P. Sassi*[2], *E. Bava*[1], *and C. Caldera*[3]

[1]Istituto Elettrotecnico Nazionale G. Ferraris, Torino, Italy
[2]Istituto di Metrologia G. Colonnetti, CNR, Torino, Italy
[3]Politecnico, Torino, Italy

The realization of a primary frequency standard at 10 μm on an OsO_4 molecular beam is here examined and the possibility of reaching a high stability level and an accuracy better than 10^{-14} is discussed.

A high accuracy frequency standard at 30 THz is interesting for the extension of the frequency measurements up to the visible, where highly precise reference transitions have been proposed [1,2] and might be available in a near future; the OsO_4 molecule can provide a standard with an intrinsically higher accuracy than CH_4, due to its higher molecular weight and in principle it appears even more suitable than SF_6 due to the absence of hyperfine structure for the most naturally abundant isotopes.

An accuracy of 3×10^{-12} has been reached with a CO_2 laser stabilized on $^{192}OsO_4$ transitions [3,4] via saturated absorption in an external cell. The Ramsey technique applied to an OsO_4 molecular beam can improve the accuracy by some orders of magnitude. In effect a very narrow transit-time limited linewidth is obtained with reduced requirements for the electromagnetic beam shape (wavefront curvature shift and residual 1st order Doppler effect), moreover the pressure-dependent shifts are eliminated; the recoil-splitting is symmetric and detection via laser-induced fluorescence is made possible in a region spatially separated from the interaction one, as it will be discussed in the following.

The capabilities of a frequency standard based on a multichannel thermal beam of OsO_4 with a Ramsey-type interaction have been evaluated [5] for real experimental conditions. Assuming for the single channels a diameter of 10 μm and a length of 2 mm, and for both the molecular and the electromagnetic beam a diameter of about 2 mm, taking into account the $^{192}OsO_4$ partition function (Z = 28,300 at 300 K) well resolved OsO_4 transitions may be observed with a line Q-factor $Q_L = 1\times10^{11}$ even with an interaction length of only 30 cm. On the other hand the molecular flux is adequate to observe the transitions with a relatively high signal-to-noise ratio directly on the power transmitted through the cavity; the expected short-term frequency stability in this case is $\sigma_y(\tau) = 5\times10^{-14}\,\tau^{-1/2}$.

A more efficient detection technique may also be used in order to further increase the signal-to-noise ratio and then the frequency stability: the strongly allowed electric dipole transitions in the first excited electronic band of OsO_4 may be induced starting from the upper level of the infrared transition and detected with high efficiency in the near UV (35,000 cm^{-1}). In the experimental conditions considered before, a frequency stability $\sigma_y(\tau) \simeq 10^{-15}\,\tau^{-1/2}$ is expected in this case, limited by the detector shot-noise of the molecular beam.

In order to discuss the accuracy limits the most important sources of uncertainty are: the second order Doppler effect giving a red-shift of the observed transitions of the order of 1×10^{-13} which may be reduced to units of 10^{-15} by taking into account the resonance lineshape and the velocity distribution which is quite

Frequency Standards and Metrology Editor: A. De Marchi

well known in this case, because the beam is not subjected to any velocity-dependent interaction before the interrogation zones; the phase-shift due to the losses and asymmetries of the interaction cavity is given by the well known relation

$$|\Delta\nu/\nu_0|_\Phi = (\alpha/2\pi\nu_0 L)\,\overline{\delta\Phi}$$

where $\overline{\delta\Phi}$ is the mean de-phasing of the elctromagnetic field between the two inter-action zones. We estimate $\overline{\delta\Phi} \simeq 4\times10^{-3}$ rad which implies $|\Delta\nu/\nu_0|_\Phi \simeq 1\times10^{-14}$; the Zeeman effect has been measured for some lines of OsO_4 and turns out $2\text{-}3\times10^{-12}$/gauss for the σ-transitions; π-transitions are symmetrically splitted only by the anomalous Zeeman effect whose amount, even if at present unknown for OsO_4, is certainly well within the transit-time limited linewidth; the wavefront curvature effects impose less severe requirements to the optics quality in the molecular beam case, with respect to a large diameter absorption cell, because of the small size of the resonator gaussian beam; moreover a strong reduction of the effect is expected in the case of the Ramsey interaction technique; the recoil splitting for the OsO_4 lines amounts to 15 Hz (5×10^{-13}); with a relative linewidth of 10^{-11}, entirely due to the transit-time broadening, the splitting, highly symmetric, is not resolved. Other uncertainty sources as cavity-pulling, servo-loop electronics, flicker limit, etc. are estimated causing effects lower than 1×10^{-14} for a well-designed system.

Cavity phase-shift, residual 1st and 2nd order Doppler effects appear the main limits at the 10^{-14} accuracy level.

References

[1] Snyder J.J. et al.: Appl. Phys. B32, 25-31(1983).

[2] Ertmer W. et al.: Progr. Quant. Electr. 8, 249-255(1984).

[3] Clairon A. et al.: IEEE Trans. Instr. Meas. IM-34, 265-268(1985).

[4] Domnin Yu. et al.: Sov. Journ. Quant. Electron. 17, 801-803(1987).

[5] Godone A. et al.: to be published.

He-Ne Laser at 612 nm Stabilized to $^{127}I_2$ Using FM Spectroscopy

F. Bertinetto, G.B. Picotto, P. Cordiale, and S. Fontana

Istituto di Metrologia G. Colonnetti, CNR, Torino, Italy

The dispersion signal, observed with sub-Doppler spectroscopic techniques using a phase modulation of the probe beam, is antisymmetric and shows a remarkable slant at the central zero-crossing. Because of these characteristics, it is particularly suitable as an error signal for stabilizing the laser frequency using as a reference an atomic or molecular transition in a low-pressure gas. At IMGC the fitness of this method for the achievement of optical frequency standards has been tested using two He-Ne lasers, independently stabilized to the hyperfine structure component "b_{15}" of the P (48) 11-3 iodine transition at 612 nm.

The optical scheme is basically the one described in /1/, but the saturating and the probe beam are orthogonally polarized, thus providing a mean to stop both beams after the iodine cell using two polarizers. A third polarizer is used to match the polarization of the light to the optical axis of the crystal in order to minimize amplitude modulation of the probe beam. For both lasers the electro-optic crystal, a commercial ADP unit cut at Brewster angle, is driven at a frequency of approximately 10 MHz; the acousto-optic crystal is driven at a 40 MHz frequency, with different alignment for the two lasers in order to select opposite velocity groups, and with a 100 kHz chopping frequency. The beams traversing the iodine cells are collimated with a diameter of 4 mm; for both lasers the saturating power is approximately 1 mW, while the power of the probe beam is 0.7 mW.

By comparison with the traditional third harmonic technique, FM spectroscopy enables the achievement of suitable signal to noise ratio at definitely lower iodine pressure (2.5 Pa against 6.7 Pa /2/) with consequently linewidth reduction (950 kHz against 1.5 MHz) and allows a larger control bandwidth (3 kHz against 30 Hz), in our case limited by the use of PZT's as frequency transducers. In present experimental conditions, changes in the modulation index from $\beta = 0.4$ to $\beta = 1.1$ and in the phase difference from the local oscillator and the detector signal from -18° to 18° cause fluctuations of the laser frequency usually well within 3 kHz. Preliminary measurements are indicating that also the power of the saturating beam, when adjusted from 0.1 mW to 1 mW, influences the laser frequency by less than 2 kHz. The pressure-dependent frequency shift of 6.98 kHz/Pa is in good agreement with previous results /2/.

The $\sigma - \tau$ plot of fig.1 shows the improvement in stability achieved with FM spectroscopy against the III harmonic method /2/. The slope of -0.2 is however a sign of not yet overcome experimental difficulties, i.e. residual optical couplings to the lasers and between optical components, rf interference signals, residual amplitude modulation. However at present time these are the most stable visible He-Ne lasers.

Frequency Standards and Metrology Editor: A. De Marchi

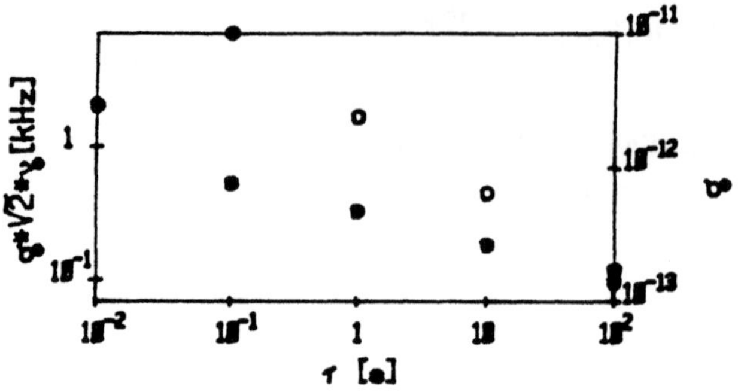

Fig.1 - σ - τ plot, both lasers beeing stabilized using FM spectroscopy (points) or the III harmonic technique (circles).

1. J.L. Hall et al., in Laser Spectroscopy V, Springer Ser.Opt.Sc., Vol.30 (Springer, Berlin, Heidelberg 1981) p.15

2. F. Bertinetto et al., IEEE Trans.Instr.Meas., IM-34, n.2, 256-261 (1985)

Frequency Stabilization of a 543.5 nm Wavelength He-Ne Laser to an Iodine Absorption Line

U. Brand and J. Helmcke

Physikalisch-Technische Bundesanstalt, Bundesallee 100,
D-3300 Braunschweig, Fed. Rep. of Germany

The introduction of commercially available He–Ne lasers emitting at 543.5 nm wavelength opens the possibility to develop a wavelength standard in the green spectral range based on a simple He–Ne laser. Recently we have developed and investigated the stabilization of a two–mode He–Ne laser to its gain profile /1/. To improve the frequency reproducibility we have stabilized the frequency of a 543.5 nm He–Ne laser to an $^{127}I_2$ absorption line. The applied sideband technique /2,3/ (modulation frequency f_{EOM}=10 MHz) allows to detect the absorption signal with high sensitivity close to the shot noise limit. The stabilization set–up shown in Fig.1 is similar to the one described in ref.3 for an I_2–stabilized dye laser.

The emission wavelength of the 543.5 nm He–Ne laser coincides with the $^{127}I_2$ absorption lines R(12)26–0 and R(106)28–0 /4/. We have observed the Doppler–free hyperfine structure (HFS) of these lines free of background. Fig.2 shows the dispersive part of the corresponding signals versus time, achieved by "thermally scanning" the laser frequency. For frequency stabilization we have chosen the HFS–component a_9 /5/ of the R(12)26–0 transition. Frequency correction was achieved by changing the discharge current of the laser.

With the available single frequency power of 100μW the best signal to noise ratio was observed at an iodine pressure of 1.44 Pa.

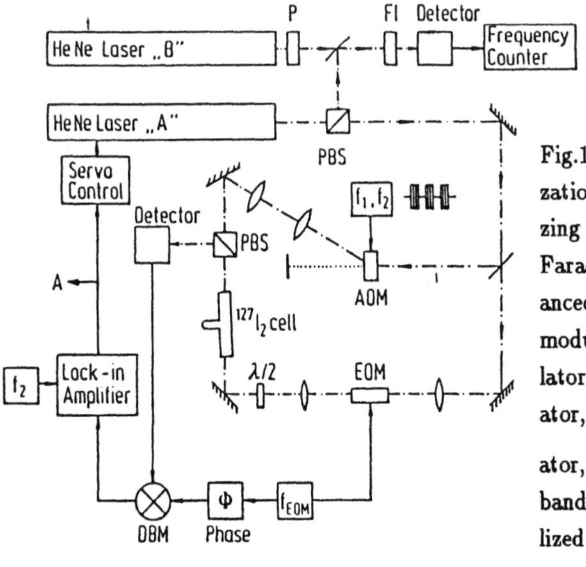

Fig.1: Block diagram of the stabilization set–up (laser "A"). (S: polarizing beam splitter, P: polarizer, FI: Faraday isolator, DBM: double balanced mixer, AOM: acoustooptic modulator, EOM: electrooptic modulator, f_1,f_2: AOM modulation generator, f_{EOM}: EOM modulation generator, A: error signal of the I_2 sideband stabilization, laser "B": stabilized to the gain profile).

Frequency Standards and Metrology Editor: A. De Marchi

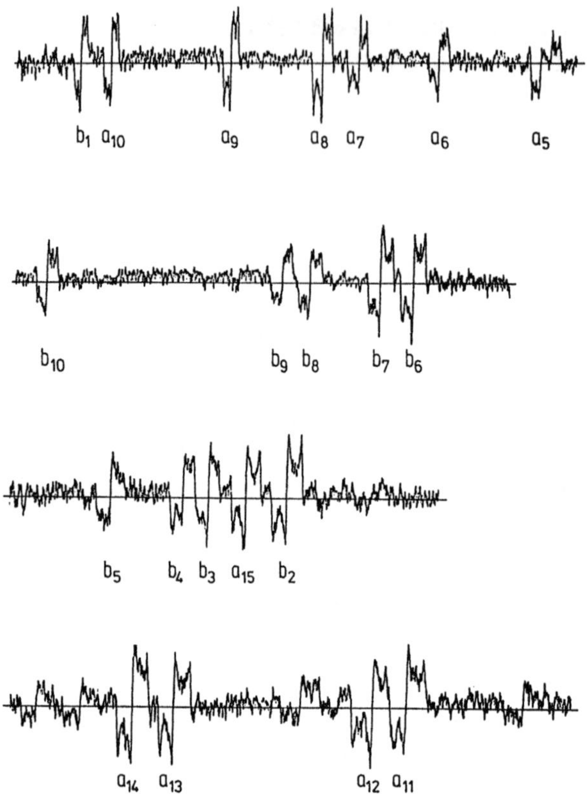

Fig.2: Dispersion signals of the $^{127}I_2$ lines R(12) 26–0 and R(106) 28–0 at 543.5 nm wavelength recorded by "thermally scanning" the laser frequency (a_i: HFS–components of the R(12) 26–0 line, b_i: HFS–components of the R(106) 28–0 line).

As a first application of the iodine stabilized laser "A" we measured the frequency behavior of a two–mode laser "B" (see Fig.1), which was frequency stabilized to its gain profile. We separated the two orthogonally polarized modes of laser "A" by a polarizing beam splitter (PBS) in front of the laser. One mode was used for frequency stabilization, the other one for beat frequency measurements. During two days of measurements we observed a drift of the beat frequency of about 1 MHz, arising from the poorer stability of laser "B". A second set–up is in preparation allowing to investigate the reproducibility of laser "A" and to measure the HFS of the I_2 lines. Precise measurements of the hyperfine splitting of the energy levels with low rotational quantum numbers [R(12)] are interesting since only few HFS data are available at low rotational quantum numbers /6/.

References

1. U.Brand, F.Mensing, J.Helmcke: to be published
2. G.C.Bjorklund: Opt. Lett. 5, 15–17 (1980)
3. J.L.Hall, L.Hollberg, T.Baer, H.G.Robinson: Appl. Phys. Lett. 39,9, 680–682 (1981)
4. J.–M.Chartier, J.L.Hall, M.Gläser: Proc. CPEM '86, ISBN 86 CH2267–3, 323 (1986)
5. M.Gläser: PTB Bericht Opt–25, 173–176 (1987)
6. A.Morinaga: private communication

Measurement of Isotope Shift in Optical Transitions of Atomic Oxygen

G.M. Tino, K. Ernst, A. Sasso, and M. Inguscio*

Dipartimento di Scienze Fisiche, Mostra d'Oltremare pad. 20,
I-80125 Napoli, Italy

The possibility of performing high resolution laser spectroscopy of atomic oxygen was demonstrated only very recently [1]. The atom was produced by dissociation of O_2 molecules in a moderate power radiofrequency discharge, substained by a buffer noble gas (Ne or Ar). A discussion of the collisional processes governing the production of O excited atoms can be found in [2]. The difficulty of working with O_2 and O in only trace amounts was overcome by the use of a very sensitive optogalvanic detection scheme, while sub-Doppler resolution of fine structures was obtained using intermodulated spectroscopy configuration.

These results now open the possibility of investigating isotope shifts in enriched samples. The measurements would be highly desirable: oxygen has three stable isotopes, 16, 17 and 18, with relative natural abundance 99.76, 0.03 and 0.20 %. No previous measurements of the shift at optical wavelengths, except for an old conventional interferometric study on a ^{18}O enriched sample [3]. ^{16}O nucleus is doubly magic and precise shift measurements on the three isotopes could evidence deviations from the "mass effect" behaviour, helping a deeper understanding of isotope effect on very light elements, now that also measurements on metrologically important transitions are available [4].

We report here the demonstration of possible precise measurements of isotope shift using the pair 16-18. Figure 1 shows intermodulated optogalvanic recordings of the $3p^5P_3$ - $5s^5S_2$ transition at 645.6 nm. In a) a natural abundance sample is used. As expected from the presence of velocity changing collisions, the lineshape is given by the superposition of a Doppler-free Lorentzian over a Gaussian pedestal. In b) a ^{18}O enriched sample (~ 50%) is used : the isotope shift is nicely resolved and the preliminary measured value is reported in Table 1, together with those of the other investigated transitions.

The Doppler broadening, always larger than the shifts to be measured, exhibits some anomalous behaviour. For the first three listed transitions is, as expected, of the order of 2 GHz, corresponding to a temperature of about 520 K. However for the last

* Permanent address: Institute of Experimental Physics, University of Warsaw, Poland.

Frequency Standards and Metrology Editor: A. De Marchi

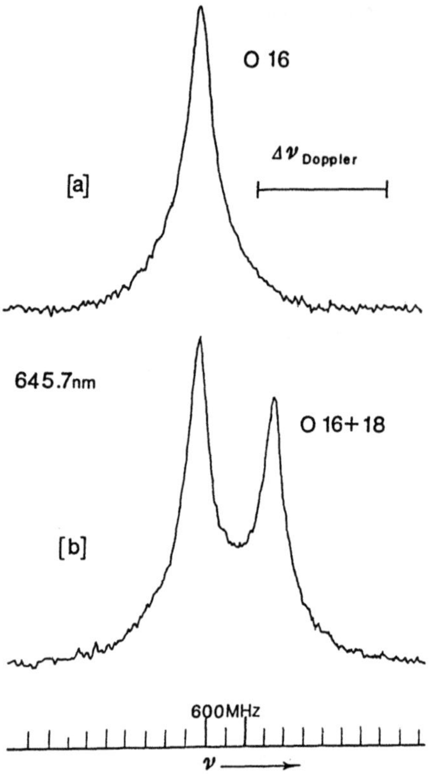

TABLE 1: Measured isotope shifts (IS) in atomic oxygen.

Transition	nm	IS (MHz)	Ref.[3]
$3p^5P_3$-$4d^5D_4$	615.8	1310(40)	<900
$3p^5P_2$-$5s^5S_2$	645.5	1160(20)	
$3p^5P_3$-$5s^5S_2$	645.6	1160(20)	<600
$3p^3P_2$-$6s^3S_1$	604.6	1300(40)	

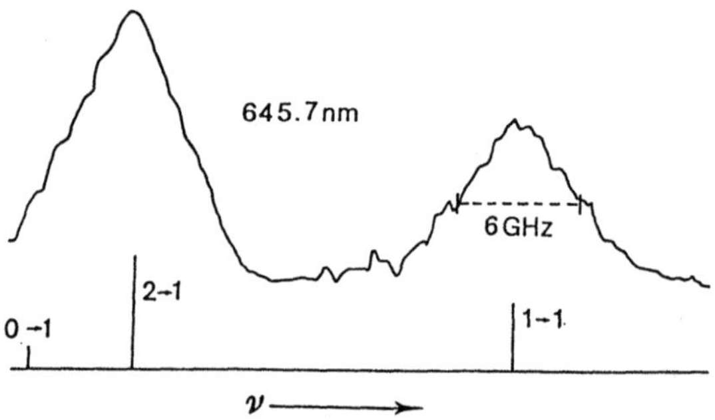

Fig.1 Sub-Doppler recording of the OI $3p^5P_3$-$5s^5S_2$ transition at 645.7nm. Partial pressures: O_2 =0.05 Torr; Ar= 2Torr. In a) is used natural oxygen and in b) an enriched sample.

Fig.2. Doppler broadened recording of the 3p $^3P_{0,1,2}$ - 6s $^3S_1$ multiplet at 604.6 nm. Pressures: O_2= 0.05 Torr (natural abundance); Ar = 3 Torr.

transition the measured Doppler width is about three times larger, corresponding to a temperature of nearly 4600 K. The Doppler broadening recording of the entire $3p^3P_{0,2,1}$ - $6s^3S_1$ multiplet is shown in Fig.2. The "anomalous" behaviour with temperature could be possibly caused by the mode of production of excited oxygen atoms, leading to a wide velocity distribution because of the energy defect in the collision with metastable noble gas, as early shown in the observation and interpretation of peculiar effects in the O laser emission at 844.6nm having $^3P_{0,2,1}$ as upper levels [5,6]. It is worth noting that, in spite of the extremely broad Doppler signal and of the relatively weaker optogalvanic signal (nearly 4 times smaller than that for the transition at 645.6nm), we were able to record sub-Doppler signals with enough signal to noise and resolution to measure the isotope shift, as shown in Fig.3.

We think that these results demonstrate that it is possible to systematically investigate isotope shift in atomic oxygen. Also, it should be possible to furthermore

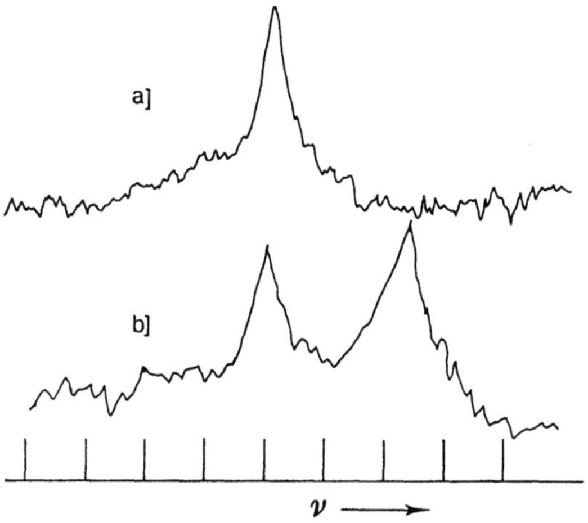

Fig.3 Intermodulated sub-Doppler recording of the $^3P_2$ - $^3S_1$ transition at 604.6 nm. In a) natural abundance oxygen is used and in b) the enriched sample . Frequency scale: 600MHz/div. Partial pressures: O_2 = 0.07 Torr; Ar = 3.7 Torr.

clarify the complex mechanisms of formation of the atoms, performing both Doppler broadened and Doppler-free investigations on different transitions and with different buffer gases.

1. M. Inguscio, P.Minutolo, A.Sasso, and G.M.Tino:Phys.Rev. A 37, 4056 (1988)
2. A.Sasso, P.Minutolo, M.I.Schisano, G.M.Tino, and M.Inguscio: J.Opt.Soc.Am. B5, (1988)
3. L.W.Parker and J.R.Holmes : J.Opt.Soc.Am. 43, 103 (1953)
4. A.Godone, E.Bava, and G.Giusfredi: Z.Phys. A318, 131 (1984)
5. W.R.Bennett, Jr., W.L.Faust, R.A.McFarlane and C.K.N.Patel: Phys.Rev.Lett. 8, 470 (1962)
6. M.S.Feld, B.J.Feldman, A.Javan, and L.H.Domash: Phys. Rev. A7, 257 (1973)

Selective Electron Attachment of SF_6 Molecules in Collision with Xe(nf) Rydberg Atoms in a R.F. Quadrupole Trap and Correlative Effects on SF_6^- Ions Lifetime

S. Rajab Pacha, G. Brincourt, R. Catella, Y. Zerega, and J. Andre

Physique des Interactions Ioniques et Moléculaires*,
Université de Provence, Centre de St Jérôme,
F-13197 Marseille Cedex 13, France
*Unité de Recherche Associée au CNRS, UA 773

Collision between SF_6 molecules and Xe(nf) Rydberg atoms gives birth to temporary excited SF_6^- ions which can lose their electron by detachment processes due to their excitation.

These ions are studied by confinement in a R.F. quadrupole trap from where they are ejected for counting after different values of confinement time.

Xenon Rydberg atoms are prepared by electron gun excitation of metastable levels followed by dye laser excitation of selected levels of the $nf(3/2)_1$ serie of Xenon. The lower the value of n, the higher the Rydberg electron binding energy. So, to catch the Rydberg electron, SF_6 molecule has to lose more energy

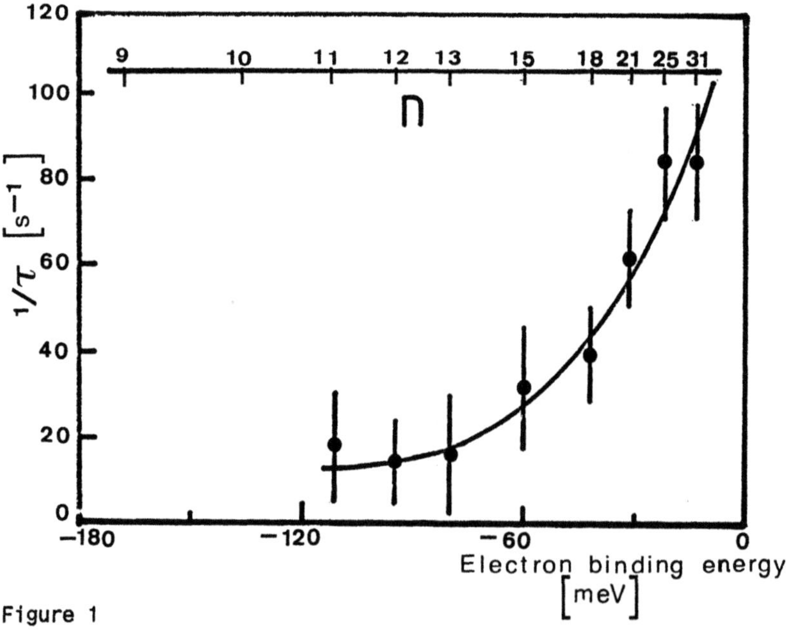

Figure 1
Reciprocal of the lifetime of SF_6^- ions versus both the principal quantum number n of the Xe(nf) Rydberg atoms and the corresponding electron binding energy.

Frequency Standards and Metrology Editor: A. De Marchi

as n decreases, resulting in an increasing of the stabilization of the molecular SF_6^- ion and then of its lifetime. Moreover, the interaction distance between the Xe^+ Rydberg core and the SF_6 molecule decreases with n, leading to a possible ion core effect by energy transfer from the molecule to the Xe^+ ion core. This effect would also contribute to enhance the lifetime of the SF_6^- ions .

As we have studied the lifetime τ of the SF_6^- ions with n in the range 11-31, ion core effect may exist for n values as low as 11 or 12.

To separate the two competitive processes, and show up the possible ion core effect, experiment is going to be conducted with krypton Rydberg atoms for comparison.

Figure 1 gives $1/\tau$ versus both electron binding energy and principal quantum number n scales, and shows clearly the increase of the SF_6^- ions lifetime when n decreases.

REFERENCES

1. G.W. Foltz, G.J. Latimer, G.F. Hildebrandt, F.G. Kellert, K.A. Smith, W.P. West, F.B. Dunning, R.F. Stebbings, J. Chem. Phys. 67 (1977) 1352

2. J.P. Astruc, R. Barbe, J.P. Schermann, J. Phys. B. : Atom. Molec. Phys. 12 (1979) L 377

3. G. Brincourt, M. Vedel, Y. Zerega, J.André, F. Vedel, Rev. Phys. Appl. 17 (1982) 813

4. J.P. Astruc, R. Barbe, A. Lagreze, J.P. Schermann, Chem. Phys. 73 (1983) 405

5. M. Vedel, J. André, G. Brincourt, Y. Zerega, G. Werth, J.P. Schermann, Appl. Phys. B 34 (1984) 229

6. B.G. Zollars, K.A. Smith, F.B. Dunning, J. Chem. Phys. 81 (1984) 3158

7. A. Chutjian, S.H. Alajajian, Phys. Rev. A 31 (1985) 2885

Closing Remarks

Before everybody leaves me alone to sweep the floors I would like to use one more time the other Chairman's privilege and grab the microphone to make a few comments.

First of all let me thank all of you again for being here. A lot of energy went into this meeting, and you supplied it. I sense we all enjoyed the intensity which it injected in this Fourth Symposium on Frequency Standards and Metrology.

We will have time in the next year or so to make a judgement on it, before we start thinking about the future, but one thing we can already tell: your intensity has confirmed in the best way that these quasi-periodic workshops are very useful to our community, and should be kept up.

The logical conclusion is that I want to voice what I think we all feel now: "there will be in a few years the need for another Symposium!" Let me then invite all prospective organizers to apply to the Governing Board. Jacques Vanier, Helmut Hellwig, Claude Audoin and myself will be happy to examine your proposals.

With this I declare this Symposium officially closed and let you go. "Arrivederci", or: see you again some place, some time.

Andrea De Marchi

List of Participants

Ouali ACEF	LPTF, PARIS	FRANCE
David ALLAN	NBS, BOULDER	USA
Roberto AMBROSINI	ISTITUTO DI RADIOASTRONOMIA DEL CNR, BOLOGNA	ITALY
Jacques ANDRE	LAB. DES INTERACTIONS IONIQUES, MARSEILLE	FRANCE
Claude AUDOIN	LHA, ORSAY	FRANCE
Sergei N. BAGAYEV	INST.OF THERMOPHYSICS, NOVOSIBIRSK	URSS
Victor I. BALYKIN	INST.OF SPECTROSCOPY, AKADEMGORODOK	URSS
Yefim BARASH	TFL, INDUSTRIAL ZONE HOLON	ISRAEL
Roland BARILLET	LHA, ORSAY	FRANCE
Andreas BAUCH	PTB, BRAUNSCHWEIG	FRG
Elio BAVA	IEN, TORINO	ITALY
Jim BERGQUIST	NBS, BOULDER	USA
Laurent-Guy BERNIER	OSCILLOQUARTZ, NEUCHATEL	SWITZERLAND
Fabrizio BERTINETTO	IMGC, TORINO	ITALY
Raimond J. BESSON	LCEP, BESANCON	FRANCE
Nicolo' BEVERINI	UNIVERSITA' DI PISA, PISA	ITALY
Francois BIRABEN	ENS, PARIS	FRANCE
Rainer BLATT	UNIVERSITY OF HAMBURG, HAMBURG	FRG
John BOLLINGER	NBS, BOULDER	USA
Christian BORDE'	LPL, VILLETANEUSE	FRANCE
Jean Simon BOULANGER	NRC, OTTAWA	CANADA
Uwe BRAND	PTB, BRAUNSCHWEIG	FRG
Christian BREANT	ETCA, ARCUEIL	FRANCE
Alain BRILLET	LHA, ORSAY	FRANCE
Giovanni BUSCA	OBSERVATOIRE DE NEUCHATEL, NEUCHATEL	SWITZERLAND
Claudio CALDERA	POLITECNICO DI TORINO, TORINO	ITALY
Jim CAMPARO	AEROSPACE CORP., LOS ANGELES	USA
Rainer CASDORF	UNIVERSITY OF HAMBURG, HAMBURG	FRG
Robert CATELLA	LAB. DES INTERACTIONS IONIQUES, MARSEILLE	FRANCE
Alberto CENCI	TELESPAZIO SpA, ROMA	ITALY
Pierre CEREZ	LHA, ORSAY	FRANCE
Andrej CHACHULSKY	IPPT, WARSAW	POLAND
Christian CHARDONNET	LPL, VILLETANEUSE	FRANCE
Andre' CLAIRON	LPTF, PARIS	FRANCE
Peter COATES	NPL, TEDDINGTON	UK
Claude COHEN TANNOUDJI	ENS, PARIS	FRANCE
Stuart B. CRAMPTON	WILLIAMS COLLEGE, WILLIAMSTOWN	USA
Leonard S. CUTLER	HP, PALO ALTO	USA
Brahim DAHMANI	LPTF, PARIS	FRANCE
John DAVIES	NPL, TEDDINGTON	UK
Emeric DE CLERCQ	LPTF, PARIS	FRANCE
Gerrit DE JONG	VAN SWINDEN LAB., DELFT	NETHERLANDS
Andrea DE MARCHI	UNIVERSITA' DI ANCONA, ANCONA	ITALY
Edoardo DE TOMA	SEPA SpA, TORINO	ITALY
Jean-Louis DEBETHUNE	BNM, PARIS	FRANCE
Hans G. DEHMELT	UNIVERSITY OF WASHINGTON, SEATTLE	USA
Michel DESAINTFUSCIEN	LHA, ORSAY	FRANCE
Noel DIMARCQ	LHA, ORSAY	FRANCE
Klaus DORENWENDT	PTB, BRAUNSCHWEIG	FRG

Robert E. DRULLINGER	NBS, BOULDER	USA
Tom ENGLISH	BALL CORP., EFRATOM DIV., IRVINE	USA
Raymond FELDER	BIPM, SEVRES	FRANCE
Cristopher FOOT	UNIVERSITY OF OXFORD, OXFORD	UK
Suzanne FREDIN PICARD	BIPM, SEVRES	FRANCE
Charles FREED	MIT - LINCOLN LABORATORY, LEXINGTON	USA
Jean Jacques GAGNEPAIN	LPMO, BESANCON	FRANCE
Jean-Marie GAGNE'	ECOLE POLITECHNIQUE, MONTREAL	CANADA
Luciano GARRAMONE	TELESPAZIO SpA, ROMA	ITALY
Patrick GILL	NPL, TEDDINGTON	UK
Aldo GODONE	IEN, TORINO	ITALY
William GOLDING	NRL, WASHINGTON	USA
Michel GRANVEAUD	LPTF, PARIS	FRANCE
Michel GROSS	ENS, PARIS	FRANCE
M.A. GUBIN	LEBEDEV PHYS.INSTITUTE, MOSCOW	URSS
Bernard GUINOT	BIH, SEVRES	FRANCE
Jan L. HALL	JILA, BOULDER	USA
Minoru HASHIMOTO	TIT, YOKOHAMA	JAPAN
Mike HAYDEN	UBC, VANCOUVER	CANADA
Helmut HELLWIG	NBS, GAITHERSBURG	USA
Jurgen HELMCKE	PTB, BRAUNSCHWEIG	FRG
Leo HOLLBERG	NBS, BOULDER	USA
Martin HURLIMANN	UBC, VANCOUVER	CANADA
Jun ISHIKAWA	PTB, BRAUNSCHWEIG	FRG
Wayne M. ITANO	NBS, BOULDER	USA
Michelle JARDINO	LHA, ORSAY	FRANCE
Ali JAVAN	MIT, CAMBRIDGE	USA
Leopoldo JETTO	UNIVERSITA' DI ANCONA, ANCONA	ITALY
Stephen JONES	UNIVERSITY OF WESTERN AUSTRALIA, NEDLANDS	AUSTRALIA
Hugh KLEIN	NPL, TEDDINGTON	UK
Vasily M.KLEMENTYEV	INST.OF THERMOPHYSICS, NOVOSIBIRSK	URSS
David J. KNIGHT	NPL, TEDDINGTON	UK
Paul A. KOPPANG	SIGMA TAU CORP., TUSCALOOSA	USA
Gunther KRAMER	PTB, BRAUNSCHWEIG	FRG
Venceslav KROUPA	URECSAV, PRAHA	CHECOSLOV.
Katsuhiko KUBOKI	TIT, YOKOHAMA	JAPAN
Steve LAMOREAUX	UNIVERSITY OF WASHINGTON, SEATTLE	USA
Philippe LAURENT	ETCA, ARCUEIL	FRANCE
Howard P. LAYER	NBS, GAITHERSBURG	USA
Paul LESAGE	LHA, ORSAY	FRANCE
Sigfrido LESCHIUTTA	POLITECNICO DI TORINO, TORINO	ITALY
Gerd LEUCHS	MAX PLANCK INSTITUT, GARCHING BEI MUNCHEN	FRG
Burghard LIPPHARDT	PTB, BRAUNSCHWEIG	FRG
Lute MALEKI	JPL, PASADENA	USA
Cipriana MANDACHE	INST.OF PHYSICS, BUCURESTI MAGURELE	ROMANIA
Guido MASETTI	UNIVERSITA' DI ANCONA, ANCONA	ITALY
Yves MILLERIOUX	CNAM, PARIS	FRANCE
Paolo MINGUZZI	UNIVERSITA' DI PISA, PISA	ITALY
Derek MORRIS	NRC, OTTAWA	CANADA
Warren NAGOURNEY	UNIVERSITY OF WASHINGTON, SEATTLE	USA
Koji NAKAGIRI	CRL, TOKYO	JAPAN
Alan C. NILSSON	STANFORD UNIVERSITY, STANFORD	USA
Carlo NOVERO	IEN, TORINO	ITALY
Bengt OEMAN	TELEWERKET RADIO, HANINGE	SWEDEN
Shin-Ichi OHSHIMA	NRLM, TSUKUBA-SHI	JAPAN
Motoichi OHTSU	TIT, YOKOHAMA	JAPAN
Bryan OWINGS	SIMA TAU CORP., TUSCALOOSA	USA
Pierre PETIT	LHA, ORSAY	FRANCE
Brian PETLEY	NPL, TEDDINGTON	UK
William D. PHILLIPS	NBS, GAITHERSBURG	USA
Francesco PIAZZA	UNIVERSITA' DI ANCONA, ANCONA	ITALY
Gianbartolo PICOTTO	IMGC, TORINO	ITALY

Jean Louis PICQUE'	LABORATOIRE AIMEE COTTON, ORSAY	FRANCE
Francine PLUMELLE	LHA, ORSAY	FRANCE
John PONSONBY	UNIV. OF MANCHESTER, JODRELL BANK	UK
Leon PROST	FEDERAL OFFICE OF METROLOGY, WABERN	SWITZERLAND
Karol RADECKI	WARSAW TECH. UNIVERSITY, WARSAW	POLAND
Norman RAMSEY	HARVARD UNIVERSITY, CAMBRIDGE	USA
Elio RESTUCCIA	ISPT, ROMA	ITALY
Friz RIEHLE	PTB, BRAUNSCHWEIG	FRG
Daniele ROVERA	THEO-TEK, PAESANA	ITALY
Enrico RUBIOLA	POLITECNICO DI TORINO, TORINO	ITALY
Yoshikazu SABURI	ANRITSU CORP., KANAGAWA	JAPAN
Christoff SALOMON	ENS, PARIS	FRANCE
Maria Paola SASSI	IMGC, TORINO	ITALY
Thomas SAUTER	UNIVERSITY OF HEIDELBERG, HEIDELBERG	FRG
Hartmut SCHWEDA	OSCILLOQUARTZ, NEUCHATEL	SWITZERLAND
Avinoam SHENHAR	HEBREW UNIVERSITY, JERUSALEM	ISRAEL
Michael SHUBERT	UNIVERSITY OF HAMBURG, HAMBURG	FRG
Howard A. SHUGART	UNIVERSITY OF CALIFORNIA, BERKELEY	USA
Peter SHUMACHER	OBSERVATOIRE DE NEUCHATEL, NEUCHATEL	SWITZERLAND
Ingo SIEMERS	UNIVERSITY OF HAMBURG, HAMBURG	FRG
Klaus SIEMSEN	NRC, OTTAWA	CANADA
Richart SLUSHER	AT&T BELL LABS, MURRAY HILL	USA
Francesco STRUMIA	UNIVERSITA' DI PISA, PISA	ITALY
Donald SULLIVAN	NBS, BOULDER	USA
Toshiharu TAKO	SCIENCE UNIVERSITY OF TOKYO, TOKYO	JAPAN
Patrizia TAVELLA	IEN, TORINO	ITALY
Joseph TAYLOR	PRINCETON UNIVERSITY, PRINCETON	USA
Harald TELLE	PTB, BRAUNSCHWEIG	FRG
Michel TETU	UNIVERSITE' LAVAL, QUEBEC	CANADA
Genevieve THEOBALD	LHA, ORSAY	FRANCE
Pierre THOMANN	OSCILLOQUARTZ, NEUCHATEL	SWITZERLAND
Claudine THOMAS	BIPM, SEVRES	FRANCE
Richard THOMPSON	IMPERIAL COLLEGE, LONDON	UK
Gaetano TIGNANELLI	ISPT, ROMA	ITALY
Guglielmo TINO	UNIVERSITA' DI NAPOLI, NAPOLI	ITALY
Peter E. TOSCHEK	UNIVERSITY OF HAMBURG, HAMBURG	FRG
Pierre TREMBLAY	UNIVERSITE' LAVAL, QUEBEC	CANADA
Claudio TURCHETTI	UNIVERSITA' DI ANCONA, ANCONA	ITALY
Robert VAN DYCK	UNIVERSITY OF WASHINGTON, SEATTLE	USA
Jacques VANIER	NRC, OTTAWA	CANADA
Nathalie VANSTEENKISTE	ENS, PARIS	FRANCE
V.L. VELICHANSKI	LEBEDEV PHYS.INSTITUTE, MOSCOW	URSS
Boudewyn VERHAAR	EINDHOVEN UNIV.OF TECHNOLOGY, EINDHOVEN	NEDERLANDS
Robert F.C. VESSOT	SAO, CAMBRIDGE	USA
Fred L. WALLS	NBS, BOULDER	USA
Jook T.M. WALRAVEN	UNIVERSITY OF AMSTERDAM, AMSTERDAM	NEDERLANDS
Harry WANG	HUGHES RES. LABORATORY, MALIBU	USA
Wang YI QIU	BEIJING UNIVERSITY, HAIDIAN, BEIJING	CHINA
Marc A. WEISS	NBS, BOULDER	USA
Gunther WERTH	UNIVERSITY OF MAINZ, MAINZ	FRG
Bradley G. WHITFORD	NRC, OTTAWA	CANADA
Dave J. WINELAND	NBS, BOULDER	USA
Cheng YU MIN	SHANGHAI BUREAU OF METROLOGY, SHANGHAI	CHINA
Jean-Jacques ZONDY	LPTF, PARIS	FRANCE

Index of Contributors

Abashev, Y.G. 374
Acef, O. 212
Aida, M. 386
Akulshin, A.M. 236
Allan, D.W. 29
Allegrini, M. 362
Ambrosini, R. 425
Andre, J. 472
Arimondo, E. 445
Aspect, A. 445
Audoin, C. 46,70,402
Aynié, P. 120
Azoubib, J. 37

Bagayev, S.N. 191,277
Balykin, V.I. 248
Barillet, R. 402
Barwood, G.P. 451
Baryshev, V.N. 374
Bauch, A. 370
Bava, E. 78,463
Bell, A.S. 451
Bergquist, J.C. 71,287
Berkhout, J.J. 102
Bernier, L.G. 398
Bertinetto, F. 180,465
Besson, R.J. 157
Beverini, N. 282
Bianco, G. 427
Biraben, F. 362
Blair, D.G. 420
Blatt, R. 300,306,453
Bollinger, J.J. 71,319,447
Bordé, C.J. 196,212
Boulanger, J.-S. 379
Brand, U. 467
Bréant, C. 441
Breton, M. 431
Brewer, L.R. 447
Brillet, A. 326
Brincourt, G. 472
Busca, G. 392,398
Byer, R.L. 225

Cagnac, B. 362
Caldera, C. 463
Camparo, J.C. 62
Casdorff, R. 306
Catella, R. 472
Cenci, A. 427
Cérez, P. 110,439
Changhua, Wu 389
Chardonnet, C. 212
Chebotayev, V.P. 191,277
Clairon, A. 120,212,441
Clercq, E., de 120
Cline, R.W. 95
Cohen-Tannoudji, C. 445
Cordiale, P. 465
Crampton, S.B. 57,86
Cyr, N. 431

Dahmani, B. 120
Dalibard, J. 137
Day, T. 225
De Marchi, A. 52,395
Dehmelt, H. 15,286,312
Depei, Wang 126
Desaintfuscien, M. 418,449
Diedrich, F. 71,287
Dimarcq, N. 110
Dinca, M. 414
Diomande, K. 439
Dmitriyev, A.K. 277
Douglas, R.J. 379
Drullinger, R.E. 116

Enders, V. 306
Ernst, K. 469

Farinas, A.D. 225
Farnham, D.L. 349
Flowers, J.L. 10
Fontana, S. 465

Gagnepain, J.-J. 150
Garramone, L. 427
Garreau, J.C. 362
Gely, B. 418
Gérard, A. 120
Gheorghiu, O. 414
Gilbert, S.L. 71,319,447
Giles, A. 420
Gilhaus, H. 300
Gill, P. 451
Giordano, V. 110,431
Giurgiu, L. 414
Glaze, D.J. 116
Godone, A. 78,463
Gould, P.L. 264
Goy, P. 356
Grangier, P. 343
Granveaud, M. 22
Gross, M. 356
Gubin, M.A. 206
Guinot, B. 37
Gustafson, E.K. 225

Haberland, H. 445
Hall, J.L. 162
Hardy, W.N. 95
Hare, J. 356
Haroche, S. 356
Hashimoto, M. 434
Hayden, M.E. 95
Hayes, R.R. 410
Heindorff, T. 370
Hellwig, H. 44
Helmcke, J. 270,467
Hils, D. 162
Hollberg, L. 116,231
Hürlimann, M.D. 95

Ikegami, T. 132
Inguscio, M. 469
Ishikawa, J. 270
Itano, W.M. 71,287,319,447

Jacques, C. 382
Jardino, M. 418
Jones, S.K. 420
Julien, L. 362

Kaiser, R. 445
Karrais, M. 445
Klein, H.A. 451
Klementyev, V.M. 191
Knight, D.J.E. 459
Koga, Y. 132
Koppang, P. 408
Kotake, N. 386
Kramer, G. 181

Kroupa, V.F. 144
Krupczak, J.J. 57
Kuboki, K. 242
Kurochkin, V.Y. 457

La Porta, A. 343
Labachelerie, M., de 439
Lamoreaux, S.K. 338
Larson, D.J. 447
Laurent, P. 441
Lesage, P. 402
Leschiutta, S. 23
Letokhov, V.S. 248
Lett, P.D. 264
Linzhen, Xie 126
Lipphardt, B. 181,436

Maccioni, E. 282
Madej, A.A. 461
Maleki, L. 174
Mandache, C. 414
Mattison, E.M 88
Meis, C. 418
Metcalf, H.J. 264
Mingshou, Li 389
Moore, F.L. 349
Morris, D. 404
Murata, M. 242

Nagourney, W. 312
Nakadan, Y. 132
Nakagiri, K. 386
Nekrasov, Y.V. 277
Neuhauser, W. 300,306,453
Nikitin, V.V. 236
Nikulchin, A.V. 206
Nilsson, A.C. 225
Novero, C. 78

Ohshima, S. 132
Ohtsu, M. 242,434
Okazawa, H. 386
Om, A.E. 277
Ovchinnikov, Y.B. 248
Owings, B. 408

Padgett, M.J. 459
Pereira, D. 282
Pernice, B. 427
Peters, H. 408
Petit, P. 402
Petley, B.W. 10
Petrovskiy, V.N. 457
Pharaoh, K.I. 459
Phillips, W.D. 264
Picotto, G.B. 465
Plumelle, F. 449
Ponsonby, J.E.B. 412
Premoli, A. 395
Protsenko, E.D. 457

Radecki, K.W. 400
Rajab Pacha, S. 472
Ramsey, N.F. 2
Reid, J. 461
Riehle, F. 270
Rolston, S.L. 264
Rovera, G.D. 395
Rurukin, A.N. 457

Salomon, C. 137
Sassi, M.P. 463
Sasso, A. 469
Sautenkov, V.A. 236
Sauter, T. 300,453
Schubert, M. 453
Schweda, H. 392,398
Schwinberg, P.B. 349
Shibuki, M. 386
Shin, C.H. 242
Shirley, J.H. 116
Shul'ga, S.V. 248
Shutong, Yao 126
Sidorov, A.I. 248
Siemers, I. 453
Siemsen, K.J. 461
Silvera, I.F. 88
Skvortsov, B.N. 277
Slusher, R.E. 343
Stoof, H.T.C. 416
Strumia, F. 282
Suzuki, M. 455
Sylos Labini, G. 427
Szekely, C. 414

Tako, T. 455
Taylor, J.H. 332
Telle, H.R. 224,436
Têtu, M. 431
Théobald, G. 110
Thériault, S. 431
Thomann, P. 392,398
Timchenko, B.A. 191
Tino, G.M. 469
Toschek, P.E. 300,306
Tremblay, P. 382
Tyurikov, D.A. 206

Van Dyck, Jr., R.S. 349
Vansteenkiste, N. 445
Vasiliev, V.V. 236
Velichansky, V.L. 236
Verhaar, B.J. 416
Vessot, R.F.C. 88
Viennet, J. 402
Villeneuve, B. 431
Vissani, G. 282
Vu, Phat 57

Walls, F.L. 145
Walraven, J.T.M. 102
Walsworth, R.L. 88
Wang, H.T.M. 410
Watts, R.N. 264
Weiss, C.O. 181
Werth, G. 293
Westbrook, C.I. 264
Whitford, B.G. 187
Wineland, D.J. 71,287,319,447

Xiaoren, Yang 389

Yakovlev, Y.N. 374
Yamashita, M. 455
Yiqiu, Wang 126
Yolkin, G.A. 374
Yu, Jin 449
Yu, Nan 312
Yumin, Cheng 406
Yurke, B. 343
Yurkin, E.K. 236

Zakharyash, V.F. 191
Zerega, Y. 472
Zibrov, A.S. 236

Symposia on Frequency Standards and Metrology Charter and By-Laws

WHEREAS three symposia with this name were organized and held in 1971 at Forêt Montmorency, Quebec, Canada, in 1976 at Copper Mountain, Colorado, USA, and 1981 at Aussois, Savoie, France,

WHEREAS the organizers of these three symposia resolve to assure continuity of these meetings in intent, scope and concept, and

WHEREAS the organizers wish to preserve the legacy and tradition established with these meetings,

the organizers resolve and agree by their signature to this document to establish this Charter and By-Laws of the Symposia on Frequency Standards and Metrology.

1. Subject of the Symposia

A symposium of this series and name is intended to serve as an international discussion-forum on precision frequency standards throughout the electromagnetic spectrum and associated metrology. It focuses on the fundamental scientific aspects of the latest ideas, results and applications in relation to these frequency standards.

2. Governing Board

The organizers of all prior symposia form a Governing Board for the next symposium. This Governing Board selects and approves the Symposium Organizer for the next meeting. The Governing Board reserves the power of decision to approve or disapprove the next symposium as proposed by the selected Symposium Organizer. Any member may resign from the Board upon notice to all other members at least 12 months before the date of a planned meeting.

3. Frequency of Meetings

There is no definite time interval between symposia. Very significant progress in the concerned field should have occurred before the next meeting is approved. A time period of no less than four (4) and no more than eight (8) years between the symposia is established in order to suit the anticipated rhythm of evolution of frequency metrology.

This document becomes null and void should a time period of more than eight (8) years occur without a symposium being held.

4. Sponsorship

The symposium must be supported by a well-established local organization. Adequate financial sponsorship must be available to allow (a) modest registration fees and (b) support, including travel, of not less than 15 invited attendees.

5. <u>Proceedings</u>

Proceedings of the symposia must be published and delivered to the attendees without additional charge.

6. <u>Attendance</u>

Attendance of the symposia is by invitation. The strong international character and focus of the symposium, as defined in 1 (one) above, must be retained. Invitations to not more than 200 scientists will be issued by the Symposium Organizer.

7. <u>Talks</u>

Most of the talks are invited after proposal by an international scientific committee which is established at the discretion of the Symposium Organizer. Some presentations may be by posters. A limited number of contributed papers may be accepted. Sessions must not be in parallel. Time for informal discussions must be made available.

A few talks by scientists of international recognition, i.e., Nobel Prize winners or people who appear to be possible candidates for such a Prize, must be included in the program.

8. <u>Welcoming Country</u>

The Symposium must be organized in a country and at a location allowing attendees to: (i) visit laboratories having an internationally leading role in the concerned field, and (ii) visit similar laboratories in neighboring countries.

9. <u>Meeting Site</u>

The meeting should be held in a relatively isolated site, favoring social interaction between attendees so as to maximize the opportunity for exchange of ideas and continuing discussions. Large cities must be avoided.

10. <u>Housing</u>

Co-location or proximity of housing and meeting facilities and arrangement for joint meals must be provided.

The Governing Board for the 5th Symposium on Frequency Standards and Metrology

Jacques VANIER, Forêt Montmorency
Quebec, (Canada) 1971

Helmut HELLWIG, Copper Mountain
Colorado (USA) 1976

Claude AUDOIN, Aussois
Savoie (France) 1981

Andrea De Marchi, Ancona
Marches (Italy) 1988

Batch number: 10838062

Printed by Printforce, the Netherlands